Carpentry
&
Construction

About the Authors

Mark R. Miller is a professor of technology and co-ordinator for the industrial management programs in the Department of Human Resource Development & Technology at The University of Texas at Tyler. He teaches technology courses for future middle managers and has authored or co-authored more than 35 technical books and numerous technical articles.

Rex Miller is professor emeritus of industrial technology at the State University of New York–College at Buffalo, where he taught technical curriculums for more than 40 years. He has authored or co-authored more than 100 texts for vocational and industrial arts programs.

Carpentry & Construction

MARK R. MILLER
Professor
The University of Texas at Tyler
Tyler, Texas

REX MILLER
Professor Emeritus
State University of New York–College at Buffalo
Buffalo, New York

5th Edition

New York   Chicago   San Francisco   Lisbon   London
Madrid   Mexico City   Milan   New Delhi   San Juan
Seoul   Singapore   Sydney   Toronto

Copyright © 2010, 2004, 1999, 1991, 1981 by The McGraw-Hill Companies, Inc. All rights reserved. Printed in the United States of America. Except as permitted under the United States Copyright Act of 1976, no part of this publication may be reproduced or distributed in any form or by any means, or stored in a data base or retrieval system, without the prior written permission of the publisher.

1 2 3 4 5 6 7 8 9 0   QPD/QPD   0 1 5 4 3 2 1 0 9

ISBN 978-0-07-162471-8
MHID 0-07-162471-6

Sponsoring Editor: Joy Bramble Oehlkers
Editing Supervisor: Stephen M. Smith
Production Supervisor: Richard C. Ruzycka
Acquisitions Coordinator: Michael Mulcahy
Project Manager: Patricia Wallenburg, TypeWriting
Copy Editor: James K. Madru
Proofreader: Teresa Barensfeld
Art Director, Cover: Jeff Weeks
Composition: TypeWriting

Printed and bound by Quebecor/Dubuque.

McGraw-Hill books are available at special quantity discounts to use as premiums and sales promotions, or for use in corporate training programs. To contact a representative, please e-mail us at bulksales@mcgraw-hill.com.

This book is printed on acid-free paper.

Information contained in this work has been obtained by The McGraw-Hill Companies, Inc. ("McGraw-Hill") from sources believed to be reliable. However, neither McGraw-Hill nor its authors guarantee the accuracy or completeness of any information published herein, and neither McGraw-Hill nor its authors shall be responsible for any errors, omissions, or damages arising out of use of this information. This work is published with the understanding that McGraw-Hill and its authors are supplying information but are not attempting to render engineering or other professional services. If such services are required, the assistance of an appropriate professional should be sought.

Contents

7 Framing the Roof

8 Roofs and Roofing

11 House Wiring

12 Plumbing

13 Insulation

14 Interior Walls and Ceilings

15 Interior Finishing

16 Special Construction Methods

17 Maintenance and Remodeling

18 The Carpenter and the Industry

19 Bathrooms

20 Construction for Solar Heating

21 Alternative Framing Methods

22 Permanent Wood Foundation System

23 Private Water Systems

24 Private Sewage Facilities

Preface

Carpentry & Construction, Fifth Edition, is written for those who want or need to know about carpentry and construction. Whether remodeling an existing home or building a new one, the rewards from a job well done are many-fold.

This text can be used by students in vocational courses, technical colleges, apprenticeship programs, and construction classes in industrial technology programs. The home do-it-yourselfer will find answers to many questions that pop up in the course of getting a job done whether over a weekend or over a year's time.

In order to prepare this text, the authors examined courses of study in schools located all over the country. An effort was made to take into consideration the geographic differences and the special environmental factors relevant to a particular area.

Notice how the text is organized. The first chapter, "Let's Get Started," presents the information needed to get construction started. The next chapter covers preparing the site. Then the footings and foundation are described. Once the roof is in place, the next step is the installation of windows and doors. When the windows and doors are in place, the exterior siding is applied. Next, heating and cooling are covered—all-important considerations for living quarters. Once the insulation is in place, the interior walls and ceilings are covered in detail before presenting interior finishing methods.

Special construction methods, maintenance and remodeling, and careers in carpentry are then described. The building of solar houses and the design of solar heating are covered to keep the student and do-it-yourselfer up-to-date with the latest developments in energy conservation. Take a closer look at the steel framing used in more abundance today for private homes. Also note the use of foam and concrete to build homes of lasting quality that are almost completely free from tornado and hurricane damage.

We trust you will enjoy using this book as much as we enjoyed writing it.

MARK R. MILLER
REX MILLER

Acknowledgments

The authors would like to thank the following manufacturers for their generous efforts. They furnished photographs, drawings, and technical assistance. Without the donations of time and effort on the part of many people, this book would not have been possible. We hope this acknowledgment of some of the contributions will let you know that the field you are working in or about to enter is one of the best. (The name given below in parentheses indicates the abbreviated form used in the credit lines for the photographs and line drawings appearing in this text.)

Abitibi Corporation; ALCOA; AFM Corporation; American Olean Tile Co.; American Plywood Association; American Polysteel Forms; American Standard, Inc.; Town of Amherst, New York (Town of Amherst, NY); Andersen Corporation (Andersen); Armstrong Cork Company (Armstrong Cork); C. Arnold and Sons; Avalon Concepts; Beaver-Advance Company (Beaver-Advance); The Bilco Company (Bilco); Bird and Son, Inc. (Bird and Son); Black & Decker Manufacturing Company (Black & Decker); Boise-Cascade; Butler Manufacturing Company (Butler); Certain-Teed Corporation (Certain-Teed); California Redwood Association (California Redwood Assoc.); Conaway Homes, Tyler, Texas; Conwed Corporation (Conwed); Corl Corporation; DeWalt, Div. of American Machine and Foundry Company (DeWalt); Dow Chemical Company (Dow Chemical); Dow-Corning Company (Dow-Corning); Duo-Fast Corporation (Duo-Fast); EMCO; Formica Corporation (Formica); General Products Corporation (General Products); Georgia-Pacific Corporation (Georgia-Pacific); Goldblatt Tools, Inc. (Goldblatt Tools); Gold Bond Building Products (Gold Bond); Grossman Lumber Company (Grossman Lumber); Gypsum Association (Gypsum); Hilti Fastening Systems (Hilti); IRL Daffin Company (IRL Daffin); Kelly-Stewart Company, Inc. (Kelly-Stewart); Kenny Manufacturing Company (Kenny); Kirch Company (Kirch); Kohler Company (Kohler); Lennox Furnace Co.; Majestic; Manco Tape, Inc. (Manco Tape); Martin Industries (Martin); Masonite Corporation (Masonite); MFS; Milwaukee Electric Tool Company (Milwaukee Electric Tool); NAHB (National Association of Home Builders); National Aeronautics & Space Administration (NASA); National Homes; National Lock Hardware Company; National Oak Flooring Manufacturers Association (National Oak Flooring Manufacturers); National Wood Manufacturers Association (National Wood Manufacturers); New York State Electric and Gas Corporation (NYSE & G); Novi; NuTone; Owens-Corning Fiberglas Corporation (Owens-Corning); Patent Scaffolding Company (Patent Scaffolding); Pella Windows and Doors, Inc. (Pella); Permograin Products; Plaskolite, Inc. (Plaskolite); Portland Cement Association (Portland Cement); Potlatch Corporation (Potlatch); Proctor Products, Inc. (Proctor Products); Red Cedar Shingle & Handsplit Shake Bureau (RCS & HSB); Reynolds Metals Products; Richmond Screw Anchor Company (Richmond Screw Anchor); Riviera Kitchens, Div. of Evans Products; Rockwell International, Power Tool Division (Rockwell); Sears, Roebuck and Company (Sears, Roebuck); Shakertown Corporation; Simplex Industries (Simplex); Stanley Tools Company (Stanley Tools); State University College at Buffalo; Southern Forest Products Association; TECO Products and Testing Corporation, Washington, D.C. 20015; U.S. Gypsum Company (U.S. Gypsum); U.S. Department of Energy (U.S. Dept. of Energy); U.S. Bureau of Labor Statistics; U.S. Forest Service, Forest Products Laboratory (Forest Products Lab); United Steel Products; Universal Fastenings Corporation; Universal Form Clamp Company (Universal Form Clamp); Valu, Inc. (Valu); Velux-American; Weiser Lock, Division of Norris Industries (Weiser Lock); Weslock; Western Wood Products Association (Western Wood Products); Weyerhauser Company (Weyerhauser); David White Instruments, Div. of Realist, Inc. (David White).

In addition, the authors would like to thank Paul Consiglio of Buffalo, NY, for some of the line drawings rendered in finished form.

AFM Corporation
American Polysteel Forms
American Standard, Inc.
Avalon Concepts
Clopay Building Products Co.
Congoleum Floors
Delta Building Products
Eljer Manufacturing Co.
Hy-Lite Block Windows
Kohler Company
L.J. Smith Stair Systems
Majestic Company
Nuconsteel Commercial Corp.
ODL, Inc.
Owens-Corning
Pittsburg-Corning
PlumbShop
Southeastern Metals Manufacturing Co.
Steel Framing, Inc.
Tri-Steel Structures, Inc.
Western Wood Products Association
Wilsonart International

1
CHAPTER

Let's Get Started

CARPENTRY INVOLVES ALL KINDS OF CHALLENGING jobs; it is an exciting industry. You will have to work with hand tools, power tools, and all types of building materials. You can become very skilled at your job, and you get a chance to be proud of what you do. You can stand back and look at the building you just helped to erect and feel great about a job well done.

One of the exciting things about being a carpenter is watching a building go up. You actually see it grow from the ground up. Many people work with you to make it possible to complete the structure. Being part of a team can be rewarding, too.

This book will help you to do a good job in carpentry, whether you are remodeling an existing building or starting from the ground up. Because it covers all the basic construction techniques, it will aid you in making the right decisions.

You may have to do something over and over again to gain skill. When you read this book, you might not always get the idea the first time. Go over it again until you understand. Then go out and practice what you just read. In this way, you can see for yourself how the instructions actually work. Of course, no one can learn carpentry merely by reading a book. You have to read, reread, and then *do*. This do part is the most important. You have to take the hammer or saw in hand and actually do the work. There is nothing like good, honest sweat from a hard day's work. At the end of the day, you can say, "I did that" and be proud that you did.

This chapter will help you to build these skills:

• Select personal protective gear

• Work safely as a carpenter

• Measure building materials

• Lay out building parts

• Cut building materials

• Fasten materials

• Shape and smooth materials

• Identify basic hand tools

• Recognize common power tools

SAFETY

Figure 1-1 shows a carpenter using one of the latest means of driving nails: a compressed-air-driven nail driver, which drives nails into wood with a single stroke. The black cartridge that appears to run up near the carpenter's leg is a part of the nailer. It holds the nails and feeds them to the nailer as needed.

Fig. 1-1 *This carpenter is using an air-driven nail driver to nail the framing members.* (Duo-Fast.)

As for safety, notice the carpenter's shoes. They have rubber soles for gripping the wood. This will prevent a slip through the joists and a serious fall. The steel toes in the shoes prevent damage to a foot from falling materials. The soles of the shoes are very thick to prevent nails from going through. The hard hat protects the carpenter's head from falling lumber, shingles, or other building materials or objects. The carpenter's safety glasses cannot be seen in Fig. 1-1, but they are required equipment for the safe worker.

Other Safety Measures

To protect the eyes, it is best to wear safety glasses. Make sure that your safety glasses are of tempered glass. They will not shatter and cause eye damage. In some instances, you should wear goggles. This prevents splinters and other flying objects from entering the eye from under or around the safety glasses. Ordinary safety glasses aren't always the best, even if they are made of tempered glass. Just become aware of the possibilities of eye damage whenever you start a new job or procedure. See Fig. 1-2 for a couple of types of safety glasses.

Fig. 1-3 *Face shield.*

Fig. 1-2 *Safety glasses.*

Sneakers are used only by roofers. Sneakers, sandals, and dress shoes do not provide enough protection for carpenters on the job. Only safety shoes should be worn on the job.

Gloves Some types of carpentry work require the sensitivity of the bare fingers. Other types do not require the hands or fingers to be exposed. In cold or even cool weather, gloves may be in order. Gloves are often needed to protect the hands from splinters and rough materials. It's only common sense to use gloves when handling rough materials.

Probably the best gloves for carpentry work are a lightweight type. A suede finish to the leather improves the gripping ability of the gloves. Cloth gloves tend to catch on rough building materials. They may be preferred, however, if you work with short nails or other small objects.

Body protection Before you go to work on any job, make sure that your entire body is properly protected. Hard hats come in a couple of styles. Under some conditions, a face shield is better protection (Fig. 1-3).

Is your body covered with heavy work clothing? This is the first question to ask before going onto a job site. Has as much of your body as practical been covered with clothing? Has your head been protected properly? Are your eyes covered with approved safety glasses or a face shield? Are your shoes sturdy, with safety toes and steel soles to protect against nails? Are gloves available when you need them?

General Safety Rules

Some safety procedures should be followed at all times. This applies to carpentry work especially:

- Pay close attention to what is being done.
- Move carefully when walking or climbing.
- Use appropriate equipment. Take a look at Fig. 1-4. This type of made-on-the-job ladder can cause trouble.
- Use the leg muscles when lifting.

Fig. 1-4 *A made-on-the-job ladder.*

- Move long objects carefully. The end of a carelessly handled 2 × 4 can damage hundreds of dollars worth of glass doors and windows. Keep the workplace neat and tidy. Figure 1-5A shows a cluttered work area. It would be hard to walk along here without tripping. If a dumpster is used for trash and debris, as in Fig. 1-5B, many accidents can be prevented.

(A)

(B)

Fig. 1-5 *A. Cluttered work site. B. A work area can be kept clean if a large dumpster is kept nearby for trash and debris.*

- Sharpen or replace dull tools.
- Disconnect power tools before adjusting them.
- Keep power tool guards in place.
- Avoid interrupting a person who is using a power tool.
- Remove hazards as soon as they are noticed.

Safety on the Job

A safe working site makes it easier to get the job done. Lost time due to accidents puts a building behind schedule. This can cost many thousands of dollars and lead to late delivery of the building. If the job is organized properly, and safety is taken into consideration, the smooth flow of work is quickly noticed. No one wants to get hurt. Pain is no fun. Safety is just common sense. If you know how to do something safely, it will not take any longer than if you did it in an unsafe manner. Besides, why would you deliberately do something that is dangerous? All safety requires is a few precautions on the job. Safety becomes a habit once you get the proper attitude established in your thinking. Some of these important habits to acquire include:

- Know exactly what is to be done before you start a job.
- Use a tool only when it can be used safely. Wear all safety clothing recommended for the job. Provide a safe place to stand to do the work. Set ladders securely. Provide strong scaffolding.
- Avoid wet, slippery areas.
- Keep the working area as neat as is practical.
- Remove or correct safety hazards as soon as they are noticed. Bend protruding nails over. Remove loose boards.
- Remember where other workers are and what they are doing.
- Keep fingers and hands away from cutting edges at all times.
- Stay alert!

Safety Hazards

Carpenters work in unfinished surroundings. While a house is being built, there are many unsafe places around the building site. You have to stand on or climb ladders, which can be unsafe. You may not have a good footing while standing on a ladder. You may not be climbing a ladder in the proper way. Holding onto the rungs of the ladder is very unsafe. You should always hold onto the outside rails of the ladder when climbing.

There are holes that can cause you to trip. They may be located in the front yard where the water or sewage lines come into the building. There may be holes for any number of reasons. These holes can cause you all kinds of problems, especially if you fall into them or turn your ankle.

The house in Fig. 1-6 is almost completed. However, if you look closely, you can see that some wood has been left on the garage roof. This wood can slide down and hit a person working below. The front porch has not been poured. This means that stepping out of the front door can be a rather long step. Other debris around the yard can be a source of trouble. Long sliv-

Fig. 1-6 *Even when a house is almost finished, there can still be hazards. Wood left on a roof could slide off and hurt someone, and without the front porch, walking out the door is a long step down.*

ers of flashing can cause trouble if you step on them and they rake your leg. You have to watch your every step around a construction site.

Outdoor work   Much of the time, carpentry is performed outdoors. This means that you will be exposed to the weather, so dress accordingly. Wet weather increases the accident rate. Mud can make a secure place to stand hard to find. Mud can also cause you to slip if you don't clean it off your shoes. Be very careful when it is muddy and you are climbing on a roof or a ladder.

Tools   Any tool that can cut wood can cut flesh. You have to keep in mind that although tools are an aid to the carpenter, they can also be a source of injury. A chisel can cut your hand as easily as it cuts wood. In fact, it can do a quicker job on your hand than on the wood it was intended for. Saws can cut wood and bones. Be careful with all types of saws, both hand and electric. Hammers can do a beautiful job on your fingers if you miss the nailhead. The pain involved is intensified in cold weather. Broken bones can be easily avoided if you keep your eye on the nail while you're hammering. Besides that, you will get the job done more quickly. And, after all, that's why you are there—to get the job done and do it right the first time. Tools can help you do the job right. They can also cause you injury. The choice is up to you.

In order to work safely with tools, you should know what they can do and how they do it. The next few pages are designed to help you use tools properly.

USING CARPENTER TOOLS

A carpenter is lost without tools. This means that you have to have some way of containing them. A toolbox

is very important. If you have a place to put everything, then you can find the right tool when you need it. A toolbox should have all the tools mentioned here. In fact, you probably will add more as you become more experienced. Tools have been designed for every task. All it takes is a few minutes with a hardware manufacturer's catalog to find just about everything that you will ever need. If you cannot find what you need, the manufacturers are interested in making it.

Measuring Tools

Folding rule   When using the folding rule, place it flat on the work. The 0 end of the rule should be exactly even with the end of the space or board to be measured. The correct distance is indicated by the reading on the rule.

A very accurate reading may be obtained by turning the edge of the rule toward the work. In this position, the marked graduations on the face of the rule touch the surface of the board. With a sharp pencil, mark the exact distance desired. Start the mark with the point of the pencil in contact with the mark on the rule. Move the pencil directly away from the rule while making the mark.

One problem with the folding rule is that it breaks easily if it is twisted. This happens most commonly when it is being folded or unfolded. The user may not be aware of the twisting action at the time. You should keep the joints oiled lightly. This makes the rule operate more easily.

Pocket tape   Beginners may find the pocket tape (Fig. 1-7) the most useful measuring tool for all types of work. It extends smoothly to full length. It returns quickly to its compact case when the return button is pressed. Steel tapes are available in a variety of lengths. For most carpentry, a rule that is 6, 8, 10, or 12 feet long is used.

Fig. 1-7   *Tape measure. (Stanley Tools.)*

Longer tapes are available. They come in 20-, 50-, and 100-foot lengths (Fig. 1-8). This tape in the figure can be extended to 50 feet to measure lot size and the location of a house on a lot. It has many uses around a building site. A crank handle can be used to wind it up once you are finished with it. The hook on the end of the tape makes it easy for one person to use it. Just hook the tape over the end of a board or nail and extend it to your desired length.

THE NUMBER OF POINTS PER INCH ON A HANDSAW DETERMINES THE FINENESS OR COARSENESS OF CUT. MORE POINTS PRODUCE A FINER CUT.

Fig. 1-8 *A longer tape measure.*

Fig. 1-9 *Using a handsaw.*

Saws

Carpenters use a number of different saws. These saws are designed for specific types of work. Many are misused. They still will do the job, but they would do a better job if used properly. Handsaws take quite a bit of abuse on a construction site. It is best to buy a good-quality saw and keep it lightly oiled.

Standard skew-handsaw This saw has a wooden handle. It has a 22-inch length. A 10-point saw (with 10 teeth per inch) is suggested for crosscutting. Crosscutting means cutting wood *across* the grain. A 26-inch-long, 5½-point saw is suggested for ripping, or cutting *with* the wood grain.

Figure 1-9 shows a carpenter using a handsaw. This saw is used in places where an electric saw cannot be used. Keeping it sharp makes a difference in the quality of the cut and the ease with which it can be used.

Backsaw The backsaw gets its name from the piece of heavy metal that makes up the top edge of the cutting part of the saw (Fig. 1-10). It has a fine tooth configuration. This means that it can be used to cut cross-grain and leave a smoother finished piece of work. This type of saw is used by finish carpenters who want to cut trim or molding.

Miter box As you can see from Fig. 1-11A, the miter box has a backsaw mounted in it. This box can

be adjusted using the lever under the saw handle (*arrow*). You can adjust it for the cut you wish. It can cut from 90 to 45 degrees. It is used for finish cuts on moldings and trim materials. The angle of the cut is determined by the location of the saw in reference to the bed of the box. Release the clamp on the bottom of the saw support to adjust the saw to any desired angle. The wood is held with one hand against the fence of

Fig. 1-10 *Backsaw.* (Stanley Tools.)

(A)

(B)

Fig. 1-11 A. Miter box. *(Stanley Tools.)* B. Powered compound miter saw. *(Delta.)*

Fig. 1-12 Coping saw. *(Stanley Tools.)*

the box and the bed. Then the saw is used by the other hand. As you can see from the setup, the cutting should take place when the saw is pushed forward. The backward movement of the saw should be made with the pressure on the saw released slightly. If you try to cut on the backward movement, you will just pull the wood away from the fence and damage the quality of the cut.

Coping saw Another type of saw a carpenter can make use of is the coping saw (Fig. 1-12). This one can cut small thicknesses of wood at any curve or angle desired. It can be used to make sure a piece of paneling fits properly or a piece of molding fits another piece in a corner. The blade is placed in the frame with the teeth pointing toward the handle. This means that it cuts only on the downward stroke. Make sure that you properly support the piece of wood being cut. A number of blades can be obtained for this type of saw. The

number of teeth in the blade determines the smoothness of the cut.

Hammers and Other Small Tools

There are a number of different types of hammers. The one a carpenter uses is the *claw* hammer. It has claws that can extract nails from wood if they have been put in the wrong place or have bent while being driven. Hammers can be bought in 20-, 24-, 28-, and 32-ounce weights for carpentry work. Most carpenters prefer a 20-ounce hammer. You have to work with a number of different weights to find out which will work best for you. Keep in mind that the hammer should be made of tempered steel. If the end of the hammer has a tendency to splinter or chip off when it hits a nail, the pieces can hit you in the eye or elsewhere, causing serious damage. It is best to wear safety glasses whenever you use a hammer.

Nails are driven by hammers. Figure 1-13 shows the gauge, inch, and penny relationships for the common box nail. The *d* after the number means "penny." This is a measuring unit inherited from the English in the colonial days. There is little or no relationship between penny and inches. If you want to be able to talk about it intelligently, you'll have to learn both inches and penny. The gauge is nothing more than the American Wire Gauge (AWG) number for the wire from which the nails were originally made. Finish nails have the same measuring unit (penny) but do not have the large, flat heads.

Nail set Finish nails are driven below the surface of the wood by a nail set. The nail set is placed on the head of the nail. The large end of the nail set is struck with a hammer. This causes the nail to go below the surface of the wood. Then the hole left by the countersunk nail is filled with wood filler and finished off with a smooth coat of varnish or paint. Figure 1-14 shows the nail set and its use.

Fig. 1-13 Nails. *(Forest Products Laboratory.)*

A carpenter would be lost without a hammer. In Fig. 1-15, the carpenter is placing sheathing on rafters to form a roof base. The hammer is used to drive the boards into place because they have to overlap slightly. Then the nails are also driven by the hammer.

In some cases, a hammer will not do the job. The job may require a hatchet (Fig. 1-16). This tool can be used to pry and to drive. It can pry boards loose when they are installed improperly. It can sharpen posts to be driven into the ground at the site. The hatchet can sharpen the ends of stakes for staking out the site. It

also can withdraw nails. This type of tool can also be used to drive stubborn sections of a wall into place when they are erected for the first time. The tool has many uses.

Scratch awl An awl is a handy tool for a carpenter (Fig. 1-17). It can be used to mark wood with a scratch mark and to produce pilot holes for screws. Once it is in your tool box, you will think of a hundred uses for it. Since it does have a very sharp point, it is best to treat it with respect.

(A)

(B)

Fig. 1-14 A. Driving a nail with a hammer. B. Finishing the job with a nail set to make sure that the hammer doesn't leave an impression in the soft wood of the window frame.

Fig. 1-15 *Putting on roof sheathing. The carpenter is using a hammer to drive the board into place.*

Fig. 1-16 *Hatchet.* (Stanley Tools.)

Fig. 1-17 *Scratch awl.* (Stanley Tools.)

Wrecking bar This device (Fig. 1-18) has a couple of names depending on which part of the country you are in at the time. It is called a *wrecking bar* in some parts and a *crowbar* in others. One end has a chisel-sharp flat surface to get under boards and pry them

Fig. 1-18 *Wrecking bars.* (Stanley Tools.)

loose. The other end is hooked so that the slot in the end can pull nails with the leverage of the long handle. This specially treated steel bar can be very helpful in prying away old and unwanted boards. It can be used to help give leverage when you are putting a wall in place and making it plumb. This tool has many uses for the carpenter with ingenuity.

Screwdrivers The screwdriver is an important tool for carpenters. It can be used for many things other than turning screws. There are two types of screwdrivers. The standard type has a straight slot-fitting blade at its end. This is the most common type of screwdriver. The Phillips-head screwdriver has a cross or X on the end to fit a screw head of the same design. Figure 1-19 shows the two types of screwdrivers.

Squares

In order to make corners meet and standard sizes of materials fit properly, you must have things square. This calls for a number of squares to check that the two walls or two pieces come together at a perpendicular angle.

Try square The *try square* can be used to mark small pieces for cutting. If one edge of a board is straight, and the handle part of the square (Fig. 1-20) is

Fig. 1-19 *Two types of screwdrivers.*

Fig. 1-20 *Use of a try square.* (Stanley Tools.)

Fig. 1-22 *Bevel.* (Stanley Tools.)

placed against this straight edge, then the blade can be used to mark the wood perpendicular to the edge. This comes in handy when you are cutting 2 × 4s and want them to be square.

Framing square The framing square is a very important tool for carpenters. It allows you to make square cuts in dimensional lumber. This tool can be used to lay out rafters and roof framing (Fig. 1-21). It is also used to lay out stair steps.

Later in this book you will see a step-by-step procedure for using a framing square. The tools are described as they are called for in actual use.

Bevel A bevel can be adjusted to any angle to make cuts at the same number of degrees (Fig. 1-22). Note how the blade can be adjusted. Now take a look at Fig. 1-23. Here, you can see the overhang of rafters. If you want the ends to be parallel with the side of the house, you can use the bevel to mark them before they are cut off. Simply adjust the bevel so that the handle is on top of the rafter and the blade fits against the soleplate below. Tighten the screw, and move the bevel down the rafter to where you want the cut. Mark the angle along the blade of the bevel. Cut along the mark, and you have what you see in Fig. 1-23. This is a good device for transferring angles from one place to another.

Fig. 1-23 *Rafter overhang cut to a given angle.*

Fig. 1-21 *Framing square.* (Stanley Tools.)

Chisel Occasionally, you may need a wood chisel (Fig. 1-24). It is sharpened on one end. When the other end is struck with a hammer, the cutting end will do its job, that is, of course, if you have kept it sharpened.

Fig. 1-24 *Using a wood chisel to complete a mortise.*

The chisel is used commonly in fitting or hanging doors. It is used to remove the area where the door hinge fits. Note how it is used to score the area in Fig. 1-24; it is then used at an angle to remove the ridges. A great deal of the work with the chisel is done by using the palm of the hand as the force behind the cutting edge. A hammer also can be used. In fact, chisels have a metal tip on the handle so that the force of the hammer blows will not chip the handle. Other applications are up to you, the carpenter. You'll find many uses for the chisel in making things fit.

Fig. 1-25 *Smooth plane.* (Stanley Tools.)

Plane Planes (Fig. 1-25) are designed to remove small shavings of wood along a surface. One hand holds the knob in front, and the other holds the handle in back. The blade is adjusted so that only a small sliver of wood is removed each time the plane is passed over the wood. A plane can be used to make sure that doors and windows fit properly. It can be used for any number of wood-smoothing operations.

Dividers and compass Occasionally, a carpenter must draw a circle. This is done with a compass. The compass shown in Fig. 1-26A can be converted to a divider by removing the pencil and inserting a straight steel pin. The compass has a sharp point that fits into the wood surface. The pencil part is used to mark the circle circumference. It is adjustable to various radii.

The dividers in Fig. 1-26A have two points made of hardened metal. They are adjustable. It is possible to use them to transfer a given measurement from a framing square or measuring device to another location (see Fig. 1-26B).

Level In order to have things look the way they should, a level is necessary. There are a number of

Fig. 1-26 A. Dividers and compass. B. Dividers being used to transfer hundredths of an inch.

Fig. 1-27 *A. Using a level to make sure that a window is placed properly before nailing.* (Andersen.) *B. A commonly used type of level.* (Stanley Tools.) *C. Plumb bobs.* (Stanley Tools.)

sizes and shapes available. The one shown in Fig. 1-27B is the most common type used by carpenters. The bubbles in the glass tubes tell you if level is obtained. In Fig. 1-27A, the carpenter is using the level to make sure that the window is in properly before nailing it into place permanently.

If the vertical and horizontal bubbles are lined up between the lines, then the window is plumb, or vertical. A plumb bob is a small, pointed weight. It is attached to a string and dropped from a height. If the bob is just above the ground, it will indicate the vertical direction by its string. Keeping windows, doors, and

frames square and level makes a difference in fitting. It is much easier to fit prehung doors into a frame that is square. When it comes to placing panels of 4- × 8-foot plywood sheathing on a roof or on walls, squareness can make a difference as to fit. Besides, a square fit and a plumb door and window look better than those that are a little off. Figure 1-27C shows three plumb bobs.

Files A carpenter finds use for a number of types of files. The files have different surfaces for doing different jobs. Tapping out a hole to get something to fit may be just the job for a file. Some files are used for sharp-

Fig. 1-28 *Wood and cabinet files. A. Half-round. B. Rasp. C. Flat. D. Triangular. E. Round.* (Millers Falls Division, a Division of Ingersol-Rand Company.)

ening saws and touching up tool cutting edges. Figure 1-28 shows different types of files. Other files may also be useful. You can acquire them later as you develop a need for them.

Clamps   C-clamps are used for many holding jobs (Fig. 1-29). They come in handy when placing kitchen cabinets by holding them in place until screws can be inserted and seated properly. This type of clamp can be used for an extra hand every now and then when two hands aren't enough to hold a combination of pieces until you can nail them.

Fig. 1-29   *C-clamp.*

Cold chisel   It is always good to have a cold chisel around (Fig. 1-30). It is very much needed when you can't remove a nail. Its head may have broken off, and the nail must be removed. The chisel can cut the nail and permit separation of the wood pieces.

Fig. 1-30   *Cold chisel.* (Stanley Tools.)

If a chisel of this type starts to "mushroom" at the head, you should remove the splintered ends with a grinder. Hammering on the end can produce a mush-

rooming effect. These pieces should be taken off because they can easily fly off when hit with a hammer. This is another reason for using eye protection when using tools.

Caulking gun   In times of energy crisis, the caulking gun gets plenty of use. It is used to fill in around windows and doors and everywhere there may be an air leak. There are many types of caulks being made today. Chapter 13 will cover the details of caulking compounds and their uses.

The caulking gun is easily operated. Insert the cartridge and cut its tip to the shape you want. Puncture the thin plastic film inside. A bit of pressure will cause the caulk to come out the end. The long rod protruding from the end of the gun is turned over. This is so that the serrated edge will engage the hand trigger. Remove the pressure from the cartridge when you are finished because caulk will continue to come out of the end. Do this by rotating the rod so that the serrations are not engaged by the trigger of the gun.

Power Tools

The carpenter uses many power tools to aid in getting the job done. The quicker the job is done, the more valuable the work of the carpenter becomes. This is called *productivity*. The more you are able to produce, the more valuable you are. This means that the contractor can make money on the job. This means that you can have a job the next time there is a need for a good carpenter. Power tools make your work go faster. They also help you to do a job without getting fatigued. Many tools have been designed with you in mind. They are portable and operate from an extension cord.

The extension cord should be the proper size to take the current needed for the tool being used (Table 1-1).

TABLE 1-1   *Size of Extension Cords for Portable Tools*

	Full-Load Rating of the Tool in Amperes at 115 Volts					
	0 to 2.0	2.10 to 3.4	3.5 to 5.0	5.1 to 7.0	7.1 to 12.0	12.1 to 16.0
Cord Length, Feet	Wire Size (AWG)					
25	18	18	18	16	14	14
50	18	18	18	16	14	12
75	18	18	16	14	12	10
100	18	16	14	12	10	8
200	16	14	12	10	8	6
300	14	12	10	8	6	4
400	12	10	8	6	4	4
500	12	10	8	6	4	2
600	10	8	6	4	2	2
800	10	8	6	4	2	1
1000	8	6	4	2	1	0

If the voltage is lower than 115 volts at the outlet, have the voltage increased or use a much larger cable than listed.

Note how the distance between the outlet and the tool using the power is critical. If the distance is great, then the wire must be larger in size to handle the current without too much loss. The higher the number of the wire, the smaller is the diameter of the wire. The larger the size of the wire (diameter), the more current it can handle without dropping the voltage.

Some carpenters run an extension cord from the house next door for power before the building site is furnished with power. If the cord is too long or has the wrong size wire, it drops the voltage below 115 volts. This means that saws or other tools using electricity will draw more current and therefore drop the voltage more. Every time the voltage is dropped, the device tries to obtain more current. This becomes a self-defeating phenomenon. You wind up with a saw that has little cutting power. You may have a drill that won't drill into a piece of wood without stalling. Of course, the damage done to the electric motor is in some cases irreparable. You may have to buy a new saw or drill. Double-check Table 1-1 for the proper wire size in your extension cord.

Portable saw This is the most often used and abused piece of carpenter's equipment. The electric portable saw, such as the one shown in Fig. 1-31, is used to cut all 2×4s and other dimensional lumber. It is used to cut off rafters. This saw is used to cut sheathing for roofs. It is used for almost every sawing job required in carpentry.

This saw has a guard over the blade. The guard always should be left intact. Do not remove the saw guard. If not held properly against the wood being cut, the saw can kick back and into your leg.

You always should wear safety glasses when using this saw. The sawdust is thrown in a number of directions, and one of these is straight up toward your eyes.

Fig. 1-31 *Portable power saw. The favorite power tool of every carpenter. Note that the blade should not extend more than ⅛ inch below the wood being cut. Also note the direction of blade rotation.*

If you are watching a line where you are cutting, you definitely should have glasses on.

Table saw If the house has been enclosed, it is possible to bring in a table saw to handle the larger cutting jobs (Fig. 1-32). You can do ripping a little more safely with this type of saw because it has a rip fence. If a push stick is used to push the wood through and past the blade, it is safe to operate. Do not remove the safety guard. This saw can be used for both crosscuts and rips. The blade is lowered or raised to the thickness of the wood. It should protrude about ¼ to ½ inch above the wood being cut. This saw usually requires a 1-hp motor. This means that it will draw about 6.5 amperes to run and over 35 amperes to start. It is best not to run the saw on an extension cord. It should be wired directly to the power source with a circuit breaker installed in the line.

Radial arm saw This type of saw is brought in only if the house can be locked up at night. The saw is ex-

Fig. 1-32 *Table saw. (Power Tool Division, Rockwell International.)*

Fig. 1-33 *Radial arm saw.* (DeWalt.)

Fig. 1-34 *The handheld router has many uses in carpentry.*

SWITCH

DEPTH ADJUSTMENT RINGS

KNOBS

DEPTH ADJUSTMENT RINGS

CHUCK

BASE

pensive and too heavy to be moved every day. It should have its own circuit. The saw will draw a lot of current when it hits a knot while cutting wood (Fig. 1-33).

In this model, the moving saw blade is pulled toward the operator. In the process of being pulled toward you, the blade rotates so that it forces the wood being cut against the bench stop. Just make sure that your left hand is in the proper place when you pull the blade back with your right hand. It takes a lot of care to operate a saw of this type. The saw works well for cutting large-dimensional lumber. It will crosscut or rip. This saw also will do miter cuts at almost any angle. Once you become familiar with it, the saw can be used to bevel crosscut, bevel miter, bevel rip, and even cut circles. However, it does take practice to develop some degree of skill with this saw.

Router The router has a high-speed type of motor. It will slow down when overloaded. It takes a beginner some time to adjust to *feeding* the router properly. If you feed it too fast, it will stall the motor. If you feed it too slowly, it may not cut the way you wish or it may burn the edge you're routing. You will have to practice with this tool for some time before you're ready to use it to make furniture. It can be used for routing holes where needed. It can also be used to take the edges off laminated plastic on countertops. Use the correct bit, though. This type of tool can be used to the extent of the carpenter's imagination (Fig. 1-34).

Saw blades There are a number of saw blades available for the portable, table, or radial arm saw (Fig. 1-35). They may be standard steel types, or they may be carbide-tipped. Carbide-tipped blades tend to last a lot longer and are better for cutting plastic laminate and hardwoods.

Fig. 1-35 *Saw blades. A. Planer blade. B. Framing rip blade. C. Carbide-tipped blade. D. Metal-cutting blade.* (Black & Decker.)

Combination blades (those that can be used for both crosscut and rip) with a carbide tip give a smooth finish. They come in 7- to 7¼-inch diameter with 24 teeth. The arbor hole for mounting the blade on the saw is ¾ to ⅝ inch. A safety combination blade is also made in 10-inch-diameter size with 10 teeth and the same arbor hole sizes as the combination carbide-tipped blade.

The planer blade is used to crosscut, rip, or miter hard- or softwoods. It is 6½ or 10 inches in diameter with 50 teeth. It too can fit anything from ¾- to ⅝-inch arbors.

If you want a smooth cut on plywood without the splinters that plywood can generate, you had better use a carbide-tipped plywood blade. It is equipped with 60 teeth and can be used to cut plywood, acrylic, or plastic laminate countertop. It also can be used for straight cutoff work in hard or soft woods. Note the shape of the saw teeth to get some idea as to how each is designed for a specific job. You can identify these after using them for some time. Until you can, mark them with a grease pencil or marking pen when you take them off. A Teflon-coated blade works better when cutting treated lumber.

Saber saw The saber saw has a blade that can be used to cut circles in wood (Fig. 1-36). It can be used to cut around any circle or curve. If you are making an inside cut, it is best to drill a starter hole first. Then insert the blade into the hole and follow your mark. The saber saw is especially useful in cutting out holes for heating ducts in flooring. Another use for this type of saw is cutting holes in roof sheathing for pipes and other protrusions. The saw blade is mounted so that it cuts on the upward stroke. With a fence attached, the saw can also do ripping.

Fig. 1-36 *Saber saw.*

Drill The portable power drill is used by carpenters for many tasks. Holes must be drilled in soleplates for anchor bolts. Using an electric power drill (Fig. 1-37A) is faster and easier than drilling by hand. This drill is capable of drilling almost any size hole through dimensional lumber. A drill bit with a carbide tip enables the

carpenter to drill in concrete as well as bricks. Carpenters use this type of masonry hole to insert anchor bolts in concrete that has already hardened. Electrical boxes have to be mounted in drilled holes in brick and concrete. The job can be made easier and can be accom-

(A)

(B)

(C)

(D)

Fig. 1-37 *A. Handheld portable drill. B. A cordless hand drill with variable torque. C. A cordless drill and a cordless saw using matching batteries. D. One charger can be used to charge saw and drill batteries of same voltage.*

plished more efficiently with a portable power hand drill.

The drill has a tough, durable plastic case. Plastic cases are safer when used where there is electrical work in progress.

Carpenters are now using cordless electric drills (see Fig. 1-37B). Cordless drills can be moved about the job without the need for extension cords. Improved battery technology has made cordless drills almost as powerful as regular electric drills. The cordless drill has numbers on the chuck to show the power applied to the shaft. Keep in mind that the higher the number, the greater is the torque. At low power settings, the chuck will slip when the set level of power is reached. This allows the user to set the drill to drive screws.

Figure 1-37C shows a cordless drill and a cordless saw. This cordless technology is now used by carpenters and do-it-yourselfers. Cordless tools can be obtained in sets that use the same charger system (see Fig. 1-37D). An extra set of batteries should be kept charging at all times and then swapped out for the discharged ones. In this way, no time is lost waiting for the battery to reach full charge. Batteries for cordless tools are rated by battery voltage. High voltage gives more power than low voltage.

As a rule, battery-powered tools do not give the full power of corded tools. However, most jobs don't require full power. Uses for electric drills are limited only by the imagination of the user. The cordless feature is very handy when mounting countertops on cabinets. Sanding disks can be placed in the tool and used for finishing wood. Wall and roof parts are often screwed in place rather than nailed. Using the drill with special screwdriver bits can make the job faster than nailing.

Sanders The belt sander shown in Fig. 1-38 and the orbital sanders shown in Fig. 1-39 can do almost any required sanding job. A carpenter needs a sander occasionally. It helps to align parts properly, especially

(A)

(B)

Fig. 1-39 *Orbital sanders: (A) dual action and (B) single action.* (Black & Decker.)

those that don't fit by just a small amount. The sander can be used to finish off windows, doors, counters, cabinets, and floors. A larger model of the belt sander is used to sand floors before they are sealed and varnished. The orbital or vibrating sanders are used primarily to put a very fine finish on a piece of wood. Sandpaper is attached to the bottom of the sander. The sander is held by hand over the area to be sanded. The operator has to remove the sanding dust occasionally to see how well the job is progressing.

Nailers One of the greatest tools carpenters have acquired recently is the nailer (Fig. 1-40). It can drive nails or staples into wood better than a hammer. The nailer is operated by compressed air. The staples and nails are especially designed to be driven by the machine. Tables 1-2 and 1-3 list a variety of fasteners used with this type of machine. The stapler or nailer can also be used to install siding or trim around a window.

Fig. 1-38 *Belt sander.* (Black & Decker.)

TABLE 1-2 *Fine Wire Staples for a Pneumatic Staple Driver*

FINE WIRE

JN / JN2330
| 3.9 5/32 A02 | 4.8 3/16 A03 | 6.4 1/4 A04 | 7.9 5/16 A05 | 9.5 3/8 A06 | 12.7 1/2 A08 | 15.9 5/8 D10 |

CROWN: 3/16 (4.8)
WIDTH: 0.030
THICKNESS: 0.0215

J / DP
| 3.1 1/8 B01 | 3.9 5/32 B02 | 4.8 3/16 B03 | 6.4 1/4 B04 | 7.9 5/16 B05 | 9.5 3/8 B06 | 12.7 1/2 B08 |

CROWN: 1/2 (12.7)
WIDTH: 0.030
THICKNESS: 0.0215

JG
| 3.9 5/32 C02 | 4.8 3/16 C03 | 6.4 1/4 C04 | 7.9 5/16 C05 | 9.5 3/8 C06 | 11.1 7/16 C07 | 12.7 1/2 C08 | 14.3 9/16 C09 | 15.9 5/8 C10 |

CROWN: 3/8 (9.5)
WIDTH: 0.030
THICKNESS: 0.0215

JN5
| 4.8 3/16 E03 | 6.4 1/4 E04 | 7.9 5/16 E05 | 9.5 3/8 E06 | 11.1 7/16 E07 | 12.7 1/2 E08 | 14.3 9/16 E09 | 15.9 5/8 E10 |

CROWN: 3/16 (4.8)
WIDTH: 0.050
THICKNESS: 0.019

J5 / DFP & F50
| 3.9 5/32 F02 | 4.8 3/16 F03 | 6.4 1/4 F04 | 7.9 5/16 F05 | 9.5 3/8 F06 | 12.7 1/2 F08 |

CROWN: 1/2 (12.7)
WIDTH: 0.050
THICKNESS: 0.019

JG45 & LG45
| 6.4 1/4 G04 | 7.9 5/16 G05 | 9.5 3/8 G06 | 12.7 1/2 G08 | 14.3 9/16 G09 | 15.9 5/8 G10 |

CROWN: 3/8 (9.5)
WIDTH: 0.045
THICKNESS: 0.023

A
| 6.4 1/4 H04 | 9.5 3/8 H06 |

CROWN: 7/16 (11.1)
WIDTH: 0.050
THICKNESS: 0.019

L75 & F75
| 9.5 3/8 J06 | 12.7 1/2 J08 | 15.9 5/8 J10 |

CROWN: 7/16 (11.1)
WIDTH: 0.075
THICKNESS: 0.023

LN2532
| 15.9 5/8 W10 | 19.1 3/4 W11 |

CROWN: 3/16 (4.8)
WIDTH: 0.032
THICKNESS: 0.025

MEDIUM WIRE

LN4450 / KN4450
| 9.5 3/8 L06 | 12.7 1/2 L08 | 15.9 5/8 L10 | 19.1 3/4 L11 | 22.2 7/8 L12 | 25.4 1 L13 | 28.6 1 1/8 L14 | 31.8 1 1/4 L15 | 38.1 1 1/2 L17 |

CROWN: 1/4 (6.4)
WIDTH: 0.050
THICKNESS: 0.044
GAGE: 18

LG4450 / KG4450
| 9.5 3/8 M06 | 12.7 1/2 M08 | 15.9 5/8 M10 | 19.1 3/4 M11 | 22.2 7/8 M12 | 25.4 1 M13 | 31.8 1 1/4 M15 | 38.1 1 1/2 M17 |

CROWN: 3/8 (9.5)
WIDTH: 0.050
THICKNESS: 0.044
GAGE: 18

LN3045
| 9.5 3/8 K06 | 12.7 1/2 K08 | 15.9 5/8 K10 | 19.1 3/4 K11 | 22.2 7/8 K12 | 25.4 1 K13 |

CROWN: 1/4 (6.4)
WIDTH: 0.045
THICKNESS: 0.030

SENCLAMPS

SC-I
| 7.9 5/16 Y05 | 11.1 7/16 Y07 | 14.3 9/16 Y09 |

CROWN: 7/16 (11.1)
THICKNESS: 0.020
GAGE: 25

KEY: 9.5 Length in millimeters
     3/8 Length in inches

TABLE 1-3 *Seven-Digit Nail Ordering System*

1st Digit: Diameter, Inches	2d Digit: Head	3d and 4th Digits: Length, Inches	5th Digit: Point	6th Digit: Wire Chem. and Finish	7th Digit: Finish
A 0.0475	**A** Brad	**08** 1/2	**A** Diam.-reg.	**A** Std. carbon-galv.	**A** Plain
D 0.072	**C** Flat	**11** 3/4	**E** Chisel	**E** Std. carb. "Weatherex" galv.	**B** Sencote
E 0.0915	**E** Flat/ring	**13** 1			**C** Painted
G 0.113	shank	**15** 1 1/4		**G** Stainless steel std. tensile	**D** Painted and sencote
H 0.120	**F** Flat/screw	**17** 1 1/2			
J 0.105	shank	**19** 1 3/4		**H** Hardened high-carbon bright basic	
K 0.131	**Y** Slight-headed	**20** 1 7/8			
U 0.080	pin	**21** 2		**P** Std. carbon bright basic	
	Z Headless pin	**22** 2 1/8			
		23 2 1/4			
		24 2 3/8			
		25 2 1/2			
		26 2 3/4			
		27 3		EXAMPLE: 10 1/4 ga. (K), flat head (C),	
		28 3 1/4		KC25AAA—2 1/2" (25), regular point (A), std. carb.	
		29 3 1/2		galvanized (A), plain, or uncoated (A) Senco-Nail	

PINS

LS-I
13.7 18.7 23.7
0.539 0.736 0.933
AZ08 AZ11 AZ13

PIN: SMOOTH SHANK, HEADLESS
GAGE: 18 (0.049 x 0.040)

LS-II
15.5 20.5 25.5
0.610 0.807 1.004
AY10 AY11 AY13

PIN: SMOOTH SHANK, SLIGHT-HEADED
GAGES: 18 (0.049 × 0.040)

FINISHING NAILS

SN-I — SFN-II B
25.4 31.8 38.1 44.5 50.8 57.2 63.5
1 1 1/4 1 1/2 1 3/4 2 2 1/4 2 1/2
DA13 DA15 DA17 DA19 DA21 DA23 UA25

NAIL: SMOOTH SHANK BRAD HEAD FINISHING
GAGES: 15 (0.072)
14 (.080) – 2 1/2 LENGTH ONLY.

RING AND SCREW SHANK NAILS

SFN-II C
38.1 44.5
1 1/2 1 3/4
EE17 EE19
0.0915
0.0915

SN-II & SN-IV
50.8 60.3
2 2 3/8
GE21 GE24
0.113
0.113

NAIL: RING SHANK, FLAT HEAD
GAGES: 13 (0.0915)
11 1/2 (0.113)

SN-IV
50.8 57.2 63.5 76.2
2 2 1/4 2 1/2 3
JF21 JF23 JF25 HF27
0.105 0.105 0.105 0.120

NAIL: SCREW SHANK, FLAT HEAD
GAGES: 12 (0.105)
11 (0.120)

FLAT HEAD NAILS

SFN-II C SFN-II C SN-II SN-II SN-IV SFN-II C SN-II SN-II SN-IV SFN-II C SN-II SN-II SN-III SN-IV SN-II SN-III SN-IV SN-III SN-IV SN-III SN-IV
38.1 47.6 50.8 53.9 57.2 60.3 63.5 76.2 82.6 88.9
1 1/2 1 7/8 2 2 1/8 2 1/4 2 3/8 2 1/2 3 3 1/4 3 1/2
EC17 EC20 GC21 EC22 GC23 EC24 GC25 HC27 KC28 KC29
0.0915 0.0915 0.113 0.0915 0.113 0.0915 0.113 0.120 0.131 0.131

HEAD HEAD
SHANK SHANK
POINT POINT

NOTE: Model numbers preceded by SN, LS, and SFN or tools drive designated fasteners.

KEY: **38.1 Length in millimeters**
1 1/2 Length in inches

NAIL: SMOOTH SHANK, FLAT HEAD
GAGES: 13 (0.0915) 11 1/2 (0.113)
11 (0.120) 10 1/4 (0.131)

Fig. 1-40 *Air-powered nailer.* (Duo-Fast.)

The tool's low air-pressure requirements (60 to 90 pounds per square inch) allow it to be moved from place to place. Nails for this machine (Fig. 1-40) are from 6d to 16d. It is magazine-fed for rapid use. Just pull the trigger.

FOLLOWING CORRECT SEQUENCES

One of the most important things a carpenter must do is follow a sequence. Once you start a job, the sequence has to be followed properly to arrive at a completed house in the least amount of time.

Preparing the Site

Preparing the site may be expensive. There must be a road or street. In most cases the local ordinances require a sewer. In most locations, the storm sewer and the sanitary sewer must be in place before building starts. If a sanitary sewer is not available, you should plan for a septic tank for sewage disposal.

Figure 1-41 shows a sewer project in progress. This shows a street being extended. The storm sewer lines are visible, as is the digger. Trees had to be removed first by a bulldozer. Once the sewer lines are in, the roadbed or street must be properly prepared. Figure 1-42 shows the building of a street. Proper drainage is very important. Once the street is in and the curbs poured, it is time to locate the house.

Fig. 1-41 *Street being extended for a new subdivision.*

Fig. 1-42 *The beginning of a street.*

Figure 1-43 shows how the curb has been broken and the telephone terminal box installed in the weeds. Note the stake with a small piece of cloth on it. This marks the location of the house.

As you can see in Fig. 1-43, the curb has been removed, and a gravel bed has been put down for the driveway. The sewer manhole sticks up in the driveway. The basement has been dug. Dirt piles around it show how deep the basement really is (Fig. 1-44). However, a closer look shows that the hole isn't too deep. This means that the dirt will be pushed back against the

Fig. 1-43 *Locating a building site and removing the curb for the driveway.*

Fig. 1-44 *Dirt from the basement excavation is piled high around a building site.*

Fig. 1-45 *Hole for a basement.*

basement wall to form a higher level for the house. This will provide drainage away from the house when it is finished. Figure. 1-45 shows the basement hole.

The Basement

Figure 1-46 shows that the columns and the foundation wall have been put up. The basement is prepared in this case with courses of block with brick on the outside. This basement appears to be more of a crawl space under the first floor than a full stand-up basement. Once the basement is finished and the floor joists have been placed, the flooring is next.

Fig. 1-46 *The columns and foundation walls will help to support the floor parts.*

The Floor

Once the basement or foundation has been laid for the building, the next step is to place the floor over the joists. Note in Fig. 1-47 that the grooved flooring is laid in large sheets. This makes the job go faster and reinforces the floor.

Fig. 1-47 *Carpenters are laying plywood subflooring with tongue-and-groove joints. This is stronger.* (American Plywood Association.)

Wall Frames

Once the floor is in place and the basement entrance hole has been cut, the floor can be used to support the wall frame. The 2 × 4s or 2 × 6s for the framing can be placed on the flooring and nailed. Once together, they are pushed into the upright position as in Fig. 1-48. For a two-story house, the second floor is placed on the first-story wall supports. Then the second-floor walls are nailed together and raised into position.

Fig. 1-48 *Wall frames are erected after the floor frame is built.*

Sheathing

Once the sheathing is on and the walls are upright, it is time to concentrate on the roof (Fig. 1-49). The rafters are cut, placed into position, and nailed firmly (Fig. 1-50). They are reinforced by the proper hori-

(A)

(B)

Fig. 1-49 *Beginning construction of the roof structure. (Georgia-Pacific.)*

Fig. 1-50 *Framing and supports for rafters.*

zontal bracing. This makes sure that they are designed properly for any snow load or other loads they may experience.

Roofing

The roofing is applied after the siding is on and the rafters are erected. The roofing is completed by applying the proper underlayment and then the shingles. If asphalt shingles are used, the procedure is slightly different from that for wooden shingles. Shingles and roofing are covered in Chapter 8. Figure 1-51 shows the sheathing in place and ready for the roofing.

Fig. 1-51 *Fiberboard sheathing over the wall frame.*

Siding

After the roofing, the finishing job will have to be undertaken. The windows and doors are in place. Finishing touches are next. The plumbing and drywall already may be in. Then the siding has to be installed. In some cases, of course, it may be brick. This calls for bricklayers to finish up the exterior. Otherwise, the carpenter places siding over the walls. Figure 1-52 shows the beginning of the siding at the top left of the picture.

Figure 1-53 shows how the siding has been held in place with a stapler. The indentations in the wood show a definite pattern. The siding is nailed to the nail base underneath after a coating of tar paper (felt paper in some parts of the country) is applied to the nail base or sheathing.

Finishing

Exterior finishing requires a bit of caulking with a caulking gun. Caulk is applied to the siding that butts the windows and doors.

Finishing the interior can be done at a more leisurely pace once the exterior is enclosed. The plumbing and electrical work has to be done before the drywall or plaster is applied. Once the wallboard has been finished, the trim can be placed around the edges of the walls, floors,

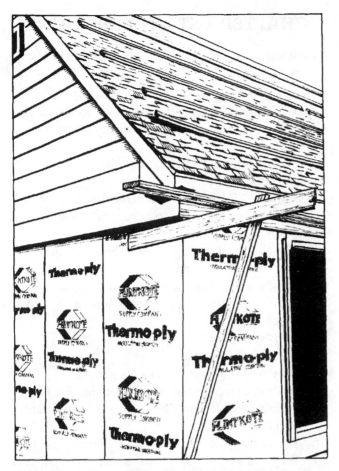

Fig. 1-52 *Siding applied on the top left side of the building.*

Fig. 1-53 *Siding applied to building. Note the pattern of the staples.*

windows, and doors. The flooring can be applied after the finishing of the walls and ceiling. The kitchen cabinets must be installed before the kitchen flooring. There is a definite sequence to all these operations.

As you can imagine, it would be impossible to place roofing on a roof that wasn't there. It takes planning and following a sequence to make sure that the roof is there when the roofing crew comes around to nail the shingles in place. The water must be there before you can flush the toilets. The electricity must be hooked up before you can turn on a light. These are rea-

sonable things. All you have to do is sit down and plan the whole operation before starting. Planning is the key to sequencing. Sequencing makes it possible for everyone to be able to do a job at the time assigned to do it.

The Laser Level

The need for plumb walls and level moldings, as well as various other straight and level points, is paramount in house building. It is difficult in some locations to establish a reference point to check for level windows, doors, and roofs, as well as ceilings and steps.

The laser level (Fig. 1-54) has eliminated much of this trouble in house building. This simple, easy-to-use tool is accurate to within ⅛ inch in 150 feet, and it has become less expensive recently so that even do-it-yourselfers can rent or buy one.

The laser level can generate a vertical reference plane for positioning a wall partition or for setting up forms (Fig. 1-55). It can produce accurate height gaug-

Fig. 1-54 *Laserspirit level moves 360 degrees horizontally and 360 degrees vertically with the optional lens attachment. It sets up quickly and simply with only two knobs to adjust.* (Stabila.)

Fig. 1-55 *The laser level can be used to align ceilings, moldings, and horizontal planes; it produces accurate locations for doorways, windows, and thresholds for precision framing and finishing.* (Stabila.)

ing and alignment of ceilings, moldings, and horizontal planes and can accurately locate doorways, windows, and thresholds for precision framing and finishing (Fig. 1-56). The laser level can aid in leveling floors, both indoors and out. It can be used to check stairs, slopes, and drains. The laser beam is easy to use and accurate in locating markings for roof pitches, and it works well in hard-to-reach situations (Fig. 1-57). The laser beam is generated by two AAA alkaline batteries that will operate for up to 16 hours.

The combination laser and spirit level quickly and accurately lays out squares and measures plumb. No protective eyewear is needed. The laser operates on a wavelength of 635 nanometers and can have an extended range of up to 250 feet.

Fig. 1-56 *The laser level can be used for indoor or outdoor leveling of floors, stairs, slopes, drains, ceilings, and moldings around the room.* (Stabila.)

Fig. 1-57 *The laser beam is used to provide easy and accurate location markings on pitches and in hard-to-reach situations.* (Stabilia.)

CHAPTER 1 STUDY QUESTIONS

1. What is carpentry involved with in terms of processes?
2. Can you learn carpentry from a book?
3. Why should a carpenter wear steel-toed shoes?
4. Why should a carpenter use safety glasses on the job?
5. Why is a hard hat suggested when a carpenter is working on site?
6. Why are gloves used on a carpentry job?
7. List at least six safety procedures that should be followed at all times.
8. List at least five important habits to acquire for safe working on the job.
9. Why would a carpenter need a folding rule?
10. What is the difference between a rip and a cross-cut saw?
11. What is a miter box?
12. What is a backsaw?
13. Why would a carpenter need a nail set?
14. What does a carpenter use a wrecking bar for?
15. What's the difference between a try square and a bevel?
16. Where does a carpenter use a caulking gun?
17. Why is the length of an extension cord important?
18. What's the difference between a table saw and a radial arm saw?
19. Why is following a sequence important?
20. How has the laser level aided in home construction?

2
CHAPTER

Site
Preparation

PLANNING IS VERY IMPORTANT TO MAKE THE FIN-ished product a success. Before you begin any construction, you must plan all aspects of the building and put it on paper. Always keep in mind that it is much cheaper to make a mistake on paper than on site.

In this chapter, you will learn how to develop these skills:

- Locate the site boundaries
- Lay out the buildings
- Use the carpenter's level
- Prepare for the start of construction

Each of these items has its importance. Locating the boundaries will allow you to properly lay out the building on the site. The carpenter's level will show you how to make sure that the building is level. No new leaning tower of Pisa is needed today. One is enough. It would be very hard to sell one today. Your first concern is a site where the soil will support the weight of the building.

There is a basic sequence to follow in building. This sequence should be followed for the benefit of those who are supposed to operate as part of the team.

BASIC SEQUENCE

The basic sequence involves

1. Cruising the site and planning the job
2. Locating the boundaries
3. Locating the building area or areas
4. Defining the needed site work
5. Clearing any unwanted trees
6. Laying out the building
7. Establishing exact elevations
8. Excavating the basement or foundation slab
9. Providing for access during construction
10. Starting the delivery of materials to the site
11. Having a crew arrive to start with the footings
12. Putting the crew to work

LOCATING THE BUILDING ON THE SITE

The proper location of a building is very important. It would be embarrassing and costly to move a building once it is built. This means that a lot of things have to be checked first.

Property Boundaries

First, a clear deed to the land should be established. This can be done in the county courthouse. Check the records, or have someone who is paid for this type of work do it. An abstract of the history of the ownership of the land is usually provided. In Iowa, for instance, the abstract traces ownership back to the Louisiana Purchase of 1803. In New York State, the history of the land is traced by owners from the days of the Holland Land Company. Alabama can provide records back to the time the Creek or other Native Americans owned the land. Each state has its own history and its own procedure for establishing absolute ownership of land. Needless to say, it is best to have proof of this ownership before starting any construction project.

Surveyors should be called in to establish the limits of the property. A plot plan is drawn by the surveyors. This can be used to locate the property. Figure 2-1 is a plot plan showing the location of a house on a lot.

Sidewalks, utilities easements, and other things have to be taken into consideration. The location of the house may be specified by local ordinance. This type of

Fig. 2-1 *Plot plan.*

Fig. 2-2 Site location must be chosen carefully.

Fig. 2-3 Rough-cleared lot. Only weeds need to be taken out as the basement is dug.

ordinance usually will specify what clearance the house must have on each side. It probably will set the limits of setback from the street. You also may want to plan around trees. Since trees increase the value of the property, it is important to save as many as possible. Figure 2-2 shows a sketch of some of these considerations.

An *easement* is the right of the utilities to use the space to furnish electric power, phone service, and gas to your location and to others nearby. This means that you have given them permission to string wires or bury lines to provide their services. Keep in mind also the rights of the city or township to supply water and sewers. These easements also may cut across the property.

Laying Out the Foundation

Layout of the foundation is the critical beginning in house construction. It is a simple but extremely important process. It requires careful work. Make sure that the foundation is square and level. You will find all later jobs, from rough carpentry through finish construction and installation of cabinetry, are made much easier.

1. Make sure that your proposed house location on the lot complies with local regulations.

2. Set the house location based on required setbacks and other factors, such as the natural drainage pattern of the lot. Level or at least rough-clear the site (Fig. 2-3).

3. Lay out the foundation lines. Figure 2-4 shows the simplest method for locating these. Locate each outside corner of the house, and drive small stakes into the ground. Drive tacks into the tops of the stakes. This is to indicate the outside line of the foundation wall. This is not the footings limit, but the outside wall limit. Next, check the squareness of the house by measuring the diagonals, corner to corner, to see that they are equal. If the structure is rectangular, all diagonal measurements will be equal. You can check squareness of any corner by measuring 6 feet down one side and then 8 feet down the other side. The diagonal line between these two end points should measure exactly 10 feet. If it doesn't, the corner isn't truly square (Fig. 2-5).

4. After the corners are located and squared, drive three 2 × 4 stakes at each corner, as shown in Fig. 2-4. Locate these stakes 3 feet and 4 feet outside the actual foundation line. Then nail 1 × 6 batter boards horizontally so that their top edges are all level and at the same grade. Levelness will be checked later. Hold a string line across the tops of opposite batter boards at two corners. Using a plumb bob, adjust the line so that it is exactly over the tacks in the two corner stakes. Cut saw kerfs ¼ inch deep where the line touches the batter boards so that the string lines may be easily replaced if they are broken or disturbed. Figure 2-6 shows how carpenters in some parts of the country use a nail instead of the saw kerf to hold the thread or string. Figure 2-7 shows how the details of the location of the stake are worked out. This one is a 3-4-5 triangle, or 9 feet and 12 feet on the sides and 15 feet on the diagonal. If you use 6, 8, and 10 feet, you get a 3-4-5 triangle also. This means that

Fig. 2-4 *Staking out a basement.* (American Plywood Association.)

Fig. 2-5 *Squaring the corner and marking the point.* (American Plywood Association.)

Fig. 2-6 *Note the location of the nails on the batter board.* (U.S. Department of Agriculture.)

Fig. 2-7 *Staking and laying out the house.* (Forest Products Laboratory.)

$6 \div 2 = 3$, $8 \div 2 = 4$, and $10 \div 2 = 5$. In the other example, 9 feet $\div 3 = 3$, 12 feet $\div 3 = 4$, and 15 feet $\div 3 = 5$. Thus you have a 3-4-5 triangle in either measurement. Other combinations can be used, but these are the most common. Cut all saw kerfs the same depth. This is necessary because the string line not only defines the outside edges of the foundation but also will provide a reference line. This ensures uniform depth of footing excavation. When you have made similar cuts in all eight batter boards and strung the four lines in position, the outside foundation lines are accurately established.

5. Next, establish the lengthwise girder location. This is usually on the centerline of the house.

CENTER
STRING LINE

STRAIGHTEDGE

FOUNDATION WALL
LINE (OUTSIDE)

TEMPORARY LEVELING
STAKE

LEVEL

BATTER BOARD

Fig. 2-8 *A. Leveling the batter boards. (American Plywood Association.)*

Double-check your house plans for the exact position. This is necessary because occasionally the girder will be slightly off the centerline to support an interior bearing wall. To find the line, measure the correct distance from the corners. Then install batter boards, and locate the string line as before.

6. Check the foundation for levelness. Remember that the top of the foundation must be level around the entire perimeter of the house. The most accurate and simplest way to check this is to use a builder's level. This tool will be explained later in this chapter. The next best approach is to ensure that the batter boards, and thus the string lines, are all absolutely level. You can accomplish this with a 10- to 14-foot-long piece of straight lumber (Fig. 2-8A). Judge the straightness of the piece of lumber by sighting along the surface. Use this straightedge in conjunction with a carpenter's level. Laser levels (Fig. 2-8B) also can be used instead of a long board. Make sure that the laser is level and at the right height. Then use the red dot to indicate the height of each leveling stake. Drive temporary stakes around the house perimeter. The distance between them should not exceed the length of the straightedge. Then place one end of the straightedge on a batter board. Check for exact levelness (see Fig. 2-8). Drive another stake to the same height. Each time a stake is driven, the straightedge and level should be reversed end for end. This should ensure close accuracy in establishing the height of each stake with reference to the batter board. The final check on overall levelness comes when you level the last stake with the

Fig. 2-8 *B. A laser level on a short tripod can be used on batter boards and leveling stakes.*

batter board where you began. If the straightedge is level here, then you have a level foundation baseline. During foundation excavation, the corner stakes and temporary leveling stakes will be removed. This stresses the importance of the level batter boards and string lines. The corners and foundation levelness must be located using the string line.

THE BUILDER'S LEVEL

Practically all optical sighting and measuring instruments can be termed *surveying instruments*. Surveying, in its simplest form, means accurate measuring. Accurate measurements have been a construction requirement ever since humans started building things.

How Does It Work?

Even during the days of pyramid building, humans recognized the fact that the most accurate distant measurements were obtained with a perfectly straight line of sight. The basic principle of operation for today's modern instruments is still the same. A line of sight is a perfectly straight line. The line does not dip, sag, or curve. It is a line without weight and is continuous.

Any point along a level line of sight is exactly level with any other point along that line. The instrument itself is merely the device used to obtain this perfectly level line of sight for measurements.

Three Main Parts of a Builder's Level

1. *Telescope* (Fig. 2-9). The telescope is a precision-made optical sighting device. It has a set of carefully ground and polished lenses. They produce a clear, sharp, magnified image. The magnification of a telescope is described as its power. An 18-power telescope will make a distant object appear 18 times closer than when viewed with the naked eye. Crosshairs in the telescope permit the object sighted on to be centered exactly in the field of view.

Fig. 2-9 *The telescope on an optical level.* (David White Instruments.)

2. *Leveling vial* (Fig. 2-10). Also called the *bubble*, the leveling vial works just like the familiar carpenter's level. However, it is much more sensitive and accurate in this instrument. Four leveling screws on the instrument base permit the user to center (level) the vial bubble perfectly and thus establish a level line of sight through the telescope. A vital first step in instrument use is leveling. Instrument vials are available in various degrees of sensitivity. In general, the more sensitive the vial, the more precise are the results that may be obtained.

Fig. 2-10 *The leveling vial on an optical level.* (David White Instruments.)

3. *Circle* (Fig. 2-11). The perfectly flat plate on which the telescope rests is called the *circle*. It is marked in degrees and can be rotated in any horizontal direction. With the use of an index pointer, any horizontal angle can be measured quickly. Most instruments have a *vernier scale*. An additional scale is subdivided. It divides degrees into minutes. There are 60 minutes in each degree. There are 360 degrees in a circle.

Fig. 2-11 *The circle on an optical level.* (David White Instruments.)

Preparing the Instrument

Figure 2-12 shows a builder's level on site. Leveling the instrument is the most important operation in preparing the instrument for use.

Leveling the instrument First, secure the instrument to its tripod, and proceed to level it as follows.

Fig. 2-12 *Using the optical (or builder's) level on the job.* (David White Instruments.)

ments if necessary. When leveling is completed, it should be possible to turn the telescope in a complete circle without any changes in the position of the bubble (Fig. 2-15).

With the instrument leveled, you know that since the line of sight is perfectly straight, any point on the line of sight will be exactly level with any other point. The drawing in Fig. 2-16 shows how exactly you can check the difference in height (elevation) between two points. If the rod reading at B is 3 feet and the reading at C is 4 feet, you know that point B is 1 foot higher than point C. Use the same principle to check if a row of windows is straight or if a foundation is level. Or you can check how much a driveway slopes.

(A)   (B)

Fig. 2-13 *A. The tripod for an optical level. B. The rod holder for use with an optical level.* (David White Instruments.)

Figure 2-13A shows the type of tripod used to support the instrument. The target pole is shown in Fig. 2-13B.

Place the telescope directly over one pair of opposite leveling screws (Fig. 2-14). Turn the screws directly under the scope in opposite directions at the same time (see step 5 in Fig. 2-14) until the level bubble is centered. The telescope then is given a quarter (90-degree) turn. Place it directly over the other pair of leveling screws (step 2 in Fig. 2-14). The leveling operation is then repeated. Recheck the other positions (steps 3 and 4 in Fig. 2-14) and make adjust-

TURNING BOTH LEVELING SCREWS
"IN" MOVES BUBBLE TO RIGHT

TURNING BOTH LEVELING SCREWS
"OUT" MOVES BUBBLE TO LEFT

SIDE VIEW

TOP VIEW

LEVEL VIAL WITH
BUBBLE CENTERED

Fig. 2-15 *Adjusting the leveling screws and watching the bubble for level.* (David White Instruments.)

1   2   3   4   5

Fig. 2-14 *Adjusting the screws on the level-transit will level it. Note how it is leveled with two screws, then moved 90 degrees, and leveled again.* (David White Instruments.)

Fig. 2-16 *Finding the elevation with a level.* (David White Instruments.)

Staking out a house Start at a previously chosen corner to stake out the house. Sight along line *AB* of Fig. 2-17 to establish the front of the house. Measure the desired distance to *B*, and mark it with a stake.

Swing the telescope 90 degrees by the circle scale. Mark the desired distance to *D*. This gives you the first corner. All the others are squared off in the same manner. You're sure all foundation corners are square, and all it took was a few minutes setup time (Fig. 2-18).

This method eliminates the use of the old-fashioned string line–tape–plumb bob methods.

The Level-Transit

There are two types of levels used for building sites. The level and the level-transit are the two instruments used. The level has the telescope in a fixed horizontal position, but it can move sideways 360 degrees to measure horizontal angles. It is usually all that is needed at a building site for a house (Fig. 2-19).

Fig. 2-17 *How to stake out a house on a building lot using a builder's level.* (David White Instruments.)

Fig. 2-18 *Squaring the other corner in laying out a building on a lot.* (David White Instruments.)

Fig. 2-19 *Builder's level.* (David White Instruments.)

A combination instrument is called a *level-transit*. The telescope can move in two directions. It can move up and down 45 degrees as well as from side to side 360 degrees (Fig. 2-20). It can measure vertical as well as horizontal angles.

A lock lever or levers permit the telescope to be securely locked in a true level position for use as a level. A full-transit instrument, in addition to the features just mentioned, has a telescope that can rotate 360 degrees vertically. The level-transit is shown in operation in Fig. 2-21.

Using the Level and Level-Transit

Reading the circle and vernier The 360-degree circle is divided in quadrants (0 to 90 degrees). The circle

Fig. 2-20 *Level-transit.* (David White Instruments.)

Fig. 2-21 *Using the level-transit on the site.* (David White Instruments.)

75 DEGREES

Fig. 2-22 *Reading the circle.* (David White Instruments.)

75 DEGREES, 45 MINUTES

Fig. 2-23 *Reading the circle and vernier.* (David White Instruments.)

Fig. 2-24 *To hang a plumb bob, attach a cord to the plumb bob hook on the tripod, and knot the cord as shown here.* (David White Instruments)

is marked by degrees and numbered every 10 degrees (Fig. 2-22).

To obtain degree readings, it is only necessary to read the exact degree at the intersection of the zero index mark on the vernier and the degree mark on the circle (or on the vertical arc of the level-transit). For more precise readings, the vernier scale is used (Fig. 2-23). The vernier lets you subdivide each whole degree on the circle into fractions, or minutes. There are 60 minutes in a degree. If the vernier zero does not line up exactly with a degree mark on the circle, note the last degree mark passed, and reading up the vernier scale, locate a vernier mark that coincides with a circle mark. This will indicate your reading in degrees and minutes.

Hanging the plumb bob   To hang the plumb bob, attach a cord to the plumb bob hook on the tripod. Knot the cord as shown in Fig. 2-24.

If you are setting up over a point, attach the plumb bob. Move the tripod and instrument over the approximate point. Be sure that the tripod is set up firmly again. Shift the instrument on the tripod head until the plumb bob is directly over the point. Then set the instrument leveling screws again to level the instrument.

Power   The power of a telescope is rated in terms of magnification. It may be 24X or 37X. The 24X means that the telescope is presenting a view that is 24 times as close as you could see it with the naked eye. Some instruments are equipped with a feature that lets you zoom in from 24X to 37X. This increases the effective reading range of the instrument by more than 42 percent. It also permits greater flexibility in matching range, image, and light conditions. Use low power for brighter images in dim light. Since it gives a wider field of view, it is also handy for locating targets. Low power also provides better visibility for sighting through heat waves (Fig. 2-25). High power is used for sighting under bright-light conditions. It is used for long-range sighting and for more precise rod readings.

24 X   37 X

Fig. 2-25 *Variable instrument power is available.* (David White Instruments.)

Rods Leveling rods are a necessary part of the transit leveling equipment. Rods provide direct reading with large graduations. All rods are equipped with a tough, permanent polyester film scale that will not shrink or expand. This is important when you consider the graduations can be 1/100 of a foot. Figure 2-26 shows a leveling rod with the target attached at 4 feet, 5¼ inches. The target (Fig. 2-27) can be moved by releasing a small clamp in the back. Figure 2-28 shows a tape and the graduations. They are ⅛ inch wide and ⅛ inch apart. The tape is marked in feet, inches, and eighths of an inch. Feet are numbered in red. A three-section rod extends to 12 feet. A two-section rod extends to 8 feet, 2 inches. The rod holder is directed by hand signals from the surveyor behind the transit. The hand signals are easy to understand because they move in the direction of desired movement of the rod.

Fig. 2-26 *Leveling rod made of wood.* (David White Instruments.)

Fig. 2-27 *Target that fits wood rods.* (David White Instruments.)

Fig. 2-28 *Tape face on the rod. This one is marked in feet, inches, and eighths.* (David White Instruments.)

Establishing Elevations

Not all lots are flat. This means that there is some kind of slope to be considered when digging the basement or locating the house. The level can help to establish what these elevation changes are. From the grade line, you establish how much soil will have to be removed for a basement. The grade line will also determine the location of the floor.

The benchmark is the place to start. A benchmark is established by surveyors when they open a section to development. This point is a reference to which the lot you are using is tied. The lot is so many feet in a certain direction from a given benchmark.

The benchmark may appear as a mark or point on the foundation of a nearby building. Sometimes it is the nearby sidewalk, street, or curb that is used as the level reference point.

The grade line is established by the person who designed the building. This line must be established accurately. Many measurements are made from this line. It determines the amount of earth removed from the basement or for the foundation footings.

Using the Leveling Rod

Use a leveling rod, and set it at any point at which you want to check the elevation. Sight through the level or transit-level to the leveling rod. Take a reading by using the crosshairs in the telescope. Move the rod to another point that is to be established. Now raise or lower the rod until the reading is the same as for the first point. This means the bottom of the rod is at the same elevation as the original point.

One person will hold the rod level. Another will move the target up or down until the crosshairs in the telescope come into alignment with the target. The difference between the two readings tells you what the elevation is.

Fig. 2-29 *Getting the elevation when the two points cannot be viewed from a single point.*

Figure 2-29 shows how the difference in elevation between two points that are not visible from a single point is determined. If point *Z* cannot be seen from point *W*, then you have to set the transit up again at two other points, such as *X* and *Y*. Take the readings at each location; then you will be able to determine how much of the soil has to be removed for a basement.

PREPARING THE SITE

Clearing

One of the first things to do in preparing a site for construction is to clear the area where the building will be located. Look over the site. Determine if there are trees in the immediate area of the house. If so, mark the trees to be removed. This can be done with a spray can of paint. Put an X on the trees to be removed or a line around them. In some cases, people have marked the trees that must go with a piece of cloth tied to a limb.

Also make sure that the trees that are staying are not damaged when the heavy equipment is brought onto the site. Scarring trees can cause them to die later. Covering their bases by more than 12 inches probably will also kill them. You have to cut off a part of the treetop. This helps it to survive the covering of its roots.

Don't dig the sewer trench or the water lines through the root system of the trees to be saved. This can cause the tops of the trees to die later and in some cases will kill them altogether.

Make a rough drawing of the location of the house and the trees to be saved. Make sure the people operating the bulldozer and digger are made aware of the effort to save the trees.

Cutting trees   Keep in mind that removing trees can also be profitable. You can cut the trees into small logs for use in fireplaces. This has become an interest of many energy conservationists. The brush and undergrowth can be removed with a bulldozer or other type of equipment. Do not burn the brush or the limbs without checking with local authorities. There is always someone who is interested in hauling off the accumulation of wood.

Stump removal   In some cases, a tree stump is left and must be removed. There are a number of ways of doing this. One is to use a winch and pull it up by hooking the winch to some type of power takeoff on a truck, tractor, or piece of heavy equipment. You could dig it out, but this can take time and too much effort in most cases.

The use of explosives to remove a stump is not permitted in some locations. It is best to check with the local police before setting off a blast.

The best way is to use the bulldozer to uproot the entire stump or tree. It all depends on the size of the tree and the size of the equipment available for the job. Anyway, be sure the lot is cleared so that the digging of the basement or footings can take place.

Excavation

A house built on a slab does not require any extensive excavation. One-piece or monolithic slabs are used on level ground and in warm climates. In cold climates, where the frost line penetrates deeper or in areas where drainage is a problem, a two-piece slab has to be used. Figures 2-30 and 2-31 both show two-piece slab foundations.

Slab footings must rest beneath the frost line. This gives stability in the soil. The amount of reinforcement needed for a slab varies. The condition of the soil and the weight to be carried determine the reinforcement. Larger slabs and those on less stable soil need more reinforcement.

The top of the slab must be 8 inches above ground. This allows moisture under the slab to drain away from the building. It also gives you a good chance to spot termites building their tunnels from the earth to the floor of the house. The slab always should rest slightly above the existing grade. This is to provide for runoff water during a rainstorm.

Fig. 2-30 *Two-piece slab with a block foundation wall.* (Forest Products Laboratory.)

Fig. 2-31 *Two-piece slab with poured wall and footer.*

Basements A basement is the area usually located underground. It provides most homes with a lot of storage space. In some, it is a place to do the laundry. It also serves as a place to locate the heating and cooling units. If a basement is desired, it must be dug before the house is started. The footings must be poured and seasoned properly. Seasoning should be done before poured concrete or concrete block is used for the wall. Some areas now use treated wood walls for a basement.

Figure 2-32 shows a basement dug for use in a colder climate. Trenches from the street to the basement must be provided for the plumbing and water. Utilities may be buried also. If they are, the electric, phone, and gas lines must also be located in trenches or buried after the house is finished. It is a good idea to notify the utility companies so that they can schedule the installation of their services when you are ready for them.

Some shovel work may have to be done to dig the basement trenches for the sewer pipes. This is done after the basement has been leveled by machines. As you can see from Fig. 2-32, the basement may also need shovel work after the digger has left. Note the cave-ins and dirt slides evident in the basement excavation in Fig. 2-32.

The basement has to be filled later. Gravel is used to form a base for the poured concrete floor. The high spots in the basement must be removed by shovel. Proper-size gravel should be spread after the sewer trench is filled. You may have to tamp the gravel to make sure that it is properly level and settled. Do this before the concrete mixer is called for the floor job.

Fig. 2-32 *Excavation for a basement.*

The footings have to be poured first. They are boxed in and poured before anything is done for the basement floor. In some instances, drain tile must be installed inside or outside the footing. The tile is allowed to drain into a sump. In other locations, no drainage is necessary because of soil conditions.

PROVIDING ACCESS DURING CONSTRUCTION

The first thing to be established is who is to be on the premises. Check with your insurance company about liability insurance. This is in case someone is hurt on the location. Also decide who should be kept out. You also have to decide how access control is to be set up. It may be done with a fence or by an alert guard or dog. These things do have to be considered before the construction gets underway. If equipment is left at the site, who is responsible? Who will pay for vandalism? Who will repair damage caused by wind, hail, rain, lightning, or tornado?

Materials Storage

Where will materials for the job be stored? In Fig. 2-33, you can see how plywood is stored. What happens if someone decides to haul off some of the plywood?

Fig. 2-33 *B. Bundled lumber delivered to the building site.*

Who is responsible? What control do you have over the stored materials after dark?

Figure 2-34 shows plywood bundles broken open. This makes it easy for single sheets to disappear. With the current price of plywood, it becomes important to plan some type of storage facility on the site.

Some of the shingles in Fig. 2-35 may be hard to find if the wind gets to the broken bundle. What's to stop children playing on the site during off hours? They also can take the shingles and spread them over the landscape.

Storage of bricks can be a problem (Fig. 2-36). They are expensive and can be removed easily by someone with a small truck. It is very important to have some type of on-site storage. It is also very important to make sure that materials are not delivered before they are needed. Some type of materials inventory has to be maintained. This may be worth a person's time. The location of the site is a major factor in

Fig. 2-33 *A. Storing plywood on the site.*

Fig. 2-34 *Broken bundle of plywood sheathing.*

Fig. 2-35 *Broken bundle of shingles.*

Fig. 2-36 *Storing bricks on site.*

the disappearance of materials. Location has a lot to do with the liability coverage needed from insurance companies.

Temporary buildings Some building sites have temporary structures to use as storage. In some cases, the plans for the building are also stored in the tool shed. Covered storage is used in some locations where rain and snow can cause a delay by wetting the lumber, sand, or cement. If you are using drywall, you will need to keep it dry. In most instances, it is not delivered to the site until the house is enclosed.

Some construction sheds are made on the scene. In other cases, the construction shed may be delivered to the site on a truck. It is picked up and moved away once the building can be locked.

Mobile homes have been used as offices for supervisors. This usually is the case when a number of houses are made by one contractor, and all are located in one row or subdivision.

A garage can be used as the headquarters for construction. The garage is enclosed. It is closed off by doors so that it can be locked at night. Since the garage is easy to close off, it becomes the logical place to take care of paperwork. It also becomes a place to store materials that should not get wet.

When building a smaller house, the carpenter takes everything home at the end of the working day. The carpenter's car or truck becomes the working office away from home. Materials are scheduled for delivery only when actually needed. In larger projects, some local office is needed, so the garage, tool shed, mobile home, or construction shack is used.

Storing construction materials Storing construction materials can be a problem. It requires a great deal of effort to make sure that the materials are on the job when they are needed. If delivery schedules are delayed, work has to stop. This puts people out of work.

If materials are stored on the job, make sure that they are neatly arranged. This prevents accidents such as tripping over scattered materials. Sand should be delivered and placed out of the way. Keep it out of the normal traffic flow from the street to the building.

Everything should be kept in some order. This means that you know where things are when you need them. Then you don't have to plow through a mound of supplies just to find a box of nails. Everything should be laid out according to its intended order of use.

Lumber should be kept flat. This prevents warpage, cupping, and twisting. Plywood should be protected from rain and snow if it is interior grade. In any case, it should not be allowed to become soaked. Keep it flat and covered.

Humidity control is important inside a house. This is especially true when you're working with drywall. It should be allowed to dry by keeping the windows open. Too much humidity can cause the wood to twist or warp.

Temporary Utilities

You will need electricity to operate your power tools. Power can be obtained by using a long extension cord from a nearby house. Or you may have to arrange for the power company to extend a line to the side and put in a meter on a nearby pole.

Water is needed to mix mortar. The local line will have to be tapped, or you may have to dig the well before you start construction. It all depends on where the building site is located. If the house is being built near another house, you may want to arrange with the neighbor to supply water with a hose from the outside faucet. Make sure that you arrange to pay for the service.

Waste Disposal

Every building site has waste. It may be human waste or paper and building-material wrappings. Human waste can be controlled by renting a porta john or a johnny-on-the-spot. This can prevent the house from smelling like a urinal when you enter. The sump basin should not be used as a urinal. It does leave an odor in the place. Besides, it is unsanitary.

Waste paper can be burned in some localities. In others, burning is strictly forbidden. You should check before you arrange to have a large bonfire for getting rid of the trash, cut lumber ends, paper, and loose shingles. There are companies that provide a trash-collecting service for construction areas. They leave the place *broom clean*. It leaves a better impression of the contractor when a building is delivered in order without trash and wood pieces lying around. If you go to the trouble of building a fine home for someone, the least you can do is deliver it in a clean condition. After all, this is going to be a home.

Arranging Delivery Routes

Damage to the construction site by delivery trucks can cause problems later. You should arrange a driveway by putting in gravel at the planned location of the drive. Get permission and remove the curb at the entrance to the driveway. Make sure that deliveries are made by this route. Pile the materials so that they are arranged in an orderly manner and can be reached when needed.

Concrete has to be delivered to the site for the basement, foundation, and garage floor. Be sure to allow room for the ready-mix truck to get to these locations. Lumber is usually strapped together. Make allowance for bundles of lumber to be dumped near the location where they will be needed.

Make sure that nearby plants or trees are protected. This may require a fence or stakes. Some method should be devised to keep the trucks, diggers, and earth movers from destroying natural vegetation.

Access to the building site is important. If this is the first house in a subdivision, or if it is located off the road, you have to provide for the delivery of materials. You may have to put in a temporary road. This should be a road that can be traveled in wet weather without the delivery trucks becoming bogged down or stuck.

As you can see, it takes much planning to accomplish a building program that will come off smoothly. The more planning you do ahead of time, the less time that will be spent trying to obtain the correct permissions and deliveries.

The key to a successful building program is planning. Make a checklist of the items that need attention beforehand. Use this checklist to keep yourself current with the delivery of materials and permissions. It is assumed here that proper financial arrangements have been made before construction begins.

Fig. 2-37 *Floor plan.*

Fig. 2-38 *Sketch of the exterior of the house.*

Check the floor plan of the proposed house before proceeding. Make sure that everything desired is included. Later changes can get very expensive (see Fig. 2-37). Also check the sketch of the proposed house to make sure that the exterior is what you want (Fig. 2-38).

CHAPTER 2 STUDY QUESTIONS

1. What is a builder's level?
2. What is the difference between a builder's level and a level-transit?
3. Why is it important to have a clear title to a piece of land before building?
4. Who establishes the correct property limits?
5. What is an easement?
6. Why should a house be laid out square and level?
7. How do you set up a builder's level?
8. What is an elevation?
9. What is a vernier?
10. What is a plumb bob?
11. How is a plumb bob used by a carpenter?
12. How is the power of a telescope on a builder's level rated?
13. What is a leveling rod?
14. What is a target on a leveling rod?
15. What methods are used to remove stumps from a building site?
16. What does excavate mean?
17. Who has access to a house under normal conditions?
18. Why is some shovel work needed after the digger has excavated the basement?
19. What is the purpose of temporary buildings at a construction site?
20. Why would you want to arrange delivery routes to a building site?

3
CHAPTER

Footings and Foundations

OOTINGS AND FOUNDATIONS ARE NOT THE SAME. A *footing* is the lowest part of a building. It carries all the weight. A *foundation* is a wall between the footing and the rest of the building.

In this chapter you will learn how to

- Design footings and foundations
- Locate corners and lines for forms
- Check the level of the footing
- Check foundation excavations
- Make forms for footings
- Make forms for foundations
- Reinforce forms as required
- Mix or select concrete for filling the forms
- Pour concrete into the forms
- Finish concrete in the forms
- Embed anchor systems in the forms
- Waterproof foundation walls, if needed
- Make necessary drainage systems

FOOTINGS

The weight of a building is borne by the footings. The footings spread the weight evenly over a wide surface. Figure 3-1 illustrates this by showing the three parts of a footing. The parts shown are

- The bearing surface
- The footing
- The foundation

The bearing surface is located beneath the frost line on firm and solid ground. The frost line is the deepest level where the ground will freeze in the win-

ter. Above the frost line, the ground moisture will freeze and thaw. When this happens, the ground will move and shift. This motion breaks or damages the footing and sometimes the foundation. Therefore, location and construction of the footing are very important. The weight of all the lumber, concrete, stone, and furniture must be supported by this layer. All this weight must be supported without sinking or moving.

Footings can be made in many ways. There are *flat footings*, *stepped footings*, *pillared or post footings*, and *pile footings*. A flat footing is shown in Fig. 3-2. It is the easiest and simplest footing to make. This is so because it is all on one level. A stepped footing is used on sides of hills, as seen in Fig. 3-3. A stepped footing is similar to a series of short, flat footings at

Fig. 3-2   *Regular flat footing.*

Fig. 3-3   *A stepped footing is used on hills or slopes.*

Fig. 3-1   *Parts of a footing and a foundation.*

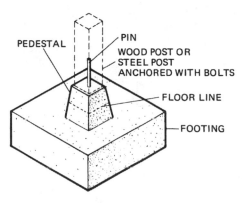

Fig. 3-4 *A pillar or post footing may be square or round.*

various levels, much like a flight of steps. By using this type of footing, no special digging (excavation) is needed. A pillard or post footing is the third type of footing (Fig. 3-4). It is used in locations where the soil is evenly packed and very little settling occurs. It consists of a series of pads or feet. Once the pads or feet are cured properly, columns are built on them, and the building rests on the columns. Buildings with either flat or stepped footings usually will have pillared footings in the center. That is so because buildings are unable to support their full weight without support in the middle areas.

A pile footing is the fourth type and is used where soil is loose, unstable, or very wet (Fig. 3-5). Long columns are put into the ground. These are long enough to reach solid soil or rock. The columns may be made of treated wood, steel, or concrete. Wooden or steel piles are driven into the ground by pounding using a pile-driver machine. Concrete piles are made by

Fig. 3-5 *Pile footings reach through water or shifting soils.*

drilling a hole and filling it with concrete. Pads or caps are put over the tops of the piles.

SEQUENCE

The order in which something is made or done is called a *sequence*. No matter what type of footing and foundation is used, a certain sequence should be followed. The sequence can change slightly based on the method involved. For example, the footing and foundation can be poured in one solid concrete piece. However, many footings are made separately, and the concrete foundation is built on top of the footings. In both cases, the sequence is similar. The basic sequence is

1. Find the amount of site preparation needed.
2. Lay out the footing and foundation shape.
3. Excavate to the proper depth.
4. Level the footing corners.
5. Build the footing forms.
6. Reinforce the forms as needed.
7. Estimate concrete needs.
8. Pour the concrete footing.
9. Build the foundation forms.
10. Reinforce the forms as needed.
11. Pour the concrete into the forms.
12. Finish the concrete and embed anchors.
13. Remove the forms.
14. Waterproof and drain as required.

LAY OUT THE FOOTINGS

Footings are at the bottom of the building and must support the weight of the building and its contents. Two factors are involved in finding the correct shape and size. The first is the strength or solidness of the soil. The second is the width and depth of the footing. This varies for the weight of the building and the type of soil.

Soil Strength

Soil strength refers to the density of the soil. This means how solid the soil is packed and how stable or unmoving the soil is. Some soils are very hard, but only when dry. Others keep their strength whether they are wet or dry. Needless to say, the soil must be dense and strong enough to support the weight of the building. When the soil is soft, the footing is made wider to spread the weight over more surface. In this way, each surface unit holds up less weight. Figure 3-6 shows

Type of Soil	Bearing Capacity (pounds per ft²)
Soft Clay loose dirt, etc.	2 000
Loose Sand hard clay, etc.	4 000
Hard Sand or Gravel	6 000
Partially Cemented Sand or Gravel soft stone, etc.	20 000

Fig. 3-6 *Bearing capacity of typical soils.*

FLOORS	BASE-MENT	ALL WOOD FRAME		WOOD FRAME WITH MASONRY VENEER	
		T	P	T	P
1	None	6"	3"	6"	4"
	Yes	6"	2"	6"	3"
2	None	6"	3"	6"	4"
	Yes	6"	4"	8"	5"

NOTE: For soil with 2000 pounds per square foot (PSF) load capacity.

Fig. 3-7 *Typical footing size.*

how much weight various soil types will support. Standard footings should not be poured on loose soil.

Footing Width

The second factor is the width of the footing. Keep in mind that the footing should be wider for soft soil. Check Fig. 3-7 for typical sizes for footings. Generally, footings are about two times as wide as they are thick. The average footing is about 8 inches thick, and the footing is about the same thickness as the foundation wall.

Locating Footing Depth

Local building codes determine the exact details on locating the footing. Usually, footings are laid out several inches below the frost line. However, in buildings with basements, the top of the footing is placed 12 inches below the frost line. For buildings that do not have basements, 4 to 6 inches below the frost line is deep enough.

Footings under Columns

Visible footings and foundations that most people see only support the outside walls. Most houses are wide, however, and support is needed in the center of a wide building. This support is from footings, pillars, or columns built in the center. Pillars or columns must have a footing just as the outside walls do. For houses with basements, the footings and pillars become part of the basement floor and walls (Fig. 3-8). Houses that do

Fig. 3-8 *Footings in a basement later became a part of the basement floor.*

Fig. 3-9 *Footings and piers must be located in crawl spaces.*

not have basements usually have a crawl space between the ground and the floor. This crawl space provides access to the pipes and utilities. Pillars or columns built on footings are used for supports in the crawl space. The footings may be any shape—square, rectangular, or round. Figure 3-9 shows a site prepared in this manner.

Basements and crawl spaces have footings that are similar. They should be below the frost line as in a regular footing. However, they carry more weight than do the outside footings. For this reason, they should be 2 to 3 feet square.

Special Strength Needs

Heavier areas of a building, such as chimneys, fireplaces, bases for special machinery, and other similar things, require wider and thicker footings. For chimneys in a one-story building, the footing should project at least 4 inches on each side. The chimneys on two-story buildings are taller and heavier. Therefore, the footing should project 6 to 8 inches on each side of the chimney (Fig. 3-10).

Fig. 3-10 *A special footing is used for fireplaces. It supports the extra weight.*

Reinforcement and Strength

In order to make the footing stronger, it is reinforced with steel rods. Then the footing is also matched or keyed so that the foundation wall will not shift or slide.

Reinforcement   In most cases, the footing should be reinforced with rods. These reinforcement rods are called *rebar*. Two or more pieces of rebar are used. The rebar should be located so that at least 3 inches of space for concrete is left around all edges (Fig. 3-11).

Fig. 3-11 *Footings may be reinforced. Note the key to keep the foundation from shifting.*

Keyed footings   The best type of separate footing is keyed, as shown in Fig. 3-11. This means that the footing has a key or slot formed in the top. The slot is filled when the foundation is formed. The key keeps the foundation from sliding or moving off the footing. Without a key, freezing and thawing of water in the ground could force the foundation walls off the footing.

EXCAVATING THE FOOTINGS

Chapter 2 explained the procedure for locating the building on the lot. Batter boards were put up, and lines were strung from them to show the location of the walls and corners. Next, the size, shape, and depth of the footings must be decided.

Finding Trench Depth

Trenches or ditches must be dug or excavated for the footings and foundation. Ground that is extremely rough and uneven should be rough graded before the excavation is begun. The topsoil that is removed can be piled at one edge of the building site. It can be used later when the ground is smoothed and graded around the building. Before the digging is started, determine how deep it is to be.

The trench at the lowest part of the site must be deep enough for the footing to be below the frost line.

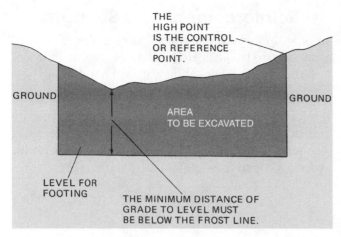

Fig. 3-12 *Footings must be below the frost line.*

If the footing is to be 12 inches below the frost line, the trench at the lowest part must be deep enough for this. Figure 3-12 shows these depths. This lowest point becomes the level line for the entire footing. Elevations are taken at each corner to find out how deep the trenches are at each corner.

Excavating for Deep Footings

Deep footings are needed in areas where the frost line is deep. Deep footings are also necessary when a basement is called for in the plans.

Draw rough lines on the ground. They do not need to be very accurate. The lines from the batter boards are used as guides. However, the rough line should be about 2 feet outside the batter-board line (Fig. 3-13). The trench for the footings is dug much wider than the footing so that there is room to work. Since the foot-

Fig. 3-13 The trench is wider than the footing and sloped for safety.

ings are made of concrete, the molds (called *forms*) for the concrete must be built. Room is needed to build or put up the forms. Work that must be done after the footing or foundation is formed includes

• Removing the forms
• Waterproofing the walls
• Making proper drainage

Measure the depth as the trench is dug. When the trench has been dug to the correct depth, the machinery is removed. Forms then are laid out. For basements, the interior ground is also excavated.

Excavating for Shallow Footings

Mark rough lines on the ground with chalk or a shovel. These lines should be marked to show the width of the footing desired. Corner stakes are removed, and lines are taken from the batter boards. Dig a trench to the correct depth. Special footings for the interior are also excavated at this point. The excavation for interior pad footings should be made to the same depth as for the outside walls.

Concrete is poured directly into these trenches. Any reinforcement is made without forms in the excavation itself. The rebar is suspended with rebar stakes or metal supports called *chairs* (Fig. 3-14).

(A)

(B)

Fig. 3-14 Chairs are used to hold rebar in place.

Types and sizes of bar supports

SYMBOL	BAR SUPPORT ILLUSTRATION	BAR SUPPORT ILLUSTRATION PLASTIC CAPPED OR DIPPED	TYPE OF SUPPORT	SIZES
SB		CAPPED	Slab Bolster	¾, 1, 1½, and 2 inch heights in 5 ft. and 10 ft. lengths
SBU*			Slab Bolster Upper	Same as SB
BB		CAPPED	Beam Bolster	1, 1½, 2, over 2" to 5" heights in increments of ¼" in lengths of 5 ft.
BBU*			Beam Bolster Upper	Same as BB
BC		DIPPED	Individual Bar Chair	¾, 1, 1½, and 1¾" heights
JC		DIPPED DIPPED	Joist Chair	4, 5, and 6 inch widths and ¾, 1 and 1½ inch heights
HC		CAPPED	Individual High Chair	2 to 15 inch heights in increments of ¼ inch
HCM*			High Chair for Metal Deck	2 to 15 inch heights in increments of ¼ in.
CHC		CAPPED	Continuous High Chair	Same as HC in 5 foot and 10 foot lengths

(C)

Fig. 3-14 *Chairs are used to hold rebar in place.*

Note how to form a key in this type of footing. Stakes are driven along the edges, as in Fig. 3-15. The board that forms the key in the footing is suspended in the center of the trench area.

In some parts of the country, concrete block wall is used on this type of footing. The blocks may be secured by inserting rebars into the footing area. The bricks or blocks are laid so that the rebar is centered in an opening in the block. The opening is then filled with mortar or concrete to secure the foundation against slipping.

Remember, it is important to know that special forms are not used with shallow footings. Also, they may not be finished smooth. As a result, they may ap-

Fig. 3-15 *Keys are made by suspending 2 × 4s in the form. A. Keys for trench forms. B. Keys for board forms.*

pear very rough or unfinished. This is not important if they are the proper shape.

Slab Footings and Basements

Slab footings are used in areas where concrete floors are made. Slabs can combine the concrete floor and the footing as one unit. Slabs, basement floors, and other large concrete surfaces are detailed in Chapter 4. Basement floors are made separately from the footings and are done after the footings and basement walls are up.

BUILDING THE FORMS FOR THE FOOTINGS

Once the excavation has been completed, check the corners. They must be relocated. After the corners are relocated, build the forms and level them. Pour the concrete, and allow it to harden. Next, remove the forms, and the foundation is erected. In many cases, the footing and the foundation are made as one piece.

Forms Layout

The excavation is completed, and then the first step is to relocate the corners and edges for the walls. Be sure to do this. The lines from the batter boards are restrung, and a plumb bob is used to locate the corner points. The corner points and other reference points are marked with a stake. The stake is driven level with the top of the footing. This level is established by using a transit or a level. See Chapter 2 for the procedure.

Nails

Use double-headed, or duplex, nails for making the forms. Forms should be nailed with the nails on the outside. This means that the nails are not in the space where the concrete will be. In this way, the nailheads do not get embedded in the concrete and are left exposed for ease in extraction later. The double head allows the nail to be driven up tight but still permits easy removal when the forms are taken apart.

Putting Up the Forms

The corner stakes are used for location and leveling the walls of the forms. The amount that the footing is to project past the wall is determined. Usually, this is one-half the thickness of the foundation wall. This dimension is needed because the corner indicates the foundation corner and not the footing corner. Stakes then are driven outside the lines so that the form will be the proper width. The carpenter must allow for the width of the stake and the width of the boards used for the forms (Fig. 3-16). Drive stakes as needed for support. As a rule, the distance between stakes is about twice the width of the footing. Nail the top board to the first stake, and level the top board in two directions. For the first direction, the top board is leveled with the corner stake. For the second direction, the top board is leveled on its length (Fig. 3-17). After the top board is leveled, nail it to all stakes. Then nail the lower boards to the stakes. Both inside and outside forms are made this way.

When 1-inch-thick boards are used to build the form, the stakes should be driven closer together. However, if boards 2 inches thick are used, the stakes may be 4 to 6 feet apart. In both cases, the stakes are braced as shown in Fig. 3-18.

Remove the loose dirt from under the footing form. It is best for the footing to be deeper than is needed. Also, keep these three points in mind:

1. Never make a footing thinner than the specifications.

CHALK LINE OR WIRE

BATTER BOARD

SAW KERF OR NAIL

CORNER STAKE

OUTSIDE FORM

Fig. 3-16 *The footing corner is located and leveled.*

Fig. 3-17 *The form is leveled all around.*

Fig. 3-18 *Bracing form boards.*

2. Never fill any irregular hole or area with loose dirt.

3. Always fill with gravel or coarse sand, and tamp it firmly in place.

The keyed notch Notice how the key or slot in the footing is made with a board. The board is nailed to a brace that reaches across the top of the forms. The brace should be nailed in place at intervals of about 4 feet apart. See Fig. 3-15 for how the key is made.

Excavation for drains and utility lines Sometimes drain pipes and utility lines are located beneath the footings in a building. When this happens, trenches are dug underneath the footing forms. These trenches are usually dug by hand. After the drain pipes or utility lines are put in place, the area is filled with coarse gravel or sand. The gravel or sand is then tightly packed in place beneath the form.

Spacing the walls of the form The weight of the concrete can make the walls spread apart. Use braces to keep the walls straight. Since the braces on the walls provide some support, special braces called *spreaders* are nailed across the top to make sure that the forms do not spread. Forms should be braced properly so that the amount of concrete ordered fills the forms. Also,

this practice ensures that excess concrete does not add extra weight to the building. The forms should be checked to make sure that there are no holes, gaps, or weak areas. These would allow the concrete to leak out of the form and thus weaken the structure. Such leaks are called *blowouts*.

WORKING WITH CONCRETE

Before the concrete is ordered and poured, several things have to be done:

• Check the forms for the proper depth and level.

• Make the openings and trenches beneath the footing area for pipes and utility lines.

• Level and fill the trenches for pipes and utility lines.

• Check the forms again to make sure that they are braced and spaced properly.

• Remove the chalk lines and corner stakes from the forms.

Reinforcement

Reinforcement rods (i.e., rebar) are placed in the footing after the forms are finished. The amount of reinforcement is usually given in the plans. The rebar is tied in place after it is laid inside the form. Soft metal wires, called *ties*, are twisted around the rebar. The carpenter must be sure that the footing conforms to the local building codes (Fig. 3-19 and Table 3-1).

Specifying Concrete

Today's concrete is made from cement, sand, and gravel mixed with water. The cement is the "glue" that hardens and holds or binds the materials. Most cement is portland cement. It is made from limestone that is heated, powdered, and mixed with certain minerals. When mixed with aggregates or sand and gravel, it becomes concrete. There are about 26 different cements used for concrete. Each has its advantages and disadvantages. Check closely for the type of cement being ordered or before being mixed.

Concrete mixes can be identified by three numbers, such as 1-2-3. This is the volume proportion of cement, sand, and gravel. 1-2-3 is the basic mix, but it is varied for strength, hardening speed, or other factors. However, it is recommended that concrete be specified by the water-to-cement ratio, aggregate size, and bags of concrete per cubic yard (Table 3-2).

Most concrete is delivered to the building site already mixed. Usually, the concrete is not mixed by the

GRADE 40 AND 50

GRADE 60 AND A 706

(A)

(B)

Fig. 3-19 *A. Identification for ASTM standard reinforcing bars. B. Various types of deformed bars.*

TABLE 3-1 *ASTM Standard Reinforcing Bars*

Bar Designation No.	Nominal Weight, lb/ft	Nominal Dimensions		
		Diameter, in	Cross-Sectional Area, in	Perimeter, in
3	0.376	0.375	0.11	1.178
4	0.668	0.500	0.20	1.571
5	1.043	0.625	0.31	1.963
6	1.502	0.750	0.44	2.356
7	2.044	0.875	0.60	2.749
8	2.670	1.000	0.79	3.142
9	3.400	1.128	1.00	3.544
10	4.303	1.270	1.27	3.990
11	5.313	1.410	1.56	4.430
14	7.65	1.693	2.25	5.32
18	13.60	2.257	4.00	7.09

carpenters. It is delivered by concrete trucks from a concrete company. Figure 3-20 shows a transit-mix truck. The concrete is sold in units of cubic yards. The carpenter may need to make the order for concrete to the concrete company. To do so, the carpenter must be able to figure how much concrete to order.

Setting Time

Setting and hardening of concrete are a continuous process. However, three points are important to consider:

• Initial setting time is the interval between the mixing of the concrete with water and the time when

Fig. 3-20 *A transit-mix truck delivers concrete to a site.*

the mix has lost plasticity and has stiffened to a certain degree. It marks roughly the end of the period when the wet mix can be molded into shape.

- Final setting time is the point at which the set concrete has acquired a sufficient firmness to resist a certain pressure. Most specifications require an initial minimum setting at ordinary temperatures of about 45 minutes and a final setting time of no more than 10 to 12 hours.

- To reach its full strength, it takes concrete 28 days.

Concrete Estimating

The volume of concrete needed is determined by a formula. Concrete is measured in cubic yards. A cubic yard is made up of 27 cubic feet ($3 \times 3 \times 3$). To convert footing sizes, use the formula

$$\frac{L'}{3} \times \frac{W''}{36} \times \frac{T''}{36} = \text{cubic yards}$$

where L is length, W is width, and T is thickness.

Example. A footing is 18 inches wide and 8 inches thick must support a building that is 48 feet long and 24 feet wide. The distance around the edges is called the *perimeter*. The perimeter is $(2 \times 48) + (2 \times 24) = 144$ feet. This would be

$$\frac{L'}{3} \times \frac{W''}{36} \times \frac{T''}{36} = \text{cubic yards}$$

$$= \frac{144}{3} \times \frac{18}{36} \times \frac{8}{36}$$

and by cancellation,

$$\frac{48}{1} \times \frac{1}{2} \times \frac{2}{9} = \frac{96}{18} = 5.33 \text{ cubic yards}$$

The minimum amount that can be ordered is 1 cubic yard. After the first cubic yard, fractions can be ordered. The estimate is 5⅓ cubic yards. Often a little more is ordered to make sure enough is delivered.

Pouring the Concrete

To be ready for the pour, the carpenter checks two things. First, the forms must be done. Then the concrete truck must have close access. The driver can move the spout to cover some distance, but it may be necessary to carry the concrete an added distance. This can be done by pumping the concrete or by carrying it. Wheelbarrows sometimes are used (Fig. 3-21). Another method is to use a dump bucket carried by a crane (Fig. 3-22).

The builder must spread, carry, and level the concrete. The truck will only deliver it to the site. The truck driver can remain only a few minutes. The driver is not allowed to help work the concrete. As the concrete is poured, it should be tamped. This is done with a board or a shovel that is plunged into the concrete

TABLE 3-2 *Concrete Use Chart*

Uses	Concrete, Bags per Cubic Yard	Sand, Pounds per Bag of Concrete	Gravel, Pounds per Bag of Concrete	Gravel Size, Average Diameter in Inches	Water, Gallons per Bag of Concrete	Consistency Slump
Footings, basement walls (8-inch), or foundation walls (8-inch thickness)	5.0	265	395	1¹/₂″	7	4 – 6 inches
Slabs, basement floors, sidewalks, etc. (4-inch thickness)	6.2	215	295	1″	6	4 – 6 inches
Basic 1-2-3 mixture (approximation only)	6.0	190	275	2″	5.5	2 – 4 inches

NOTES: 1. All figures are for slight to moderate ground water and medium-fineness sand.
2. All figures vary slightly.

Fig. 3-21 *Sometimes the concrete must be carried from the truck to the work site.*

Fig. 3-22 *A dump bucket is used to dump concrete into forms that trucks can't reach.*

Fig. 3-23 *Concrete is tamped into forms to get rid of air pockets.* (Portland Cement.)

(Fig. 3-23). Tamping helps get rid of air pockets. This makes the concrete solidly fill all the form.

For shallow footings, no smoothing or "finishing" need be done. For deep footings, the surface should be roughly leveled. This is done by resting the ends of a board across the top of the form. The board is then used to scrape the top of the concrete smooth and even with the form.

Slump Test

The slump test is an easy test to perform. Just follow the six steps outlined below:

1. Fill one-third of the cone with concrete and rod 25 times (Fig. 3-24A).

2. Fill two-thirds of the cone and rod the second layer 25 times (Fig. 24B).
3. Fill the cone to overflow and rod 25 times (Fig. 3-24C).
4. Remove the excess from the top of the cone and at the base (Fig. 3-24D).
5. Lift the cone vertically with a slow, even motion (Fig. 3-24E).
6. Invert the cone and place it next to the concrete. Measure the distance from the top of the cone to the top of the concrete. (Fig. 3-24F).

BUILDING THE FOUNDATION FORMS

The wall between the footing and the floor of the building is called the *foundation*. It is sometimes made of concrete. However, it also may be made of concrete blocks, bricks, or stone. In some regions, foundation walls are made of treated plywood as well. See Chapter 22 for permanent plywood foundations.

Special forms are available commercially for pouring concrete foundations. These forms are easily

Fig. 3-24 *Slump test. (Courtesy of Portland Cement Association.)*

put up and taken down. Often the footing and the foundation are made in one solid piece.

Builders also make the foundation forms in much the same way as they make the forms for the footings. After the form is removed, the lumber is used in framing the house.

Form Spacing

In either the one- or the two-piece method, the form should be spaced for the correct width. It also must be spaced to prevent the weight of the concrete from spreading the forms. The width of the form is important. A form that is not wide enough will not carry the weight of the building safely. A form that is too wide uses too much concrete. Too much concrete costs more and adds weight to the building. This weight can cause settling problems. However, spreading forms also can cause errors in pouring the concrete. Frequently, just enough concrete is ordered to fill the forms. If the forms are allowed to spread, more concrete is used. Thus enough concrete might not be delivered to the site.

Constructing the Forms

Several things must be considered when making forms. First, sections of a form must fit tightly together. This prevents leaks at the edges. Leaks can cause bubbles and air pockets in the concrete. This is called *honeycombing*. A honeycomb weakens the foundation wall.

When special forms are used and assembled to make the total form, they must be braced properly. Forms up to 4 feet wide are braced on the back side with studs. These forms are made from metal sheets or from plywood sheathing ¾ inch thick or thicker. For building walls higher than 4 feet, special braces (called *wales*) are used (Fig. 3-25).

Sheathing is nailed to the studs and wales from the inside. The studs are laid out flat on the ground. The sheathing then is laid on the studs and nailed down. The assembled form is then erected and placed into position. It is spaced properly, and wales and braces are added. Erect braces every 4 to 6 feet. However, for extra weight or wall height, braces may be closer.

Fig. 3-25 *Special braces called* wales *are sometimes needed.*

Joining the Forms Together

Edges and corners should be joined tightly so that no concrete leaks occur. When using plywood forms, join the edges together by nailing the plywood sheathing to the studs. Use 16d nails, as in Fig. 3-26. When nailing the corners together, use the procedure also shown in the figure. When special metal forms are used, the manufacturer's directions should be followed carefully.

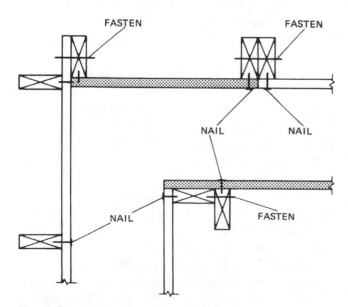

Fig. 3-26 *Nailing plywood panel forms.*

Spreaders

All forms use spreaders to hold the walls apart evenly. Several types may be used. The spreaders may be made of metal at the site. Metal straps may be nailed in between the sections of the forms. However, most builders use spacers that have been made at a factory for that type of form.

After the concrete has hardened, the forms are removed. The spreaders are broken off when the form is removed. Special notches are made in the rods to weaken them so that they will break at the required place. Figure 3-27 shows a typical spreader.

Panel Forms

A panel form is a special form made up in sections. The forms may be used many times. Most are specially made by manufacturers. Each style has special connectors that enable the forms to be erected quickly

Fig. 3-27 *Special braces called* spreaders *keep the forms spaced apart. These rods are later broken off at the notches, called* breakbacks. *The rest of the spreader remains in the concrete.*

and easily. By use of standard sizes, such as 2- × 4- or 4- × 8-foot sections, walls of almost any size and shape can be erected quickly. The forms are made of metal or wood. The advantages are that they are quick and easy to use, they may be used many times, and they may be used on almost any size or shape of wall.

Panel forms must be braced and spaced just as forms constructed on the site are. If the builder is making just one building, there is little advantage to using such forms. They must be purchased, and they are not cheap. However, when they are used many times, the savings in time make them economical. Figure 3-28 shows an example. There are also companies that specialize in renting concrete forms; usually these are made of metal.

One-Piece Forms

When the same style of footing and foundation is often used in the same type of soil, a one-piece form is used. This combines the footings and the foundation as one piece (Fig. 3-29). Several versions may be used. Some types allow a footing of any size to be cast with a foundation wall of any thickness. Some incorporate the footing and the foundation wall as a stepped figure. Others use a tapered design (Fig. 3-30).

The one-piece form saves operational steps. Casting is quicker, easier, and done in one operation. Two-

Fig. 3-29 Panel forms can combine the footing and the foundation.

Fig. 3-30 Tapered form.

Fig. 3-28 Reusable forms are made from panels of plywood or metal. These special panel forms are assembled with special fasteners.

piece forms and the conventional processes require that the footing be cast and allowed to harden. The footing forms then are removed and the foundation forms erected, cast, allowed to harden, and removed. This takes several days and many hours of work. The one-piece form offers many advantages in the savings of time and cost of labor.

Special Forms

Certain special types of forms are used specific shapes are more useful. A round form, for instance, such as that shown in Fig. 3-31 is used commonly. This type of form may be used to cap pilings. It also may be used for the footings under central foundation pillars.

Fig. 3-31 *Pier form.*

Other special forms include forms made of steel and cardboard. Steel forms may be used to cast square or round columns. They are normally used in construction of large projects such as bridges and dams. They are also used on large business buildings. Cardboard forms are made of treated paper and fibers. They are used one time and destroyed when they are removed. Figure 3-32 shows a pillar made with a cardboard form.

Special forms allow time and labor to be saved. Little labor and time are needed to set up special forms. Reinforcement may be added as required. Also, such forms are available in a variety of shapes and sizes.

Openings and Special Shapes

Openings for windows and doors are frequently required in concrete foundation walls. Also, special keys or notches are often needed. These hold the ends of support frames, joists, and girders. At times, utility and sewer lines run through a foundation. Special openings also must be made for these. It is very expensive to try to cut such openings into concrete once it is hardened and cured.

Fig. 3-32 *A round form made of cardboard.*

However, if portions of the forms are blocked off, concrete cannot enter these areas. In this way, almost any shape can be built into the wall before it is formed. This shape is called a *block-out* or a *buck*. The concrete then is poured and moves around these blocked-out areas. When the concrete hardens, the shape is part of the wall. This is quicker, cheaper, and easier than cutting. It also provides better strength to the wall and makes the forms used more versatile.

Of course, a carpenter can build a block-out of almost any shape in a form. First, one wall of the form is sheathed. The shape then can be framed out on that side. The inside of the shaped opening may be used for nails and braces. The outside, next to the concrete, should be kept smooth and well finished. However, it is expensive to pay a carpenter to frame special openings if they have to be repeated many times. It is better to use a form that can be used over again. Figure 3-33 shows an example.

When building a buck or block-out, first check the plans. Sometimes bucks are removed; other times they are left to form a wooden frame around the opening. In either case, the size of the opening is the important thing. To determine rough opening sizes for windows in walls, see Chapter 6.

Buck keys Strips of wood are used along the sides of openings in concrete. These are used as a nail base to hold frames or units in the opening. These strips are called *keys* (Fig. 3-34). If the buck is removed, the key is left in place. If the buck is left in place, the key holds the buck frame securely. Note how the key is undercut. The undercut prevents the key from being pulled out. Bucks should be made from 2-inch lumber. The key

Fig. 3-33 *Openings may be made with special forms that can be reused.*

Fig. 3-34 *Keys are placed along the sides of openings as a nail base. Note the undercut so that key cannot be removed.*

Fig. 3-35 *The buck may be left in the wall as the frame. In this case, the opening in the buck is the desired size.*

Fig. 3-36 *The buck may be removed. In this case, the buck frame is the size desired.*

can be made of either 1- or 2-inch lumber. The key needs to be only 1 or 2 inches wide. Usually, only the sides of the bucks are keyed.

Buck left as a frame   First, find the size of the opening. Next, cut the top and bottom pieces longer than the width. These pieces are usually 3 inches longer than the width (Fig. 3-35). Next, cut the two sides to the same height as the desired opening. Nail with two or three 16d nails as shown. Note that the top piece goes over the sides. The desired size of the opening is the same as the size of the opening in the buck.

Buck removed   First, find the opening size. Next, cut the top and bottom pieces to the exact width of the opening. Then cut the two sides shorter. The amount is usually 3 inches (twice the thickness of the lumber used). Nail the frame together with 16d nails as shown in Fig. 3-34.

Note that the opening size is the same as the outside dimensions of the buck. Also, the outside faces are oiled. This keeps the concrete from sticking to the sides of the buck. It also makes it easier to remove the buck.

Buck braces   When the opening is large, the weight of the concrete can bend the boards. If the boards bend, the opening will not be the right size or shape. To prevent this, braces are placed in the opening (Fig. 3-36).

Note that the braces can run from side to side or from top to bottom.

Another type of form is used for porches, sidewalks, or overhangs (Fig. 3-37). This allows porch supports to be part of the foundation. The earth is filled in later.

Reinforcing Concrete Foundations

Concrete is very strong when compressed. However, it does not support weight without cracking. Even though it is very hard, it is also very brittle. In order to resist shifting soil, concrete should be reinforced. It is not a matter of whether the soil will shift. It is more a matter of how much the soil will shift.

It should be noted that sometimes the reinforcement is added before the forms are done. When the forms are tall, very narrow, or hard to get at, reinforcement is done first.

Concrete reinforcement is done in two basic ways. The first is to use concrete reinforcement bars. The second is to use mesh. Mesh is similar to a large screen made with heavy steel wire. The foundation may be reinforced by running rebar lengthwise across the top

(A)

(B)

(C)

(D)

Fig. 3-37 *A. Special forms used for an overhang on a basement wall. B. The overhang after the forms are removed. C. The basement walls are coated and waterproofed. D. Finally, the concrete porch is poured.*

Fig. 3-38 *Foundations should be reinforced at the top.*

and at intervals up and down. Figure 3-38 shows a typical reinforcement.

The amount and type of reinforcement used are determined by the soil and geographic location. The reinforcement is spaced and tied. There should be at least 3 inches of concrete around the reinforcement. The carpenter should be sure that the foundation and reinforcement conform to local building codes.

Estimating Concrete Volume

When the forms are complete, the amount of concrete needed is computed. The same formula as for footings is used:

$$\frac{L'}{3} \times \frac{W''}{36} \times \frac{T''}{36} = \text{cubic yards}$$

However, W in inches is replaced by the wall height H in feet, so we get

$$\frac{L'}{3} \times \frac{H'}{3} \times \frac{T''}{36} = \text{cubic yards}$$

If the foundation is to be 8 inches thick and 8 feet high and the perimeter is 144 feet, then

$$\frac{144}{3} \times \frac{8}{3} \times \frac{8}{36} = \text{cubic yards}$$

and by cancellation,

$$\frac{48}{1} \times \frac{8}{3} \times \frac{2}{9} = \frac{16}{1} \times \frac{8}{1} \times \frac{2}{9}$$

$$= \frac{256}{9}$$

$$= 28.4 \text{ cubic yards}$$

Delivery and Pouring

Once the needs are estimated and the concrete has been ordered or mixed, the wall should be poured. If a transit-mix truck is used, the concrete is mixed and delivered to the site. The concrete truck should be backed as close as possible to the forms. As in Fig. 3-39, the concrete should be poured into the forms, tamped, and spread evenly. By doing this, air pockets and honeycombs are avoided.

Fig. 3-39 *Concrete is poured or pumped into the finished form.*

Finishing the Concrete

Two steps are involved in finishing the pouring of the concrete foundation wall. First, the tops of the forms must be leveled. Sometimes the concrete is poured to within 2 or 3 inches of the top. The concrete then is allowed to partially cure and harden. A concrete or grout with a finer mixture of sand may be used to finish out the top of the foundation.

Anchors are embedded in the concrete before it hardens completely. One end of each anchor bolt is threaded, and the other is bent (Fig. 3-40). The threaded end sticks up so that a sill plate may be bolted in place. As the concrete begins to harden, the bolts are slowly worked into place by being twisted back and forth and pushed down. Once they are embedded firmly, the concrete around them is troweled smooth. Figure 3-41 shows an anchor embedded in a foundation wall.

The concrete cures and hardens, and the forms are removed. Low spots are filled. A small spot can be

Fig. 3-40 *Anchor bolt.*

Fig. 3-41 *Anchor bolts are embedded in foundation walls and protrude from them.*

shimmed with a wooden shingle. However, a larger area should be filled in with grout or mortar.

CONCRETE BLOCK WALLS

Concrete blocks are often used for basement and foundation walls. When the foundation is exposed, it may be faced with brick (Fig. 3-42). Blocks need no form work and go up more quickly than brick or stone. The most common size is 7⅝ inches high, 8 inches wide, and 15⅝ inches long. The mortar joints are ⅜ inch wide. This gives a finished block size of 8 × 16 inches and a wall 8 inches thick.

Fig. 3-42 *Concrete block foundations may be faced with brick or stone.*

The footing is rough and unfinished. This is so because the mortar for the block is also used to smooth out the rough spots. No key is needed, but reinforcement rod should be used. Figure 3-43 shows a footing for a block wall.

Block walls should be capped with either concrete or solid block. Anchor bolts are mortared in the last row of hollow block. They then pass through the mortar joint of the solid cap (Fig. 3-44).

Fig. 3-43 *A footing for a concrete block foundation.*

Fig. 3-44 *Concrete block basement wall.*

NOTE: Wood and plywood are treated.

Fig. 3-45 *A typical plywood foundation panel.* (American Plywood Association.)

A special pattern is sometimes used to lay the block. This is done when the wall also will be the visible finished wall. This pattern is called a *stack bond* (see Fig. 3-44). It is reinforced with small rebar.

PLYWOOD FOUNDATIONS

Plywood may be used for foundations or for basement walls in some regions. There are several advantages to using plywood. First, it can be erected in even the coldest weather. It is fast to put up because no forms or reinforcement are required. For the owner, plywood makes a wall that is warmer and easier to finish inside. It also conserves on the energy required to heat the building.

The frame is formed with 2-inch studs located on 12- or 16-inch centers. The frame then is sheathed with plywood. Insulation is placed between the studs. The exterior of the wall is covered with plastic film, and building paper is lapped over the top part. The building paper is laid over the top of the rock fill and helps drainage (Figs. 3-45 and 3-46). It is important to note that all lumber and plywood must be pressure-treated with preservative.

DRAINAGE AND WATERPROOFING

A foundation wall should be drained and waterproofed properly. If a wall is not drained properly, the water may build up and overflow the top of the foundation. Unless the wall is waterproofed, water may seep through it and cause damage to the foundation and

BASEMENT WALL

Fig. 3-46 *Cross section of plywood basement wall.* (American Plywood Association.)

Fig. 3-47 *Basement walls should be coated and drained.*

footings. It also may cause damage to the interior of a basement. Proper drainage is ensured by placing drain pipes or drain tile around the outside edges of the footing. A gravel fill is used to place the tile slightly below the level of the top of the footing. Figure 3-47 shows the proper location of the drain pipe. The pipe then is covered with loose gravel and compacted slightly. If the house has a basement, the foundation walls should be waterproofed. If the house has only a crawl space, no waterproofing is needed.

Waterproofing Basement Walls

Three types of walls are commonly used today for basements. The most common is the solid cast-concrete wall. However, concrete blocks are also used, and so is plywood.

Concrete block foundations Concrete blocks should be plastered and waterproofed. Figure 3-44 shows the processes involved. First, the concrete wall is coated with a thin coat of plaster. This is called a *scratch coat.* After this has hardened, the surface is scratched so that the next coat will adhere more firmly. The second coat of plaster is applied thickly and smoothly over both the wall and the top of the footing. As shown in Fig. 3-44, this outside layer is then covered with a waterproof coating. Such a coating could include layers of bitumen, builder's felt, or plastic.

Waterproofing concrete walls No plaster is needed over a cast-concrete wall. The wall may be quickly coated with bitumen. However, plastic sheeting also may be applied to cover both the footing and the foundation in one piece. The most common process, as shown in Fig. 3-47, involves a bitumen layer. This bi-

tumen layer is sometimes reinforced by a plastic panel that is then coated with another layer of bitumen.

Basement Walls Coatings

For many years, asphalt-based coatings have been applied to basement walls to waterproof them. However, this coating is often damaged with the backfill and careless workmanship on the part of those responsible for cleaning up the work site. A more permanent and damage-resistant coating now has become commonly available. Rub-R-Wall foundations are coated with a 100 percent polymer membrane that is guaranteed not to leak for a lifetime. Conventional asphalt-based coatings, such as cutback and emulsified asphalts, are susceptible to leaching. The Rub-R-Wall polymer membrane is virtually leachproof. Because of its nonhazardous ingredients, it will not contaminate the groundwater.

Protection board ranging in thickness from ⅜ inch (which is standard) to 2 inches can be put on to protect the membrane (Fig. 3-48). In addition, plastic drain board may be used.

The basic use for this waterproofing membrane is for vertical and horizontal building elements above and below grade, such as walls, slabs, decks, and underground structures. Rub-R-Wall prevents the passage of water under hydrostatic, dynamic, or static pressure. When used on basement walls and floors inside the basement, it is possible to eliminate the threat from radon gas that plagues some areas of the United States. The material is make of rubber polymer and is applied in liquid form by spraying it onto walls and floors. The more complicated the surface to be waterproofed, the more reason there is to use a liquid-applied membrane. The material can conform to all irregular shapes. It can be applied at low temperatures (15°F) provided that substrates are dry and frost-free. Once the membrane is applied, it is impermeable to water. The excellent and tenacious bond of the membrane to substrates and protection course prevents the lateral movement of water between the membrane and substrate.

The membrane is installed in a minimum of 40 mils thick, averaging 50 mils thick for waterproofing purposes. The only limitation is that the waterproofing must be protected from ultraviolet light rays (sun) and mechanical damage and should not be left permanently exposed. However, do not backfill sooner than 24 hours after the membrane is applied. It will dry to the touch and not be sticky within 20 minutes. This coating does not require maintenance. Damaged areas can be repaired easily by spraying over the affected

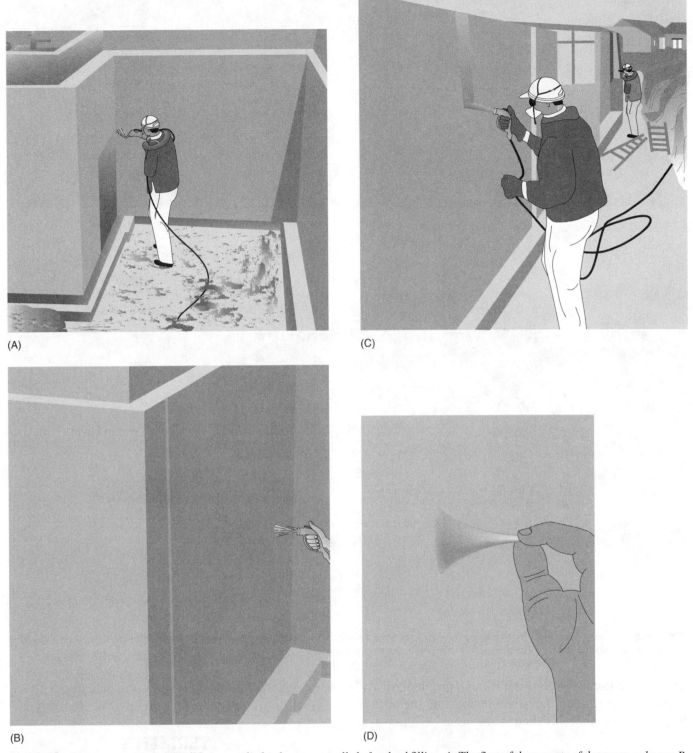

(A)

(C)

(B)

(D)

Fig. 3-48 *Green-coated Rub-R-Wall being applied to basement walls before backfilling. A. The first of three coats of the green polymer. B. The second coat of three. C. The third coat of the membrane being applied with a spray gun. D. The fresh membrane already can have much flexibility.* (Rubber Polymer Company)

(E)

(G)

(F)

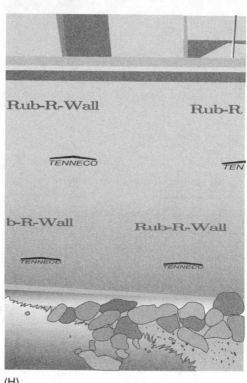

(H)

Fig. 3-48 *Green-coated Rub-R-Wall being applied to basement walls before backfilling. E. Plastic flashing on corners and footer. F. Flashing above the finished board. G. Finished job with protective wall coating. H. Closer look at the finished job.* (Rubber Polymer Company)

areas. Cold joints or recoating is not a problem; newly applied material blends easily with existing membrane to provide a monolithic membrane.

Gray Wall

Gray Wall is a 100 percent gray rubber coating that outperforms all asphalt-based damp proofing on the market today. It has 1400 percent elongation that allows the material to bridge small cracks in the foundation. It is applied at about half the thickness and amount of products such as Rub-R-Wall. Protection board also may be used on this product. Gray Wall offers seamless protection at an economical price (Fig. 3-49).

TERMITES

Termite is the common name applied to white ants. They are neither all white nor ants. In fact, they are closely related to cockroaches. Termites do, however, live in colonies somewhat in the same manner as ants do.

Most termite colonies are made up of three castes. The highest caste is the royal or reproductive group (Fig. 3-50A). The middle caste is the soldier (Fig. 3-50C). The worker is at the bottom of the social groupings

Fig. 3-49 *Basement walls coated with a gray-coat polymer coating.* (Rub-R-Wall.)

(Fig. 3-50B). In every mature colony of termites, a group of young winged reproductives leaves the parent nest, mates, and sets out to found new colonies. Their wings are used only once; then they are broken off just before they seek a mate.

The worker caste is made up of small, blind, and wingless termites (see Fig. 3-50B). They have pale or whitish soft bodies. Only their feet and heads are covered by a hard coating. The worker caste makes up the largest group within a colony.

The soldiers have very long heads in proportion to their bodies and are responsible for protecting the

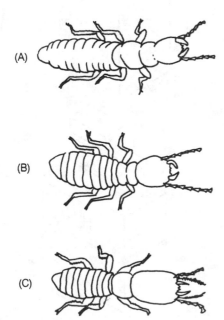

Fig. 3-50 *A. The supplementary queen leaves the colony to mate and then sets up a new colony. She breaks off her wings before mating. B. The worker termite is small and has a pale, soft body. Workers gather food for the colony. C. The soldier termite is extremely large compared with the worker. The soldier has a hard head and defends the colony.*

colony against its enemies, usually ants. The soldiers are also blind and wingless. Because of this, the workers do all the work. They enlarge the nest, search for food and water, and make tunnels. They also take care of the soldiers because they have to be fed individually.

Termite castes contain both females and males. The kings live as long as the queens. The queens become as long as the males—3 inches—when they are full of eggs. They can lay many thousands of eggs a day for many days. The eggs for all castes appear the same.

Termites live in warm areas such as Africa, Australia, and the Amazon. They build nests as high as 20 feet, with the inside divided into chambers and galleries. They keep the king and queen separated in closed cells. The workers carry the eggs away as fast as the queen lays them. The workers care for the eggs until they hatch and then take care of the young until they grow up. Termites digest wood, paper, and other materials. They use cellulose for food. Most of the damage they do to homes and furniture is through tunneling. They also have been known to destroy books in search of cellulose to digest. They also do great damage to sugar cane and orange trees. They are considered a serious pest in many parts of the United States, where they damage houses.

The best way to provide protection for a home is to follow some suggestions that have been developed through the years. There are about 2,000 different species of termites known. About 40 species live in the United States and 2 species in Europe. They do not build large mounds in the United States or Europe but do most of their damage out of sight.

Types of Termites

Three groups of termites exist in the United States. They are grouped according to their habits. The *subterranean* (underground) termites are the smallest and most destructive, for they nest underground. They extend their habitat for long distances into wood structures. The *damp wood* termites live only in very moist wood. This type causes trouble only on the Pacific Coast. The *dry wood* termites need very little moisture. They are found to be destructive in the Southwest. The damp wood and dry wood types do not have a distinctive working caste (Fig. 3-51).

Once termites get into a house, they eat books, cloth, and furniture, as well as the wood of the house. They also attack bridges, trestles, and other wooden structures. They do more damage each year in the United States than fire.

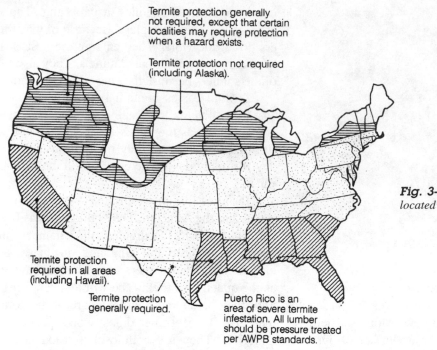

Fig. 3-51 *Note where the termites are located in the United States.*

Termite protection generally not required, except that certain localities may require protection when a hazard exists.

Termite protection not required (including Alaska).

Termite protection required in all areas (including Hawaii).

Termite protection generally required.

Puerto Rico is an area of severe termite infestation. All lumber should be pressure treated per AWPB standards.

Termite Protection

Most termites cannot live without water. The best approach is to eliminate their source of moisture. This can be done by applying chemicals to the soil around the foundation of a building or impregnating the wood with chemicals that repel or kill termites. Creosote and other types of chemicals are used around the footings of buildings, and termite shields are used to keep them from reaching the wood sill (Fig. 3-52). Metal shields are also placed on copper water pipes or soil pipes to prevent their climbing up the metal pipes to reach the wood of the building (Fig. 3-53). Builders must take every precaution to prevent infestation.

Infestation can be prevented by using minimum space between joists and the soil in the crawl space (Fig. 3-54). Keep the space to a minimum of 18 inches. The minimum recommended space between girders and the soil is 18 inches. The lowest wood member of the exterior of a house should be placed 6 inches or more above the grade. Metal termite shields should be used on each side of a masonry wall (see Fig. 3-52A). Metal shields should be at least 24-gauge galvanized iron. Concrete should be compacted as it is placed in the forms. This makes sure that rock pockets and honeycombing are eliminated.

After a house is completed, make sure that all wood scraps are removed. Any scraps buried in backfilling or during grading become possible pockets for termite colonies. From there they can tunnel into a house (Fig. 3-55). Make sure that there are no fine

Fig. 3-52 *A. A metal shield is placed over the concrete foundation before the wood sill is applied. B. Even masonry walls should have a termite shield in areas infested with termites.*

SILL

SILL SEALER

METAL SHIELD

FOUNDATION

(A)

(B)

Fig. 3-53 *Note the placement of shield over pipes to protect the wooden floors from infestation.*

cracks in the concrete foundation or loose mortar in the cement blocks. Termites have been known to build clay tunnels around metal shields.

Termites and Treated Wood

Decay and termite attack can occur when all four of the following conditions prevail: a favorable temperature (approximately 50 to 90°F), a source of oxygen, a moisture content above 20 percent, and a source of food (wood fiber). If any one of these conditions is re-

Fig. 3-54 *Note the minimum spacings for wood in reference to the grade.*

Fig. 3-55 *Termites are well known for their tunneling ability.*

moved, infestation will not occur. Chemical preservatives eliminate wood as a food source.

PRESSURE-TREATED WOOD

Treatment of wood adds to its versatility. Treatment with chemical preservatives protects wood that is exposed to the elements, is in contact with the ground, or is used in areas of high humidity. The treatment of wood allows you to build a project for outside exposure that will last a lifetime.

Properly processed pressure-treated wood resists rot, decay, and termites and provides excellent service, even when exposed to severe conditions. It is wood that has been treated under pressure and under controlled conditions with chemical preservatives that penetrate deeply into the cellular structure.

Just because a wood is green in color doesn't mean that it is pressure-treated wood. And just because it is pressure-treated doesn't mean that there is enough of the chemical deep enough in the wood to prevent decay and keep insects at bay. Only in recent years has treated wood been readily available from local lumber yards.

In order to obtain the treated wood properly, you should use the American Wood Preservers Association recommendations. First, you should consider the wood. Not all wood is created equal. Most wood species do not readily accept chemical preservatives. To assist preservative penetration, the American Wood Preservers Association standards require incising for all species except southern pine, ponderosa pine, and red pine. Incising is a series of little slits along the grain of the wood that assist chemical penetration and uniform retention. Depth of penetration is important in providing a chemical barrier that is thick enough that any checking or splitting won't expose untreated wood to decay or insect attack.

Southern Pine is the preferred lumber species for building the PermanentWood Foundation. Southern Pine lumber is readily available in a wide range of grades and sizes. In structural terms, it is one of the strongest softwoods.

Quality Southern Pine lumber is graded in accordance with the grading rules of the Southern Pine Inspection Bureau (SPIB).

Fig. 3-56 *Two typical lumber grade marks. (Southern Forest Products Association.)*

To make sure that the wood is strong enough for your intended use, you always should insist that the lumber you buy bears a lumber grade mark. Figure 3-56 shows the typical symbols (grade marks) found on southern pine lumber.

Preservatives

Three types of wood preservatives are used. *Waterborne preservatives* are used in residential, commercial, recreational, marine, agricultural, and industrial applications. *Creosote* and mixtures of creosote and coal tar in heavy oil are used for railroad ties, utility poles, piles, and similar applications. *Pentachlorophenol* (Penta) in various solvents is used in industrial applications and utility poles and for some farm applications.

Waterborne preservatives are used most commonly in home construction owing to their clean, odorless appearance. Wood treated with waterborne preservatives can be stained or painted when dry. These preservatives also meet stringent Environmental Protection Agency health guidelines. The preservative used most commonly in residential construction lumber is known by a variety of brand names, but in the trade it is known simply as *CCA*. The CCA stands for chromates-copper-arsenate. It also accounts for the green (copper) color in the treated wood.

CCA is a waterborne preservative. The chromium salts combine with the wood sugar to form an insoluble compound that renders the CCA preservative non-leachable. Some people estimate the life of properly treated wood to exceed 100 years because the chemicals do not leach out in time or with exposure to the elements. However, in order to make sure that the wood is treated properly after it has been cut, you must treat the exposed area with chemicals. It is a good idea to avoid field cuts and drilling in portions of treated wood that will be submerged in water.

Above-Ground and In-Ground Treatment

The standard retentions found most commonly in lumber yards are 0.25 (pounds of chemical per cubic foot of wood) for above-ground use and 0.40 for below-ground use (or in-ground contact with the ground). The only difference is a slightly higher concentration of chemical in 0.40 below-ground treatments. Thus, if the above-ground (0.25) material is not available, below-ground (0.40) material can be used instead. However, you should not use 0.25 material in contact with the ground.

Retentions (lb/ft.^{3})	Uses
0.25	Above ground
0.40	Ground contact
0.60	Qood foundation
2.50	In saltwater

Only foundation (FDN) treatment should be used for wood foundation lumber and plywood applications. All pieces of lumber and plywood used for wood foundation applications must be identified by the FDN stamp (0.60 retention). That means that the foundation should have a life longer than that of the rest of the structure.

All lumber used in the foundation must be pressure-treated with CCA in accordance with the requirements of the American Wood Preservers Bureau (AWPB) Foundation (FDN) Standard. Each piece must bear the stamp of an approved agency certified to inspect preservative-treated lumber. The mark AWPB-FDN indicates that the lumber is suitable for foundations. It has the required 0.60 pound of preservative per cubic foot of wood. It has been kiln dried to 19 percent moisture content. All cut or drilled lumber must be field treated with preservative. Any cut end 8 inches or higher above grade or top plates, headers, or upper course of plywood need not be treated. Extend footing plates beyond the corners to avoid cutting them.

Nails and Fasteners

Hot-dipped galvanized or stainless steel nails and fasteners should be used to ensure maximum performance in treated wood. Such fasteners ensure permanence and prevent corrosion, which stains both the wood and its finishes. In structural applications, where a long service life is required, stainless steel, silicon, bronze, or copper fasteners are recommended. Smaller nails may be used with southern pine because of its greater nail-holding ability. Use 10d nails to fasten 2-inch dimensional lumber, 8d nails for fastening 1-inch boards to 2-inch dimensional lumber, and 6d nails for fastening 1-inch boards to 1-inch boards. The use of treated lumber adhesives can be considered for attaching deck boards to joists, in building fences, or in applications where the appearance of nailheads is not desired. For deck applications, fastening boards bark side up will help to reduce surface checking and cupping (Fig. 3-57).

Fig. 3-57 *The bark side of the lumber is exposed when using treated wood for a deck or other exposed surface.* (Southern Forest Products Association.)

Handling and Storing Treated Wood

When handled properly, pressure-treated wood is not believed to be a health hazard. Treated wood should be disposed of by ordinary trash collection or burial. It should not be burned. Prolonged inhalation of sawdust from untreated and treated wood should be avoided. Sawing should be performed outdoors while wearing a dust mask. Eye goggles should be worn when power-sawing or machining. Before eating, drinking, or using tobacco products, areas of skin that have come in contact with treated wood should be washed thoroughly. Clothes accumulating preservatives and sawdust should be laundered before reuse and washed separately from other household clothing.

Care should be taken to prevent splitting or excessively damaging the surface of the lumber because this could permit decay organisms to get past the chemical barrier and start deterioration from within. Treated lumber should be stacked and stored in the same manner as untreated wood. Treated wood also will weather. If it is stored outside and exposed to the sun and elements, the green color eventually will turn to the characteristic gray, just as natural brown or red-colored wood does.

CHAPTER 3 STUDY QUESTIONS

1. What is the lowest part of a building?
2. How is footing size determined?
3. What types of footings are used?
4. What is a form?
5. How deep should footings be for basements?
6. How deep should footings be for crawl spaces?
7. How are forms leveled?
8. What is a spreader?
9. How much concrete would be needed for a one-piece footing and foundation with these dimensions:

Footing thickness	8 inches
Width	18 inches
Perimeter	150 feet
Foundation thickness	8 inches
Height	9 feet

10. What should be used to fill holes for utility lines?
11. How is a footing key made?
12. How are anchor bolts embedded?
13. What two materials are commonly used to make foundation walls?

14. How should rebar be spaced?
15. How is rebar tied together?
16. How are foundations drained?
17. How are foundation walls waterproofed?

18. Why are double-headed nails preferred for building concrete forms?
19. What is a block-out?
20. What are the advantages of using panel forms?
21. What is a "basic" concrete volume mix?

4
CHAPTER

Concrete Slabs and Floors

CONCRETE SLABS MAY BE USED FOR MANY THINGS. Slabs often combine footings, foundations, and subfloors into one piece. Concrete floors are commonly found in basements and bathrooms. Concrete is used outdoors to form stairs, driveways, patios, and sidewalks. Carpenters are the form builders. In some cases, they also aid those who pour and finish the concrete.

Once you have studied this chapter you should be able to

• Excavate

• Construct forms

• Prepare subsurfaces

• Lay drains and utilities

• Lay reinforcement

• Determine concrete needs

• Ensure correct pouring and surfacing

SLABS

Concrete slabs are often combinations of footings, foundations, and subfloors, all as one piece. Slabs are easier to plan and build than basements. Formerly, basements were used as storage areas for furnaces, fuels, and ashes. Basements also held cooling, heating, and ventilation units, as well as laundry areas. These things are usually installed at ground level. However, in most cold climates, people prefer basements. Basements keep pipes from freezing and add warmth to the upper floors. They also provide storage space, play areas, and sometimes living areas.

Slabs are best when placed on level ground and in warmer climates. They can be used where ground hardness is uneven. Slabs are good with split-level houses or houses on hills, where the slab is used for the lower floor (Fig. 4-1).

Modern building methods can solve most of the problems encountered in the past when slabs did not make comfortable floors. They were very cold in the winter, and water would condense on them. Water would also seep up through the slab from the ground.

With concrete slabs, most of the heat-energy loss occurs at the edges. Energy loss can be reduced by using rigid insulation (Fig. 4-2). In extreme cases, where a warmer floor is needed, heating ducts may be built in the slab flooring itself. This is an efficient method and provides an even temperature in the building. A warmer type of floor can also be built over the concrete floor and is discussed later in this chapter.

The Slab Sequence

Many types of buildings use slabs for several types of outdoor surfaces. In most cases, the general procedure is about the same. After the site is prepared and the building is located, the following sequence is used most often:

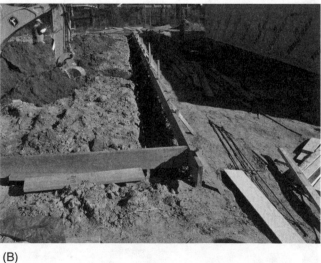

Fig. 4-2 A. Insulation for a slab. B. Forms for a slab.

Fig. 4-1 A typical split-level home. The lower floor is a slab; the middle floor is a frame floor.

SHEATHING
STUD
WOOD BLOCK OR
RESILIENT TILE
SILL CAULK
ADHESIVE
8" MINIMUM
WIRE MESH
CONCRETE SLAB
VAPOR BARRIER
GRAVEL
REINFORCING RODS

Fig. 4-3 *Combined slab and foundation (thickened-edge slab).*

1. Excavate.
2. Construct forms.
3. Prepare subsurface.
   a. Spread sand or gravel and level.
   b. Install drains, pipes, and utilities.
   c. Install moisture barrier.
4. Install reinforcement bar.
5. Construct special forms for lower levels, stairs, walks, etc.
6. Level tops of forms.
7. Determine concrete needs.
8. Pour concrete.
9. Tamp, level, and finish.
10. Embed anchors.
11. Cure and remove forms.

Types of Slabs

Slabs have two basic designs:

- *Monolithic slabs* are poured in one piece and are used on level ground in warm climates. In cold climates, the frost line penetrates deeper into the ground. This means that the footing must extend deeper into the ground to be below the frost line. Another method for slabs involves building the footing separately.

- *Two-piece slabs* are best suited for cold or wet climates. Figures 4-2 and 4-3 show both types of slabs.

All slab footings must rest beneath the freeze line. This gives the slab stability in the soil. Slabs are reinforced, but the amount needed varies. It depends on soil conditions and weights to be carried. On dry, stable soil, slabs need little reinforcement. Larger slabs or slabs on less stable soils need more reinforcement.

The top of the slab should be 8 inches above ground. If the slab is above the rest of the ground, moisture under the slab can drain away from the building. The ground around the slab should also be sloped for the best drainage (Fig. 4-4).

SLOPE IS BELOW
SLAB FOR
8" DRAINAGE
MINIMUM

SLOPE IS AWAY
FROM SLAB

Fig. 4-4 *Slab drainage.*

Excavate

At this point, the building has already been located, and batter boards are in place. Lines on the batter boards show the locations of corners and walls.

Trenches or excavations are made for the footings, drains, and other floor features. These features must be deeper than the rest of the slab. Footings are, of course, around the outside edges. However, a slab big enough for a house should also have central footings.

Locating and digging for slab footings are the same as for foundation footings. There is no difference at all when a two-piece slab is to be made. The main difference for a monolithic slab is how the forms are made.

Trenches now can be dug. Rough lines are used for guides. As a rule, inner trenches are dug first. Trenches for the outside footings are done last. This is easier when machines are moved around the site.

Excavations are checked for depth level. Remember from Chapter 3 that the lowest point determines the depth. Trenches are also dug for drains and sewer lines inside the slab. Trenches are dug from the slab to the main utility lines. Sewer lines must connect the slab to the main sewer line, and so forth.

Construct the Forms

Once the excavation is done, the corners are relocated. Lines are restrung on batter boards, and the corners are plumbed (see Fig. 3-16). The footing forms for two-piece slabs are made like standard footings or foundation forms (see Chapter 3).

Lumber is brought to the place where it is needed, and forms are constructed (Fig. 4-5). Monolithic forms are made like footing forms. The top board is placed and leveled first (see Fig. 3-16). It is leveled with the corner first. Then its length is leveled, and the ends are nailed to stakes. As before, double-headed nails should be used from the outside. The remaining form boards are then nailed in place.

Shallow footings may be another method used for monolithic slabs. It is a very fast and inexpensive method. The form boards are put up before excavating (Fig. 4-6). Be sure to check the plans carefully. Next, the sand or gravel is dumped inside the form area (Fig. 4-7). Sand is spread evenly over the form area, as in

Fig. 4-6 *The forms are erected.*

Fig. 4-8. The outside footings are then dug by hand. Chalk lines are strung from the forms.

The chalk lines are then used as guides to dig the central footings (Fig. 4-9). Trenches for drains and sewers are then dug.

(A)

(B)

Fig. 4-5 *Forms are erected.*

Fig. 4-7 *Sand is dumped and spread in the form area.*

Fig. 4-8 *Excavations are made for outside footings. Plumbing is roughed in.*

Fig. 4-9 *Excavations are also made for footings in the slab.*

Prepare the Subsurface

Ground under either type of slab must be prepared for moisture control. Water must be kept from seeping upward through the slab. The water must also be drained from under the slab. This preparation is necessary because water from rain and snow will seep under a slab.

Outside moisture is reduced by using good siding methods. The edges of the slab can be stepped for brick. The sheathing can overlap the slab edge just as on other types of siding. As mentioned previously, heat energy can be lost easily from slabs. The main area of loss is around the edges. Rigid foam insulation can be put under the slab's edges to reduce the heat-energy losses.

Subsurface preparation After the excavation, drains, water lines, and utilities are roughed in. These should be placed for areas such as the kitchen, baths, and laundry. Water lines may be run in ceilings or beneath the slab. If they are to be beneath the slab, soft copper should be used. An extra length of the copper tubing should be coiled loosely to allow for slab movement. Metal or plastic pipe may be used for the drains. Conduit (metal pipe for electrical wires) should be laid for any electrical wires that go under the slab. Wires should never be laid without conduit. All openings in the pipes are then capped (Fig. 4-10). This keeps dirt and concrete from clogging them.

One thing to keep in mind: *Once the price of copper exceeds a certain point, there are thieves who ruin days of work by the plumbing crew by stealing the copper tubing during the time the building site is left unguarded.* Also, electrical wiring is damaged by thieves looking for copper and aluminum wire for sale to salvage operators. It is becoming more evident that night guards have to be employed to watch buildings during construction, thus adding to the cost of construction and delaying building progress. The old practice of leaving materials stacked up around a building site and

Fig. 4-10 *All pipe ends are covered. This prevents them from being clogged with dirt or concrete.*

inside unlocked buildings probably will disappear as the economy changes for the worse.

Various pipes are then covered with sand or gravel. Sand or gravel is dumped in the slab area and carefully smoothed and leveled. Chalk lines are strung across the forms to check the level (Fig. 4-11). It is also wise to install a clean-out plug between the slab and the sewer line (Fig. 4-12).

Lay a vapor seal   A moisture barrier is laid after the sand is leveled. The terms *vapor barrier*, *moisture barrier*, *vapor seal*, and *membrane* mean the same thing. As a rule, plastic sheets are used for moisture barriers. Lay the moisture barrier so that it covers the whole subsurface area. To do this, several strips of material can be used. The strips should overlap at least 2 inches at the edges. Check for overlaps once the whole surface has been covered with the plastic to make sure that there is no obvious overlap mistake.

Insulate the edges   Go back to Fig. 4-2, which shows the insulation for a slab. Insulation is laid after placement of the vapor barrier. The insulation is placed around the outside edges. This is called *perimeter insulation*. The insulation should extend to the bottom of the footing. It also should extend into the floor area at least 12 inches. A distance of 24 inches is recommended.

Fig. 4-12 *Trenches for drains and sewers are also excavated. Clean-out plugs are usually outside the slab.*

Rigid foam at least 1 inch thick is used. Perimeter insulation is not always used in warm climates. However, the moisture barrier always should be used.

Reinforcement   Always use reinforcement. Use the same amount of reinforcement rods (rebar) as used in the footing. Be sure to conform to local codes. Also reinforce the slab with mesh. This mesh is made of 10-gauge wire. The wires are spaced 4 to 6 inches apart. Figure 4-13 shows the reinforced form ready for pour-

Fig. 4-11 *Sand is leveled. Use chalk lines as guides.*

Fig. 4-13 *The vapor barrier is laid, and the reinforcement is also laid.*

(A)

Fig. 4-14 *Rebar is tied together with soft wire ties.*

(B)

Fig. 4-15 *Short stakes hold mesh and rebar in place during pouring.*

ing. Where the soil is unstable, more reinforcement is needed. The amount is usually given on the plans.

Rebar is laid and tied in place. Soft metal ties are twisted around the rebar (Fig. 4-14). The mesh also may be held off the bottom by metal stakes (Fig. 4-15). Mesh also may be lifted into place as the concrete is poured. Chairs hold rebar in place during pouring (Fig. 4-16).

Different Shapes

There are several reasons for some shapes depending on location and purpose. A stepped edge, for instance, as in Fig. 4-17, helps drainage. It prevents rain water from flowing onto the floor surface. Lower surface areas are also common. They are used for garages, entryways, and so forth.

Stepped edges are formed easily. Use a 2 × 4 or 2 × 6, and nail it to the top of the form (Fig. 4-18). To lower a larger surface, an extension form may be used. The first form is used as a nailing base. The extension

form is built inside the outer form. The lower area can then be leveled separately (Fig. 4-19).

Pouring the Slab

The corners of the forms are leveled first (Fig. 4-20). Use a transit for large slabs, but small areas can be leveled with a carpenter's level. See Chapter 2 for the leveling process using a transit. Diagonals are checked for squareness, and all dimensions are checked.

In modern building, the concrete is delivered to the building site already mixed. Usually, the concrete is not mixed by the carpenters. It is delivered by transit-mix trucks from a concrete specialty company. The concrete is sold by the cubic yard. Before the concrete

Fig. 4-16 *Chairs hold rebar in place for pouring.*

Fig. 4-17 *A stepped form aids drainage. The step also forms a base for brick siding.*

(A)

(B)

Fig. 4-18 *A stepped edge is formed by nailing a board to the form.*

$$\frac{30}{3} \times \frac{48}{3} \times \frac{6}{36} = \frac{8640}{324}$$

$$\frac{10}{1} \times \frac{16}{1} \times \frac{1}{6} = \frac{160}{6}$$

$$= 26\frac{2}{3}$$

But the perimeter footings are 18 inches deep. They are not just 6 inches deep. Thus a portion 12 inches thick (18 minus 6) must be added. This additional amount of concrete is calculated as follows: The linear distance around the slab is

$$48 + 48 + 30 + 30 = 156$$

Then

$$\frac{L}{3} \times \frac{W}{3} \times \frac{T}{36} = \frac{1}{3} \times \frac{156}{3} \times \frac{12}{36}$$

$$= \frac{1}{3} \times \frac{52}{1} \times \frac{1}{3} = \frac{52}{9} = 5.7 \text{ cubic yards}$$

is ordered, the amount of concrete needed must be determined.

Estimating the volume needed   A formula is used to estimate the volume of concrete needed:

$$\frac{L'}{3} \times \frac{W'}{3} \times \frac{T''}{36} = \text{cubic yards}$$

For instance, a slab has a footing 1 foot wide. The slab is to be 30 feet wide and 48 feet long. The slab is to be 6 inches thick. From the formula, 26.6666 cubic yards are required. That is,

(A)

(B)

(C)

(D)

(E)

Fig. 4-19 *A lower surface is used for a garage. (See top left through bottom right.)*

Fig. 4-20 *Corners of forms are leveled before pouring the slab.*
(Portland Cement.)

Thus, to fill the slab, the two elements are added:

$$\begin{array}{r} 26.7 \\ + 5.7 \\ \hline 32.4 \text{ cubic yards} \end{array}$$

Pouring In order to be prepared for the truck to pour the ready-mixed concrete into the forms, the carpenter checks two things. First, the forms must be complete. Second, the concrete truck must have close access.

It is the responsibility of the builder to spread, carry, and level the concrete. The truck only delivers it to the site. The truck driver can remain only for a few minutes. The driver is not allowed to help work the concrete. As the concrete is poured, it should be spread and tamped. This is done with a board or shovel (Fig. 4-21). The board or shovel is plunged into the con-

(A)

(B)

Fig. 4-21 *Concrete is spread evenly in the form.*

crete. Be careful not to cut the moisture barrier. Tamping helps to get rid of air pockets. This makes the concrete solid as it fills all the form.

After tamping, the concrete is leveled with a long board called a *strike-off* (Fig. 4-22). The ends rest across the top of the forms. The board is moved back and forth across the top. A short back-and-forth motion is used. If the board is not long enough, special supports are used. These are called *screeds*. A screed may be a board or pipe supported by metal pins. The screed is leveled with the tops of the forms. It is removed after the section of concrete is leveled. Any holes left by the screeds are patched.

Once leveled, the surface is treated with a "jitterbug" to remove trapped air in the concrete. Then it is floated. This is done after the concrete has been allowed to get stiff. However, the concrete must not have hardened. A finisher uses a float to tamp the surface gently. During tamping, the float is moved across the surface. Large floats called *bull floats* are used (Fig. 4-23). Floating lets the smaller concrete particles float to the top. The large particles settle. This gives a smooth surface to the concrete. Floats may be made of wood or metal.

After floating, the finish is done. A rough, lined surface can be produced. A broom is pulled across the top to make lines. This surface is easier to walk on in bad weather (Fig. 4-24). For most flooring surfaces, a smooth surface is desired. A smooth surface is made by troweling. For small surfaces, a hand trowel is used, as in Fig. 4-25. However, for larger areas, a power trowel is used (Fig. 4-26).

Fig. 4-22 *A. As the form is filled, it is leveled. This is done with a long board called a strike-off. B. Using a "jitterbug" to remove trapped air from the concrete.*

Fig. 4-23 *After leveling, the surface is smoothed by floating with a bull float.*

Fig. 4-24 *A broom can be used to make a lined surface.*

Fig. 4-25 *Hand trowels may be used to smooth small surfaces.*

Fig. 4-26 *A power trowel is used to finish large surfaces.*

Once the surface has been finished, the anchors for the walls are embedded. Remember, the concrete has not yet hardened. Do not let the anchors interfere with joist or stud spacing. The first anchor is embedded at about one-half the distance of the stud spacing. This would be 8 inches for 16-inch spacings. Anchors are placed at 4- to 8-foot intervals. Only two or three anchors are needed per wall.

The anchors are twisted deep into the concrete. The anchors are moved back and forth just a little. This settles the concrete around them. After the anchors are embedded, the surface is smoothed using a hand trowel (Fig. 4-27).

Fig. 4-27 *Anchor bolts are embedded into the foundation wall.*

The concrete is then allowed to harden and cure. Afterwards, the form is removed. It takes about three days to cure the concrete slab. During this time, work should not be done on the slab. When boards and lumber are used for the forms, they are recycled. These boards and lumber are used later in framing the house. Lumber is not thrown away if it is reasonably clean and nail-free. It is used where the concrete stains do not matter.

Expansion and Contraction

Concrete expands and contracts even when it is hardened. With alternate heat and cooling long after installation, it can expand and crack or contract, leaving a weakened slab under pressure. To compensate for this movement, expansion joints are needed. Expansion joints are used between sections. The expansion joint is made with wood, plastic, or fiber. Joint pieces are placed before pouring. Such pieces are used between foundations and basement walls or between driveways and slabs (Fig. 4-28). Most commonly noticed expansion joints are in sidewalks. Often, the screed is made of wood. This can be left for the expansion joint.

Fig. 4-28 *Expansion joints are used between large, separate pieces.*

Joints

Other joints are also used to control cracking. These, however, are shallow grooves troweled or cut into the concrete. They may be troweled in the concrete as it hardens. The joints may also be cut with a special saw blade after the concrete is hard. The joints help to offset and control cracking. Concrete is very strong against compression. However, it has little strength against bending or twisting. Note, for example, how concrete sidewalks break. The builder only tries to control breaks. They cannot be prevented. The joints form a weak place in concrete. The concrete will break at the weak point. But the crack will not show in the joint. The joint gives a better appearance when the concrete cracks. Figure 4-29 shows a troweled joint. Figure 4-30 shows a cut joint.

Fig. 4-29 *Joints may be troweled into a surface.*

CONCRETE FLOORS

Concrete is the most often used material for floors in basements and commercial buildings. Footings and foundations are built before the floor is made. An entire house can be built before the basement floor is poured (Fig. 4-31). The concrete can be

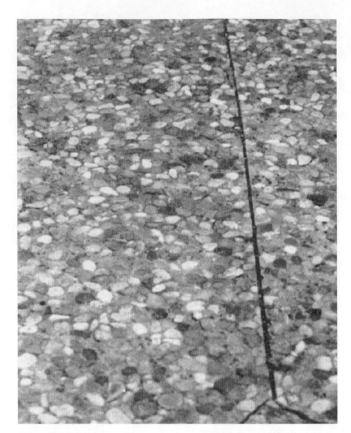

Fig. 4-30 *Some joints may be cut with saws.*

poured through basement windows or first-floor deck openings.

Concrete floors are made like a slab. First, the ground is prepared. Drains, pipes, and utilities are placed into position and covered with gravel. Next, coarse sand or gravel is leveled and packed. A plastic-film moisture barrier is placed to reach above the floor level. Perimeter insulation is laid as indicated in Fig. 4-32. Reinforcement is placed and tied. Rigid foam in-

Fig. 4-31 *Footings in a basement later become part of the basement floor.*

Fig. 4-32 Rigid foam should be laid at the edges of the basement floor.

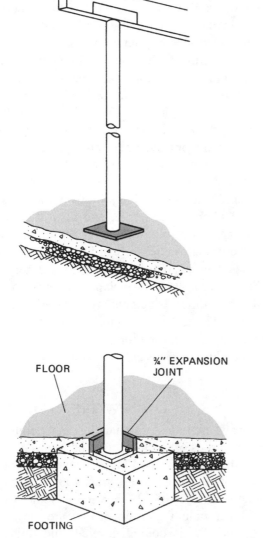

Fig. 4-33 Floors should not be major supports. Separate footings should be used.

sulation can also be used as an expansion joint. Finally, the floor is poured, tamped, leveled, and finished. Separate footings should be used to support beams and girders. Figure 4-33 shows this procedure.

Stairs

Entrances to the house are made as a part of the main slab. But stairs used with slabs are poured separately. Separate stairs are typically used where slabs serve as an access in steep areas. They are also used with steep lawns. Forms for steps may be made by two methods. The first method uses short parallel boards (Fig. 4-34). The second method uses a long stringer. Support boards are added as in Fig. 4-35. The top of the stair tread is left open in both types of forms. This lets the concrete step be finished. The bottoms of the riser forms are beveled. This lends a troweled finish to the full surface. The stock used for building supports should be 2-inch lumber. The heavier lumber keeps the weight of the concrete from bulging the forms.

Fig. 4-34 A form for steps.

Fig. 4-35 *A form for steps against a wall.*

SIDEWALKS AND DRIVEWAYS

Sidewalks and driveways are not a part of a slab. They float free of the building. An expansion joint is needed where they touch. Sidewalks and driveways should slope away from the building to allow water to drain away from the building. Then special methods are used for drainage. Often, a ledge about 1½ inches high will be used. In this way, the floor of the building is higher than the driveway. One side of the driveway is also raised. Water will then stop at the ledge and flow off to the side. This keeps water from draining into the building.

Two-inch-thick lumber should be used for the forms. It should be 4 or 6 inches wide. The width determines slab thickness. Commercial forms may also be used (Fig. 4-36).

Fig. 4-36 *Special forms may be used for sidewalks and driveways.* (Proctor Products.)

Sidewalks

Sidewalks are usually 3 feet wide and 4 inches thick. Main walks or entryways may be 4 feet or wider. No reinforcement is needed for sidewalks on firm ground, although it may be used for sidewalks on soft ground. A sand or gravel fill is used for support of sidewalks on wet ground. The earth and fill should be tamped solid. Figure 4-37 shows a sidewalk form. The slab is poured, leveled, and finished like other surfaces.

In making sidewalk joints, it is interesting to note that new devices are constantly reaching the market in an affordable size and shape. In Fig. 4-37B, notice how the joint holder is made of galvanized wire and welded. This is in order for it to hold the joint material, whether it is plastic, wood, or creosoted tar paper, as shown in Fig. 4-37D. Sometimes it is inserted in the sidewalk so that the finisher can make a groove of an inch or so deep along the line where an expansion joint would normally be placed. The properly welded metal wires allow for cracking of the concrete and keep it in check. The metal inserted in the concrete at pouring time aids in containing the concrete once it hardens. Figure 4-37C shows joint holders installed in the sidewalk without an expansion board. This type of joint holder is ideal for use in flatwork where expansion joints are required. The properly positioned holder expansion joint eliminates the need for backup formwork and therefore speeds up installation time. This device is reasonably priced and packaged for the convenience of the installer. It installs easily, holds the expansion joint firmly in place while keeping it plumb and true, and also allows you to place expansion joints wherever required. The joint holder eliminates checkerboard pours as well as bulk heading. It is available in two standard heights: 2½ inches high for 3½- to 4-inch pavement and 4½ to 5½ inches for 6-inch pavement. Both heights are available in standard 48-inch nominal length, packed five pieces per bundle, for a total of approximately 20 linear feet. Custom sizes are available.

Driveways

Driveways are required to handle great weights. Reinforcement rods or mesh are used in driveways. Slabs 4 inches thick may be used for passenger cars. Six-inch-thick slabs are used where trucks are expected. The standard driveway is 12 feet wide. Double driveways are 20 feet wide. Other dimensions are shown in Fig. 4-38. The slab is poured, leveled, and finished like other surfaces.

(A)

SPECIFY
EXPANSION
JOINT
THICKNESS

4½" FOR
5½" OR 6" SLABS

2½" FOR 3½"
OR 4" SLABS

(B)

(C)

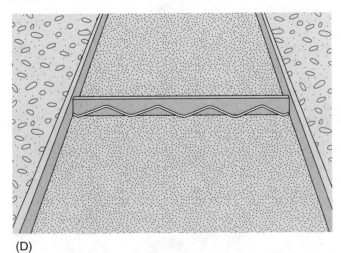

(D)

Fig. 4-37 *A. Sidewalk forms are made with 2 x 4 lumber. B. Expansion joint holder. C. Joint holders installed in sidewalks without the expansion board. D. Expansion joint ready for concrete.*

DRAIN FROM
CENTER OUTWARD

8' MINIMUM
9' PREFERABLE

CROSS SECTION

EXPANSION JOINT

SIDEWALK

CURB

5' RADIUS

(A)

5' O.C.

3' WIDE WHEN
USED AS WALK

2'   3'   2'

(B)

Fig. 4-38 *Driveway details: (A) single-slab driveway; (B) ribbon-type driveway.* (Forest Products Laboratory.)

SPECIAL FINISHES AND SURFACES

Concrete is finished in several ways. Various surface textures are used for better footing or appearance. Also, concrete may be colored and is often combined with other materials.

Surface Textures

Different surface textures may be used for appearance. However, the most common purpose is for better footing and tire traction, especially in bad weather. Several methods may be used to texture the surface.

Floating   The surface simply may be floated. This gives a smooth, slightly roughened surface. Floated surfaces are often used on sidewalks.

Brushing   The surface also may be brushed. Brushing is done with a broom or special texture brush. The pattern may be straight or curved. For a straight pattern, the brush is pulled across the entire surface. Refer again to Fig. 4-24. For swirls, the brush can be moved in circles.

Pebble finish   Pebbles can be put in concrete for a different appearance (Fig. 4-39). The pebble finish is not difficult. As the concrete stiffens, pebbles are poured on the surface. The pebbles are then tamped into the top of the concrete. The pebbles in the surface are leveled with a board or a float. Some hours later, the fine concrete particles may be hosed away.

Fig. 4-39 *A pebbled concrete surface.*

Color additives Color may be added to concrete for better appearance. The color is added as the concrete is mixed and is uniform throughout the concrete. The colors used most are red, green, and black. The surface of colored concrete is usually troweled smooth. Frequently, the surface is waxed and polished for indoor use.

Terrazzo A terrazzo floor is made in two layers. The first layer is plain concrete. The second layer is a special type of white or colored concrete. Chips of stone are included in the second, or top, layer of concrete. The top layer is usually about ½ to 1 inch thick. It is leveled but not floated. The topping is then allowed to set. After it is hardened, the surface is finished. Terrazzo is finished by grinding it with a power machine. This grinds the surface of both the stones and the cement until smooth. Metal strips are placed in the terrazzo to help control cracking. Both brass and aluminum strips are used. Figure 4-40 shows this ef-

fect. The result is a durable finish with natural beauty. The floor may be waxed and polished.

Terrazzo will withstand heavy foot traffic with little wear. It is often used in buildings such as post offices and schools. It is also easy to maintain. However, it can be slippery when wet. That is one of the reasons it is used for indoor service. Snow and rain make its slick surface very dangerous.

Ceramic tile, brick, and stone Concrete may also be combined with ceramic tile or brick. The result is a better-appearing floor. The floor contrasts the concrete and brick or tile. Also, the concrete may be finished in several ways. This gives more variety to the contrast. For example, a pebble finish on the concrete may be used with special brick. Bricks are available in a variety of shapes, colors, and textures.

Ceramic tile comes in many sizes. The largest is now 18 × 18 inches, with 12- × 12-inch tiles still being used. Several shapes are available. Glazed tile is used in bathrooms because the glaze seals water from the tile. Unglazed tile also has many uses, but it is not waterproof.

To set tile, stone, or bricks in a concrete floor, you must have a lowered area. The pieces are placed in the lowered area. The area between the pieces is filled with concrete, grout, or mortar.

Stone, tile, and brick are useful. They add contrast and beauty and resist wear. They also are easy to clean and resist oil, water, and chemicals.

Concrete over wood floors Concrete is also used for a surface over wood flooring. Figure 4-41 shows this kind of floor being made. A concrete topping on a floor has several advantages. These floors are harder and more durable than wood. They resist water and chemicals and may be used in hallways and rest rooms.

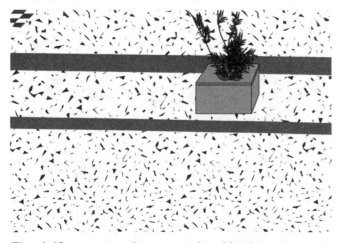

Fig. 4-40 *A terrazzo floor is smooth and hard. It is frequently used in schools and public buildings.*

Fig. 4-41 *A concrete floor is being placed over a plywood floor. The concrete will make the floor more durable.*

Wood over concrete   Wood may be used for a finish floor over a concrete floor. The wooden floor is warmer in cold climates. Because it does not absorb heat as does concrete, energy can be saved. The use of wood can also improve the appearance of the floor.

Two methods are used for putting wood over a concrete floor.

1. The first method is the older. A special glue, called *mastic*, is spread over the concrete. Then strips of wood are laid on the mastic. A wooden floor may be nailed to these strips. Figure 4-42 shows a cross section.

2. For better energy savings, a newer method is used. In this method, rigid insulation is laid on the

NOTE: Vapor barrier may be omitted if an effective vapor barrier is in place beneath the slab.

Fig. 4-43   *A wood floor may be laid over insulation. The insulation is glued to the concrete floor.*

concrete. Mastic may be used to hold the insulation to the floor (Fig. 4-43). The wooden floor is laid over the rigid insulation. Plywood or chipboard underlayment can also be used. The floor may then be carpeted. If desired, special wood surfaces can be used.

ENERGY FACTORS

Two ways used to save energy with concrete floors are

- Insulate around the edges of the slab.
- Cover the floor with insulated material.

Both methods have been mentioned previously.

CHAPTER 4 STUDY QUESTIONS

1. What does *monolithic* mean?
2. What are two ways of saving energy with concrete floors?
3. What is the sequence for pouring a slab?
4. Why are slab forms leveled just before pouring?
5. How is moisture beneath a slab drained?
6. How is moisture drained away from a building?
7. What thickness of lumber is best for making concrete forms?
8. What should be the thickness of a sidewalk?
9. What should be the thickness of a driveway for trucks?
10. How is energy saved when concrete floors are built?
11. How are step forms made?
12. What is the width of a single driveway?

Fig. 4-42   *A. Base for wood flooring on a concrete slab with a vapor barrier under slab. B. Base for wood flooring on a concrete slab with no vapor barrier under slab. (Forest Products Laboratory.)*

13. Where should footings for a slab be located?

14. How is a large surface floated?

15. How is a large surface troweled?

16. How can concrete be finished for better footing?

17. How is a pebble surface made?

18. How much concrete would be needed for a plain slab floor 6 inches thick, 21 inches wide, and 36 feet long?

19. What happens to lumber used for forms?

20. How is cracking controlled? Is it really stopped?

5
CHAPTER

Floor Frames

THIS CHAPTER COVERS HOW TO BUILD FLOOR frames—floors over basements and over crawl spaces. Openings for stairs and other things are also covered.

In this chapter, you will learn how to

- Connect the floor to the foundation
- Place needed girders and supports
- Lay out the joist spacings
- Measure and cut the needed parts
- Put the floor frame together
- Lay the subflooring
- Install sills
- Build special framing
- Alter a standard floor frame to save energy

FLOORS

Needless to say, floors are an important and integral part of a house or any building. Floor frames are built over basements and crawl spaces. Houses built on concrete slabs do not have floor frames. However, some multilevel buildings can make use of slabs and floor frames.

Framing Methods

It is true that no two buildings are put together in exactly the same manner. Disagreement exists among architects and carpenters as to which method of framing will prove most satisfactory for a given condition. Light-framed construction may be classified into the following three distinct types:

- Balloon frame
- Post and beam
- Platform frame

Balloon-Frame Construction

The principal characteristic of balloon framing is the use of studs extending in one piece from the foundation to the roof (Fig. 5-1). The joists are nailed to the studs and supported by a ledger board set into the studs. Diagonal sheathing may be used instead of wallboard to eliminate corner bracing.

Fig. 5-1 *Details of balloon-frame construction.*

Post-and-Beam Construction Post-and-beam construction is the oldest method of framing used in the United States. It is still used, particularly where the builder wants large open areas in a house. This type of framing is characterized by heavy timber posts, often with intermediate posts between (Fig. 5-2).

Platform-Frame Construction

Platform (or western) framing is characterized by platforms framed independently, the second or third floor being supported by the studs from the first floor (Fig. 5-3). The chief advantage in this type of framing and in all-lumber construction lies in the fact that if there is any settlement caused by shrinkage, it will be uniform throughout and will not be noticeable.

The foundation is laid. Then the floor frame is made using posts, beams, sill plates, joists, and a subfloor. When these are put together, they form a level platform. The rest of the building is held up by this platform. The first wooden parts are called the *sill plates*. The sill plates are laid on the edges of the foundations. Often, additional supports are needed in the middle of the foundation area (Fig. 5-4). These are called *midfloor supports* and may take several forms. These supports may be made of concrete or masonry. Wooden posts and metal columns are also used. Wooden timbers called *girders* are laid across the central supports. Floor joists then reach, or span, from the sill on the foundation to the central girder. The floor joists support the floor surface. The joists are supported by the sill and girder. These, in turn, rest on the foundation.

Multistory buildings use two types of floor framing. Most common is the *platform type*. In platform construction, each floor is built separately. The other type is called the *balloon frame*. In balloon frames, the wall studs reach from the sill to the top of the second floor. Floor frames are attached to the long wall studs. The two differ on how the wall and floor frames are connected. These are covered in detail later.

Fig. 5-2 *Post-and-beam frame construction. Heavy solid timbers are fastened together with pegged mortise-and-tenon joints.*

RAFTER HIP CROSS BRIDGING
ROUGH FLOOR
PLATE
STUD
SOLE
HEADER
GIRT
ROUGH FLOOR
STUD
SOLE
HEADER
SILL
DIAGONAL BRACING SET INTO FACE OF STUD
SHEATHING MASONRY WALL

PARTITION CAP
STUD
BRIDGING
JOIST
SOLE
SOLID BRIDGING
PARTITION CAP
BRIDGING
JOIST
SOLE
GIRDER
LEDGER OR SPIKING STRIP
CROSS BRIDGING
ROUGH FLOOR
CORNER POST

Fig. 5-3 *Details of platform, or western, construction.*

Fig. 5-4 *The piers and foundation walls will help to support the floor frame.*

SEQUENCE

The carpenter should build the floor frame using this sequence:

1. Check the level of foundation and supports.
2. Lay sill seals, termite shields, etc.
3. Lay the sill.
4. Lay girders.
5. Select joist style and spacing.
6. Lay out joists for openings and partitions.
7. Cut joists to length and shape.
8. Set joists.
9. Lay in place.
10. Nail opening frame.
11. Nail regular joists.
12. Cut scabs and trim joist edges.
13. Nail bridging at tops.
14. Lay subfloor.
15. Nail bridging at bottom.
16. Trim floor at ends and edges.
17. Cut special openings in floor.

PLACEMENT OF THE SILL

The first wooden part attached to the foundation is the sill. However, other things must be done before the sill is laid. When the anchors and foundation surface are

Fig. 5-5 *Perimeter forms for slab are installed, now trenches are dug for cables and plumbing.*

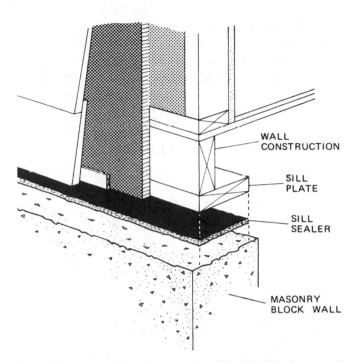

Fig. 5-6 *A seal fills in between the top of the foundation wall and the sill. It helps to conserve energy by making the sill more weather-tight.* (Conwed.)

adequate, a seal must be placed on the foundation. The seal may be a roll of insulation material or caulking. If a metal termite shield is used, it is placed over the seal. Next, drill anchor holes in the sill so that it will fit over the anchor bolts that protrude out of the foundation wall (Fig. 5-5).

The seal forms a barrier to moisture and insects and prevents heat loss during cold weather. Roll insulation-type material, as in Fig. 5-6, may be used. The roll should be laid in one continuous strip with no joints. At corners, the rolls should overlap about 2 inches.

Two things that can help to protect against termites are often done. Pour a solid masonry top or use metal shields. Some foundation walls are built of brick or concrete block that have hollow spaces. Seal these hollow spots with mortar or concrete on the top. A solid

Fig. 5-7 *Metal termite shield used to protect wood over a foundation.*

concrete foundation provides the best protection from termite penetration unless it cracks from shifting soil.

Termites can penetrate cracks in masonry. They can enter a crack as small as an inch in width. Metal termite shields are used in many parts of the country. Figure 5-7 shows a termite shield that has been installed.

Anchor the Sill

The sill is anchored to the foundation, and the anchors keep the frame from sliding from the foundation. They also keep the finished building from lifting in high winds. Three methods are used to anchor the sill to the foundation. The first uses bolts embedded in the foundation, as in Fig. 5-8. Sill straps are also used, and so

STEP 1: DRILL   **STEP 2: INSERT**   **STEP 3: ANCHOR**

Fig. 5-10 *Anchor holes may be drilled after the concrete has set.* (Hilti-Fastening Systems.)

Fig. 5-8 *Anchor bolt in foundation.*

Fig. 5-9 *Anchor straps or clips can be used to anchor the sill.*

are special drilled bolt anchors (Figs. 5-9 and 5-10). Special masonry nails are also sometimes used; however, they are not recommended for anchoring exterior walls. It is not necessary to use many anchors per wall. Anchors should be used about every 4 feet depending on local codes and prevailing winds. Anchors may not be required on walls shorter than 4 feet.

Anchor bolts must fit through the sill. The holes are located first. Washers and nuts are taken from the anchor bolts. The sill board is laid next to the bolts (Fig. 5-11). Lines are marked using a framing square as a guide. The sheathing thickness is subtracted from one-half the width of the board. This distance is used to find the center of the hole for each anchor. The centers for the holes then are marked. As a rule,

Fig. 5-11 *Locate the holes for anchor bolts.*

the hole is bored ¼ inch larger than the bolt. This leaves some room for adjustments and makes it easier to place the sill.

Place the sill over the anchors, and check the spacing and locations. All sills are fitted and then removed. Sill sealer and termite shields are laid, and the sills are replaced. The washers and nuts are put on the bolts and tightened. The sill is checked for levelness and straightness. Low spots in the foundation can be shimmed with wooden wedges. However, it is best to use grout or mortar to level a foundation.

Special masonry nails may be used to anchor interior walls on slabs. These are driven by sledge hammers or by nail guns. The nail mainly prevents side slippage of the wall. Figure 5-12 shows a nail-gun application.

Fig. 5-12 *A nail gun can be used to drive nails in the slab and for toenailing.* (Duo-Fast.)

Setting Girders

Girders are used to support the joists on one end. Usually the girder is placed halfway between the outside walls. The total distance between the supports is called the *span*. The span on most houses is too great for joists to reach from wall to wall. In that case, central support is provided by girders.

Determine girder location   Plans give the general spacing for supports and girders. Spans up to 14 feet are common for 2- × 10-inch or 2- × 12-inch lumber. The girder is laid across the leveled girder supports. A chalk line may be used to check the level. The support may be shimmed with mortar, grout, or wooden wedges. Supports are placed to equalize the span. The supports also help to lower expense. The piers shown in Fig. 5-4 must be leveled for the floor frame.

Fig. 5-13   *Built-up wood girder.* (Forest Products Laboratory.)

Girders can be built by nailing boards together. Figure 5-13 shows a built-up girder. Girders are often made of either 2- × 10-inch or 2- × 12-inch lumber. Joints in the girder are staggered. The girder and joist sizes are also specified on the plans.

There are several advantages to using built-up girders. First, thin boards are less expensive than thick ones. The lumber is more stable because it is drier. There is less shrinkage and movement of this type of girder. Wooden girders are also more fire-resistant than steel girders. Solid or laminated wooden girders take a long time to burn through. They do not sag or break until they have burned nearly through. Steel, on the other hand, will sag when it gets hot. It only takes a few minutes for steel to get hot enough to sag.

There are several ways to support the ends of girders. Figure 5-14 shows two methods. Ends of girders set in walls should be cut at an angle (Fig. 5-15). In a fire, the beam may fall free. If the ends are cut at an angle, they will not break the wall.

The metal girder should have a wooden sill plate on top. This board forms a nail base for the joists. Basement girders are often supported by post jacks (Fig. 5-16). Post jacks are used until the basement floor is finished. A support post, called a *lolly column*, may be built beneath the girder. It is usually made from 2 × 4 lumber. Walls may be built beneath the girder. In many areas, this is done so that the basement may be finished out as rooms later.

When a girder is supported by a wall or column, it must be remembered that such a member delivers a large, concentrated load to a small section of the wall or column. Therefore, care must be taken to see that the wall or column is strong enough to carry the load imposed on it by the girder. Girders generally are used only where the joist will not span the distance safely. The size of a girder is determined by the span length and the load to be carried.

(A)

(B)

Fig. 5-14 *Two methods of supporting girder ends: A. projecting post; B. recessed pocket.*

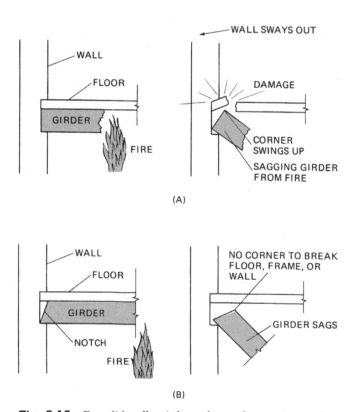

(A)

(B)

Fig. 5-15 *For solid walls, girder ends must be cut at an angle.*

Fig. 5-16 *A post jack supports the girder until a column or a wall is built.*

JOISTS

Joists are used to support the floor. They span from the sill to the girder. The subfloor is laid on the joists.

Lay Out the Joists

There are two basic ways to build joists. The *platform method* is the most common method. The other method is *balloon framing*. It is used for two-story buildings in some areas. However, the platform method is more common for multistory buildings.

Joist spacing   The most common spacing for joists is 16 inches. This makes a stronger floor with more support. It also allows the carpenter to use standard sizes. However, 12- and 24-inch spacing are also sometimes used. The spacing depends on the weight the floor must carry. Weight comes from people, furniture, and snow, wind, and rain. Local building codes often will specify what joist spacings are to be used.

Joist spacings are measured by the distance from the center of one board to the center of the next. This is called the distance *on centers*. For a 16-inch spacing, it is written *16 inches O.C.*

Modular spacings are 12, 16, or 24 inches O.C. These modules allow the carpenter to use standardized sheet materials easily. The standard-size sheet is 48 × 96 inches (4 × 8 feet). Any of the modular sizes divides evenly into the standard sheet size. By using modules, the amount of cutting and fitting is greatly reduced. This is important because sheet materials (such as plywood) are used on subfloors, floors, outside walls, inside walls, roof decks, and ceilings.

Joist layout for platform frames   Mark the position of the floor joists on a board called a *header*. Then nail the header across the ends of the joists (Fig. 5-17).

Fig. 5-17   *The positions of the floor joists may be marked on the header.*

The spacing of joists is from center to center or the distance between centers. Because the center of the board is a hard mark to use, it is much easier to mark the edge of a board. After all, if the centers are spaced right, the edges will be too!

A header is laid flat on the foundation. The end of the header is even with the end joist. The distance from the end of the header to the edge of the first joist is marked because this distance is not the same as the O.C. spacing (Fig. 5-18). The first distance is always ¾ inch less than the spacing. This lets the edge of the flooring rest flush, or even, with the outside edge of the joist on the outside wall. This makes laying the flooring quicker and easier.

The rest of the marks are made at the regular O.C. spacing (see Fig. 5-18). As shown, an X indicates on which side of the line to put the joist.

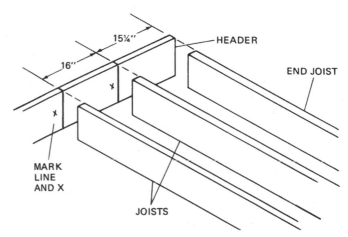

Fig. 5-18   *The first joist must be spaced ¾ inch less than the O.C. spacing used. Mark from end of header.*

Mark a pole   It is faster to transfer marks than to measure each one. The spacing can be laid out on a board first. This board then can be used to transfer spacings. This board is called a *pole*. A pole will save time because measurements are done only once. Just transfer the marks using the pole. It is laid next to a header. Use a square to project the spacing from the pole to the header. A square may be used to check the squareness of the line and the mark.

Joists under walls   Double the joists under walls. There are two ways of building a double joist. When the joist supports a wall, the two joists are nailed together (Fig. 5-19). Pipes or vents sometimes go through the floors and walls. Then a different method is used (Fig. 5-20). The joists are spaced approximately 4 inches apart. This space allows the passage of pipes or vents. Figure 5-21 shows the header layout

Fig. 5-19   *Joists are doubled under partitions.*

Fig. 5-20   *Double joists under a partition are spaced apart when pipes must go between them.*

pole with a partition added. Special blocking should be used in the double joist. Two or three blocks are used. The blocking serves as a fire stop and as bracing.

Joists for openings   Vents for the plumbing must have room and not create cut or reduced-strength joists. There also must be openings in the floors for stairs and chimneys. Double joists are used on the sides of these openings. They are called *double trimmers*. These double trimmers are placed without re-

Fig. 5-22   *Frame parts for a floor opening.*

gard for regular joist spacings. Regular spacing is continued on each end of the opening. Short, or "cripple," joists are used (Fig. 5-22). A pole can show the spacing for the openings (Fig. 5-23).

Girder spacing   Joists are also located on the girders. Remember, marks do not show centerlines of the joists. Centerlines are hard to use, so marks show the edge of a board. These edge marks are easily seen.

Balloon layout   In the balloon type of framing, there are some differences (Fig. 5-24). Wall studs rest on the sill. Joists and studs are nailed together as shown. However, the end joists are nailed to the end wall studs.

Locate the first joist back from the edge. Keep the distance the same as the wall thickness. Then locate the second joist by the first wall stud.

Locate the wall stud first. The first edge of the stud is ¾ inch less than the O.C. spacing. For 16 inches O.C., the stud is 15¼ inches from the end. A 2-inch

Fig. 5-21   *Header-joist layout. Next, add the joists for partitions.*

STEP 2

STAIRWELL OPENING

DOUBLE TRIMMER   CRIPPLE   CRIPPLE   TRIMMER ADDED

STEP 3

Fig. 5-23 *Add the trimmers for the opening to the layout pole.*

stud will be 1½ inches thick. Thus the edge of the first joist will be 16¾ inches from the edge.

Engineered Wood Joists

An alternative to solid dimensional lumber, which can warp as the humidity in your house changes throughout the year, are engineered wood joists. Various types and sizes are illustrated in Fig. 5-25. Engineered wood products can be used for floor joists and rafters. They are made in the shape of an I-beam. The top and bottom sections are made of laminated or solid lumber

Fig. 5-25 *Types of engineered wood joists.*

that is grooved to allow for the thinner middle section. This middle sections is typically composed of ⅜-inch-thick oriented strand board (OSB) The OSB is glued and pressed into the top and bottom sections.

The rim joists used for this type of construction resemble solid dimensional lumber; however, they are manufactured from laminated veneer lumber for extra strength and increased dimensional stability. The main advantages of engineered wood products over standard solid dimensional lumber are the following:

1. They are lighter in weight.
2. They have better strength-to-weight ratios.
3. Knockouts can be removed for cross-ventilation and wiring.
4. They are more stable, with no shrinkage-related callbacks.
5. They can be manufactured in continuous 30-foot lengths.
6. They are less expensive when reduced labor and material costs are considered.

SECOND FLOOR JOISTS
FIRE STOP
FIRE STOP
RIBBON (LET-IN)
STUD
DIAGONAL BRACE
FIRST FLOOR JOIST
FIRE STOP
ANCHORED SILL

Fig. 5-24 *Joist and stud framing used in balloon construction.*

Fig. 5-26 Sizes of engineered wood joists.

Fig. 5-28 Joists may overlap on the girder. The overlap may be long or short.

Fig. 5-27 Installing engineered wood joists.

Engineered wood joists are made in several sizes depending on load and span. Sizes range from 9½ inches tall with 1½-inch-wide top and bottom flanges to 16 inches tall with 3-inch-wide top and bottom flanges (Fig. 5-26). Installation of floor sheathing is done in the same manner as if you were using traditional solid-lumber floor joists (Fig. 2-27).

Because of the irregular I shape of the joists, they typically are not used around the perimeter as rim joists. The solid laminated veneer rim joists create a smooth side on which sheathing or siding can be attached easily. In addition, metal joist hangers also can be attached easily to the solid laminated veneer board. Because the sizes and types of joist vary among manufacturers, charts, tables, and building-technique brochures are provided wherever you purchase these products. Building techniques for joist are quite similar to those used when framing with traditional solid dimensional lumber, which will be discussed in depth later in this chapter.

Cut Joists

Joists span, or reach, from the sill to the girder. Note that joists do not cover the full width of the sill. Leave space on the sill for the joist header (see Fig. 5-5). For lumber 2 inches thick, the spacing would be 1½ inches. Joists are cut so that they rest on the girder. Four inches of the joist should rest on the girder.

The quickest way to cut joists is to cut each end square. Figure 5-28 shows square-cut ends. The joists overlap across the girder. The ends rest on the sill with room for the joist header. In this way, the header fits even with the edge.

It is sometimes easier to put the joist header on after the joists are toenailed and spaced. Or the joist header may be put down first. The joists then are just butted next to the header. It is very important, though, to carefully check the spacing of the joists before they are nailed to either the sill or the header.

Ends of joists may be cut in other ways. The ends may be aligned and joined. Ends are cut square for some systems. For others, the ends are notched. Metal girders are sometimes used. Then joists are cut to rest on metal girders.

End-joined joists The ends of the joists are cut square to fit together. The ends then are butted together as in Fig. 5-29. A gusset is nailed (10d) on each side to hold the joists together. Gussets may be made of either plywood or metal. This method saves lumber. Builders use it when they build several houses at one time.

Fig. 5-29 Joists may be butt-joined on the girder. Gussets may be used to hold them. Plywood subfloor may be used to stabilize the joists.

Fig. 5-30 *Joists can be notched and lapped on a girder.* *(Forest Products Laboratory.)*

Fig. 5-31 *Joists may be butted against the girder. Note that the girder may be spaced for pipes, etc.* *(Forest Products Laboratory.)*

Notched and lapped joists   Girders may be notched and lapped (Fig. 5-30). This connection has more interlocking but takes longer and costs more. First, a notch is cut on the end of the joist. Next a 2- × 4-inch joist support is nailed (16d) on the girder. Nails should be staggered 6 or 9 inches apart. The joist is then laid in place. The end overlaps across the girder.

Joist-girder butts   This is a quick method. With it, the top of the joist can be even with the top of the girder. A 2- × 4-inch ledger is nailed to the girder with 16d common nails (Fig. 5-31). The joist rests on the ledger and not on the girder. This method is not as strong.

Also, a board is used to join the girder ends. The board is called a *scab*. The scab also makes a surface for the subfloor. It is a 2- × 4-inch board. It is nailed with three 16d nails on each end.

Joist hangers   Joist hangers are metal brackets (Fig. 5-32). These brackets hold up the joists. They are nailed (10d) to the girder. The joist ends are cut square. Then the joist is placed into the hanger. It is also nailed with 10d nails, as in Fig. 5-32. Using joist hangers saves time.

Fig. 5-32 *Using metal joist hangers saves time.*

The carpenter need not cut notches or nail up ledgers.

Joists for metal girders   Joists must be cut to fit into metal girders (Fig. 5-33). A 2- × 4-inch board is first bolted to the metal girder. The ends of the joist then are beveled. This lets the joist fit into the metal girder. The joist rests on the board. The board also is a nail base

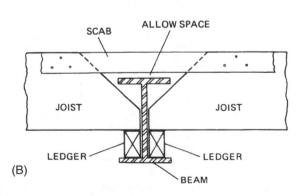

Fig. 5-33 *Systems for joining joists to metal girders.* *(Forest Products Laboratory.)*

for the joist. The tops of the joists must be scabbed. The scabs are made of 2- × 4-inch boards. Three 16d nails are driven into each end.

Setting the Joists

Two jobs are involved in setting joists. The first is laying the joists in place. The second is nailing the joists. The carpenter should follow a given sequence.

Lay the joists in place   First, the header is toenailed in place. Then the full-length joists are cut. Then they are laid by the marks on the sill or header. Each side of the joist then is toenailed (10d) to the sill (Fig. 5-34). Joists next to openings are not nailed. Next, the ends of the joists are toenailed (10d) on the girder. Then the overlapped ends of the joists are nailed (16d) together (see Fig. 5-28). These nails are driven at an angle, as in Fig. 5-35.

Nail opening frame   A special sequence must be used around the openings. The regular joists next to the opening should not be nailed down. The opening joists are nailed (16d) in place first. These are called *trimmers.*

Then the first headers for openings are nailed (16d) in place. Note that two headers are used. For 2- × 10-inch joists, three nails are used. For 2- × 12-inch

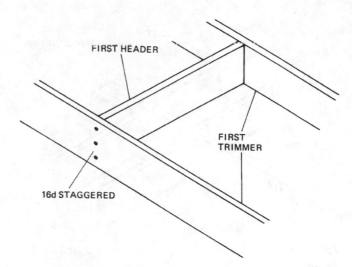

Fig. 5-36 *Nailing the first parts of an opening.*

lumber, four nails are used. Figure 5-36 shows the spacing of the nails.

Next, short cripple or tail joists are nailed in place. They span from the first header to the joist header. Three 16d nails are driven at each end. Then the second header is nailed (16d) in place (Fig. 5-37).

Fig. 5-37 *Add the second header and trimmer joists.*

The double-trimmer joist now is nailed (16d) in place. These pieces are nailed next to the opening. The nails are alternated top and bottom (Fig. 5-38). This finishes the opening. The regular joists next to the opening are nailed in place. Finally, the header is nailed to the joist ends. Three 16d nails are driven into each joist.

Fire and Draft Stops

Fire stops are short pieces nailed between joists and studs (Fig. 5-39). They are made of the same boards as the joists. Fire stops keep fire from spreading between walls and floors. They also help to keep joists from twisting and spreading. Fire stops are usually put at or

Fig. 5-34 *The joists are toenailed to the girders.*

Fig. 5-35 *When nailing joists together, drive nails at an angle. This holds better, and the ends do not stick through.*

Fig. 5-38 *Stagger nails on double trimmer. Alternate nails on top and bottom.*

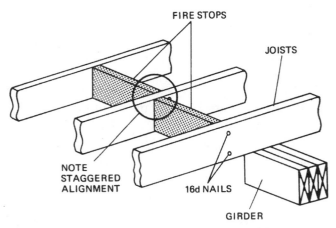

Fig. 5-39 *Fire stops are nailed in. They keep fire from spreading between walls and floors.*

near the girder. Two 16d nails are driven at each end of the stop. Stagger the boards slightly as shown in the figure. This makes it easy to nail them in place.

It is well known that many fires originate on lower floors. It is therefore important that fire stops be provided to prevent a fire from spreading through the building by way of air passages between the studs. Similarly, fire stops should be provided at each floor level to prevent flames from spreading through the walls and partitions from one floor to the next. Solid blocking should be provided between joists and studs to prevent fire from passing across the building.

In platform and post-and-beam framing, the construction itself provides stops at all levels. In this construction, therefore, fire stops are needed only in the floor space over the bearing partitions. Masonry is sometimes used for the stopping but is usually adaptable in only a few places. Generally, obstructions in the air passages may be made of 2-inch lumber, which will effectively prevent the rapid spread of fire. Precautions should be taken to ensure the proper fitting of fire stops throughout the building.

Bridging

To prevent joists from springing sideways under load (which would reduce their carrying capacity), they are tied together diagonally by 1 × 3 or 2 × 3 strips in a process called *bridging*. The 1 × 3 ties are used for small houses, and the 2 × 3 stock is used on larger work. Metal bridging also may be used.

Rows of bridging should not be more than 8 feet apart. Bridging pieces may be cut all in one operation with a miter box, or bridging may be cut to fit. Bridging is put in before the subfloor is laid, and each piece is fastened with two nails at the end. The subfloor should be laid before the bottom end is nailed.

A more rigid (less vibrating) floor can be made by cutting in solid 2-inch joists of the same depth. They should be cut perfectly square and a little full (say, 1/16 inch longer than the inside distance between the joists). First, set a block in every other space; then go back and put in the intervening ones. This keeps the joists from spreading and allows the second ones to be driven in with the strain the same in both directions. This solid blocking is much more effective than crossbridging. The blocks should be toenailed and not staggered and nailed through the joists.

Bridging is used to keep joists from twisting or bending. Bridging is centered between the girder and the header. For most spans, center bridging is adequate. For joist spans longer than 16 feet, more bridging is used. Bridging should be put in every 8 feet. This must be done to comply with most building codes.

Most bridging is cut from boards. It may be cut from either 1- or 2-inch lumber. Use the framing square to mark the angles, as in Fig. 5-40. With this method, the angle may be found.

A radial arm saw may be used to cut multiple pieces (Fig. 5-41). Also, a jig can be built to use with a portable power saw (Fig. 5-42).

Fig. 5-40 *Use a square to lay out bridging.*

STEP 1. Set saw at angle and place stop.

BOARD (ANY LENGTH)   SAW DIRECTION

STEP 2. Make first cut.

STOP

STEP 3. Reverse short board and make second cut.

STEP 4. Reverse long board and make cut.

Fig. 5-41 *Radial arm saw set up for bridging.*

Steel bridging is also used. Figure 5-43 shows an example. Often, only one nail is needed in each end. Steel bridging meets most local codes and standards.

The bridging pieces should be cut first. Nails are driven into the bridging before it is put up. Two 8d or 10d nails are driven into each end. Next, a chalk line is strung across the tops of the joists. This provides a line for the bridging.

Bridging is nailed at the top first. This lets the carpenter space the joists for the flooring when it is laid.

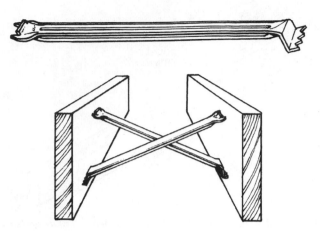

Fig. 5-43 *Most building codes allow steel bridging.*

The bottoms of the bridging are nailed after the flooring is laid.

The bridging is staggered on either side of the chalk line. This prevents two pieces of bridging from being nailed at the same spot on a joist. To nail them both at the same place would cause the joist to split (Fig. 5-44).

SUBFLOORS

Subflooring is the last step in making a floor frame. Subflooring most often is called *underlayment*. The subfloor is the platform that supports the rest of the structure. It is covered with a finished floor material in the living spaces. This may be of wood, carpeting, tile, or stone. However, the finished floor is added after the exterior of the house is completed.

The most common material used for subflooring is plywood. Plywood for subfloors is graded C–D with waterproof or exterior glues. Materials such as chipboard, fiberboard, and other type boards are also used for subflooring.

STOP BLOCK FOR THE LENGTH

BRIDGING PIECE

SAW ANGLED AS NEEDED

SAW BASE AGAINST GUIDE BOARD

Fig. 5-42 *A jig may be built to cut bridging.*

CHALK LINE

ALTERNATE LOCATION TO PREVENT SPLITTING

(A)

NAIL TOP NAILS FIRST

JOISTS

(B)

NOTE: Nails are partially driven into bridging before it is put in place.

Fig. 5-44 *Bridging pieces are staggered.*

TABLE 5-1 *Minimum Flooring Standards*

Single-Layer (Resilient) Floor			
Joists, Inches O.C.	Minimum Thickness, Inches	Common Thickness, Inches	Minimum Index
12	19/32	5/8	24/12
16	5/8	5/8 or 3/4	32/16
24	3/4	3/4	48/24
Subflooring with Finish Floor Layer Applied			
Joists, Inches O.C.	Minimum Thickness, Inches	Common Thickness, Inches	Minimum Index
12	1/2	1/2 or 5/8	32/16
16	1/2	5/8	32/16
24	3/4	3/4	48/24

NOTES: 1. C-C grade underlayment plywood.
2. Each piece must be continuous over two spans.
3. Sizes can vary with span and depth of joists in some locations.

Plywood Subfloor

Plywood and chipboard are ideal subflooring materials. They are quickly laid, and the job takes little cutting and trimming. The sheets may be nailed or glued to the joists. Plywood and chipboard are very flat and smooth. This makes the finished floor smooth and easy to lay. Builders use plywood and chipboard thicknesses from ½ to ¾ inch. The most common thicknesses are ½ and ⅝ inch. The Federal Housing Administration (FHA) minimum is ½-inch-thick plywood.

Plywood, when used as subflooring, has fewer squeaks than boards. This is so because fewer nails are required. The squeak in a floor is caused when nails work loose. Table 5-1 shows minimum standards for plywood use.

Chipboard and Fiberboard

As a rule, plywood is stronger than other types of underlayment. However, both fiberboard and chipboard are also used. Chipboard underlayment is used more often. The minimum thickness for chipboard or fiberboard is ⅝ inch. This thickness also must be laid over 16-inch joist spacing. Both chipboard and fiberboard are laid in the same manner as plywood. In any case, the ends of the large sheets are staggered (Fig. 5-45).

Laying Sheets

Use the same method for any sheet subflooring. Nails are used most often, but glue also can be used. Start at an outside corner. The long grain or sheet length is laid across the joists (Fig. 5-46). The ends of the different courses are staggered. This prevents the ends from all lining up on one joist. If they did, it could weaken the floor. By staggering the end joints, each layer adds strength to the total floor. The carpenter must allow for expansion and contraction. To do this, the sheets are spaced slightly apart. A paper matchbook cover may be used for spacing. Its thickness is about the correct space.

Nailing The outside edges are nailed first with 8d nails. Special "sinker" nails may be used. The outside nails should be driven about 6 inches apart. Nails are driven into the inner joists about 10 inches apart (Fig. 5-47). Power nailers can be used to save time, cost, and effort (Fig. 5-48). Nailers can be plugged directly into a wall outlet or portable generator for operation on 120 volts ac, or they can be the compressed air type, which uses a long piece of tubing from the compressor to the nailer. Both types are cumbersome to use in comparison with the battery-operated nailers presently available at Home Depot or Lowe's.

Fig. 5-45 *The ends of subfloor sheets are staggered.*

Fig. 5-46 *The long grain runs across the joists.*

Fig. 5-47 *Flooring nails. Note the "sinker" head on the first nail.*

Fig. 5-48 *Using power nailers saves time and effort.* (Duo-Fast.)

Gluing Gluing is now used widely for subflooring. Modern glues are strong and durable. Glues, also called *adhesives*, are applied quickly. Glue will not squeak, as will nails. Figure 5-49 shows glue being applied to floor joists. Floors are also laid with tongue-and-groove joints (Fig. 5-50). Buffer boards are used to protect the edges of the boards as the panels are put in place (Fig. 5-51).

Caution: Gluing the subfloor to the joists produces a stiffer floor, but not one that is much stronger than a nailed-only floor. Strength is not the same as stiffness. The span of floor joists framed with weaker wood, such as spruce-pine-fir (SPF), is controlled by the strength of the wood. Plywood glued to SPF joists will stiffen them, but the small increase in strength will not allow an increase in span length.

Board Subflooring

Boards are also used for subflooring. There are two ways of using boards. The older method lays the

Fig. 5-49 *Subflooring is often glued to the joists. This makes the floor free of squeaks.*

Fig. 5-50 *Plywood subflooring also may have tongue-and-groove joints. This is stronger.*

Fig. 5-51 *A buffer board is used to protect the edges of tongue-and-groove panels.*

Fig. 5-52 *Diagonal board subfloors are still used today.*

boards diagonally across the joists. Figure 5-52 shows this. This way takes more time and trimming. It takes a longer time to lay the floor, and more material is wasted in trimming. However, diagonal flooring is still used. It is preferred where wood board finish flooring will be used. In this way, the finish flooring may be laid at right angles to the joists. Having two layers that run in different directions gives greater strength.

Board subflooring is often laid at right angles to the joists. This is appropriate when the finish flooring will be sheets of material.

Either way, two kinds of boards are used. Plain boards are laid with a small space between them. This allows for expansion. End joints must be made over a joist for support (Fig. 5-53). Grooved boards are also used (Fig. 5-54). End joints may be made at any point with grooved boards.

Fig. 5-53 *Plain board subflooring may be laid across joists. Joints must be made over a joist.*

Nailing Lay the boards from an outside edge toward the center. The first course is laid and nailed with 8d nails. Two nails are used for boards 6 inches wide or less. Three nails are used for boards wider than 6 inches.

Nail the boards down untrimmed. The ends stick out over the edge of the floor. This is done for both

Fig. 5-54 *Grooved flooring is laid across joists. Joints can be made anywhere.*

(A)

(B)

Fig. 5-55 *A. A bay window rests on overhanging floor joists. B. A different type of bay. Both types rest on overhanging floor joist systems.*

grooved and plain boards. After the floor is done, the ends are sawed off even with the floor edges.

SPECIAL JOISTS

Carpenters should know how to make special joists. Several types of joists are used in some buildings. Special joists are used for overhangs and sunken floors. A sunken floor is any floor lower than the rest. Sunken floors are used for special flooring such as stone. Floors also may be lowered for appearance. Special joists are also used to recess floors into foundations. This is done to make a building look lower, often referred to as a *low-profile building*.

Overhangs

Overhangs are called *cantilevers*. They are used for special effects. Porches, decks, balconies, and projecting windows are all examples. Figure 5-55A shows an example of projecting windows. Figure 5-55B is a different type of bay. However, both rest on overhanging floor joist systems. Overhangs are also used for garrison-style houses. When a second floor extends over the wall of the first, it is called a *garrison style* (Fig. 5-56).

The longest projection without special anchors is 24 inches. Windows and overhangs seldom extend 24 inches. A balcony, however, would extend more than 24 inches. This means that it calls for special anchors.

Overhangs with joist direction Some overhangs project in the same direction as the floor joists. Little extra framing is needed for this. This is the easiest way to build overhangs. In this method, the joists are simply made longer. Blocking is nailed over the sill with

16d nails. Figure 5-57 shows blocking and headers for this type of overhang. Here, the joists rest on the sill. Some overhangs extend over a wall instead of a foundation. Then the double top plate of the lower wall supports the joists.

Overhangs at angles to joist direction Special construction is needed to frame this type of overhang.

(A)

(B)

Fig. 5-56 *A. Joist protections from the overhang for a garrison-type second story. B. The finished house.*

Fig. 5-57 *Some overhangs simply extend the regular joist.*

It is similar to framing openings in the floor frame. Stringer joists form the base for the subflooring. Stringer joists must be nailed to the main floor joists (Fig. 5-58). They must be inset twice the distance of the overhang. Two methods of attaching the stringers are used. The first method uses a wooden ledger. However, this ledger is placed on the top (Fig. 5-59A). The other method uses a metal joist hanger. Special anchors are needed for large overhangs such as rooms or decks.

Fig. 5-58 *Frame for an overhang at an angle to the joists.*

Fig. 5-59 *A. A top ledger "let in" is a good anchor.*

Cantilevered In-Line Joist System

When two joists ends are secured together with a metal or plywood gusset, the spliced joint does not have to be located directly over the girder. Such joists can extend beyond the girder, cantilevered, as shown in Fig. 5-59B.

The cantilevered system is more structurally efficient than the simple two-joist span. There is less stress on the suspended joist because the span is shorter. It runs from the foundation wall to the girder. Off-center spliced joists act as though they were a single unit. A reduction in joist size may be possible—a 2

Fig. 5-59 *B. Cantilevered joist diagram from APA.* (Courtesy American Plywood Associiaton.)

× 8 rather than a 2 × 10. However, even if no reduction is allowed, the cantilevered joist combined with a plywood subfloor is stiffer than the simple two-joist span. The offset joist can be preassembled, and each joist simply can be dropped in place.

Sunken Floors

Subfloors are lowered for two main reasons. A finish floor may be made lower than an adjoining finish floor for appearance. Or the subfloor may be made lower to accommodate a finish floor of a different material. The different flooring could be stone, tile, brick, or concrete. These materials are used for appearance or to drain water. However, they are thicker than most finish floors. To make the floor level, special framing is done to lower the subfloor.

The sunken portion is framed like a special opening. First, header joists are nailed (16d) in place (Fig. 5-60). The headers are not as deep or wide as the main joists. This lowers the floor level. To carry the load with thinner boards, more headers are used. The headers are added by spacing them closer together. Double joists are nailed (16d) after the headers.

Low Profiles

The lower-profile home has a regular size frame. However, the subfloor and walls are joined differently. Figure 5-61 shows the arrangement. The sill is below the top of the foundation. The bottom plate for the wall is attached to the foundation. The wall is not nailed to the subfloor. This makes the joists below the common foundation level. The building will appear to be lower than normal.

Fig. 5-60 *Details of framing for a sunken floor.*

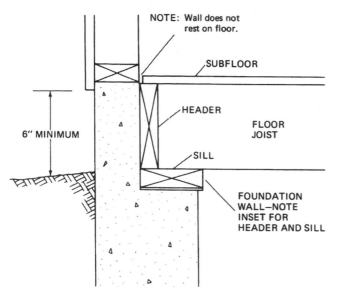

Fig. 5-61   *Floor detail for a low-profile house.*

ENERGY FACTORS

More energy (heat) is lost through ceilings because heat rises. However, energy can be saved by insulating the ceiling and the floor. In the past, most floors were not insulated. Floors over basements need not be insulated if the basement is heated or finished for a living space. Floors over enclosed basements are also good energy savers, even if not insulated.

Floors over crawl spaces, however, should be insulated. The crawl space also should be totally enclosed. The foundation should have ventilation ports. However, they should be closed in winter. More energy is saved by insulating certain areas. Floors under overhangs and bay windows should be insulated because they are exposed to the outdoors.. Floors next to the foundation also should be insulated. The insulation should start at the sill or header and extend 12 inches into the floor area (Fig. 5-62). Outer corners are the

Fig. 5-62   *Insulate the outside floor edges.*

Fig. 5-63   *Insulation between floor joists should be supported.*

most critical spots. However, for the best results, the whole floor should be insulated. Roll or bat insulation is placed between the joists and supported, usually by a wire mesh. Supports can be made of wood strips or wire. Nail (6d) wire or wood strips to the bottom of the joists to support the insulation material (Fig. 5-63).

Moisture Barriers

Slabs as well as basements must have moisture barriers beneath them. Moisture barriers are not needed under a floor built over a basement. However, floors over crawl spaces should have moisture barriers. The moisture barrier is laid over the subfloor (Fig. 5-64). A moisture barrier may be added to older floors below the joists. This is usually held in place by either wooden strips or wires. Usually, 6-mil plastic or builder's felt is used (Fig. 5-65).

Energy Plenums

What is a plenum? A *plenum* is a space for controlled air. The air is usually slightly pressurized. Plenum systems over crawl spaces allow air to circulate beneath floors. This maximizes the heating and cooling effects. Figure 5-66 shows how the air is circulated. Doing this keeps the temperature more even. Even temperatures are not only more efficient but also more comfortable.

The plenum must be carefully built. Insulation is used in special areas (Fig. 5-67). A hatch is needed for plenum floors. The hatch gives access to the plenum area. Access is needed for inspection and servicing. There are no outside doors or vents to the plenum. The hot-air furnace has a plenum sitting on top above the burners, and it serves to collect the heat generated by the burners for distribution as part of a central heating system. Ducts carry the hot air or cool conditioned air throughout the house.

A plenum arrangement–style of house offers an advantage to the builder: The plenum house can be

FINISH FLOOR

VAPOR BARRIER

SUBFLOOR

INSULATION

SCREENED VENT

FLOOR JOIST

SOIL COVER (VAPOR BARRIER)

Fig. 5-64 *A moisture barrier is laid over the subfloor above a crawl space.*

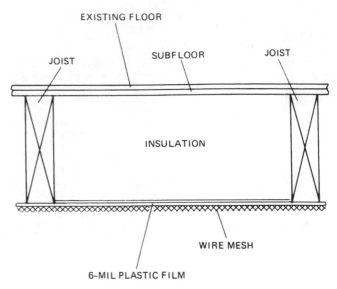

EXISTING FLOOR

JOIST

SUBFLOOR

JOIST

INSULATION

WIRE MESH

6-MIL PLASTIC FILM

Fig. 5-65 *A moisture barrier also may be added beneath floors.*

built inexpensively. There are several reasons for this. The circulation system is simpler. No ducts are built beneath floors or in attics. Common vents are cut into the floors of all the rooms. The system forces air into the sealed plenum (Fig. 5-68). The air does not lose

energy in the insulated plenum. The forced air then enters the various rooms from the plenum. The blower unit is in a central portion of the house. The blower can send air evenly from a central area. The enclosed louvered space permits the air easy return to the blower.

Similar to most houses, the rough plumbing is brought into the crawl space first. Then the foundation

Fig. 5-66 *The air-circulation system for an energy plenum.*

Fig. 5-67 *Section view of an energy plenum.*

is laid. Fuel lines and cleanouts are located outside the crawl space. The minimum clearance in the crawl space should be 18 inches. The maximum should be 24 inches. This size gives the greatest efficiency for air movement.

Foundation walls may be masonry or poured concrete. Specially treated plywood foundations are also used (see Chapter 22). Proper drainage is essential. After the foundation is built, the sill is anchored. Standard sills, seals, and termite shields may be used. The plenum area must be covered with sand. Next, a vapor barrier is laid over the ground and extends up and over the sill (see Fig. 5-67). This completely seals the plenum area (see Fig. 5-68). Next comes the insulation. Either rigid or batt insulation may be used. It should extend from the sill to about 24 inches inside the plenum. The most energy loss occurs at foundation corners. Be sure to insulate corners. Then the floor joists are nailed to the sill. The joist header is nailed (16d) on and insulated. Subflooring then is nailed to

the joists. This completes the plenum. After this, the building is built as a normal platform.

Note: Most of the line drawings for this chapter were furnished by American Wood Products, the Forest Products Laboratory, the American Plywood Association, and the Western Wood Products Association.

CHAPTER 5 STUDY QUESTIONS

1. What is the procedure for building a floor?
2. List the parts of a frame floor.
3. Why are regular joists by openings nailed last?
4. What does a girder do?
5. What is a crawl space?
6. What holds the wooden frame to the foundation?
7. Where is joist spacing marked?
8. Why is the first joist spaced differently?
9. What is a post jack?
10. What are the two joist (framing) methods?
11. What are common joist spacings?
12. What size lumber is used for joists?
13. What is the easiest way to cut joists?
14. What are the advantages of using joist hangers?
15. What sequence is used to nail the opening frames?
16. Why are fire stops and bridging used?
17. What is done to frame overhangs?
18. What is done to frame sunken floors?
19. How can floors be built to save energy?
20. What is an energy plenum?

Fig. 5-68 *Conditioned air is circulated through the plenum.*

6
CHAPTER

Framing Walls

UILDING A WALL FRAME IS THE TOPIC OF THIS chapter. How to cut the parts for the wall is covered, then we look at how to connect the wall to the rest of the building. Why the parts are made as they are is also explained. You will learn how to

- Lay out wall sections
- Measure and cut the parts
- Assemble and erect wall sections
- Join wall sections together
- Change a standard wall frame to save energy and materials

FRAMING

The two ways of framing a building are

- The western platform method
- The balloon method

Most contemporary buildings use the western platform method of construction. Residential housing uses the western platform method, where walls are put up after the subfloor has been laid. Walls are started by making a frame. The frame is made by nailing boards to the tops and bottoms of other boards. The top and bottom boards are called *plates*. The vertical boards are called *studs*. The frame must be made very strong because it holds up the roof. After the frame is put together, it is raised and nailed in place. The wall frames are put up one at a time. The roof can be built next. Then the walls are covered, and windows and doors are installed. Figure 6-1 shows a wall frame in place. Note that the roof is not yet on the building.

Sometimes the first covering for the wall is added before the wall is raised. This first covering is called *sheathing*. It is very easy to nail the sheathing to the frame while the frame is flat on the floor. Doing this

makes the job quicker and reduces problems with keeping the frame square. Another advantage is that no scaffolds will be needed to reach all the areas. This reduces the possibility of accidents and saves time in moving scaffolds. However, most builders still nail sheathing on after the wall is raised.

The wall is attached to the floor in most cases. When the floor is built over joists, the wall is nailed to the floor (Fig. 6-2). When the wall is built on a slab, it is anchored to the slab (Fig. 6-3). As a rule, walls for both types of floors are made the same way.

Walls that help to hold up the roof or the next floor are made first. As a rule, all outside, or exterior, walls do this. These walls are also referred to as *load-bearing walls*.

Inside walls are called *partitions*. They can be load-bearing walls, too. However, not all partitions carry loads. Interior walls that do not carry loads may be built after the roof is up. Interior walls that do not carry loads are also called *curtain walls*.

After the walls are put up, the roof is built. Then the walls are covered. The first cover is the sheathing. Putting on the sheathing after the roof lets a builder get the building waterproof or weatherproof a little sooner. The siding is put on much later.

Wall sections are made one at a time. The longest outside walls are made first. The end walls are made next. However, the sequence can be changed to fit the

Fig. 6-2 *Section view of a wall on a frame floor.*

Fig. 6-1 *Wall frames are put up after the floor is built.*

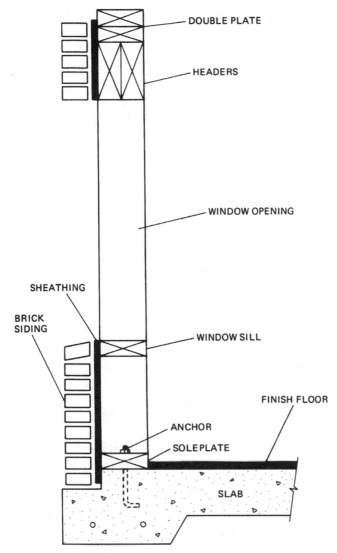

Fig. 6-3 *Section view of a wall on a slab.*

job. There are many ways to make walls. This chapter will show the most common method.

SEQUENCE

Follow this general sequence for making wall frames:

1. Lay out the longest outside wall section.
2. Cut the parts.
3. Nail the parts together.
4. Raise the wall.
5. Brace it in place.
6. Lay out the next wall section.
7. Repeat the process.
8. Join the walls.
9. Do all outside walls.
10. Do all inside walls.

11. Build the roof.
12. Sheath the outside walls.
13. Install outside doors and windows.
14. Cut soleplates from inside door openings.

WALL LAYOUT

All the parts of the wall must be planned. The carpenter must know beforehand what parts to cut and where to nail. The carpenter plans the construction by making a wall layout on boards. One layout is done on the top and bottom boards (plates). Another layout is done for the wall studs.

Plate Layout

Soleplates   First, select pieces of 2-inch lumber for the bottom of the wall. The bottom part of the wall is called a *soleplate*. The soleplate is laid along the edges of the floor where the walls will be. No soleplate is put across large door openings. Sometimes the soleplate can be made across small doors. After the wall is put up, the soleplate is sawed out for the doorway.

Top plate   After a soleplate has been laid, another piece is laid beside it. It will be the top part of the wall. This piece is called a *top plate*. The soleplate and the top plate are laid next to each other with a flat side up. Figure 6-4 shows how the soleplate and the top plate are laid so that you can measure and mark both the top and bottom plates at the same time. Because it keeps both top and bottom locations aligned, this method ensures accuracy.

Soleplates and top plates are often spliced. A splice must occur over the center of a full stud (Fig. 6-5). Otherwise, the wall section will be weakened.

Stud Layout

Several types of studs are used in walls. The studs that run from the soleplate to the top plate are called *full studs*. Studs that run from the soleplate to the top of a rough opening are called *trimmer studs*. Short studs that run from either plate to a header or a sill are called *cripple studs*. Figure 6-6 shows a part of a wall section.

Spacing full studs   Most full studs are spaced a standard distance apart. This standard distance is an even part of the sizes of plywood, sheathing, and other building materials. The studs act both as roof support and as a nailing base for the sheathing. The most common standard distance is 16 inches. Studs spaced 16 inches apart are said to be 16 inches on center (O.C.). Another common spacing is 24 inches O.C.

100' TAPE

1" SCRAP

CORNER STUDS

TAPE MARKED 16 INCHES O.C.

NAIL A PIECE OF 1 INCH SCRAP AT THE END. HOOK
THE TAPE OVER THE END. MARK THE INTERVAL.
THEN MAKE A SMALL X TO THE RIGHT OF THE MARK.

Fig. 6-4 *Lay soleplates and top plates for marking.*

16"   16"   16"

PLACE THIS EDGE
OF TONGUE ON MARK

MARK ALL STUDS →

16"   16"

USE A 2 INCH SCRAP TO MARK WIDTHS. THEN
MARK STUD LOCATIONS.

Fig. 6-5 *Top plates are spliced over a stud.*

Lines on centers are not used to show where boards are placed. An easier way is shown in Fig. 6-4. The stud is located by using the left end of the wall as a starting point. Nail a 1-inch scrap piece there. Then the distance between centers is measured from the block on the outside corner of the wall. A mark is made, and an X is made to the right of the mark.

After all the marks are made, a square is used to line in the locations. This method is easier because the measurement is taken from the side of the stud. The distance is the same whether it is center-to-center or side-to-side.

Space the rough openings The next step in spacing is to find where the openings for windows and doors are to be made. The openings for windows and doors are framed no matter what the stud spacing is. The openings are called *rough openings* (ROs). The size of rough openings may be shown in different ways. The actual size may be shown in the plan. However, many plans show window schedules, which often list the window as a number, such as 2442. This means that the window sash opening is 2 feet, 4 inches wide and 4 feet, 2 inches high. The rough openings are larger. Wooden windows require larger ROs than metal ones. It is best to refer to the specifications.

However, when the specs are not available, carpenters can use a rule of thumb: Wooden windows are usually written, for example, 24 × 42. The RO should be 3 inches wider (2 feet, 7 inches) and 4 inches higher (4 feet, 6 inches). Metal frames are usually written without the X. The RO is 2 inches wider and 3 inches higher. For a 2442 metal window, the RO would be 2 feet, 6 inches wide and 4 feet, 5 inches high.

Fig. 6-6 *A wall section and parts. Note the large header over the opening. It is sometimes larger than needed to eliminate the need for top cripples. This saves labor costs because it takes longer to cut and nail cripples.*

To find the locations of ROs, measure the distance from the corner or end of the wall to the center of the RO. Make a mark called a *centerline* on the soleplate. From the centerline, measure half the RO width on each side. Make another mark at each side of the centerline for the RO. Mark a line, as in Fig. 6-7, to show the thickness of a stud. This thickness goes on the outside of the RO. Note that the distance between.

The lines must be the width of the opening. Mark a T in the regular stud spaces. This tells you that a trimmer stud is placed there. On the outside of the space for the trimmer stud, lay out another thickness.

Mark an X from one corner of the opening to the other, as shown in the figure.

The X is used to show where a full stud is placed. The T is used to show where a trimmer stud is placed. The trimmer stud does not extend from the soleplate to the top plate. Thus it is shown only on the soleplate.

Corner Studs

A strong way to nail the walls together is needed. To accomplish this, a double stud is used for one corner.

Fig. 6-7 *Locate rough openings.*

Corner studs on the first wall The spacing of studs starts at one corner. Another stud space is marked at the second corner. The regular spacing is not used at corners. A stud is laid out at each corner, or end, of the wall.

To make the corner stronger, another stud is placed in the corner section (Fig. 6-8A). The second corner stud is spaced from the first with spacer blocks. This is called a *built-up corner*. It is done to give the corner greater strength and to make a nail base on the inside. A nail base is needed on the inside to nail the inside wall covering in place. After the end stud is marked on the corner, mark the thickness of one more stud on the plate. Mark it with an S for spacer. Next, lay out the stud as shown in Fig. 6-8B. Mark it with an X to indicate a full stud.

Corner studs for the second wall The second wall does not need double studs. It is laid out in the regular way. The walls are nailed together as in Fig. 6-9.

Fig. 6-9 *The second wall is nailed to the double corner.* (Forest Products Laboratory.)

Partition Studs

As mentioned earlier, inside walls are called *partitions*. The partitions must be solidly nailed to the outside walls. To make a solid nail base, special studs are built into the exterior walls (Fig. 6-10). The most common method is to place two studs as shown in the figure. The studs are placed 1½ inches apart. The space is made just like the corners on the first wall. The corner of the partition now can rest on the two studs. The two special studs act as a nail base for the wall. Also, ¾ inch of each stud is exposed. This makes a nail base for the interior wall covering. Other methods of joining partitions to the outside walls are shown in Figs. 6-11 and 6-12.

Find Stud Length

Before cutting any studs, find the proper lengths. There are two ways of doing this. One way lets the carpenter measure the length. In the other way, the carpenter must compute it.

Fig. 6-8 *Stud layout for corners: A. the assembly; B. the layout.*

Fig. 6-10 *Partition walls are nailed to special studs in the outside walls.* (Forest Products Laboratory.)

Fig. 6-11 *Another way to join partitions to the outside walls.*
(Forest Products Laboratory.)

Fig. 6-12 *A third way to join partitions to outside walls.*

Make a master stud pattern A carpenter can make a master stud pattern. This is the way that stud lengths are measured. A 2-inch board just like the studs is used. A side view of the wall section is drawn. The side view is full size and will show the stud lengths.

Start by laying out the distance between the floor and the ceiling. The distances in between are then shown. Rough openings are added. This then shows the lengths of trimmers and cripples. The pattern also lets the carpenter check the measurements before cutting the pieces. The stud pattern is also called a *story pole* or *rod*. Generally, it is done for only one floor of the house. Figure 6-13 shows a master stud pattern.

Compute stud length Another way to find stud lengths is to compute them. The carpenter must do some arithmetic and check it very carefully. This method is shown in Fig. 6-14.

The usual finished floor-to-ceiling distance is 8 feet, ½ inch. The thicknesses of the finish floor and ceiling material are added to this dimension. The

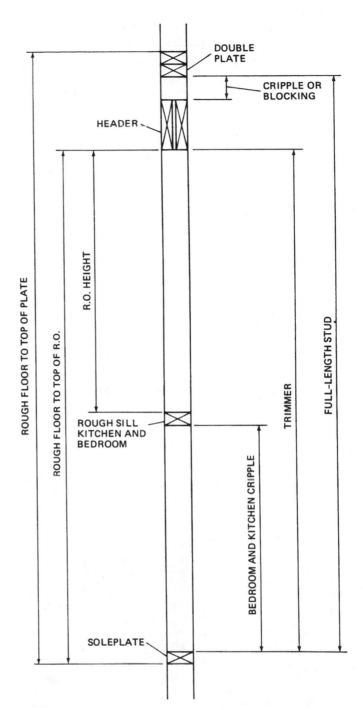

Fig. 6-13 *Make a master stud pattern by drawing it full size on a board.*

floor and ceiling thicknesses are commonly ½ inch each. The distance is written all in inches and adds up to 97½ inches. The thickness of the soleplate and top plate is subtracted. For one soleplate and a double top plate, this thickness is 4½ inches (3 times 1½ inches). The remainder is the stud length. For this example, the stud length is 93 inches. This is a commonly used length.

Fig. 6-14 *Computing stud length.*

Using precut studs Sometimes the studs are cut to length at the mill and delivered to the site. When this is done, the carpenter does not make any measurements or cuts for full studs. The standard length for such precut studs is 93 inches. The carpenter who orders such materials should be very careful to specify *precision end trim* (P.E.T.) for the lumber (Fig. 6-15).

Fig. 6-15 *Precut studs ready at a site.*

Frame Rough Openings

The locations of full and trimmer studs for the RO are shown on the soleplates. The lengths of these can be found from the size of the RO. The story pole is used for reference. The width of the RO sets the distance between the trimmers (see Fig. 6-6). A full stud is used on the outside of the trimmer.

Trimmer Studs
Trimmer studs extend from the soleplate to the top of the RO. They provide support for the header. The header must support the wall over the RO. It is important that the header be solidly held. The trimmers give solid support to the ends of the headers. The length of trimmers is the distance from the soleplate to the header.

Header Size

Size of the header is determined by the width of the RO. Table 6-1 shows the size for a typical RO width. As in Fig. 6-3, two header pieces are used over the RO. In some cases, the headers may be large enough to completely fill the space between the RO and the top plate (Fig. 6-16). However, doing this will make the wall section around the header shrink and expand at a different rate than the rest of the wall. A solid header section is also harder to insulate.

The sill The bottom of the RO is framed in by a sill. The sill does not carry a load. Thus it simply can be

Fig. 6-16 *Solid headers may be used instead of cripples.*

Table 6-1 *Size of Lumber for Headers*

Width of Rough Opening	Minimum Size Lumber	
	One Story	Two or More
3'0"	2 × 4	2 × 4
3'6"	2 × 4	2 × 6
4'0"	2 × 6	2 × 6
4'6"	2 × 6	2 × 6
5'0"	2 × 6	2 × 6
6'0"	2 × 6	2 × 8
8'0"	2 × 8	2 × 10
10'0"	2 × 10	2 × 12
12'0"	2 × 12	2 × 12

nailed between the trimmer studs at the bottom of the RO. The sill does not require solid support beneath the ends; however, it is common to use another short trimmer for this (Fig. 6-17).

(A)

(B)

Fig. 6-17 *A trimmer may be used under both the header and the sill.*

Cripples Cripple studs are short studs. They join the header to the top plate. They also join the sill to the soleplate. They continue the regular stud spacing. This makes a nail base for sheathing and wallboard.

CUTTING STUDS TO LENGTH

After the wall parts are laid out, carpenters must find how many studs are needed. They must do this for each length. A great deal of time can be saved if all the studs are cut at one time. Saws are set, and full studs are cut first. Then the settings are changed, and all trimmer studs are cut. Next, the cripple studs are cut. Headers and sills should be cut last.

Cutting Tips

Most cuts to length are done with power saws. Two types of saws are used. Radial arm saws can be moved to the site (Fig. 6-18). Portable circular saws can be used almost anyplace (Fig. 6-19).

Fig. 6-18 *A radial arm saw can be used on the site.*

With either type of saw, special setups can be used. Pieces can be cut to the right length without measuring each one.

Portable circular saw Sawing several pieces to the same length is done with a special jig (Fig. 6-20). Two pieces of stud lumber are nailed to a base. Enough space is left between them for another stud. A stop block is nailed at one end. A guide board then is nailed across the two outside pieces. Care should be taken because the guide is for the saw frame. The blade cuts a few inches away.

Radial arm saw A different method is used on a radial arm saw. No marking is done. Figure 6-21 shows how to set the stop block. This is quicker and easier.

Fig. 6-19 *Portable circular saws can be used almost anywhere.*

Fig. 6-20 *A jig for cutting studs with a portable circular saw.*

The piece to be cut is simply moved to touch the stop block. This sets the length. The piece is held against the back guide of the saw. The saw is then pulled through the work.

WALL ASSEMBLY

Once all pieces are cut, the wall may be assembled. Headers should be assembled before starting (Figs. 6-22 and 6-23). The soleplate is moved about 4 inches away from the edge of the floor. It is laid flat with the stud markings on top. Then it is turned on edge. The markings should face toward the middle of the house. Then the top plate is moved away from the soleplate. The distance should be more than the length of a full stud. The soleplate and top plate markings should be aligned. The must point toward each other.

Fig. 6-21 *A stop block is used to set the length with a radial arm saw. In this way, all the studs can be cut without measuring. This saves much time and money.*

Fig. 6-22 *Use ½-inch plywood for spaces between headers.*

Fig. 6-23 *Nailers can be used to assemble headers.*

Fig. 6-24 *Laying the studs in place on the subfloor.*

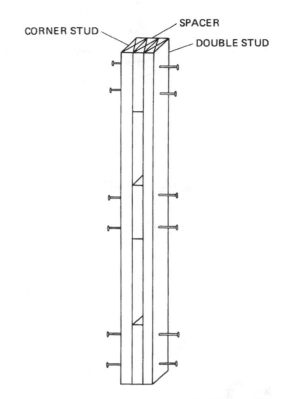

Fig. 6-25 *Nailing spacers to corner studs. Use two 16d nails on each side of a spacer.*

The straightest studs are selected for the corners. They are put at the corners just as they will be nailed. Next, a full stud is laid at every X location between the soleplate and top plate. Figure 6-24 shows carpenters doing this.

Nailing Studs to Plates

All studs are laid in position. The soleplate and top plate are tapped into position. Make sure that each stud is within the marks.

Corner studs   Before nailing, corner spacers are cut. Three spacers are used at each corner. Each spacer should be about 16 inches long. Spacers are put between the two studs at the corners. One spacer should be at the top, one should be at the bottom, and one should be in the center. Two 16d nails should be used on each side (Fig. 6-25). After the corner studs are nailed together, they are nailed to the plates. Two 16d nails are used for each end of each stud.

Full studs   All the full studs are nailed in place at the X marks. Two 16d nails are used at each end. Figures 6-26 and 6-27 show how.

Trimmer studs   Next, the trimmer studs are laid against the full studs. The spacing of the ROs is checked. The trimmers then are nailed to the studs from the trimmer side (Fig. 6-28). Use 10d nails. The nails should be staggered and 16 inches apart. *Staggered* means that one nail is near the top edge, and the next nail is near the bottom edge (Fig. 6-29).

Fig. 6-26 *Nailing full studs to the plates.*

Fig. 6-27 Trimmers are nailed into the rough opening.

Fig. 6-28 Nailing studs for wall frames.

Fig. 6-29 In staggered nailing, one nail is near the top and the other is near the bottom.

Headers and sills The headers are nailed in place next. The headers are placed flush with the edges of the studs. The headers are nailed in place with 16d nails. The nails are driven through the studs into the ends of the headers.

The sills are nailed in place after the headers. Locate the position of the sill. Toenail the ends of the sill

to the trimmers. Also, another trimmer may be used (Fig. 6-30).

Cripple studs Next, the cripples are laid out. They are nailed in place from the soleplate with two 16d nails. Then two 16d nails are used to nail the sill to the ends of the cripples. Finally, the top cripples are nailed into place (Fig, 6-31).

CORNER BRACES

Corner braces should be put on before the wall is raised. The bracing prevents the wall frame from swaying sideways. Two methods are commonly used.

Plywood Corner Braces

The first method uses plywood sheets, as shown in Fig. 6-32. It is best to nail the plywood on before the wall is erected. Plywood bracing costs more but takes little time and effort. If used, it should be the same thickness as the wall sheathing. Plywood should not be used where "energy" sheathing is used. This will be discussed later.

Diagonal Corner Braces

The second method is to make a board brace. The brace is made from 1- × 4-inch lumber. It is "let in," or recessed into, the studs. The angle may be any angle, but 45 degrees is common. Braces are recessed into the studs so that the outside surface of the wall will remain flat (Fig. 6-33).

To make a diagonal brace, select the piece of wood to be used, and nail it temporarily in place across the

Fig. 6-30 *Another trimmer may be used to support the sill.*

Fig. 6-31 *Nail cripples through soleplate and sill.*

(A)

(B)

Fig. 6-32 *Plywood can be used for corner braces.*

outside. Mark the layout on each stud. Also mark the angle on the ends of the brace. Remove the brace, and gauge the depth of the cut across each stud. Make the cuts with a saw. Knock out the wood between the cuts

with a chisel. Then trim off the ends of the brace. Place the brace in the "let in" slots. Check for proper fit. If the fit is good, nail the brace in place. Use two 8d nails at each stud.

TOP PLATES

LAP
TOP PLATES
AND NAIL

WINDOW
HEADER

TEMPORARY
BRACE

LET-IN
CORNER BRACE

STUD

WINDOW SILL

HEADER
JOIST

SOLEPLATE

SUBFLOOR

STRINGER JOIST

FOUNDATION WALL

ANCHORED SILL PLATE

SPACER BLOCK

Fig. 6-33 *Boards also can be used for corner bracing.* (Forest Products Laboratory.)

ERECT THE WALLS

Once the wall section is assembled, it is raised upright. The wall may be raised by hand (Fig. 6-34). Special wall jacks also may be used (Fig. 6-35). Figure 6-36 shows another method, raising the wall with a forklift unit. This is common when a large wall must be placed over anchor bolts on a slab. For slabs, the anchor holes in the soleplate should be drilled before the wall is raised (Fig. 6-37; see also the section on anchoring sills in Chapter 5). Walls that have been raised must be braced. The brace is temporary. The brace is used to hold the wall upright at the proper angle. Wind will easily blow walls down if they are not braced. Bracing is usually kept until the roof is on (Figs. 6-38 and 6-39).

Fig. 6-36 *Raising a wall section with a fork-lift unit.*

Fig. 6-34 *Raising a wall section by hand.*

Fig. 6-37 *Wall sections are anchored to slabs after being erected.*

Fig. 6-35 *Raising a wall section with wall jacks.*

Fig. 6-38 *Once raised, walls are held in place with temporary braces.*

Fig. 6-39 *Nailing a temporary brace.*

Wall Sheathing

Sheathing may be put on a wall before it is raised. The advantage is that this reduces the amount of lifting and holding. However, it slows down getting the roof up. It is also harder to make extra openings for vents and other small objects. Wall sheathing may be nailed or stapled. Manufacturer's recommendations should be followed. Sheathing is made from wood, plywood, fiber, fiberboard, plastic foam, or gypsum board.

To Raise the Wall

Before the wall is raised, a line should be marked. The line should show the inside edge of the wall. It is made with the chalk line on the subfloor. This shows the position of the wall.

Before the wall is raised, all needed equipment should be ready. Enough people or equipment to raise the wall should be ready. Also, the tools to plumb and brace the wall should be ready.

When all are ready, the wall may be raised. The top edge of the wall is picked up first. Lower parts are grasped to raise it into a vertical position (see Figs. 6-34 through 6-36). The wall is held firmly upright. As it is held, the wall is pushed even with the chalk line on the floor.

Some walls are built erect and not raised into position. As a rule, these walls are built on slabs. The layout and cutting are the same. However, the soleplate is bolted to the slab first. Next, the studs are toenailed to the soleplate. The top plate can be put on next. Then the wall is braced. Openings and partition bases then may be built where needed.

Put Up a Temporary Brace

After the wall is positioned on the floor, it is nailed in place (Fig. 6-40). Then one end is plumbed for vertical alignment. A special jig can be used for the level (Fig. 6-41). Once the wall is plumb, a brace is nailed (see Fig. 6-39).

Fig. 6-40 *Once the walls have been pulled into place, they are nailed to the subfloor.*

Each wall is treated in the same manner. After the first wall, each added wall will form a corner. The corners are joined by nailing, as in Fig. 6-42. It is important that the corners be plumb.

Once several walls have been erected, the double plate is added at the top. Figure 6-42 shows a double plate in place. Note that the double plate overlaps on corners to add extra strength.

INTERIOR WALLS

Interior walls, or partitions, are made in much the same manner as outside walls. However, the carpenter must remember that studs for inside walls may be longer. This is so because inside walls are often curtain walls. A curtain wall does not help to support the load of the roof. Because of this, a double plate might not be used. But the top of the wall must be just as high.

TOP PLATE

1-INCH BLOCKS NAILED TO A STRAIGHT 2 × 4

2 × 4 ON EDGE

CARPENTER'S LEVEL

WALL FRAME

CENTER BUBBLE BETWEEN MARKS

Fig. 6-41 *Use a spanner jig to plumb the corners with the level.*

WALL DOUBLE PLATE

10d NAILS

PARTITION DOUBLE PLATE

10d NAILS

Fig. 6-42 *Double plates are added after several walls have been erected.*

Most builders wish to make the building weather tight quickly. Therefore, the first partitions that are made are load-bearing partitions.

When roof trusses are used, load-bearing partitions may not be necessary. Trusses distribute loads so that inside support is not needed. However, many carpenters make all walls alike. This lets them cut all studs the same. It also lets them use any wall for support.

Locate Soleplates for Partitions

To locate the partitions, the centerlines are determined from the plans. The centerline is then marked on the floor with the chalk line. Plates are then laid out.

Studs

Studs and headers are cut. The partition walls are done in the same way as outside walls. As a rule, the longer partitions are done first. Then the short partitions are done. Last are the shortest walls for the closets.

Corners

Corners and wall intersections are made just as for outside walls. The size and amount of blocking can be reduced. The purpose of blocking is to provide nail surfaces. These are needed at inside and outside corners. They are a base for nailing wall covering.

Headers and Trimmers

Headers are not required for rough openings in curtain walls. Often, openings are framed with single boards (Fig. 6-43). The header for inside walls is much like a

Fig. 6-43 *Single boards are also used as headers on inside partitions.*

sill. Trimmers are optional. They provide more support. They are recommended when single-board headers are used.

Soleplate

The soleplate is not cut out for door openings. It is made as one piece. It is cut away after the wall is raised (Fig. 6-44).

Special Walls

Several conditions may call for special wall framing. Walls may need to be thicker to enclose plumbing. Drain pipes may be wider than the standard 3½-inch-thick wall. Thickness may be added in two ways: Wider studs may be used, or extra strips may be nailed to the edges of studs. Other special needs are soundproofing and small openings.

Soundproofing

Most household noise is transmitted by sound waves vibrating through the air. Blocking the air path with a standard interior wall or ceiling reduces the sound somewhat, but not completely.

The reason? Vibrations still travel through solids, particularly when the materials provide a continuous path. Standard interior walls and metal air ducts allow sound waves to continue their transmission.

Therefore, an effective sound-control system must block not only the sound path but also the vibration

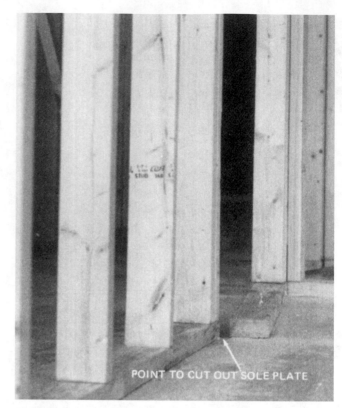

Fig. 6-44 *At door openings, the partition soleplates are cut out after the wall is up.*

path. Further noise reduction is accomplished with a system that absorbs the sound.

The Sound Transmission Class (STC) rating is a measurement—expressed in decibels—that describes

WALL DETAIL	DESCRIPTION	STC RATING
⊢— 16″ —⊣ 2 × 4	1/2-INCH GYPSUM WALLBOARD	45
2 × 4	5/8-INCH GYPSUM WALLBOARD (DOUBLE LAYER EACH SIDE)	53
2 × 4 BETWEEN OR "WOVEN"	1/2-INCH GYPSUM WALLBOARD 1 1/2-INCH FIBROUS INSULATION	60
2 × 4	1/2-INCH SOUND DEADENING BOARD (NAILED) 1/2-INCH GYPSUM WALLBOARD (LAMINATED)	51

Fig. 6-45 *Sound insulation of double walls.*

a structure's ability to resist sound transmission. The higher the STC rating, the greater is the structure's ability to limit the transmission of sound.

By incorporating combinations of all three elements for proper sound control, it is possible to raise the STC rating of a standard interior wall from 4 to 12 decibels depending on the wall's construction. Special insulation is available for sound control. When installed inside interior walls, the acoustic batts provide extra material to help block the sound path. The fiberglass insulation has tiny air pockets that absorb sound energy.

To break the vibration path, the insulation manufacturer recommends two options. One approach is to install staggered wall studs. The soleplate and top plate are made wider. Regular studs are used, but the studs are staggered from one side to the other (Fig. 6-45). Insulation now can be woven between the studs. Sound is transmitted better through solid objects. With staggered studs, though, there is no solid part from wall to wall. In this way, there is no solid bridge for easy sound passage.

Another alternative is to install resilient channels between the drywall and the studs. For floors and ceilings, resilient channels should be installed in addition to the insulation batts. Figure 6-46A illustrates this using wood studs, 16 inches O.C., with ½-inch drywall on each side and one thickness of fiberglass acoustic batt, which is 3½ inches thick. Figure 6-46B shows a better option. Single wood studs are used 16 inches O.C. with a resilient channel and ½-inch drywall on each side. A 3½-inch-thick insulation batt is used. Resilient channels help to reduce noise by dissipating sound energy and decreasing sound transmission through the framing. For the wall, install resilient channels spaced 24 inches horizontally over 16 inches O.C. framing. Another, better option is shown in Fig. 6-46C. Here, staggered wood studs placed 16 inches O.C. with ½-inch gypsum board on each side with one thickness of fiberglass acoustic batt, 3½ inches thick. Keep in mind that the insulation batts inserted between

the walls are not intended as a thermal barrier but are for sound control only.

There are other sources of sound or noise transmission that also must be addressed if the problem is to be minimized. Figure 6-47 shows that the electrical wiring can be a source of sound transmission if the holes for the wire are not caulked. Do not place light switches and outlets back to back. Place wall fixtures a minimum of 24 inches apart. Note also how elastic caulk is used to fill in around the electrical receptacles.

Fig. 6-47 *Soundproofing light switches and receptacles.* (Owens-Corning.)

Wood joist floors (Fig. 6-48) usually have a standard carpet and pad with a ⅜-inch plywood or particle board subfloor. A single-layer ½-inch gypsum ceiling mounted on resilient channels spaced every 24 inches also improves the sound-deadening quality of the ceiling and floor. This can be improved more by adding 3.5-inch-thick fiberglass acoustic batts.

Fig. 6-48 *Standard carpet and pad ⅜-inch particle board surface, ⅜-inch plywood subfloor, and a single sheet of ½-inch gypsum board mounted on resilient channels spaced 24 inches.* (Owens-Corning.)

Doors with a solid-wood core will improve noise control (Fig. 6-49). Install a threshold closure at the bottom of the door to reduce sound transmission. Ducts can be insulated with fiberglass batts to aid in maintaining constant air temperature while minimizing sound transmission. Ductwork featuring a smooth acrylic coating on the inner surface for easy cleaning and maintenance also controls the sound-transmission qualities (Fig. 6-50).

Caulking also can decrease the sound level. Caulking around the perimeter of drywall panels, plumbing fixtures, pipes, and wall plates can reduce the incidence of unwanted noises considerably (Fig. 6-51).

(A)   (B)   (C)

Fig. 6-46 *A. Single wood studs, 16 inches O.C.* (Owens Corning.) *B. Single wood studs, 16 inches O.C., with a resilient channel. C. Staggered wood studs, 16 inches O.C.* (Owens-Corning.)

Fig. 6-49 *Doors need a threshold closure to keep out noise.*
(Owens-Corning.)

Fig. 6-50 *Installing fiberglass on ductwork.*

Small openings   Small openings are often needed in walls. They are needed for air ducts, plumbing, and drains. Recessed cabinets, such as bathroom medicine cabinets, also need openings. These openings are framed for strength and support (Fig. 6-52).

Fig. 6-51 *Caulking around plumbing helps to seal out noise.*

Fig. 6-52 *Framing for small openings in walls.*

SHEATHING

Several types of sheathing are used. Sheathing is the first layer put on the outside of a wall. Sheathing makes the frame still and rigid and provides insulation. It also will keep out the weather until the building is finished.

Five main types of sheathing are presently in use:

- Fiberboard
- Gypsum board
- Boards
- Plywood
- Rigid foam

Corner bracing is needed for all types except plywood and boards.

Fiberboard Sheathing

The most common type of sheathing is treated fiberboard. It should be applied vertically for best bracing and strength characteristics. It is usually ½ inch thick. Plywood ½ inch thick can be used for corner bracing. The ½-inch fiberboard sheathing and plywood fit flush for a smooth surface. The exterior siding can easily be attached. Fiber sheathing should be fastened with roofing nails. Nails 1½ inches long are spaced 3 to 6 inches apart. Nails always should be driven at least ⅜ inch from the edge. If plywood is not used, diagonal corner braces are required (Fig. 6-53).

Gypsum Sheathing

Sheathing made from gypsum material is also used (Fig. 6-54). This gypsum sheathing is not the same as the sheets used on interior walls. The gypsum sheathing is treated to be weather resistant. The most common thickness is ½ inch. A ½-inch-thick plywood corner brace may be used.

The sheathing should be fastened with roofing nails. Nails 1¾ inches long should be used and should be spaced 4 inches apart.

Plywood Sheathing

Plywood is also used for sheathing. When it is used, no corner bracing is required. Plywood is both strong and fire resistant. It can be nailed up quickly with little cutting. A moisture barrier must be added when plywood is used. However, plywood and board sheathing are both expensive.

(A)

(B)

(C)

Fig. 6-53 *Corner bracing: A. fiberboard sheathing with plywood corner braces; B. fiberboard sheathing with inlet board corner braces—seen from the outside; C. fiberboard sheathing with inlet board corner braces—seen from the inside.*

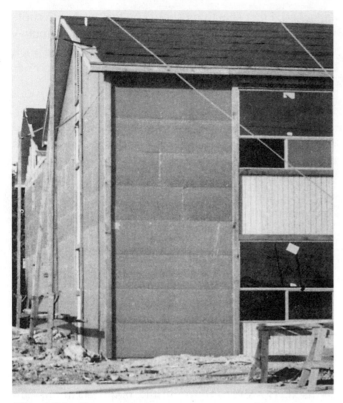

Fig. 6-54 *Gypsum sheathing is inexpensive and fire resistant.*

Fig. 6-55 *Rigid polystyrene foam sheathing provides more insulation.* (Dow Chemical.)

Fig. 6-56 *Rigid polystyrene foam may be coated with reflective foil to increase its effectiveness as an insulation sheathing.*

When plywood is used, it should be at least ⅜ inch thick. Half-inch thickness is recommended. Exterior siding can be nailed directly to ½-inch plywood. These plywood sheets can be applied vertically or horizontally. The sheets should be fastened with 6d or 8d nails. Nails should be 6 to 12 inches apart.

"Energy" Sheathing

Two types of "energy" sheathing are used commonly. The first is a special plastic foam called *rigid foam* (Fig. 6-55). Its use can greatly reduce energy consumption. It is roughly equivalent to 3 more inches of regular wall insulation. It is grooved on the sides and ends and is fitted together horizontally. It is nailed in place with 1¼-inch nails spaced 9 to 12 inches apart. Joints may be made at any point. Rigid foam also may be covered with a shiny foil on one or both sides (Fig. 6-56). The foil surface further reduces energy losses. It does so because the shiny surface reflects heat. Also, the foil prevents air from passing through the foam. However, foam burns easily. A gypsum-board interior wall should be used with the foam. The gypsum wall reduces the fire hazard.

The second type is a special fiber. It is also backed on both sides with foil. Its fibers do not insulate as well as foam. It does prevent air movement better than plain foam, though. The foil surface is an effective reflector. Also, it is more fire resistant and costs less (Fig. 6-57).

Fig. 6-57 *Another type of energy-efficient sheathing is similar to rigid foam but is coated with reflective foil.* (Simplex.)

Boards

Boards are still used for sheathing. However, diagonal boards are seldom used. Plain boards may be used. Boards with both side and end tongue-and-groove joints are used. For grooved boards, joints need not occur over a stud. This saves installation time. Carpenters need not cut boards to make joints over studs. However, for plain boards, joints should be made over studs. A special moisture barrier should be added on the outside. Builder's felt is used most often. It is nailed in place with 1-inch nails through metal disks. The bottom layers are applied first (Fig. 6-58).

FACTORS IN WALL CONSTRUCTION

A carpenter may learn the procedure for making a wall. However, the carpenter may not know why walls are made as they are. Each part of a wall frame has a specific role. The corner pieces help to tie the walls together. The double top plates also add strength to the corners. The top plate is doubled to help support the weight of the rafters and ceiling. A rafter or a ceiling joist between studs is held up by the top plate. A single top plate eventually could sag and bend.

Standard Spacing

Spacing of wall members is important. Standard construction materials are 4 feet wide and 8 feet long. Standardized finish floor-to-ceiling heights are 8 feet, ½ inch. The extra ½ inch lets pieces 8 feet long be used without binding.

Further, using even multiples of 4 or 8 means that less cutting is needed. Buildings are often designed in multiples of 4 feet. Two-foot roof overhangs can be used. The overhang at each end adds up to 4 feet. This reduces the cutting done on siding, floors, walls, and ceilings, thereby reducing time and cost.

Notching and Boring

When a hole or notch is cut into a structural member, the structural capacity of the piece is weakened, and a portion of the load supported by the cut member must be transferred properly to other joists. It is best to design and frame a project to accommodate mechanical

Fig. 6-58 *Board sheathing must have a separate moisture barrier.* (Forest Products Laboratory.)

Fig. 6-59 *Placement of cuts in floor joists. (Western Wood Products Association.)*

systems from the outset because notching and boring should be avoided whenever possible. However, unforeseen circumstances can arise during construction.

If it is necessary to cut into a framing member, the Figs. 6-59 through 6-61 provide a guide for doing so in the least destructive manner. These drawings comply with the requirements of the major model building codes:

- Uniform (UBC)
- Standard (SBC)
- National (BOLA)
- CABO

These apply to one- and two-family dwellings. Figure 6-59 shows the placement of cuts in floor joists. Notches and holes in 2 × 4 studs are shown in Figs. 6-60 and 6-61. Table 6-2 presents maximum sizes for cuts in floor joists.

The Western Wood Products Association provides information on the woods harvested in their area and

Fig. 6-60 *Notches in 2 × 4 studs.*

Fig. 6-61 *Bored holes in 2 × 4 studs. (Western Wood Products Association.)*

TABLE 6-2 *Maximum Sizes for Cuts in Floor Joists (Source: Western Wood Products Association.)*

Joist Size	Max. Hole	Max Notch Depth	Max. End Notch
2x4	none	none	none
2x6	1-1/2"	7/8"	1-3/8"
2x8	2-3/8"	1-1/4"	1-7/8"
2x10	3"	1-1/2"	2-3/8"
2x12	3-3/4"	1-7/8"	2-7/8"

used for studs and floor joists. Notching and boring holes in wood are a common practice for use as a means of installing plumbing, electrical wiring, security systems, and sound systems. However, there are some very good reasons for not indiscriminately drilling a hole where you want. There can be serious results when the lumber being used for its strength is weakened by the placement of a hole or notch.

For instance, when structural wood members are used vertically to carry loads in compression, the same engineering procedure is used for both studs and columns. However, differences between studs and columns are recognized in the model building codes for conventional light-frame residential construction.

The term *column* describes an individual major structural member subjected to axial compression loads, such as columns in timber-frame or post-and-beam structures. The term *stud* describes one of the members in a wall assembly or wall system carrying axial compression loads, such as 2 × 4 studs in a stud wall that includes sheathing or wall board. The difference between columns and studs can be further described in terms of the potential sequences of failure.

Columns function as individual major structural members; consequently, failure of a column is likely to result in partial collapse of a structure (or complete collapse in extreme cases owing to the domino effect). However, studs function as members in a system. Owing to the system effects (i.e., load sharing, partial composite action, redundancy, load distribution, etc.), studs are much less likely to fail and result in a total collapse than are columns.

Notching or boring into columns is not recommended and rarely acceptable; however, model codes established guidelines for allowable notching and boring into studs used in a stud-wall system. Figures 6-60 and 6-61 illustrate the maximum allowable notching and boring of 2 × 4 studs under all model codes except BOLA. BOLA allows a hole one-third the width of the stud in all cases. Bored holes should not be located in the same cross section of a stud as a cut or a notch.

It is important to recognize the point at which a notch becomes a rip, such as when floor joists at the entry of a home are ripped down to allow underlayment for a floor. Ripping wide-dimension lumber lowers the grade of the material and is unacceptable under all building codes.

When a sloped surface is necessary, a nonstructural member can be ripped to the desired slope and fastened to the structural member in a position above the top edge. Do not rip the structural member.

Modular Standards

Currently, a new modular system of construction is being used more often. The modular system uses stud and rafter systems 24 inches O.C. The wider spacing does not provide as much support for wall sheathing. However, the ability to support the roof is not reduced much. A building with studs 16 inches O.C. is very strong. Yet the difference in strength between 16- and 24-inch centers is very small. The advantage of using the 24-inch modular system is that it saves on costs. For a three-bedroom house, the cost of wall studs may be reduced about 25 percent. Moreover, the cost of labor is also reduced. This saves the time that would be used for cutting and nailing that many more studs.

The savings are even more in terms of money and wages paid. Fewer resources are used with almost the same results. In the modular system, several ways of reducing material use are employed. Floor joists are butted and not lapped. Building and roof size are exact multiples of 4 or 8 feet. Also, single top plates are used on partitions.

Energy

Energy is also a matter of concern. Insulated walls save energy used for heating and cooling. The amount of insulation helps to determine the efficiency. A 6-inch-thick wall can hold more insulation than a 4-inch wall. It will reduce the energy used by about 20 percent. However, it takes more material to build such a wall. Canadian building codes often specify 6-inch walls.

The double-wall system of frame buildings is a better insulator than a solid wall. Wood is good insulation compared with other materials. But the best insulation is a hollow space filled with material that does not conduct heat. A wall should have three layers. An outside weatherproof wall is needed. This layer stops rain and wind. A thick layer of insulation is next. This helps to keep heat energy from being conducted through the wall. The inside wall is the third layer. It helps to reduce air movement. It also aids in making a tight seal and holds the insulation in place.

More of the insulation effect also can be added by using a shiny surface (see Figs. 6-56 and 6-57). "Energy" sheathing adds insulation in two ways. It insulates in the same way as regular insulation, and it also covers the studs. In this way, the stud does not conduct heat directly through the wall (Fig. 6-62). The use of headers also affects wall construction. Headers reduce time and construction costs, but they are also difficult to insulate. Truss headers or a single header is better. This allows more insulation to be used.

The builder and the carpenter can alter a wall frame to save energy. Walls can be made thicker so that more insulation can be used. Thicker walls also let insulation cover studs in the same way as for sound insulation. "Energy" sheathing can be used to cover the studs. It adds insulation without requiring thick walls. Reflective foil also makes it more efficient (Fig. 6-63).

WITHOUT INSULATION   WITH ALUMINUM FOIL INSULATION   WITHOUT INSULATION   WITH ALUMINUM FOIL INSULATION

Fig. 6-62 *Reflected heat makes the house cooler in summer and warmer in winter.* (US Gypsum.)

(A)   (B)

Fig. 6-63 *A. Solid sheathing, stud, and wall are a solid path. This conducts energy loss straight through a wall. B. "Energy" sheathing forms a barrier. There is no solid path for energy.*

CHAPTER 6 STUDY QUESTIONS

1. Which is usually built first, the wall or the floor?
2. On what piece are studs spaced?
3. What size nail is used to join studs to plates?
4. What size nail is used to join headers to studs?
5. How are headers supported?
6. When are soleplates cut for interior doors?
7. When are headers made?
8. What is a curtain wall?
9. Why is corner bracing done?
10. How are corners braced?
11. How are corners plumbed?
12. Why are walls braced temporarily?
13. What is used for sheathing?
14. How can you tell when a wall is in place?
15. How can walls be changed to save energy?
16. What is a story pole?
17. What is needed for sound control in a house?
18. What kind of door improves sound control in a house?
19. What happens to a structural member when a hole is drilled in it?
20. What happens to dimensional lumber when it is ripped?

7
CHAPTER

Framing the Roof

FRAMING A ROOF IS DETERMINED BY THE CHOICE OF roof style. There are a number of types of rooflines. Some of these types of roofs include

- Gable
- Mansard
- Shed
- Hip
- Gambrel

Each style presents unique problems. Rafters and sheathing will be shaped according to the roofline desired. In this chapter you will learn how to

- Build roof frames
- Make roofs of different types
- Make openings for chimneys and soil pipes
- Design a particular type of roof rafter
- Mark off and cut a roof rafter
- Put roof trusses in place
- Lay sheathing onto a roof frame
- Build special combinations of framing for roofs
- Identify the type of plywood needed for a particular roof

ROOFS

A roof is needed to keep out the weather and to control the heat and cooling provided for human comfort inside a house. There are many types of roofs. Each serves a purpose. Each one is designed to keep the inside of the house warm in the winter and cool in the summer. The roof is designed to keep the house free of moisture, whether rain, snow, or fog.

The roof is made of rafters and usually supported by ceiling joists. When all the braces and forms are put together, they form a roof. In some cases, roofs have to be supported by more than the outside walls. This means that some of the partitions must be weight bearing. However, with truss roofs, it is possible to have a huge open room without supports for the roof.

Framing Lumber

The basic construction material in carpentry is lumber. There are many kinds of lumber that vary greatly in structural characteristics. Here we deal with the lumber common to construction carpentry, its application, the standard sizes in which it is available, and the methods of computing lumber quantities in terms of board feet.

Standard Sizes of Bulk Lumber

Lumber is usually sawed into standard lengths, widths, and thicknesses. This permits uniformity in planning structures and in ordering material. Table 7-1 lists the common widths and thicknesses of wood in rough and dressed dimensions in the United States. Standards have been established for dimensional differences between nominal size and the standard size (which is actually the reduced size when dressed). It is important that these dimensional differences be taken into consideration when planning a structure. A good example of the dimensional difference may be illustrated by the common 2- $\times$ 4-foot board (commonly known as a 2 $\times$ 4). As may be seen in the table, the familiar quoted size (2 $\times$ 4) refers to a rough or nominal dimension, but the actual standard size to which the lumber is dressed is 1½ inches $\times$ 3½ inches.

TABLE 7-1 *Your Guide to Lumber Sizes*

What You Order	What You Get Dry or Seasoned*	Green or Unseasoned†	What You Used to Get, Seasoned or Unseasoned
1 × 4	¾ × 3½	25/32 × 3 9/16	25/32 × 3⅝
1 × 6	¾ × 5½	25/32 × 5⅝	25/32 × 5½
1 × 8	¾ × 7¼	25/32 × ½	25/32 × 7½
1 × 10	¾ × 9¼	25/32 × 9½	25/32 × 9½
1 × 12	¾ × 11¼	25/32 × 11½	25/32 × 11½
2 × 4	1½ × 3½	1 9/16 × 3 9/16	1⅝ × 3⅝
2 × 6	1½ × 5½	1 9/16 × 5⅝	1⅝ × 5½
2 × 8	1½ × 7¼	1 9/16 × 7½	1⅝ × 7½
2 × 10	1½ × 9¼	1 9/16 × 9½	1⅝ × 9½
2 × 12	1½ × 11¼	1 9/16 × 11½	1⅝ × 11½
4 × 4	3½ × 3½	3 9/16 × 3 9/16	3⅝ × 3⅝
4 × 6	3½ × 5½	3 9/16 × 5⅝	3⅝ × 5½
4 × 8	3½ × 7¼	3 9/16 × 7½	3⅝ × 7½
4 × 10	3½ × 9¼	3 9/16 × 9½	3⅝ × 9½
4 × 12	3½ × 11¼	3 9/16 × 11½	3⅝ × 11½

*Up to 19 percent moisture content.
†Over 19 percent moisture content.

Grades of Lumber

Lumber as it comes from the sawmill is divided into three main classes:

- Yard lumber
- Structural material
- Factory or shop lumber

In keeping with the purpose of this book, only yard lumber will be considered. Yard lumber is manufactured and classified on a quality basis into sizes, shapes, and qualities required for ordinary construction and general building purposes. It is then further subdivided into classifications of select lumber and common lumber.

Select lumber Select lumber is of good appearance and finished or dressed. It is identified by the following grade names:

- *Grade A.* This is suitable for natural finishes of high quality and is *practically* clear.
- *Grade B.* This is suitable for natural finishes of high quality and is *generally* clear.
- *Grade C.* This is adapted to high-quality paint finish.
- *Grade D.* This suitable for paint finishes and is between the higher finishing grades and the common grades.

Common lumber Common lumber is suitable for general construction and utility purposes and is identified by the following grade names:

- *No. 1 common.* This is suitable for use without waste. It is sound and tight knotted and may be considered watertight material.
- *No. 2 common.* This is less restricted in quality than No. 1 but of the same general quality. It is used for framing, sheathing, and other structural forms where the stress or strain is not excessive.
- *No. 3 common.* This permits some waste, with prevailing grade characteristics larger than in No. 2. It is used for footings, guardrails, and rough subflooring.
- *No. 4 common.* This permits waste and is of low quality, admitting the coarsest features (such as decay and holes). It is used for sheathing, subfloors, and roof boards in the cheaper types of construction. The most important industrial outlet for this grade is for boxes and shipping crates.

Framing lumber The frame of a building consists of the wooden form constructed to support the finished members of the structure. It includes such items as posts, girders (beams), joists, subfloor, soleplate, studs, and rafters.

Softwoods are usually used for light wood framing and all other aspects of construction carpentry considered in this book. One of the classifications of softwood lumber cut to standard sizes is called *yard lumber*, and it is manufactured for general building purposes. It is cut into the standard sizes required for light framing, including 2×4, 2×6, 2×8, 2×10,

and 2×12, as well as all other sizes required for framework, with the exception of sizes classed as structural lumber.

Although No. 1 and No. 3 common woods sometimes are used for framing, No. 2 common is used most often and is, therefore, most often stocked and available at retail lumber yards in the common sizes for various framing members. However, the size of lumber required for any specific structure will vary with the design of the building (such as light- or heavy-frame construction) and the design of the particular members (such as beams or girders).

Exterior walls traditionally consisted of three layers—sheathing, building paper, and siding. Plywood sheathing has replaced board sheathing. Tyvek and other breathable air retarders have replaced building paper. Rigid nonstructural foam boards are also used as sheathing.

Roof Shapes

Figure 7-1 shows various roof shapes. The shape can affect the type of roofing material used. The various shapes call for some special details. Roofs have to withstand high winds as well as ice and snow. The weight of snow can cause a roof to collapse unless it is designed properly. It is necessary to make sure that the load on the roof can be supported. This calls for properly sized rafters and decking.

Shingles on a roof protect it from rain, wind, and ice. They have to withstand many years of exposure to all types of weather conditions.

The hip roof with its hip and hip-jack rafters is of particular concern. It is one of the most popular roof

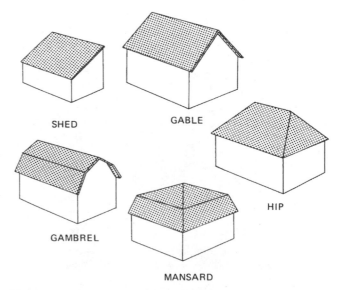

Fig. 7-1 *Various roof shapes or styles.*

types. The gable roof with its common rafter is the simplest. The mansard roof and the gambrel roof present some interesting problems with some very interesting solutions. All this will be explained in detail later in this chapter. Various valleys, ridge boards, and cripple rafters also will be described in detail here.

SEQUENCE

The following steps should be followed in building roof frames:

1. Check the plans to see what type of roof is desired.
2. Select the ceiling joist style and spacing.
3. Lay out the ceiling joists for openings.
4. Cut the ceiling joists to length and shape.
5. Lay out regular rafter spacing.
6. Lay out the rafters and cut to size.
7. Set the rafters in place.
8. Nail the rafters to the ridge board.
9. Nail the rafters to the wall plate.
10. Figure out the sheathing needed for the roof.
11. Apply the sheathing according to specifications.
12. Attach the soffit.
13. Put in braces or lookouts where needed (Fig. 7-2).
14. Cut special openings in the roof decking.

There may be a need for cutting dormer rafters after the rest of the roof is finished. Structural elements may be installed where needed. Truss roof rafters will need special attention as to spacing and nailing. Double-check to make sure that they fit the manufacturer's specifications.

Fig. 7-2 *Braces (lookouts) are put in where they are needed.* (Duo-Fast.)

TRUSS ROOFS
Truss Construction

Trussed rafters are used commonly in light-frame construction. They are used to support the roof and ceiling loads. Such trusses are designed, fabricated, and erected to meet the design and construction criteria of various building codes. They efficiently use the excellent structural properties of lumber. They are spaced either 16 inches on center (O.C.) or in some cases 24 inches O.C. (Fig. 7-3).

Fig. 7-3 *W-type truss roof and metal plates used to make the junction points secure.*

Truss Disadvantages

Keep in mind that the truss type of construction does have some disadvantages. The attic space is limited by the supports that make up the truss. Truss roofs may need special equipment to construct them. In some instances, it is necessary to use a crane to lift the trusses into position on the job site.

Roof framing   The roof frame is made up of rafters, ridge board, collar beams, and cripple studs (Fig. 7-4). In gable roofs, all rafters are precut to the same length and pattern. Figure 7-5 shows a gable roof. Each pair of rafters is fastened at the top to a ridge board. The ridge board is commonly a 2 × 8 for 2 × 6 rafters. This board provides a nailing area for rafter ends (see Fig. 7-4).

Getting started   Getting started with erection of the roof framing is the most complicated part of framing a house. Plan it carefully. Make sure that you have all

Fig. 7-4 *Gable-type rafter with cripple studs and ridge board. Note the notched top in the cripple stud.* (American Plywood Association.)

Fig. 7-5 *Two-story house with a gable roof. Note the bay windows with a hip roof.*

materials on hand. It is best to make a "dry run" at ground level. The erection procedure will be much easier if you have at least two helpers. A considerable amount of temporary bracing will be required if the job must be done with only one or two persons.

Steps in framing a roof   Take a look at Fig. 7-6. It shows two persons tipping up trusses. They are tipped up and nailed in place one at a time. Two people, one working on each side, will get the job done quickly. This is one of the advantages of trussed-roof construction. The trusses are made at a lumber yard or some other location. They are usually hauled to the construction site on a truck. Then they are lifted to the roof of the building with a crane or by hand. In some cases, the sheet-metal bracket shown in Fig. 7-7 holds the truss to the wall plate. Figure 7-8 shows how the metal bracket is used to fasten the truss in place. In Fig. 7-9, toenailing is used to fasten the truss to the wall plate.

Fig. 7-6 *Roof trusses are placed on the walls and then tipped up and nailed in place.*

Fig. 7-7 *Rafter truss attached to the wall plate with a sheet-metal bracket.*

Fig. 7-8 *Using a nailing bracket to attach a truss to the wall plate and nailing into the gusset to attach the rafter.*

Advantages of trusses   Manufacturers point out that the truss has many advantages. Figure 7-10 shows conventional framing. Note how bearing walls are required inside the house. With a truss roof, you can

UPPER CHORD

GUSSET

TOENAIL

Fig. 7-9 *Toenailing the truss rafter to the wall plate.*

place the partitions anywhere. They are not weight-bearing walls. With conventional framing, it is possible for ceilings to sag. This causes cracks. The truss has supports to prevent sagging ceilings (Fig. 7-11). Notice how the triangle shape is obvious in all the var-

ious parts of the truss. The triangle is a very strong structural form.

Details of trusses   There are a number of truss designs. A W-type is shown in Fig. 7-12. Note how the 2 × 4s are brought together and fastened by plywood that is both nailed and glued. In some cases, as shown in Fig. 7-3, steel brackets are used. Whenever plywood is used for the gussets, make sure that the glue fits the climate. In humid parts of the country, where the attic may be damp, the glue should be able to take the humidity. The manufacturer will inform you of the glue's use and how it will perform in humid climates.

The glue should not lose its strength when the weather turns humid. Most glue containers have this information on them. If not, check with the manufacturer before making trusses. In most instances, the manufacturer of trusses is very much aware of the glue requirements for a particular location.

Figure 7-13 shows how the three suggested designs are different. The truss in Fig. 7-13A is known as a W-type truss. Note the W in the middle of the truss. Figure 7-13B shows a simple king-post truss. It is used for a low-pitch roof. The scissors-type truss is shown

Sagging rafter—unsightly and possible source of leaks

Sagging ridge—creates outward thrust on walls as well as unsightliness

Ceiling joist sags—causes cracks

Bearing wall required—limits planning and room alignment. Heavy footing required

Outward thrust on walls created by sagging ridge—possible cause of sticking and ill-fitting windows and doors, source of cracks

Fig. 7-10 *Disadvantages of conventional roof framing.*

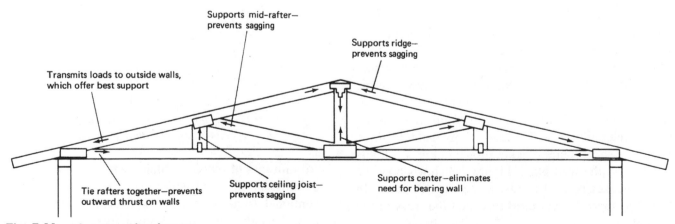

Supports mid-rafter—prevents sagging

Supports ridge—prevents sagging

Transmits loads to outside walls, which offer best support

Tie rafters together—prevents outward thrust on walls

Supports ceiling joist—prevents sagging

Supports center—eliminates need for bearing wall

Fig. 7-11 *Advantages of roof trusses.*

Fig. 7-12 *Construction of a W-type truss.*

in Fig. 7-13C. The W-type truss is the most popular. It can be used on long spans. It also can be made of low-grade lumber. The scissor-type truss is used in houses with sloping living-room ceilings. It also can be used for a cathedral ceiling. This truss is cheaper to make than the conventional type of construction for cathedral ceilings. King-post trusses are very simple. They are used for houses. This truss is limited to about a 26-foot span—that is, if 2 × 4s are used for the members of the truss. Compare this with a W-type truss, which could be used for a span of 32 feet. The king-post truss is economical for use in medium spans. It is also useful in short spans (Fig. 7-14).

The type of truss used depends on the wind and snow. The weight applied to a roof is an important factor. Make sure that the local codes allow for truss roofs. In some cases, the Federal Housing Authority (FHA) has to inspect them before insuring a mortgage on a house that has them.

Lumber to use in trusses   The lumber used in construction of trusses must be that which is described in Table 7-2. The moisture content should be from 7 to 16 percent. Glued surfaces have to be free of oil, dirt, and

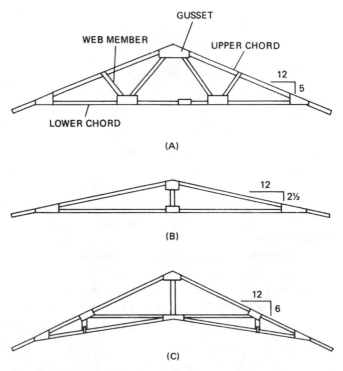

Fig. 7-13 *Wood trusses: A. W type; B. king-post type; C. scissors type.*

Fig. 7-14 *Construction of a king-post truss.*

any foreign matter. Each piece must be machine finished but not sanded.

Lumber with roughness in the gusset area cannot be used. Twisted, cupped, or warped lumber should not be used either. This is especially true if the twist or cup is in the area of the gusset. Keep the intersecting surfaces of the lumber within ⅟₃₂ inch.

Glue for trusses   For dry or indoor locations, use a casein-type glue. It should meet federal specification MMM-A-125, Type 11. For wet conditions, use a resorcinol-type glue. It should neet military specifica-

tion MIL-A-46051 for wet locations. If the glue joint is exposed to the weather or used at the soffit, use the resorcinol-type glue.

Load conditions for trusses   Table 7-3 shows the loading factors needed in designing trusses. Note the 30 and 40 pounds per square foot columns. Then look up the type of gusset—either beveled-heel or square-heel type. Standard sheathing or C-C Ext APA grade plywood is the type used here for the gussets and sheathing (Table 7-4). C-D plywood is often used for roof sheathing.

TABLE 7-2 *Chord Code Table for Roof Trusses*

Chord Code	Size	Grade and Species Meeting Stress Requirements	Grading Rules	f	t//	c//
1	2 × 4	Select structural light framing WCDF No. 1 dense kiln-dried Southern pine 1.8E	WCLIB SPIB WWPA	1950 2000 2100	1700 2000 1700	1400 1700 1700
2	2 × 4	1500f industrial light framing WCDF 1500f industrial light framing WCH No. 1 2-inch dimension Southern pine 1.4E	WCLIB WCLIB SPIB WWPA	1500 1450 1450 1500	1300 1250 1450 1200	1200 1100 1350 1200
3	2 × 4	1200f industrial light framing WCDF 1200f industrial light framing WCH No. 2 2-inch dimension Southern pine	WCLIB WCLIB SPIB	1200 1150 1200	1100 1000 1200	1000 900 900
4	2 × 6	Select structural J&P WCDF Select structural J&P Western larch No. 1 dense kiln-dried Southern pine 1.8E	WCLIB WWPA SPIB WWPA	1950 1900 2000 2100	1700 1600 2000 1700	1600 1500 1700 1700
5	2 × 6	Construction grade J&P WCDF Construction grade J&P WCH Structural J&P Western larch No. 1 2-inch dimension Southern pine 1.4E	WCLIB WCLIB WWPA SPIB WWPA	1450 1450 1450 1450 1500	1300 1250 1300 1450 1200	1200 1150 1200 1350 1200
6	2 × 6	Standard grade J&P WCDF Standard grade J&P WCH Standard structural Western larch No. 2 2-inch dimension Southern pine	WCLIB WCLIB WWPA SPIB	1200 1150 1200 1200	1100 1000 1100 1200	1050 950 1050 900
7	2 × 4	Select structural light framing WCDF Select structural light framing Western larch No. 1 dense kiln-dried Southern pine 1.8E	WCLIB WWPA SPIB WWPA	1900 1900 2050 2100	1900 1900 2050 1700	1400 1400 1750 1700
8	2 × 4	1500f industrial light framing WCDF Select structural light framing WCH Select structural light framing WH 1500f industrial light framing Western larch No. 1 2-inch dimension Southern pine 1.4E	WCLIB WCLIB WWPA WWPA SPIB WWPA	1500 1600 1600 1500 1500 1500	1500 1600 1600 1500 1500 1200	1200 1100 1100 1200 1350 1200
9	2 × 4	1200f industrial light framing WCDF 1200f industrial light framing Western larch 1500f industrial light framing WCH 1500f industrial light framing WH No. 2 2-inch dimension Southern pine	WCLIB WWPA WCLIB WWPA SPIB	1200 1200 1500 1500 1200	1200 1200 1500 1500 1200	1000 1000 1000 1000 900
10	2 × 6	Select structural J&P WCDF Select structural J&P Western larch No. 1 dense kiln-dried Southern pine 1.8E	WCLIB WWPA SPIB WWPA	1900 1900 2050 2100	1900 1900 2050 1700	1500 1500 1750 1700
11	2 × 6	Construction grade J&P WCDF Construction grade J&P Western larch Select structural J&P WCH Select structural J&P WH No. 1 2-dimension Southern pine 1.4E	WCLIB WWPA WCLIB WWPA SPIB WWPA	1500 1500 1600 1600 1500 1500	1500 1500 1600 1600 1500 1200	1200 1200 1200 1200 1350 1200
12	2 × 6	Standard grade J&P WCDF Standard grade J&P Western larch Standard grade J&P WCH Standard grade J&P WH No. 2 2-inch dimension dense Southern pine	WCLIB WWPA WCLIB WWPA SPIB	1200 1200 1200 1200 1400	1200 1200 1200 1200 1400	1000 1000 1000 1000 1050

Covering trusses Once the trusses are in place, cover them with sheathing or plywood. This will make a better structure once the sheathing is applied. The underlayment is applied, and the roofing is attached properly. The sheathing or plywood makes an excellent nail base for shingles. See Chapter 8 for applying shingles.

THE FRAMING SQUARE

In carpentry, it is necessary to be able to use a framing square. This device has a great deal of information stamped on its body and tongue. Figure 7-15 shows what a square can do in terms of a right angle. Figure 7-16 shows the right angle made by a framing square.

TABLE 7-3 *Designs When Using Standard Sheathing or C-C Ext Plywood*

Loading Condition, Total Roof Load, psf	Span	Beveled-Heel Gusset Dimensions, Inches					Chord code		Square-Heel Gusset Dimensions, Inches					Chord code	
		A	B	C	H	O	Upper	Lower	A	B	C	H	O	Upper	Lower
30 psf (20 psf live load, 10 psf dead load) on upper chord and 10 psf dead load on lower chord. Meets FHA requirements.	20'8"	32	48	12	45⅛	44	2	3	19	32	12	48¾	44	2	3
	22'8"	32	48	12	49⅛	48	1	2	19	32	12	52¾	48	1	3
	24'8"	48	60	16	53⅛	48	1	2	24	48	12	56¾	48	1	3
	26'8"	48	72	16	57⅛	48	1	2	32	60	16	60¾	48	1	2
40 psf (30 psf live load, 10 psf dead load) on upper chord and 10 psf dead load on lower chord.	20'8"	32	48	12	45⅛	43	8	9	19	32	12	48¾	48	7	9
	22'8"	32	60	16	49⅛	48	7`	8	19	48	12	52¾	48	7	9
	24'8"								32	60	16	56¾	48	7	9

TABLE 7-4 *Plywood Veneer Grades*

Grade	Description
N	Special order "natural finish" veneer. Select all heartwood or all sapwood. Free of open defects. Allows some repairs.
A	Smooth and paintable. Neatly made repairs permissible. Also used for natural finish in less demanding applications.
B	Solid surface veneer. Circular repair plugs and tight knots permitted.
C	Knotholes to 1″. Occasional knotholes ½″ larger permitted providing total width of all knots and knotholes within a specified section does not exceed certain limits. Limited splits permitted. Minimum veneer permitted in Exterior type plywood.
C Plugged	Improved C veneer with splits limited to ⅛″ in width and knotholes and borer holes limited to ¼″ by ½″.
D	Permits knots and knotholes to 2½″ in width, and ½″ larger under certain specified limits. Limited splits permitted.

The steel square or carpenter's square is made in the form of a right angle. That is, two arms (the body and the tongue of the square) make an angle of 90 degrees.

Note in Fig. 7-15 how a right triangle is made when points A, B, and C are connected. Figure 7-16 shows the right triangle. A right triangle has one angle that is 90 degrees.

Fig. 7-15 *The steel square has a 90-degree angle between the tongue and the body.* (Stanley Tools.)

Fig. 7-16 *The parts of a right triangle.* (Stanley Tools.)

Parts of the Square

The steel square consists of two parts—the *body* and the *tongue*. The body is sometimes called the *blade* (Fig. 7-17).

Body The body is the longer and wider part. The body of the Stanley standard steel square is 24 inches long and 2 inches wide.

Tongue The tongue is the shorter and narrower part and usually is 16 inches long and 1½ inches wide.

Heel The point at which the body and the tongue meet on the outside edge of the square is called the

Fig. 7-17 *Parts of the steel square—face side.* (Stanley Tools.)

heel. The intersection of the inner edges of the body and tongue is sometimes also called the *heel.*

Face The face of the square is the side on which the manufacturer's name, Stanley in this case, is stamped or the visible side when the body is held in the left hand and the tongue is held in the right hand (see Fig. 7-17).

Back The back is the side opposite the face (Fig. 7-18). The modern scale usually has two kinds of marking: scales and tables.

Fig. 7-18 *Backside of the steel square labeled.* (Stanley Tools.)

Scales The scales are the inch divisions found on the outer and inner edges of the square. The inch graduations are in fractions of an inch. The Stanley square has the following graduations and scales:

Face of body, outside edge	Inches and sixteenths
Face of body, inside edge	Inches and eighths
Face of tongue, outside edge	Inches and sixteenths
Back of tongue, inside edge	Inches and eighths
Back of body, outside edge	Inches and twelfths
Back of body, inside edge	Inches and sixteenths
Back of tongue, outside edge	Inches and twelfths
Back of tongue, inside edge	Inches and tenths

Hundredths scale This scale is located on the back of the tongue, in the corner of the square, near the brace measure. The hundredths scale is 1 inch divided into 100 parts. The longer lines indicate 25 hundredths, whereas the next shorter lines indicate 5 hundredths. With the aid of a pair of dividers, fractions of an inch can be obtained. Figure 7-19 shows the location of the hundredths scale.

Fig. 7-19 *Location of the hundredths scale on a square.* (Stanley Tools.)

One inch graduated in sixteenths is below the hundredths scale on current squares so that the conversion from hundredths to sixteenths can be made at a glance without the need to use a calculator. This comes in handy when you are determining rafter lengths using rafter tables, where hundredths are given.

Rafter scales These tables will be found on the face of the body and will help you to determine rapidly the lengths of rafters as well as their cuts. The rafter tables consist of six lines of figures. Their use is indicated on the left end of the body. The first line of figures gives the lengths of common rafters per foot of run. The second line gives the lengths of hip-and-valley rafters per foot of run. The third line gives the length of the first jack rafter and the differences in the lengths of the others centered at 16 inches. The fourth line gives the length of the first jack rafter and the differences in the lengths of the others centered at 24 inches. The fifth line gives the side cuts of jacks rafters. The sixth line gives the side cuts of hip-and-valley rafters.

Octagon scale The octagon, or "eight square," scale is found along the center of the face of the tongue. Using this scale, a square timber may be shaped into one having eight sides, or an octagon.

Brace scale This table is found along the center of the back of the tongue and gives the exact lengths of common braces.

Essex board measure This table is on the back of the body and gives the contents of any size lumber.

Steel Square Uses

The steel square has many applications. It can be used as a try square or to mark a 90-degree line along a piece of lumber (Fig. 7-20).

THE STEEL SQUARE
USED AS A TRY
SQUARE

MARKING WORK
AT 90°

Fig. 7-20 *Uses of the steel square.* (Stanley Tools.)

The steel square also can be used to mark 45-degree angles and 30- to 60-degree angles (Fig. 7-21). In some instances, you may want to use the square for stepping off the lengths of rafters and braces (Fig. 7-22). Another use for the square is in laying out of stair steps. Figure 7-23 shows how this is done.

However, one of the most important roles the square plays in carpentry is in the layout of roof framing. It is used to make sure that the rafters fit the proper angles. The rafter lengths and angles to be cut can be

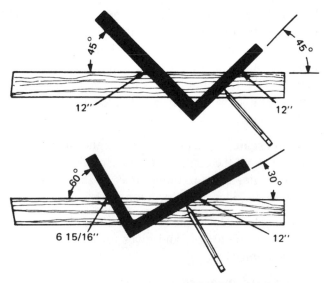

Fig. 7-21 *Using the steel square for marking angles.* (Stanley Tools.)

Fig. 7-22 *Using the steel square to step off the lengths of rafters and braces.* (Stanley Tools.)

RISE

FLOOR LINE

RUN

Fig. 7-23 *Using the square to lay out stairs.* (Stanley Tools.)

determined by using the framing square. Rafter cuts are shown in the following sections.

ROOF FRAMING

As we've seen, there are a number of types of roofs. A great many shapes are available, as you can see in any neighborhood. Some of the most common types will be identified and worked with here.

Shed roof   The shed roof is the most common type. Easy to make, it is also sometimes called the *lean-to roof*. It has only a single slope (Fig. 7-24).

Fig. 7 24   *Lean-to or shed roof.* (Stanley Tools.)

Gable or pitch roof   This is another type of roof that is used commonly. It has two slopes meeting at the center or ridge, forming a gable. It is a very simple roof and perhaps the easiest to construct (Fig. 7-25).

Fig. 7 25   *Gable roof.* (Stanley Tools.)

Gable-and-valley or hip-and-valley roof This is a combination of two intersecting gable or hip roofs. The valley is the place where the two slopes of the roof meet. The slopes run in different directions. There are many modifications of this roof, and the intersections usually are at right angles (Figs. 7-26 and 7-27).

Fig. 7-26 *Gable-and-valley roof. (Stanley Tools.)*

Fig. 7-27 *Hip-and-valley roof. (Stanley Tools.)*

Hip roof This roof has four sides, all sloping toward the center of the building. The rafters run up diagonally to meet the ridge, into which the other rafters are framed (Fig. 7-28).

Fig. 7-28 *Hip roof. (Stanley Tools.)*

Roof Terms

There are a number of terms with which you should be familiar to work with roof framing. Each type of roof has its own particular terms, but some of the common terms include

Span The span of a roof is the distance over the wall plates (Fig. 7-29).

Run The run of a roof is the shortest horizontal distance measured from a plumb line through the center of the ridge to the outer edge of the plate (see Fig. 7-29). In equally pitched roofs, the run is always equal to half the span, or generally half the width of the building.

Rise The rise of a roof is the distance from the top of the ridge and of the rafter to the level of the foot. In figuring rafters, the rise is considered as the vertical distance from the top of the wall plate to the upper end of the measuring line (see Fig. 7-29).

Deck roof When rafters rise to a deck instead of a ridge, the width of the deck should be subtracted from the span. The remainder divided by 2 will equal the run. Thus, in Fig. 7-30, the span is 32 feet, and the deck is 12 feet wide. The difference between 32 and 12 is 20 feet, which divided by 2 equals 10 feet. This is the run of the common rafters. Since the rise equals 10 feet, this is a half-pitch roof.

Pitch The pitch of a roof is the slant or slope from the ridge to the plate. It may be expressed in several ways: (1) The pitch may be described in terms of the ratio of the total width of the building to the total rise of the roof. Thus the pitch of a roof having a 24-foot span with an 8-foot rise will be 8 divided by 24, which equals one-third pitch (Fig. 7-31). (2) The pitch of a roof also may be expressed as so many inches of vertical rise to each foot of horizontal run. A roof with a 24-foot span and rising 8 inches to each foot of run will have a total rise of $8 \times 12 = 96$ inches, or 8 feet. Eight divided by 24 equals ⅓. Therefore, the roof is one-third pitch (see Fig. 7-31). Note that in Fig. 7-32, the building is 24 feet wide. It has a roof with a 6-foot rise. What is the pitch of the roof? The pitch is 6 divided by 24, or ¼.

Fig. 7-29 *Span, run, rise, and pitch. (Stanley Tools.)*

Fig. 7-30 *Location of the deck roof.* (Stanley Tools.)

Fig. 7-31 *One-third pitch.* (Stanley Tools.)

Fig. 7-32 *One-fourth pitch.* (Stanley Tools.)

Principal roof pitches Figure 7-33 shows the principal roof pitches. They are called *one-half pitch*, *one-third pitch*, or *one-fourth* pitch, as the case may be, because the height from the level of the wall plate to the ridge of the roof is one-half, one-third, or one-quarter of the total width of the building.

Keep in mind that roofs of the same width may have different pitches depending on the height of

Fig. 7-34 *Different types of rafters used in a roof frame.*

the roof. Take a look at Fig. 7-34. This will help you interpret the various terms commonly used in roof construction.

Principal Roof Frame Members

The principal members of the roof frame are the plates at the bottom and the ridge board at the top. Various rafters are fastened to them (see Fig. 7-34).

Plate The plate is the roof member to which rafters are framed at their lower ends. The top, *A*, and the outside edge of the plate, *B*, are the important surfaces from which rafters are measured in Fig. 7-29.

Ridge board The ridge board is the horizontal member that connects the upper ends of the rafters on one side to the rafters on the opposite side. In cheap construction, the ridge board is usually omitted. The upper ends of the rafters are spiked together.

Common rafter A common rafter is a member that extends diagonally from the plate to the ridge.

Hip rafter A hip rafter is a member that extends diagonally from the corner of the plate to the ridge.

Fig. 7-33 *Principal roof pitches.* (Stanley Tools.)

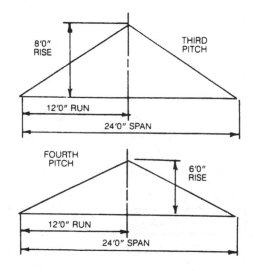

Valley rafter   A valley rafter is one that extends diagonally from the plate to the ridge at the line of intersection of two roof surfaces.

Jack rafter   Any rafter that does not extend from the plate to the ridge is a jack rafter. There are different kinds of jacks. According to the position they occupy, they can be classified as hip jacks, valley jacks, or cripple jacks.

Hip jack   A hip jack is a jack rafter with the upper end resting against a hip and the lower end against the plate.

Valley jack   A valley jack is a jack rafter with the upper end resting against the ridge board and the lower end against the valley.

Cripple jack   A cripple jack is a jack rafter with a cut that fits in between a hip-and-valley rafter. It touches neither the plate nor the ridge.

All rafters must be cut to proper angles so that they will fit at the points where they are framed. The different types of cuts are described below.

Top or plumb cut   The cut of the rafter end that rests against the ridge board or against the opposite rafter.

Bottom or heel cut   The cut of the rafter end that rests against the plate. The bottom cut is also called the *foot* or *seat cut*.

Side cut   Hip-and-valley rafters and all jacks, besides having top and bottom cuts, also must have their sides at the end cut to a proper bevel so that they will fit into the other members to which they are to be framed. These are called *side cuts* or *cheek cuts*. All rafters and their cuts are shown in Fig. 7-35.

Fig. 7-35   *Rafter cuts. (Stanley Tools.)*

RAFTERS
Layout of a Rafter

The measuring line of a rafter is a temporary line on which the length of the rafter is measured. This line runs parallel to the edge of the rafter and passes through the point *P* on the outer edge of the plate. Point *P* is where cut lines *A* and *B* converge. This is the point from which all dimensions are determined (see Fig. 7-29).

Length   The length of a rafter is the shortest distance between the outer edge of the plate and the center of the ridge line.

Tail   The tail is that portion of the rafter extending beyond the outside edge of the plate. In some cases it is referred to as the cave. The tail is figured separately and is not included in the length of the rafter, as mentioned earlier (see Fig. 7-29).

Figure 7-36 shows the three variations of the rafter tail. Figure 7-36A shows the flush tail or no tail. The rafter butts against the wall plate with no overhang. Figure 7-36B shows the full tail. Note the overhang. Figure 7-36C shows a separate curved tail. It is nailed onto the no-tail or flush rafter. All the cuts for the various types of common rafters are made at right angles to the sides of the rafter.

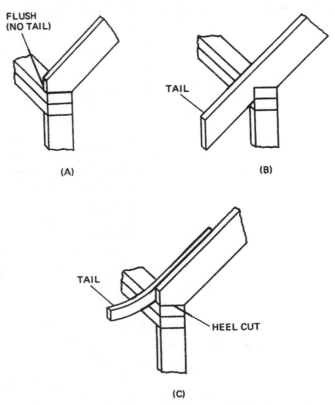

Fig. 7-36   *Tails or overhangs of rafters: A. flush tail (no tail); B. full tail; C. separate curved tail.*

Fig. 7-37 *Using the steel square to mark off the top or plumb cut.*

Figure 7-37 shows how the framing square is used to lay out the angles. Find the 12-inch mark on the square. Note how the square is set at 12 for laying out

the cut. The distance 12 is the same as 1 foot of run. The other side of the square is set with the edge of the stock to the rise in inches per foot of run. In some cases, the tail is not cut until after the rafter is in place. Then it is cut to match the others and aligns better with the fascia board.

Figure 7-38 shows a method of using the square to lay out the bottom and lookout cuts. If there is a ridge board, you have to deduct one-half the thickness of the ridge from the rafter length. The figure shows how to place the square for marking the rafter lookout. Scribe the cut line as shown in Fig. 7-38A. The rise is 4, and the run is 12. Then move the square to the next position and mark from *C* to *E*. The distance from *B* to *E* is equal to the length of the lookout. Move the square up to *E* (with the same setting). Scribe line *CE*. On this line, lay off *CD*. This is the length of the vertical side of the bottom cut. Now apply the same setting to the bottom edge of the rafter. This is done so that the edge of the square cuts *D*. Scribe *DF* (see Fig. 7-38B). This is the horizontal line of the bottom cut. In making this cut, the wood is

Fig. 7-38 *Using the steel square to lay out the rafter lookout and bird's mouth.*

(A)

(B)

Fig. 7-39 *A. Notice the saw cuts past the bird's mouth. B. Rafters in place. They are nailed to the ridge board, the wall plate, and the ceiling joists. The other side of the roof is already covered with plywood sheathing.*

Fig. 7-40 *Common rafters in place.*

Fig. 7-41 *Length per foot of run.* (Stanley Tools.)

cut out to the lines *CD* and *DE*. Figure 7-39A provides an example of rafters cut this way. Note how the portable handsaw makes cuts beyond the marks.

In Figs. 7-39B and 7-40 you can see the rafters in place. Note the 90-degree angle. The rafter and the ridge should meet at 90 degrees. The rafter in Fig. 7-40 has not had the lookout cut. The overhang is slightly different from the type just shown. Can you find the difference?

Length per foot of run The rafter tables on Stanley squares are based on the rise per foot of run, which means that the figures in the tables indicate the length of rafters per 1-foot run of common rafters for any rise of roof. This is shown in Fig. 7-41.

The roof has a 6-foot span and a certain rise per foot. The figure may be regarded as a right triangle *ABC* having for its sides the run, the rise, and the rafter. The run of the rafter has been divided into three equal parts, each representing a 1-foot run.

Note that by drawing vertical lines through each division point of the run, the rafter also will be divided into three equal parts *D*. Since each part *D* represents the length of rafter per 1-foot run and the total run of the rafter equals 3 feet, it is evident that the total length of the rafter will equal length *D* multiplied by 3.

The reason for using this per-foot-run method is that the length of any rafter may be easily determined for any width of building The length per foot of run will be different for different pitches. Therefore, before you can find the length of a rafter, you must know the rise of the roof in inches or the rise per foot of run.

Rule: To find the rise per foot of run, multiply the rise by 12, and divide by the length of the run.

The factor 12 is used to obtain a value in inches. The rise and run are expressed in feet (Figs. 7-42 and 7-43).

The rise per foot run is always the same for a given pitch and can be easily remembered for all ordinary pitches:

RISE = 10 FT.
RUN = 15 FT.
PITCH = 10 ÷ 30 = 1/3
RISE PER FOOT RUN = $\frac{10 \times 12}{15}$ = 8″

Fig. 7-42 *Finding the rise per foot of run.* (Stanley Tools.)

RISE = 8 FT
RUN = 16 FT
PITCH = 8 ÷ 32 = 1/4
RISE PER FOOT RUN = $\frac{8 \times 12}{16}$ = 6″

Fig. 7-43 *Finding the rise per foot of run.* (Stanley Tools.)

Pitch	$\frac{1}{2}$	$\frac{1}{3}$	$\frac{1}{4}$	$\frac{1}{6}$
Rise per foot of run, inches	12	8	6	4

The members of a firmly constructed roof should fit snugly against each other. Rafters that are not cut properly make the roof shaky and the structure less stable. Therefore, it is very important that all rafters are the right length and that their ends are cut properly. This will provide a full bearing against the members to which they are connected. Correct length, proper top and bottom cuts, and the right-side or cheek cuts are very important features to be observed when framing a roof.

Lengths of Rafters

The lengths of rafters also may be obtained in other ways. There are three in particular that can be used:

1. Mathematical calculations
2. Measuring across the square
3. Stepping off with the square

The first method, while absolutely correct, is very impractical on the job. The other two are rather unreliable and quite frequently result in costly mistakes.

The tables on the square have eliminated the need for using the three methods just mentioned. These tables let the carpenter find the exact length and cuts for any rafter quickly and thus save time and avoid the possibility of errors.

Common Rafters

The common rafter extends from the plate to the ridge. Therefore, it is evident that the rise, the run, and the rafter itself form a right triangle. The length of a common rafter is the shortest distance between the outer edge of the plate and a point on the centerline of the ridge. This length is taken along the measuring line. The measuring line runs parallel to the edge of the rafter and is the hypotenuse, or longest side, of a right triangle. The other sides of the triangle are the run and the rise (Fig. 7-44). The rafter tables on the face of the body of the square include the outside-edge graduations on both the body and the tongue, which are in inches and sixteenths of an inch.

Length of common rafters The lengths of common rafters are found on the first line, indicated as "Length of main rafters per foot run." There are 17 of these ta-

Fig. 7-44 *Measuring the line for a common rafter.* (Stanley Tools.)

Fig. 7-45 *Note the labels for the tables on the steel square.*
(Stanley Tools.)

Fig. 7-47 *Finding the length of the rafter.* (Stanley Tools.)

bles, beginning at 2 inches and continuing to 18 inches. Figure 7-45 shows the square being used.

> **Rule:** To find the length of a common rafter, multiply the length given in the table by the number of feet of run.

For example, if you want to find the length of a common rafter where the rise of the roof is 8 inches per foot of run, or one-third pitch, and the building is 20 feet wide, you first find where the table is. Then, on the inch line on the top edge of the body, find the figure that is equal to the rise of the roof, which in this case is 8. On the first line under the figure 8 will be found 14.42. This is the length of the rafter in inches per foot run for this particular pitch (Fig. 7-46). The building is 20 feet wide. Therefore, the run of the rafter will be 20 divided by 2, which equals 10 feet.

Fig. 7-46 *Finding the proper number on the square.* (Stanley Tools.)

Since the length of the rafter per 1-foot run equals 14.42 inches, the total length of the rafter will be 14.42 multiplied by 10, which equals 144.20 inches, or 144.20 divided by 12, which equals 12.01 feet or, for all practical purposes, 12 feet. Check Fig. 7-46.

Top and bottom cuts of the common rafter The top, or plumb, cut is the cut at the upper end of the rafter, where it rests against the opposite rafter or against the ridge board. The bottom, or heel, cut is the cut at the lower end, which rests on the plate (Fig. 7-47).

The top cut is parallel to the centerline of the roof, and the bottom cut is parallel to the horizontal plane of the plates. Therefore, the top and bottom cuts are at right angles to each other.

> **Rule:** To obtain the top and bottom cuts of a common rafter, use 12 inches on the body and the rise per foot run on the tongue. Twelve inches on the body will give the horizontal cut, and the figure on the tongue will give the vertical cut.

To illustrate this rule, we will examine a large square placed alongside the rafter, as shown in Fig. 7-48. Note

Fig. 7-48 *Using the steel square to check plumb and heel cuts.* (Stanley Tools.)

that the edge of the tongue coincides with the top cut of the rafter. The edge of the blade coincides with the heel cut. If this square were marked in feet, it would show the run of the rafter on the body and the total rise on the tongue. Line *AB* would give the bottom cut, and line *AC* would give the top cut.

However, the regular square is marked in inches. Since the relation of the rise to 1 foot of run is the same as that the total rise bears to the total run, we use 12 inches on the blade and the rise per foot on the tongue to obtain the respective cuts. The distance 12 is used as a unit and is the 1-foot run, whereas the figure on the other arm of the square represents the rise per foot of run (Figs. 7-49 and 7-50).

Fig. 7-49 *Using the steel square to lay out the heel cut. (Stanley Tools.)*

Fig. 7-50 *Using the steel square to lay out the plumb cut. (Stanley Tools.)*

Actual length of the common rafter The rafter lengths obtained from the tables on the square are to the centerline of the ridge. Therefore, the thickness of half the ridge board always should be deducted from the total length obtained before the top cut is made (Fig. 7-51). This deduction of half the thickness of the ridge is measured at right angles to the plumb line and is marked parallel to this line.

Figure 7-52 shows the wrong and right ways of measuring the lengths of rafters. Diagram *D* shows the measuring line at the edge of the rafter, which is the case when there is no tail or eave.

Fig. 7-51 *How to find the difference between the actual length and the length obtained. (Stanley Tools.)*

Cutting the rafter After the total length of the rafter has been established, both ends should be marked and allowance made for a tail or eave. Don't forget to allow for half the thickness of the ridge. Both cuts are obtained by applying the square so that the 12-inch mark on the body and the mark on the tongue that represents the rise are at the edge of the stock. All cuts for common rafters are made at right angles to the side of the rafter.

For example, a common rafter is 12 feet, 6 inches long, and the rise per foot of run is 9 inches. Obtain the top and bottom cuts (Fig. 7-53).

Points *A* and *B* are the ends of the rafter. To obtain the bottom, or seat, cut, take 12 inches on the body of the square and 9 inches on the tongue. Lay the square on the rafter so that the body will coincide with point *A*, or the lower end of the rafter. Mark along the body of the square and cut. To obtain the top cut, move the square so that the tongue coincides with point *B*. This is the upper end of the rafter. Mark along the tongue of the square.

Deduction for the ridge The deduction for half the thickness of the ridge should be measured now. Half

Fig. 7-52 *Right and wrong ways of measuring rafters. (Stanley Tools.)*

Fig. 7-53 *Applying the square to the layout cuts.* (Stanley Tools.)

the thickness of the ridge is 1 inch. One inch is deducted at right angles to the top cut mark or plumb line, which is point *C*. A line then is drawn parallel to the top-cut mark, and the cut is made. You will notice that the allowance for half the ridge measured along the measuring line is 1¼ inches. This will vary according to the rise per foot of run. It is therefore important to measure for this deduction at right angles to the top-cut mark or plumb line.

Measuring rafters The lengths of rafters having a tail or eave also can be measured along the back or top edge instead of along the measuring line, as shown in Fig. 7-54. To do this, it is necessary to carry a plumb line to the top edge from *P*, and the measurement is started from this point.

Occasionally, in framing a roof, the run may have an odd number of inches; for example, a building might have a span of 24 feet, 10 inches. This would mean a run of 12 feet, 5 inches. The additional 5 inches can be added easily without mathematical division after the length obtained from the square for 12 feet of

Fig. 7-55 *Adding extra inches to the length of a rafter.* (Stanley Tools.)

run is measured. The additional 5 inches is measured at right angles to the last plumb line. Figure 7-55 illustrates the procedure.

Hip-and-Valley Rafters

The hip rafter is a roof member that forms a hip in the roof. This usually extends from the corner of the building diagonally to the ridge. The valley rafter is similar to the hip, but it forms a valley or depression in the roof instead of a hip. It also extends diagonally from the plate to the ridge. Therefore, the total rise of hip-and-valley rafters is the same as that of common rafters (Fig. 7-56).

The relation of hip-and-valley rafters to common rafters is the same as the relation of the sides of a right triangle. Therefore, it will be well to explain here one of the main features of right triangles. In a right triangle, if the sides forming the right angle are 12 inches each, the hypotenuse, or the side opposite the right angle, is equal to 16.97 inches. This is usually taken as 17 inches (Fig. 7-57).

The position of the hip rafter and its relation to the common rafter is shown in Fig. 7-58, where the hip

A—RAFTER MEASURED ALONG
  MEASURING LINE
B—RAFTER MEASURED ALONG
  TOP EDGE

Fig. 7-54 *Two places to measure a rafter.* (Stanley Tools.)

Fig. 7-56 *Hip-and-valley rafters. (Stanley Tools.)*

Fig. 7-57 *In a right triangle, a 12-inch base and 12 inches of altitude produce an isosceles triangle. This means that the hypotenuse is 16.97 inches. (Stanley Tools.)*

rafter is compared with the diagonal of a square prism. The prism (as shown in the figure) has a base of 5 square feet, and its height is 3 feet, 4 inches.

D is the corner of the building.

BC is the total rise of the roof.

AC is the common rafter.

DB is the run of the hip rafter.

DC is the hip rafter.

Note that the figure *DAB* is a right triangle whose sides are the portion of the plate *DA*, the run of the common rafter *AB*, and the run of the hip rafter *DB*. The run of the hip rafter is opposite right angle *A*. The hypotenuse is the longest side of the right triangle.

If we should take only 1 foot of run of the common rafter and a 1-foot length of the plate, we would have right triangle *H*. The triangle's sides are each 12 inches long. The hypotenuse is 17 inches, or more accurately 16.97 inches (see Figs. 7-57 and 7-59).

The hypotenuse of the small triangle *H* in Fig. 7-59 is a portion of the run of the hip rafter *DB*. It corresponds to a 1-foot run of the common rafter. Therefore, the run of the hip rafter is always 16.97 inches for every 12 inches of run of the common rafter. The total run of the hip rafter will be 16.97 inches mul-

tiplied by the number of feet of run of the common rafter.

Lengths of hip-and-valley rafters Lengths of the hip-and-valley rafters are found on the second line of the rafter table. It is entitled "Length of hip or valley rafters per foot run." This means that the figures in the table indicate the lengths of hip-and-valley rafters per foot run of the common rafters (see Fig. 7-45).

> **Rule:** To find the length of a hip or valley rafter, multiply the length given in the table by the number of feet of the run of the common rafter.

For example, find the length of a hip rafter where the rise of the roof is 8 inches per foot of run, or one-third pitch. The building is 10 feet wide (see Fig. 7-58). Proceed as in the case of common rafters. That is, find on the inch line of the body of the square the figure corresponding to the rise of roof, which is 8. On the second line under this figure is found 18.76. This is the length of the hip rafter in inches for each foot of run of the common rafter for one-third pitch (see Fig. 7-59).

The common rafter has a 5-foot run. Therefore, there are also five equal lengths for the hip rafter, as

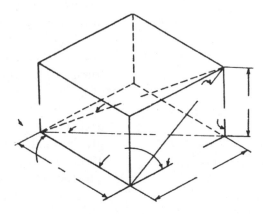

Fig. 7-58 *Relative position of the hip rafter and common rafter. (Stanley Tools.)*

Fig. 7-59 Finding the length of the hip rafter. Note the location of the 18.76 on the square. (Stanley Tools.)

may be seen in Fig. 7-59. We have found the length of the hip rafter to be 18.76 inches per 1 foot of run. Thus the total length of the hip rafter will be 18.76 × 5 = 93.80 inches. This is 7.81 feet or, for practical purposes, 7 feet, 9¹³⁄₁₆ inches, or 7 feet, 9⅝ inches.

Top and bottom cuts   The following rule should be followed for top and bottom cuts:

> **Rule:** To obtain the top and bottom cuts for hip or valley rafters, use 17 inches on the body and the rise per foot run on the tongue. Seventeen on the body will give

the seat cut, and the figure on the tongue will give the vertical, or top, cut (Fig. 7-60).

Measuring hip-and-valley rafters   The lengths of all hip-and-valley rafters always must be measured along the center of the top edge or back. Rafters with a tail or eave are treated like common rafters, except that the measurement or measuring line is at the center of the top edge.

Deduction from hip or valley rafter for ridge   The deduction for the ridge is measured in the same way as for the common rafter (see Fig. 7-53), except that half

Fig. 7-60 Using the square to lay out top and seat cuts on a hip rafter. (Stanley Tools.)

the diagonal thickness (45 degrees) of the ridge must be used.

Side cuts In addition to the top and bottom cuts, hip-and-valley rafters also must have side, or check, cuts at the point where they meet the ridge. These side cuts are found on the sixth or bottom line of the rafter tables. It is marked "Side cut hip or valley—use." The figures given in this line refer to the graduation marks on the outside edge of the body (see Fig. 7-45).

The figures on the square have been derived by determining the figure to be used with 12 on the tongue for the side cuts of the various pitches by the following method: From a plumb line, the thickness of the rafter is measured and marked, as at *A* in Fig. 7-61. A line then is squared across the top of the rafter, and the diagonal points are connected as at *B*. The line *B*, or side, cut is obtained by marking along the tongue of the square.

STEP 1. Shorten for ridge thickness.

STEP 2. Mark a line parallel to the plumb cut, equal in distance to the thickness of the rafter.

STEP 3. Square a line across top of hip rafter. The diagonal is the side cut.

Fig. 7-61 *Making side cuts so that the hip will fit into the intersection of rafters.* (Stanley Tools.)

Rule: To obtain the side cut for hip or valley rafters, take the figure given in the table on the body of the square and 12 inches on the tongue. Mark the side cut along the tongue where the tongue coincides with the point on the measuring line.

For example, find the side cut for a hip rafter where the roof has 8 inches rise per foot of run or pitch. Figure 7-62 shows the position of the hip rafter on the roof. The rise of the roof is 8 inches to the foot. First, locate the number 8 on the outside edge of the body. Under this number in the bottom line you will

Fig. 7-62 *Hip-rafter cuts.* (Stanley Tools.)

find 10⅞. This figure is taken on the body, and 12 inches is taken on the tongue. The square is applied to the edge of the back of the hip rafter. The side-cut *CD* comes along the tongue.

In making the seat cut for the hip rafter, an allowance must be made for the top edges of the rafter. They would project above the line of the common and jack rafters if the corners of the hip rafter were not removed or backed. The hip rafter must be slightly lowered. Do this by cutting parallel to the seat cut. The distance varies with the thickness and pitch of the roof.

Note that on the Stanley square the 12-inch mark on the tongue is always used in all angle cuts—top, bottom, and side. This leaves the worker with only one number to remember when laying out side or angle cuts. This is the figure taken from the fifth or sixth line in the table.

The side cuts always come on the right-hand or tongue side of the rafters. When you are marking boards, these can be reversed for convenience at any time by taking the 12-inch mark on the body and using the body references on the tongue.

Obtain additional inches in run of hip or valley rafters by using the explanation given earlier for common rafters. However, use the diagonal (45 degrees) of the additional inches. This is approximately 7¹⁄₁₆ inches for 5 inches of run. This distance should be measured in a similar manner.

Jack Rafters

Jack rafters are discontinued common rafters. They are common rafters cut off by the intersection of a hip or valley before reaching the full length from plate to ridge. Jack rafters lie in the same plane as common rafters. They usually are spaced the same and have the same pitch. Therefore, they also have the same length per foot of run as common rafters.

Jack rafters are usually spaced 16 or 24 inches apart. Because they rest against the hip or valley equally spaced, the second jack must be twice as long

Fig. 7-63 *Location of jack rafters. (Stanley Tools.)*

as the first one, the third must be three times as long as the first, and so on (Fig. 7-63).

Lengths of jack rafters The lengths of jacks are given in the third and fourth lines of the rafter tables on the square. They are indicated as follows:

Third line: "Difference in length of jacks—16-inch centers"

Fourth line: "Difference in length of jacks—2-foot centers"

The figures in the table indicate the length of the first or shortest jack, which is also the difference in length between the first and second jacks, between the second and third jacks, and so on.

> **Rule:** To find the length of a jack rafter, multiply the value given in the tables by the number indicating the position of the jack. From the length obtained, subtract half the diagonal (45 degree) thickness of the hip or valley rafter.

For example, find the length of the second jack rafter. The roof has a rise of 8 inches to 1 foot of run of common rafter. The spacing of the jacks is 16 inches. On the outer edge of the body, find the number 8, which corresponds to the rise of the roof. On the third line under this figure, find 19¼. This means that the first jack rafter will be 19¼ inches long. Since the length of the second jack is required, multiply 19¼ by 2, which equals 38½ inches. From this length, half the diagonal (45 degrees) thickness of the hip or valley rafter should be deducted. This is done in the same manner as the deduction for the ridge made on the hip rafter.

Proceed in the same manner when the lengths of jacks spaced on 24-inch centers are required. It should be borne in mind that the second jack is twice as long as the first one. The third jack is three times as long as the first jack, and so on.

Top and bottom cuts for jacks Since jack rafters have the same rise per foot of run as common rafters,

the method of obtaining the top and bottom cuts is the same as for common rafters. That is, take 12 inches on the body and the rise per foot run on the tongue. Twelve inches will give the seat cut. The figure on the tongue will give the plumb cut.

Side cut for jacks At the end where the jack rafter frames to the hip or valley rafter, a side cut is required. The side cuts for jacks are found on the fifth line of the rafter tables on the square. It is marked "Side cut of jacks—use" (see Fig. 7-45).

> **Rule:** To obtain the side cut for a jack rafter, take the figure shown in the table on the body of the square and 12 inches on the tongue. Mark along the tongue for the side cut.

For example, find the side cut for jack rafters of a roof having 8 inches of rise per foot of run, or one-third pitch (Figs. 7-64 and 7-65). Under the number 8 in the fifth line of the table, find 10. This number taken on the outside edge of the body and 12 inches taken on the tongue will give the required side cut.

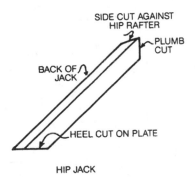

Fig. 7-64 *Hip jack-rafter cuts. (Stanley Tools.)*

Fig. 7-65 *Valley jack-rafter cuts. (Stanley Tools.)*

BRACE MEASURING

In all construction there is the need for some braces to make sure that certain elements are held securely (see Fig. 7-120A). The brace measure table is found along the center of the back of the tongue of the carpenter's square. It gives the lengths of common braces (Fig. 7-66).

For example, find the length of a brace whose run on post and beam equals 39 inches (Fig. 7-67). In the brace table, find the following expression:

$$\frac{39}{39} \quad 55.15$$

This means that with a 39-inch run on the beam and a 39-inch run on the post, the length of the brace will be 55.15 inches. For practical purposes, use 55⅛ inches.

Fig. 7-67 *Cutting a brace using a square table.* (Stanley Tools.)

Braces may be regarded as common rafters. Therefore, when the brace run on the post differs from the run on the beam, their lengths as well as top and bottom cuts may be determined from the figures given in the tables of the common rafters.

ERECTING THE ROOF WITH RAFTERS

Rafters are cut to fit the shape of the roof. Roofs are chosen by the builder or planner. The design of the rafter is determined by the pitch, span, and rise of the type of roof chosen. The gable roof is simple. It can be made easily with a minimum of difficult cuts. In this example, we start with the gable type and then look at other variations of rooflines. Figure 7-68 provides an example of a gable roof.

Fig. 7-68 *Gable roof.*

Rafter Layout

One of the most important tools used to lay out rafters is the carpenter's square. All rafters must be cut to the proper angle or bevel. They fasten to the wall plate or to the ridge board. In some cases, there is an overhang. This overhang must be taken into consideration when the rafter is cut. Gable siding, soffits, and overhangs are built together (Fig. 7-69).

Fig. 7-69 *Gable siding, soffits, and overhangs can be built together.*

Raising Rafters

Mark rafter locations on the top plate of the sidewalls. The first rafter pair will be flush with the outside edge of the end wall (Fig. 7-70). Note the placement of the gable end studs. The notch in the gable end stud is made to fit the 2 × 4 or whatever thickness of rafter you are using. Space the first interior rafter 24 inches, measured from the end of the building to the center of the rafter. In some cases, 16 inches O.C. is used for spacing. All succeeding rafter locations are measured 24 inches center to center. They will be at the sides of ceiling joist ends (Fig. 7-71).

Next, mark rafter locations on the ridge board. Allow for the specified gable overhang. To achieve the required total length of ridge board, you may have to splice it (Fig. 7-72). Do not splice at this time. It is eas-

Fig. 7-70 *Placement of the end rafter.* (American Plywood Association.)

Fig. 7-71 *Spacing of the first interior rafter.* (American Plywood Association.)

Fig. 7-72 *Method of splicing a ridge board.* (American Plywood Association.)

ier to erect in shorter sections and then splice after it is in place. Check your house plan for roof slope. For example, a 4-inch rise in 12 inches of run is common. It is usually considered the minimum for asphalt or wood shingles.

Lay out one pair of rafters as shown previously. Mark the top and bottom angles and seat-cut location. Make the cuts, and check the fit by setting the rafters up at floor level. Mark this set of rafters for identification, and use them as a pattern for the remainder.

Cut the remaining rafters. For a 48-foot house with rafters spaced 24 inches O.C., you will need 24 pairs cut to the pattern (25 pairs counting the pattern). In addition, you will need two pairs of fascia rafters for the ends of the gable overhang (Fig. 7-73). Since they cover the end of the ridge board, they must be longer than the pattern rafters by half the width of the ridge board. Fascia rafters have the same cuts at the top and bottom as regular rafters. However, they do not have a seat cut.

Build temporary props of 2 × 4s to hold the rafters and ridge board in place during roof framing. The props should be long enough to reach from the top plate to the bottom of the ridge board. They should be fitted with a plywood gusset at the bottom. When the

Fig. 7-73 *Placement of rafters, ridge board, and collar beam.* (American Plywood Association.)

Fig. 7-74 *Placement of angle braces and vertical props. (American Plywood Association.)*

props are installed, the plywood gusset is nailed temporarily to the top plate or to a ceiling joist. The props are also braced diagonally from about midpoint in both directions to maintain true vertical (check with a plumb bob) (Fig. 7-74).

Move the ridge-board sections and rafters onto the ceiling framing. Lay plywood panels over the ceiling joists for a safe footing. First, erect the ridge board and the rafters nearest its ends (see Fig. 7-74). If the ridge of the house is longer than the individual pieces of ridge board, you'll find it easier to erect each piece separately rather than splice the full length of the ridge board first. Support the ridge board at both ends with temporary props. Toenail the first rafter pair securely to the ridge board using at least two 8d nails on each side. Then nail it at the wall. Install the other end-rafter pair in the same manner.

Make the ridge-board joints, using plywood gussets on each side of the joint. Nail them securely to the ridge board. Check the ridge board for level. Also check the straightness over the centerline of the house.

After the full length of the ridge board is erected, put up the remaining rafters in pairs. Nail them securely in place. Check them occasionally to make sure that the ridge board remains straight. If all rafters are cut and assembled accurately, the roof should be self-aligning.

Toenail the rafters to the wall plate with 10d nails. Use two per side. Also nail the ceiling joists to the rafters. For a 24-foot-wide house, you will need four 16d nails at each lap. In high-wind areas, it is a good idea to add metal-strap fasteners for extra uplift resistance (Fig. 7-75).

Cut and install 1 × 6 collar beams at every other pair of rafters (4 feet O.C.) (see Fig. 7-73). Nail each end with four 8d nails. Collar beams should be in the upper third of the attic crawl space. Remove the temporary props.

Square a line across the end wall plate directly below the ridge board. If a vent is to be installed, mea-

Fig. 7-75 *Metal framing anchors.*

sure half its width on each side of the center mark to locate the first stud on each side. Mark the positions for the remaining studs at 16 inches O.C. Then measure and cut the studs. Notch the top end to fit under the rafter so that the stud bottom will be flush with the top plate. Cut the cripple studs and headers to frame in the vent opening (Fig. 7-76).

Fig. 7-76 *Vent openings should be blocked in. (American Plywood Association.)*

Cut and install fascia board to the correct length of the ridge board. Bevel the top edge to the roof slope. Nail the board to the rafter ends. Then install the fascia rafters. Fascia rafters cover the end of the ridge board (see Fig. 7-73). Where the nails will be exposed to weather, use hot-dipped galvanized or other nonstaining nails.

SPECIAL RAFTERS

Special rafters are used to make special roof shapes. The mansard roof and the hip roof both call for special rafters. Jack rafters are needed for the hip roof. This type of roof also may have valleys and have to be treated especially well. Dormers call for some special rafters, too. For bay windows and other protrusions, some attention may have to be given to rafter construction.

Dormers

Dormers are protrusions from the roof. They stick out from the roof. They may be added to allow light into an upstairs room. Or they may be added for architectural effect (Fig. 7-77).

Fig. 7-78 *Shed-type dormer.*

Fig. 7-77 *A dormer.*

Dormers may be made in three types:

1. Dormers with flat, sloping roofs that have less slope than the roof in which they are located. This is called a *shed-type dormer* (Fig. 7-78).
2. Dormers with roofs of the gable type at right angles to the main roof (Fig. 7-79). There is no slope in this one, called a *gable dormer*.
3. The two types can be combined. This is called a *hip-roof dormer.*

Bay Windows

Bay windows are mostly for decoration. They add to the architectural qualities of a house. They stick out from the main part of the house. This means that they have to have special handling. The floor joists are extended out past the foundation the required amount. A band is then used to cap off the joist ends (Fig. 7-80). Take a closer look at the ceiling joists and rafters for the bay. The rafters are cut according to the rise called for on the plans. Cuts and lengths already have been discussed. No special problems should be presented by this method of framing. In order to make it easier, it is best to lay out the rafter plan at first so that you know which are the common and which are the hip rafters. In some cases, you may need a jack rafter or two depending on the size of the bay (Fig. 7-81).

CEILING JOISTS

Ceiling joists serve a number of purposes. They keep the wall from falling inward or outward. They anchor the rafters to the top plate. Ceiling joists also hold the finished ceiling in place. The run of the joist is important. The distance between supports for a joist determines its size. In some cases, the ceiling joist has to be

Fig. 7-79   *Gable dormer.*

(A)

(B)

Fig. 7-80   *A. Bay window framing. B. Two bay windows stacked for a two-story house. The metal cover does not require rafters.*

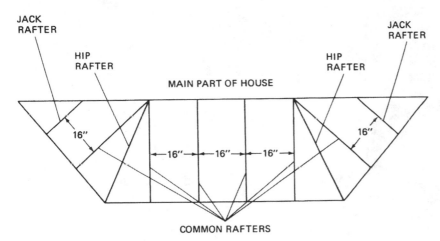

JACK RAFTER
HIP RAFTER
MAIN PART OF HOUSE
HIP RAFTER
JACK RAFTER

16"
16"
16" — 16" — 16"

COMMON RAFTERS

Fig. 7-81 *Rafter layout for a bay window.*

spliced. Figure 7-82 shows one method of splicing. Note how the splice is made on a supporting partition. Figure 7-83 shows how the joists fit on top of the plate. In some cases, it is best to tie the joist down to the top plate by using a framing bracket. Figure 7-84 shows how one type of bracket is used to hold the joist. This helps in high-wind areas.

The ceiling joists in Fig. 7-85 have been trimmed to take the angle of the rafter into consideration once the rafter is attached to the top plate and joist. The size of the joist is determined by the local code. However, there are charts that will give you some idea of what size piece of dimensional lumber to use. Table 7-5 in-

NAIL JOISTS TO EACH OTHER
FILLER BLOCKS
CEILING JOIST

Fig. 7-82 *Ceiling joist splices are made on a supporting partition.*

Fig. 7-83 *Looking up toward the ceiling joists. Notice how these are supported on the wall plate and are not cut or tapered. Two nails are used to toenail the joists to the plate.*

Fig. 7-84 *Steel bracket used to hold ceiling joist to the top plate. Note how it is bent to fit.*

dicates some allowable spans for ceiling joists. These are given using lumber graded nonstress.

There are some special arrangements for ceiling joists. In some cases it is necessary to interrupt the free flow of lines represented by joists. For example, a chimney may have to be allowed for. An attic opening may be called for on the drawings. These openings have to be reinforced to make sure that the joists maintain their ability to support the ceiling and some weight in the ceiling at a later time (Fig. 7-86).

Figure 7-87 shows how framing anchors are used to secure the joists to the double header. This method can be used for both attic openings and fireplace openings.

OPENINGS

As mentioned earlier, fireplaces do come out through the roof. This must be allowed for in the construction process. The floor joists have to be reinforced. The area around the fireplace has to be strong enough to hold the hearth. However, what we're interested in

(A)

(B)

Fig. 7-85 *A. Ceiling joists in place on the top plate. Note the cuts on the ends of the joists. B. Set the first ceiling joist on the inside of the end of the wall. This will allow the ceiling drywall to be nailed to it later.*

TABLE 7-5 *Ceiling Joists*

Size of Ceiling Joists, Inches	Spacing of Ceiling Joists, Inches	Maximum Allowable Span			
		Group I	Group II	Group III	Group IV
2 × 4	12	11'6"	11'0"	9'6"	5'6"
	16	10'6"	10'0"	8'6"	5'0"
2 × 6	12	18'0"	16'6"	15'6"	12'6"
	16	16'0"	15'0"	14'6"	11'0"
2 × 8	12	24'0"	22'6"	21'0"	19'0"
	16	21'6"	20'6"	19'0"	16'6"

Fig. 7-86 *Blocking in the joists to allow an opening for a chimney.*

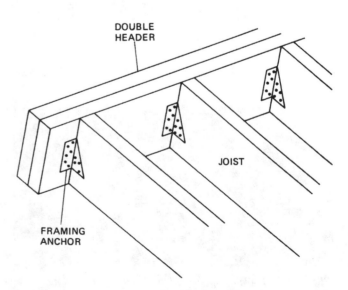

Fig. 7-87 *Using framing anchors to hold the tail joists to the header.*

Fig. 7-88 Arrows show the openings in the roof for the chimney.

here is how the fireplace opening comes through the rafters (Fig. 7-88). The roofing here has been boxed off to allow the fireplace to come through. The chimney at the top has the flashing ready for installation as soon as the bricks are laid around the flue.

Other openings are for soil pipes. These are used for venting the plumbing system. A hole in the plywood deck is usually sufficient to allow their exit from the inside of the house (Fig. 7-89).

Fig 7-89 Soil pipe stack coming through the roof.

DECKING

A number of types of roof decks are used. One is plywood. It is applied in 4- × 8-foot sheets. Plywood adds structural strength to the building. Plywood also saves time because it can be placed on the rafters rather quickly.

Boards of the 1- × 6- or 1- × 8-inch size can be used for sheathing. This decking takes a longer time to apply. Each board has to be nailed and sawed to fit. This type of decking also adds strength to the roof. Another type of decking is nothing more than furring

strips applied to the rafters. This is used as a nail base for cedar shingles.

Plywood Decking

Roof sheathing, the covering over rafters or trusses, provides structural strength and rigidity. It makes a solid base for fastening the roofing material. A roof-sheathing layout should be sketched out first to determine the amount of plywood needed to cover the rafters (Fig. 7-90).

NOTE: For "open soffits" all panels marked with an "x" must be EXT-DFPA "soffit" panels.

Fig. 7-90 Roof sheathing layout for a plywood deck. (American Plywood Association.)

Draw your layout. It may be a freehand sketch, but it should be relatively close to scale. The easiest method is to draw a simple rectangle representing half the roof. The long side will represent the length of the ridge board. Make the short side equal to the length of your rafters, including the overhangs. If you have open soffits, draw a line inside the ends and bottoms. Use a dotted line, as shown in Fig. 7-90. This area is to be covered by exterior plywood. Remember that this is only half the roof. Any cutting of panels on this side can be planned so that the cut-off portions can be used on the other side. If your eave overhang is less than 2 feet and you have an open soffit, you may want to start with a half-panel width of soffit plywood.

Figure 7-91 shows the open soffit. Figure 7-92 shows the boxed soffit. Otherwise, you probably will start with a full 4- × 8-foot sheet of plywood at the bottom of the roof and work upward toward the ridge. This is where you may have to cut the last row of panels. Stagger panels in succeeding rows.

Complete your layout for the whole roof. The layout shows panel size and placement as well as the number of sheathing panels needed. See all this in Fig. 7-90.

If your diagram shows that you will have a lot of waste in cutting, you may be able to reduce scrap by slightly shortening the rafter overhang at the eave or the gable overhang. An example is shown in Fig. 7-90,

SHIM AT EACH RAFTER FOR FLUSH JOINT AT CHANGE OF PLYWOOD THICKNESS

PLYWOOD SHEATHING

LEAVE 1/16" SPACE AT ALL PANEL END AND EDGE JOINTS

NOTE: Any appropriate grade of exterior plywood of adequate thickness (½" or more) prevents protrusion of roofing nails or staples at exposed underside, and carries design roof load.

Fig. 7-91 *Open soffit. (American Plywood Association.)*

NOTE: Protect edges of interior plywood sheathing against exposure to weather.

PLYWOOD SHEATHING

APPROPRIATE GRADE OF EXTERIOR PLYWOOD

CONTINUOUS SCREENED VENT OR LOUVERED VENT

Fig. 7-92 *Boxed soffit. (American Plywood Association.)*

where nearly half the panels are "soffit" panels. In such a case, rather than using shims to level up soffit and interior sheathing panels, you may want to use interior sheathing panels of the same thickness as your soffit panels, even though they might then be a little thicker than the minimum required.

Cut panels as required, marking the cutting lines first to ensure square corners. Begin panel placement at any corner of the roof. If you are using special soffit panels, remember to place them best, or textured, side down.

Fasten each panel in the first course (row), in turn, to the shank with spiral-threaded nails. Space nails 6 inches O.C. along the panel ends and 12 inches O.C. at intermediate supports. Leave a ¹⁄₁₆-inch space at panel ends and ⅛ inch at edge joints. In areas of high humidity, double the spacing.

Apply the second course using a soffit half-panel in the first (overhang) position. If the main sheathing panels are thinner than the soffit sheathing, install small shims to ease the joint transition. See Fig. 7-91 for location of the shims. Apply the remaining courses as above. Note that if your plans show closed soffits,

Fig. 7-93 *Using a stapler to fasten plywood decking to the rafters. (American Plywood Association.)*

(A)

(B)

Fig. 7-94 *Plywood clips reinforce the surface area where the sheathing butts.*

Fig. 7-95 *Erection of sidewalls to a house.*

Fig. 7-96 *Plywood sheathing placed on one portion of rafters. Note the ladder made for plywood lifting.*

the roof sheathing all will be the same grade thickness. To apply plywood to the underside of closed soffits, use nonstaining-type nails.

Figure 7-93 shows plywood decking being applied with a stapler. Figure 7-94 shows an H clip for plywood support along the long edges. This gives extra support for the entire length of the panels.

Figure 7-95 shows the erection of the sidewalls of a house. In Fig. 7-96, the plywood sheathing has been placed on one portion of the rafters. Note the rig (*arrow*) that allows a sheet of plywood to be passed up from the ground to roof level. The sheet is first placed on the rack. Then it is taken by the person on the roof and moved over to the needed area.

Figure 7-97 shows a boxed soffit. Note the louvers already in the board. The temporary supports hold the

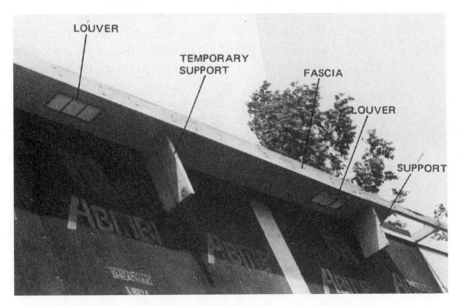

Fig. 7-97 *Louvered soffit held in place with temporary braces.*

Fig. 7-98 *Different sizes and shapes of roof decking made of lumber: A. regular V-jointed decking; B. strait decking; C. grooved type; D. eased joint or bull-nosed type; E. single tongue and groove with a V joint.*

soffit in place until final nailing is done and it can be supported by the fascia board. The fascia has a groove along its entire length to allow the soffit to slide into it.

Boards for Decking

Lumber may be used for roof decking. In fact, it is necessary in some special-effects ceilings. It is needed where the ceiling is exposed and the underside of the decking is visible from below or inside the room.

Fig. 7-99 *A 2-degree angle is cut in the lumber decking ends. This ensures a tight face joint on the exposed ceiling below.*

Fig. 7-100 *Note the locations of the nails in the lumber decking.*

Roof decking comes in a variety of sizes and shapes (Fig. 7-98). A 2-degree angle is cut in the lumber decking ends to ensure a tight joint (Fig. 7-99). This type of decking usually is nailed, so the nails must be concealed. This requires nailing as shown in Fig. 7-100. Eight-inch spikes usually are used for this type of nailing. Note the chimney opening in Fig. 7-101.

Figure 7-102 illustrates the application of 1 × 6 or 1 × 8 sheathing to the rafters. Note the two nails used to hold the boards down. The common nail is used here. For concealed nailing, it takes a finishing nail to be completely concealed.

Once the decking is in place, it is covered by an underlayment of felt paper. This paper then is covered by shingles, as selected by the builder.

Shingle Stringers

For cedar shingles, the roof deck may be either spaced or solid. The climatic conditions determine if it is a solid deck or one that is spaced. In areas where there are blizzards and high humidity, the spaced deck is not used. In snow-free areas, spaced sheathing is practical.

Use 1 × 6s spaced on centers equal to the weather exposure at which the shakes are to be laid. However, the spacing should not be over 10 inches. In areas where wind-driven snow is encountered, solid sheathing is recommended. Figure 7-102 shows spaced sheathing.

Roof pitch and exposure   Hand-split shakes should be used on roofs where the slope or pitch is sufficient to ensure good drainage. The minimum recommended pitch is one-sixth, or 4-in-12 (4 inches of vertical rise for each 12 inches of horizontal run), although there have been satisfactory installations on lesser slopes.

(labels: TRIMMER, RAFTER, CHIMNEY OPENING, HEADER, TRIMMER, SHEATHING, VALLEY, RIDGE BOARD, PLATE, STUD)

Fig. 7-101 *Note the chimney opening in this lumber sheathing.*

Climatic conditions and the technique and skill of application are modifying factors.

The maximum recommended weather exposure is 10 inches for 24-inch shakes and 7½ inches for 18-inch shakes. A superior three-ply roof can be achieved at slight additional cost if these exposures are reduced to 7½ inches for 24-inch shakes and 5½ inches for 18-inch shakes.

Figure 7-103 shows the shakes in place on spaced sheathing. Note how the amount of exposure

Fig. 7-103 *Hand-split shakes should be used on roofs where the slope is sufficient to ensure good drainage. Two different exposures to the weather are shown. Note the spacing of the sheathing under the shakes. (Red Cedar Shingle & Hand-Split Shake Bureau.)*

Fig. 7-102 *Spacing of sheathing for wood shingles. (Red Cedar Shingle & Hand-Split Shake Bureau.)*

(A)

(B)

Fig. 7-104 *A. Roof ready for application of shingles. B. Cedar shingles being applied to prepared sheathing.* (Red Cedar Shingles & Hand-Split Shake Bureau.)

to the weather makes a difference in the spacing of the sheathing. Figure 7-104 gives a better view of the spaced sheathing and how the roofing is applied to it.

CONSTRUCTING SPECIAL SHAPES

The gambrel shape is familiar to most people because it is the favorite shape for barns. It consists of a double-slope roof. This allows for more space in the attic or upper story. More can be stored there. In modern home designs, this type of roof has been used to advantage. It gives good headroom for an economical structure with two stories. This design was brought to the United States by Germans in the early days of the country.

Gambrel-Shaped-Roof Storage Shed

A storage shed will give you an idea of the simplest way to use the gambrel-shaped roof. Examine the details and then obtain the equipment and supplies needed. The bill of materials lists the supplies that are needed (Fig. 7-105). Now take a look at Figs. 7-106, 7-107, and 7-108. They show different ways of finishing the shed for different purposes. For instance, the structure can be covered with glass, Plexiglas, or polyvinyl as in Fig. 7-107 and made into a greenhouse.

Fig. 7-106 *Rustic shed design.* (TECO.)

BILL OF MATERIALS			
QUAN.	DESCRIPTION	QUAN.	DESCRIPTION
28	2″ x 4″ x 8 FT. LONG	40	TECO C-7 PLTS.
9	4′ x 8′ x ½″ PLYWD.	30	TECO JOIST HGR.
2	1″ x 4″ x 6 FT. LONG	12	TECO ANGLES
1 ROLL	ROOFING FELT	30	TECO A-5 PLTS.
1 GAL	ROOF. CEMENT	3	3 BUTT HINGES
1 GAL	BARN-RED PAINT	1	HASP & LOCK
5 #	6d COM. NAILS	10 BG.	90# CONC. MIX
2 #	12d COM. NAILS		OR
2 #	½″ ROOF. NAILS	4	6″ x 8″ x 8′ RAIL TIE

Fig. 7-105 *Bill of materials for a tool shed.* (TECO.)

Fig. 7-107 *A greenhouse can be made by covering the frame with plastic. (TECO.)*

Fig. 7-108 *Contemporary shed design. (TECO.)*

Fig. 7-109 *Frame layout for the shed. (TECO.)*

Then there is the rustic look shown in Fig. 7-106 and the contemporary look shown in Fig. 7-108.

Frame layout Note the dimensions of the shed. It is 7 feet high and 8 feet wide. A frame-to-sill detail is given in Fig. 7-109. Note the spacing of the slopes of the roof. Figure 7-110 indicates how the vertical stud members and the roof members are attached with metal plates. Figure 7-111 indicates how an 18-degree angle is used for cutting the studs and roof members.

Frame-cutting instructions Mark off 18-degree angles on the 2 × 4s (see Fig. 7-111). Cut to length. Note the exact length required for the roof member and the vertical stud. When you have cut the required number of studs and roof members, place the sections on a hard, flat surface.

Fig. 7-110 *Detail of connection of vertical stud member and roof member. (TECO.)*

Fig. 7-111 *Frame-cutting instructions. (TECO.)*

A driveway or sidewalk can be used. Nail the metal plates equally into each member, and flip the frame over. Then nail the metal plates on this side. Make four frames for a shed with an 8-foot depth. You can change the number of frames to match your length or depth requirements. Just add a frame for every 2 feet, 8 inches of additional depth.

Roof-framing plan Figure 7-112 shows the roof-framing layout. Such a layout will eliminate any problems you might have later if you did not plan your project properly. The purlin is extended in length every time you extend the depth of the shed by 2 feet, 6 inches. Note how the purlins are attached to the vertical stud member and the roof member (Fig. 7-113). Once you have attached all the purlins, you have the standing frame. It is time to consider other details. Figure 7-114 shows how a vent is built into the rear elevation. It can be a standard window or constructed from 1 × 2s with a polyvinyl backing.

Fig. 7-112 *Roof-framing plan for the shed.* (TECO.)

Fig. 7-114 *Rear elevation of the shed. Note how the vent is built in.* (TECO.)

Door details You have to decide on the design of the doors to be used. Figure 7-115 shows how the door is constructed for both rustic and contemporary styles. Cut the angles at 18 degrees when making the door for the rustic style. Use the same template as was used for the studs and roof members.

Fig. 7-113 *Attachment of purlins to vertical stud member and roof member.* (TECO.)

Fig. 7-115 *Door details for the shed.* (TECO.)

PLYWOOD
SHEATHING,
ROLL ROOFING,
AND HAND-SPLIT
SHAKES

HASP

HINGE
(TYP. 3 PLCS)

LAG SCREW
½" X 5" AT 2'
CENTERS

ANCHOR
BOLT ½" X
5½' CTRS.

6' X 8" X 6' RAIL TIE OR 4' X 8" X 8' CONCRETE

Fig. 7-116 *Front elevation of the shed.* (TECO.)

The contemporary door is nothing more than squared-off corners. Figure 7-116 shows the placement of the hasps used on the door.

Mansard Roofs

The mansard roof has its origins in France. The mansards usually made in the United States are slightly different from the French ones. Figure 7-117 shows the American-style mansard. The roof uses a rafter with a steep slope for the side portion and one with a very low slope for the top. Standard 2 × 4 or 2 × 6 framing lumber is used to make these rafters. Plywood sheathing is applied over the rafters, and roofing is usually applied over the entire surface. This is supposed to make the house look lower. It effectively low-

Fig. 7-117 *Mansard roof.*

ers the "beltline" and makes the roof look closer to the ground. Shingles have to be chosen for a steep slope so that they are not blown off by high winds.

Figure 7-118 shows a mansard roof with cedar shakes. The steep slope portion adds to the effect when covered by cedar shingles. It is best to use the wooden shakes for the low slope as well; however, in some cases, the slope may be too low, and other shingles may be needed to do the job properly.

Fig. 7-118 *Hand-split shakes used on a mansard roof.* (Red Cedar Shingle & Hand-Split Shake Bureau.)

Figure 7-119 shows how the French made the mansard roof truss. It was used on hotels and some homes. This style was popular during the nineteenth century. Note the elaborate framing used in those days to hold the various angles together. At that time, builders used wrought-iron straps instead of today's metal (steel) brackets and plates.

Post-and-Beam Roofs

The post-and-beam type of roof is used for flat or low-slope roofs. This type of construction can use the roof decking as the ceiling below (Fig. 7-120A). The exposed ceiling or roof decking has to be finished. This means that the wood used for the roof deck has to be surfaced and finished on the inside surface. Post-and-beam methods don't use regular rafters (Fig. 7-120B). The rafters are spaced at a greater distance than in conventional framing. This calls for larger dimensional lumber in the rafters. These are usually also exposed and need to be finished according to plan.

Fig. 7-119 *Truss construction for an old-type mansard roof.*

Fig. 7-120 *A. Post-and-beam framing. B. Conventional framing.*

ROOF LOAD FACTORS

Plywood roof decking offers builders some of their most attractive opportunities for saving time and money. Big panels go down really fast over large areas. They form a smooth, solid base with a minimum of joints. Waste is minimal, contributing to the low in-place cost. It is frequently possible to cut costs still further by using fewer rafters with a somewhat thicker panel for the decking. For example, use ¾-inch plywood over framing 4 feet O.C. Plywood and trusses are often combined in this manner. For recommended spans and plywood grades, see Table 7-6.

Plywood roof sheathing with conventional shingle roofing Plywood roof sheathing under shingles provides a tight deck with no wind, dust, or snow infiltration and high resistance to racking. Plywood has stood up for decades under asphalt shingles, and it has performed equally well under cedar shingles and shakes. Plywood sheathing over roof trusses spaced 24 inches O.C. is widely recognized as the most economical construction for residential roofs and has become the industry standard.

Design Plywood recommendations for plywood roof decking are given in Table 7-6. They apply for the following grades: C-D INT APA, C-C EXT APA, Structural II and II C-D INT APA, and Structural I and II C-C EXT APA. Values assume 5 pounds per square foot dead load. Uniform load deflection limit is 1/180 of the span under live load plus dead load or 1/120 under live load only. Special conditions, such as heavy, concentrated loads, may require constructions in excess of these minimums. Plywood is assumed to be continuous across two or more spans and applied face grain across supports.

Application Provide adequate blocking, tongue-and-groove edges, or other edge support such as ply-clips when spans exceed maximum length for unsupported edges. Figure 7-121 shows the installation of ply-clips. Use two ply-clips for 48-inch or greater spans and one for lesser spans.

Space panel ends 1/16 inch apart and panel edges ⅛ inch apart. Where wet or humid conditions prevail, double the spacings. Use 6d common smooth ring-shank or spiral-thread nails for plywood ½ inch thick or less. Use 8d nails for plywood to 1 inch thick. Use 8d ring-shank or spiral-thread nails or 10d common smooth nails for 1-, 1⅛-, and 1¼-inch panels. Space nails 6 inches apart at panel edges and 12 inches apart at intermediate supports, except where spans are 48 inches or more. Then space nails 6 inches apart at all supports.

Plywood nail holding Extensive laboratory and field tests, reinforced by more than 25 years of experi-

TABLE 7-6 *Plywood Roof Decking*

Identification Index	Plywood Thickness, Inches	Maximum Span, Inches	Unsupported Edge—Max. Length, Inches	Allowable Live Loads, psf									
				Spacing of Supports Center to Center, Inches									
				12	16	20	24	30	32	36	42	48	60
12/0	5/16	12	12	150									
16/0	5/16, 3/8	16	16	160	75								
20/0	5/16, 3/8	20	20	190	105	65							
24/0	3/8, 1/2	24	24	250	140	95	50						
32/16	1/2, 5/8	32	28	385	215	150	95	50	40				
42/20	5/8, 3/4, 7/8	42	32		330	230	145	90	75	50	35		
48/24	3/4, 7/8	48	36			300	190	120	105	65	45	35	
2•4•1	1⅛	72	48				390	245	215	135	100	75	45
1⅛" Grp. 1 and 2	1⅛	72	48				305	195	170	105	75	55	35
1¼" Grp. 3 and 4	1¼	72	48				355	225	195	125	90	85	40

LEAVE 1/16" SPACE AT ALL PANEL
END JOINTS AND 1/8" SPACE AT ALL
PANEL EDGE JOINTS. WHEN WET OR
HUMID CONDITIONS CAN BE EXPECTED
DOUBLE THESE SPACINGS.

DETAIL
PLYCLIP-S

PLYCLIPS OR TONGUE-
AND-GROOVED EDGES
IF REQUIRED

ASPHALT, ASBESTOS,
OR WOOD SHINGLES.
FOLLOW ROOFING MFR'S.
RECOMMENDATIONS FOR
ROOFING FELT.

PLYWOOD
SHEATHING

EXTERIOR PLYWOOD
STARTER STRIP OR
PROTECT EDGES OF
DECKING AGAINST
EXPOSURE TO WEATHER

EXTERIOR
PLYWOOD SOFFIT

Fig. 7-121 *Using a ply-clip to reinforce plywood decking.* (American Plywood Association.)

ence, offer convincing proof that even 5/16-inch plywood will hold shingle nails securely and permanently in place, even when the shingle cover is subjected to hurricane-force winds.

The maximum high wind pressure or suction is estimated at 25 pounds per square foot, except at the southern tip of Florida, where wind pressures may attain values of 40 to 50 pounds per square foot. Because of shape and height factors, however, actual suction or lifting action even in Florida should not exceed 25 pounds per square foot up to 30-foot heights. Thus any roof sheathing under shingles should develop at least that much withdrawal resistance in the nails used.

Plywood sheathing provides more than adequate withdrawal resistance. A normal wood-shingled room will average more than 6 nails per square foot. Each nail need carry no more than 11 pounds. Plywood sheathing only 5/16 inch thick shows a withdrawal resistance averaging 50 pounds for a single 3d shingle nail in laboratory tests and in field tests of wood shingles after 5 to 8 years of exposure. In addition, field experience shows that asphalt shingles consistently tear through at the nail before the nail pulls out of the plywood.

Figure 7-122 shows the markings found on plywood used for sheathing. Note the interior and exterior

Fig. 7-122 *Plywood grades identified.* (American Plywood Association.)

glue markings. APA stands for the American Plywood Association.

LAYING OUT A STAIR

So far, you have used the framing square to lay out rafters. There is also another use for this type of instrument. It can be used to lay out the stairs going to the basement or going upstairs in a two-story house.

Much has been written about stairs. Here we only lay out the simplest and most useful of the types available. The fundamentals of stair layout are offered here:

1. Determine the height, or rise. This is from the top of the floor from which the stairs start to the top of the floor on which they are to end (Fig. 7-123).

2. Determine the run, or distance measured horizontally.

NOTE: The stairs shown with approximately the above dimensions of the riser and tread are considered easy or comfortable to climb.

Fig. 7-123 *Laying out stairs with the steel square.* (Stanley Tools.)

3. Mark the total rise on a rod or a piece of 1- × 2-inch furring to make a so-called story pole. Divide the height or rise into the number of risers desired. A simple method is to lay out the number of risers wanted by spacing off the total rise with a pair of compasses. It is common to have this result in fractions of an inch. For example, a total rise of 8 feet, 3¾ inches, or 99¾ inches divided by 14 = 7.125, or a 7⅛-inch riser. This procedure is not necessary in the next step because the horizontal distance, or run, is seldom limited to an exact space, as is the case with the rise.

4. Lay out or space off the number of treads wanted in the horizontal distance, or run. There is always one less tread than there are risers. If there are 14 risers in the stair, there are only 13 treads. For example, if the tread is 10 inches wide and the riser is 7 inches, the stair stringer would be laid out or "stepped off "with the square, ready for cutting, as shown in Fig. 7-123. The thickness of the tread should be deducted from the first riser as shown. This is done to have this first step the same height as all the others.

ALUMINUM SOFFIT

So far, the soffit has been mentioned as the covering for the underside of the overhang. This has been shown to be covered with a plywood sheathing of ¼-inch thickness or with a cardboard substance about the thickness of the plywood suggested. The cardboard substitute is called *Upson board* because it is made by the Charles A. Upson Company. If installed properly and painted, it will last for years. However, it should not be used in some climates.

One of the better materials for soffit is aluminum. More and more homes are being fitted with this type of maintenance-free material. This particular manufacturer no longer makes roll-type soffit material. It is included here in case you need to know how it was installed so that you can either repair or replace it.

Material Availability

Aluminum soffit material was available in 50-foot rolls with various widths. The rolls could be obtained in widths of 12, 18, 24, 30, 36, and 48 inches (Fig. 7-124). These were pushed or pulled into place as shown in Fig. 7-125. The hip roof with an overhang all around the house would require soffit material pulled in as shown in Fig. 7-126. The runners supporting the material are shown in Fig. 7-127. Covering the ends is important to ensure a neat job. Corner trim and fascia closure were available to help give the finished job a look not unlike that of an all-wood soffit.

Fig. 7-124 *Rolls of soffit material.*

SOFFIT COIL MATERIAL

Fig. 7-125 *Method of inserting the aluminum soffit material.* (Reynolds Metal Products.)

Fig. 7-126 *Steps for installing aluminum soffit in a hip roof with an overhang all around.* (Reynolds Metal Products.)

Fig. 7-127 *Closing off the ends of the soffit with aluminum.* (Reynolds Metal Products.)

Figure 7-128 shows how the fascia runner, the frieze runner, and corner trim were located for ease of installation for the soffit coil. After the material had been put in place and the end cuts had been made, the last step was to insert a plastic liner to hold the aluminum in place. This prevented rattling when the wind blew. The material could be obtained with a series of holes prepunched. This served as ventilation for the attic (Fig. 7-129).

Figure 7-130 gives more details on the installation of the runners that supported the soffit material. The fascia runner was notched at points *b* for about 1½ inches maximum. Then the tab was bent upward and against the inside of the fascia board. Here it was nailed to the board for support. Take a look at *c* in Fig. 7-130 to see how the tab was bent up. Note how the width of the channels was the soffit coil width plus at least ⅜ inch and not more than ⅞ inch. This allowed for expansion by the aluminum. Aluminum expands in hot weather.

Cutting the runner to desired lengths could be done by cutting the channel at *a* and *b* of Fig. 7-131. Then the metal back was bent back and forth along a line such as at *c* until it broke. Of course, you could use a pair of tin snips to make a clean cut.

Figure 7-132 shows how the soffit was installed on a brick-veneer house. Figure 7-132A shows how the frieze runner was located along the board. Figure 7-132B shows how the runner was nailed to the board.

When the fascia board was more than 1 inch wide, it was necessary to place 1-inch aluminum strips as shown in Fig. 7-133. The tabs were hooked in between the fold in the runner. Then the runner was brought under the fascia board and bent back and nailed. This allowed the runner to expand when it was hot. The overlapping runners were not nailed to one another.

In some instances, it was necessary to use a double-channel runner. This was done so that there would be no sagging of the soffit material (Fig. 7-134). Note

CORNER TRIM

FASCIA RUNNER

SOFFIT COIL

FRIEZE RUNNER

Fig. 7-128 *Installing soffit coil in the over-hang space. Note the corner trim to give a finished appearance.* (Reynolds Metal Products.)

Fig. 7-129 *Finishing up the job with a vinyl insert to hold the aluminum in place.* (Reynolds Metal Products.)

1" THICKNESS

FASCIA BOARD

FASCIA RUNNER

2' MAX.

c

1-1/2" MAX.

b

b

W

FRIEZE RUNNER

W = SOFFIT COIL WIDTH + 3/8" TO 7/8"
(3/8" MINIMUM) (7/8" MAXIMUM)

Fig. 7-130 *Method used to support the runner on the fascia.* (Reynolds Metal Products.)

how the frieze runner and double-channel runner were located. Note the gravel stop on this flat roof. In some parts of the country, more overhang is needed because that gives more protection from the sun.

Figure 7-135 shows how the H-molding joint works to support the two soffit materials as they are unrolled into the channel molding. Note the location of the vent strip, when needed.

As was mentioned earlier, the aluminum soffit made for a practically maintenance-free installation.

BEND

c

b

a

Fig. 7-131 *Bending and breaking the runner material.* (Reynolds Metal Products.)

Aluminum Soffit **187**

Fig. 7-132 Soffit on a brick-veneer house: A. locating the frieze runner along the board; B. using the quarter-round type of frieze runner. (Reynolds Metal Products.)

Fig. 7-133 Installing a tab to keep the runner free to move as aluminum expands on hot days. (Reynolds Metal Products.)

Fig. 7-134 Using a double-channel runner and a double row of aluminum soffit material. (Reynolds Metal Products.)

Fig. 7-135 Installing the H-molding joint. (Reynolds Metal Products.)

METAL CONNECTORS

Every year, many houses are destroyed when the force of high winds causes roofs to fly off and walls to collapse. One of the methods used by builders located in high-wind areas such as along the shores of lakes, bays, oceans, and gulfs is to use metal connectors. Var-

ious fasteners are designed to increase the structural strength of homes built to withstand hurricanes and, in some instances, tornadoes and earthquakes.

Figure 7-136 shows how the use of metal connectors can increase the chances for a house to escape a hurricane's full force. The connectors are illustrated in the next few pages. Take a close look at the encircled

Fig. 7-136 *Location of metal connectors. (SEMCO.)*

number, and then refer to the following pages for an illustration of where and how the connector is used to its best advantage.

High winds can be fought by building a house that will withstand the forces of Mother Nature. The only successful way to combat these forces is to employ proper construction techniques ahead of time to ensure the integrity of the structure. The best method is to use an uninterrupted load path from the roof members to the foundation. Metal connectors are engineered to satisfy the necessary wind uplift load requirements (Fig. 7-137).

Figure 7-138 shows how foundations are prepared with metal connectors to cause the soleplate and studs to be permanently and solidly anchored to prevent damage by high winds. Figure 7-139 illustrates how connectors are used for a stem-wall system. A wood-to-wood type of construction enhanced by metal connectors is shown in Fig. 7-140. Note the allowable load tables that give the nail size and uplift in pounds per square inch (psi).

A poured masonry header has the rafters attached by hangers anchored in the concrete (Fig. 7-141). Second-floor problems can be solved by tying the first and second floors together with metal connectors, as shown in Fig. 7-142. A number of fasteners are illustrated. Regular truss-to-top-plate construction is shown reinforced in Fig. 7-143, whereas the top-plate-to-stud reinforcement is illustrated by Fig. 7-144.

These metal fasteners or connectors are standard requirements in areas buffeted by high winds, hurricanes, and earthquakes. The additional costs often can be offset by lower insurance rates on the finished house.

There are charts that show the expected winds in a given area. Figure 7-145 is a map of the United States with the wind speeds indicated. The wind loads placed on a structure are determined by a number of factors. Chief among these is the wind-speed ratings for the location of the structure.

Fig. 7-136 (Continued)

⑧ UHP

⑨ UHPKG

⑩ **L-Post Strap** S-345

⑪ **T-Post Strap** S-320

HP

DOUBLE

⑫ HCPRS

⑬ ABJ

⑭ ADS2 & ADS5

⑮ FAP18 FAP18S FAP18F

Fig. 7-136 (Continued)

Fig. 7-136 (Continued)

192 Framing the Roof

① Hurricane Anchor   This tie adds increased resistance to wind uplift. The tie eliminates toenailing using correctly located nail holes for fast, easy, and strong attachment of rafters and trusses to plates and studs. These ties are made of 20- to 18-gauge galvanized steel. They can be installed on each side of the rafter for twice the loads when the rafter thickness is a minimum of 2.5 inches or diagonally when rafter is 1.5 inches.

② 90-Degree Bracket   Used for tying trusses to non-load-bearing walls.

③ Rafter Clip/Stud Plate   A fast, economical tie to secure rafters or trusses to wall studs and top plates and from studs to sill plate.

④ Girder Truss Strap   Ideal for girder truss connections when there is a high uplift load requirement. Can be used for wood-to-wood application or concrete-to-wood application. A washer plate adds increased resistance to wind uplift.

⑤ A. Beam Support   These face-mount supports are designed to provide full support of the top chord, preventing joist rotation. Bolt holes are also provided for additional load capacity.

B. Floor Tie Anchor   These ties are designed especially for use with floors constructed above grade as a connection between first- or second-floor level to studs. They are designed for engineered floor systems with larger clear spans of 21 and 24 inches. They can be fastened with bolts or nails.

⑥ Heavy-Duty Post Anchor   Post base is to elevate post above concrete to allow for ventilation. Heavy-duty design permits higher uplift loads and simplified installation with 0.5-inch anchor bolts that fit through prepunched holes, slotted to permit adjustment to align for off-position anchor bolts.

⑦ Staircase Bracket   These brackets are designed to simplify and reinforce stair construction.

⑧ Joist Support   The regular joist support is made of 18-gauge steel, and the heavy-duty joist hanger is made to support headers, joists, and trusses. It is made from 14-gauge galvanized steel.

⑨ Kwik Grip Joist Support   These supports are designed for quick installation in place for easy nailing. The offset nail hole placement is for secure positive nailing. Precision-formed high-strength 18-gauge galvanized is used for long life.

⑩ L-Post Strap   Can be used to tie the window framing together.

⑪ T-Post Strap   Can be used at perpendicular junction points for cripple studs and the rough sill.

⑫ Stud Plate J Rafter Clip   Used for tying studs to soleplate and top plates or rafters to top plates.

⑬ Anchor Bolt   A 6-inch minimum embedment with 3,000 psi concrete will resist 1,635 lb. Wind unlift loads are based on the shear capacity of No. 2 southern pine. Compression perpendicular to grain = 565 psi.

⑭ Hold Down Anchor   Ideal for shear walls and vertical posts.

⑮ Universal Framing Anchor   This is a multipurpose anchor for almost any wood-connection task. It anchors rafters and roof trusses to plates, and it anchors floor and ceiling joists to headers and solid blocking to plates. The 90-degree framing angles can be used to join posts to beams and make other right-angle connections.

⑯ Post Cap   Simplifies installing 4 × 4 wood posts to wood beams and trusses. Makes a full-strength, positive connection between the post top and lateral beams or trusses. It is considerably stronger than random toenailing and spiking and is less time-consuming than drilling and lag bolting. Post caps are of the split design to offer maximum flexibility in application. Direct-load path-side plates maximize load capacities. Toenail slots speed plum and level adjustments.

⑰ Floor Tie Anchor   As noted previously, these ties are designed especially for use with floors constructed above grade as a connection between first or second floor level to studs. Nails or bolts can be used.

⑱ Foundation Tie   There is a prepunched hole in the foot of the tie to increase concrete grip and allow alternate rebar rod support. Prepunched holes in the bend are there for nails to hold it in the form board. Ties should have a minimum of 3 inches in the concrete.

⑲ Heavy-Duty Masonry Beam Hanger   Designed to work with standard block wall or concrete tie beam construction. Eliminates the need of constructing special seats to support door joists.

⑳ Rafter Tie   Tie straps meet a variety of application and design-load conditions and specifications. Use rafter ties when tying rafters to plate, anchoring studs to sill, or framing over girders and bearing partitions.

㉑ Lateral Truss Anchor   This anchor is designed to meet the lateral and uplift load demands for hurricane-resistant construction. It provides a custom connection to wood for trusses or rafters. An attached beam seat-plate eliminates treated sill or moisture-barrier installation. The riveted plate on the seat-plate design prevents truss movement parallel and perpendicular to the wall.

㉒ Truss Anchor   Accommodates diverse design requirements for concrete-to-wood installation and allows a 4-inch embedment in concrete. You can use two anchors installed, one on each side of the rafter, for twice the load per single-rafter thickness. Minimum edge distance is 2 inches.

㉓ Galvanized Plywood Clip   This clip is designed for easy and fast application to sheathing edges. It gives a snug self-lock that keeps the clip firmly in position throughout panel placement. It eliminates unreliable wood edge blocking. This type of clip is also available in aluminum. The clip gives an automatic spacing to the plywood or sheathing.

㉔ Truss Tie Down Strap (Gun Tie)   This strap provides additional increased resistance to wind uplift to secure rafters or trusses to top plates. It eliminates fastening through the truss nail plates and requires no truss nailing. It is designed to allow gun nailing for quick installation. The clip is made of 20-gauge steel, and the installer should wear eye protection. Fasteners are placed on each side of the truss into the top and bottom layers of the double top-plate members in equal quantity. Fasteners should be no less than 0.25 inch from the edge of the strap and placed no less than 0.375 inch from the edge of the framing member. True Tie is the same share as the Gun Tie, but it has 18-gauge galvanized metal.

Fig. 7-136   (Continued)

PRODUCT CODE	GAUGE	FASTENER SCHEDULE		ALLOWABLE LOADS	
		HEADER / PLATE	STUD	WIND / EARTHQUAKE	
				UPLIFT 133%	UPLIFT 160%
ABJBL10W	---	---	---	1635	1635
FOP41	12	12-16d	---	2190	2465
HCPFA	16	8-16d	8-16d	1200	1415
HCPSA	18	---	16-16d	1200	1415
HCPRS	18	5-8d	6-8d	540	540
CLP5W	18	11-8d	6-8d	530	540
SRP121630F	12	---	18-16d	2815	3380
RS150	16	---	11-10d	1645	1645
FAP18F	18	6-8d	6-8d	765	915
ADS2	12	(1) 5/8"	(2) 5/8"	2775	3330

PRODUCT CODE	GAUGE	FASTENER SCHEDULE		ALLOWABLE LOADS	
		STUD	PLATE	WIND / EARTHQUAKE	
				UPLIFT 133%	UPLIFT 160%
HCPLR	18	4-8d	4-8d	510	520
FAP18	18	6-8d	6-8d	745	745
HCPRS	18	6-10d	5-10d	540	540
CLP5W	18	6-10d	11-10d	540	540
TPP4	20	8-10d	8-10d	1335	1335

PRODUCT CODE	GAUGE	FASTENER SCHEDULE		ALLOWABLE LOADS			
		TRUSS / RAFTER	SEAT PLATE OR BEAM	LATERAL		WIND / EARTHQUAKE	
				PERP. TO WALL	PARAL. TO WALL	UPLIFT 133%	UPLIFT 160%
SGP2	14	14-16d	---	---	---	1455	1455
TAPL12	14&20	11-16d	4-10dx1 1/2"	1405	1405	1950	1950
TAP16	14	11-16d	---	595	210	1950	1950
TAPR216	14&20	11-16d	---	595	210	1950	1950
HDA6	1/4"	(2) 3/4"	(2) 3/4"	---	---	4256	4256

PRODUCT CODE	GAUGE	FASTENER SCHEDULE			ALLOWABLE LOADS			
		TRUSS / RAFTER	PLATE	STUD	LATERAL		WIND / EARTHQUAKE	
					PERP. TO WALL	PARAL. TO WALL	UPLIFT 133%	UPLIFT 160%
HDPT2	18	---	12-16d	---	450	450	1915	2300
RT10	20	5-8dx1 1/2"	8-8d	5-8dx1 1/2"	95	115	555	555
HCPLR	18	4-8d	4-8d	4-8d	95	145	510	520
HCPRF	18	6-10d	6-10d	6-10d	395	235	540	540
RTPGA818T	14	9-16d	---	9-16d	---	---	1360	1635
HCPFA	16	---	8-16d	8-16d	---	---	1200	1415
TPP4	20	---	---	8-10d	---	---	1290	1335

PRODUCT CODE	GAUGE	FASTENER SCHEDULE		ALLOWABLE LOADS	
		STUD / TOP	SILL	WIND / EARTHQUAKE	
				UPLIFT 133%	UPLIFT 160%
TAP18	14	(12) 16d		1950	1950
ADS2	12	(2) 5/8"	(1) 5/8"	2775	3330
FA3	16	(4) 8dX1 1/2"	(2) 8dX1 1/2"	1155	1155
FTP42*	12	(22) 16d		4050	4050
FAS118	18	(4) 8dX1 1/2"	(4) 8dX1 1/2"	755	755
SGP2	14	(14) 16d		1455	1455
ABJBL10W	---			1635	1635

Fig. 7-137   *Characteristics of metal connectors.* (SEMCO.)

Fig. 7-138 *Wood-to-concrete foundation connections.* *(SEMCO.)*

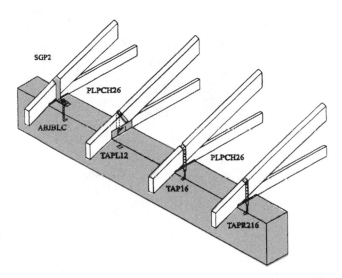

Fig. 7-141 *Poured masonry heading with connectors holding the trusses in place.* *(SEMCO.)*

Fig. 7-139 *Stem-wall-system connectors.* *(SEMCO.)*

Fig. 7-142 *Wood frame second-floor connections.* *(SEMCO.)*

Fig. 7-140 *Wood-to-wood connectors.* *(SEMCO.)*

Fig. 7-143 *Top plate, truss, or rafter connections with connectors making the truss to top plate a little more secure.* *(SEMCO.)*

Fig. 7-144 *Various connectors used in the top-plate-to-stud construction. (SEMCO.)*

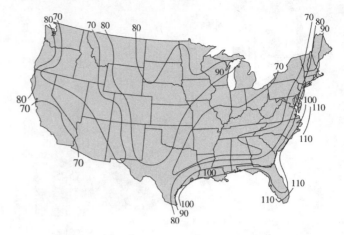

Fig. 7-145 *Basic wind-speed map for continental United States.*

CHAPTER 7 STUDY QUESTIONS

1. List at least five types of roof lines.
2. What is the difference between a hip roof and a mansard roof?
3. What are the parts of a roof frame?
4. Where are trussed rafters commonly used?
5. What is a disadvantage of a truss roof?
6. What is an advantage of the truss type of construction?
7. What is used to cover trusses?
8. What is a framing square?
9. Identify the following parts of a steel square:
    a. Body
    b. Tongue
    c. Heel
    d. Face
10. How can you make use of the hundredths scale on a square?
11. What is the difference between the octagon and brace scales on a square?
12. Identify the following terms:
    a. Shed roof
    b. Gable or pitch roof
    c. Valley roof
13. What do the following terms mean:
    a. Span
    b. Run
    c. Rise
    d. Pitch
14. Identify the following roof frame members:
    a. Plate
    b. Ridge board
    c. Common rafters
    d. Hip rafters
    e. Jack rafters
    f. Valley rafters
15. What part of a rafter is the tail?
16. What is the difference between a valley rafter and a hip rafter?
17. How do you describe a jack rafter?
18. What is a bay window?
19. What method does a builder use to minimize high wind damage to a house?
20. In what areas of the country are metal fasteners or connectors standard requirements?

8
CHAPTER

Roofs and Roofing

IN THIS CHAPTER YOU WILL LEARN HOW THE CARPENTER covers roofs and how the roof is prepared for shingles. Also, you will learn how to place shingles on a prepared surface. In addition, details for applying an asphalt shingle roof are given. Skills covered include how to

• Prepare a roof deck for shingles

• Apply shingles to a roof deck

• Estimate the number of shingles needed for a job

• Figure slope of a roof

ROOFING

Roofing shingles are made in different sizes and shapes. Roofs have different angles and shapes. Pipes stick up through the roof. Roofing has to be fitted. Crickets also need to be fitted. (A cricket fits between a chimney and the roof.) A number of fine details are presented by roof shapes. All these have to be considered by the roofer.

Some industrial and commercial buildings use roll shingles. This material requires a slightly different approach. Asphalt shingles are safer than wooden shingles. The asphalt shingles will resist fire longer. Because of fire regulations, wooden shingles are not allowed in some sections of the country.

SEQUENCE

The carpenter should apply a roof in this order:

1. Check the deck for proper installation.
2. Decide which shingles to use for the job.
3. Estimate the amount needed for the job.
4. Apply drip strips.
5. Place the underlayment.
6. Nail the underlayment.
7. Start the first course of shingles.
8. Continue other courses of shingles.
9. Cut and install flashings.
   a. Valleys:
      (1) Open
      (2) Closed-cut
      (3) Woven
   b. Soil pipe flashing
   c. Chimney flashing
   d. Other flashings
10. Cover ridges.
11. Cover all nailheads with cement.
12. Glue down tabs, if needed.

FLAT ROOF SHED OR LEAN-TO ROOF GABLE ROOF

HIP ROOF GABLE ROOF AND DORMER

"DUTCH" HIP

GAMBREL ROOF

BUTTERFLY ROOF

GABLE AND VALLEY ROOF HIP-AND-VALLEY ROOF MANSARD ROOF

Fig. 8-1 *Various types of roof shapes.*

Types of Roofs

There are a number of roof types; each is classified according to its shape. Figure 8-1 shows the different types. Each type presents roofing problems, and different methods are used to cover the decking, ridges, and drip areas.

The *mansard roof* presents some unique problems. Figure 8-2 shows a mansard roof. Note that the dark area is covered with shingles. The angles presented by the various vertical and sloping sections need special bracing. Attaching the shingles also requires attention to detail on the vertical sections. (Refer back to Figure 7-119 for an illustration showing truss construction on a mansard-style roof.) The attached garage in Fig. 8-2 has a hip roof. Figure 8-3 also shows a hip roof. Note how the entrance is also a hip but shorter.

Fig. 8-2 *Mansard roof with hip on garage.*

Fig. 8-3 *Hip-and-valley roof.*

The *gable roof* is a common roof type (Fig. 8-4). It is a simple roof that is easy to build. Figure 8-5 shows a variety of gable roofs. Each is a complete unit. The garage shows the angles of this type of roof very well.

Fig. 8-4 *Gable roof.*

Fig. 8-5 *Gable roof with add-ons.*

Drainage Factors

The main purpose of a roof is to protect the inside of a building. This is done by draining the water from the roof. The water goes onto the ground or into the storm sewer. Some parts of the country allow the water to be dropped onto the earth below. Other sections require the collected water to be moved to a storm sewer. The main idea is to prevent water seepage. Roof water should not seep back into a basement.

Ice is frequently a problem in colder climates. Ice forms and makes a dam for melting snow (Fig. 8-6A). Water backs up under the shingles and leaks into the ceiling below. This problem can be caused by insufficient insulation. The lack of soffit ventilation also will cause leaks. Heating cables can be installed to prevent ice dams. Leaking can be prevented by adding ventilation. If insulation can be added, this too should be done (Fig. 8-6B).

The point where rooflines come together is called a *valley* (Fig. 8-7). Valleys direct water to the drain. This keeps it out of the house. Valleys need special attention during roofing.

Fig. 8-6 *A. Water leakage caused by an ice dam. B. Using ventilation and insulation to prevent leaks.*

Eave troughs and downspouts drain water from the roof. It is drained into gutters. Downspouts carry the water to the ground. Downspouts connect to other pipes. That piping sometimes goes to the street storm drain. This eliminates seepage into the basement or under the slab (Fig. 8-8A). In most cases, a splash block is located under the downspout to disperse the water (Fig. 8-8B).

Roofing Terms

A number of roofing terms are used by roofers and carpenters. You should become familiar with these terms; you then will be able to talk with roofing salespersons.

Square Shingles needed to cover 100 square feet of roof surface. This means 10 feet *square*, or 10 × 10 feet.

Exposure Distance between exposed edges of overlapping shingles. Exposure is given in inches (Fig. 8-9). Note the 5- and 3-inch exposures in the figure.

Head lap Distance between the top of the bottom shingle and the bottom edge of the one covering it (Fig. 8-10).

Top lap Distance between the lower edge of an overlapping shingle and the upper edge of the lapping shingle (see Fig. 8-10). Top lap is measured in inches.

Side lap Distance between adjacent shingles that overlap. Measured in inches.

Valley Angle formed by two roofs meeting. The internal part of the angle is the valley.

Rake On a gable roof, the inclined edge of the surface to be covered.

Flashing Metal used to cover chimneys and around things projecting through the roofing. Used to keep the weather out.

Underlayment Usually No. 15 or No. 30 felt paper applied to a roof deck. It goes between the wood and the shingles.

Ridge The horizontal line formed by the two rafters of a sloping roof being nailed together.

Hip The external angle formed by two sides of a roof meeting.

Roofing is part of the exterior building. The carpenter is called on to place the covering over a frame. This frame is usually covered by plywood. Plywood comes in 4- × 8-foot sheets and can be installed quickly onto the rafters. Sheathing may be 1- × 6-inch or 1- × 10-inch boards. Sheathing takes longer to install than plywood. The frame is covered with a tar or felt paper. This paper goes on over the sheathing. The paper allows moisture to move from the wood upward. Moisture then escapes under the shingles. This prevents a buildup of moisture. If the weather is bad, moisture can freeze and form frost under the shingles.

Shingles of wood, asphalt, asbestos, and Fiberglas are used for roofing. Tile and slate were once commonly used, but they are rather expensive to install nowadays. Copper, galvanized iron, and tin are also used as roof coverings.

Commercial buildings may use a built-up roof, which has a number of layers. This type uses a gravel topping or cap sheet. Asphalt-saturated felt is mopped down with hot asphalt or tar (see Fig. 8-8C). Choice of roofing is determined by three things: cost, slope, and life expected. In some local climates (e.g., where there is wind, rain, and snow), flat roofs may have to be rejected.

Fig. 8-7 *A. Closed-cut valley. B. Woven valley. C. Open valley.*

In certain applications, such as homes, appearance is another important consideration. Shingles are used most frequently for homes (see Fig. 8-8D). In some areas, cedar shingles are not permitted because wood burns too easily. Once aged, however, cedar becomes more fire resistant.

Pitch

Drainage of water from a roof surface is essential. This means that pitch should be considered. The pitch or slope of a roof deck determines the choice of shingles. It also determines drainage.

Fig. 8-8 *A. Eave trough and downspout. B. Downspout elbow turns water away from the basement. C. Preparing a flat roof. Asphalt-saturated felt is mopped down with hot asphalt or tar. D. Applying a shingle roof. The shingles are packaged so that they can be placed in a location convenient for the roofer.*

Fig. 8-9 *Exposure is the distance between the exposed edges of overlapping shingles.* (Bird and Son.)

Fig. 8-10 *Head lap and top lap.* (Bird and Son.)

Pitch limitations are shown in Fig. 8-11. Any shingle may be used safely on roofs with normal slopes. Normal is 4 inches of rise or more per horizontal foot. An exception exists for square-butt strip shingles. They may be used on slopes included in the shaded area in Fig. 8-11.

When the pitch is less than 4 inches per foot, it is best to use roll roofings. In the range of 4 inches down to 1 inch per foot, the following rules apply:

• Roll roofing may be applied by the exposed-nail method if the pitch is not lower than 2 inches per foot.

• Roll roofings applied by the concealed-nail method may be used on pitches down to, but not below, 1 inch per foot. This is true if (1) they have at least 3 inches of top lap, and (2) they have double-coverage roofing with a top lap of 19 inches.

Either of the preceding may be applied on a deck with a pitch steeper than the stated minimum. Pitch is given as a fraction. For example, a roof has a rise of 8 feet and a run of 12 feet. Then its pitch is

$$\frac{8}{2 \times 12} \quad \text{or} \quad \frac{8}{24} \quad \text{or} \quad \frac{1}{3}$$

Pitch is equal to the rise divided by two times the run or

$$\text{Pitch} = \frac{\text{rise}}{2 \times \text{run}}$$

Slope

Slope is how fast the roof rises from the horizontal (Fig. 8-12). Slope is equal to the rise divided by the run or

$$\text{Slope} = \frac{\text{rise}}{\text{run}}$$

Slope and *pitch* are often used interchangeably. However, you can see that there is a difference. Some roofers' manuals use them as if they were the same.

Before a roof can be applied, you have to know how many shingles are needed. This calls for estimating the area to be covered. First, determine the number of square feet. Then divide the number of square feet by 100 to produce the number of squares needed.

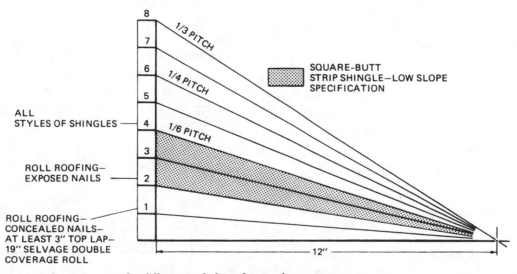

Fig. 8-11 *Minimum pitch requirements for different asphalt roofing products.* (Bird and Son.)

SLOPE	PITCH
2 in 12	1/12
3 in 12	1/8
4 in 12	1/6
5 in 12	5/24
6 in 12	1/4
7 in 12	7/24
8 in 12	1/3
10 in 12	5/12
12 in 12	1/2

Assume:
Rise = 4'; Run = 12'

Slope: 4/12 or 4 in 12

Pitch: $\dfrac{4}{2 \times 12} = \dfrac{4}{24} = \dfrac{1}{6}$

Fig. 8-12 *Slope, pitch, and run of a roof.*

ESTIMATING ROOFING QUANTITIES

Roofing is estimated and sold in squares. A *square* of roofing is the amount required to cover 100 square feet. To estimate the required amount, you have to compute the total area to be covered. This should be done in square feet. Then divide the amount by 100. This determines the number of squares needed. Some allowance should be made for cutting and waste. This allowance is usually 10 percent. If you use 10 percent for waste and cutting, you will have the correct number of shingles. A simple roof with no dormers will require less than 10 percent. Complicated roofs will require more than 10 percent for cutting and fitting.

Estimating Area

The areas of simple surfaces can be computed easily. The area of the shed roof in Fig. 8-13 is the product of the eave line and the rake line ($A \times B$). The area of the simple gable roof in Fig. 8-13 equals the sum of the two rakes B and C multiplied by the eave line A. A gambrel roof is estimated by multiplying rake lines A, B, C, and D by eave line E (see Fig. 8-13).

Complications arise in roofs such as the one in Fig. 8-14. Ells, gables, or dormers can cause special problems. Obtain the lengths of the eaves, rakes, valleys, and ridges from drawings or sketches. Measuring calls for dangerous climbing. You may want to estimate without climbing. To do this,

1. The pitch of the roof must be known or determined.
2. The horizontal area in square feet covered by the roof must be computed.

Pitch is shown in Fig. 8-15. The pitch of a roof is stated as the relationship between rise and span. If the

Fig. 8-13 *Simple roof types and their dimensions: shed, gable, and gambrel.* (Bird and Son.)

Fig. 8-14 *Complicated dwelling roof shown in perspective and plan views.* (Bird and Son.)

span is 24 feet, 0 inches and the rise is 8 feet, 0 inches, the pitch will be 8/24, or 1/3. If the rise were 6 feet, 0 inches, then the pitch would be 6/24, or 1/4. The one-third pitch roof rises 8 inches per foot of horizontal run. The one-quarter pitch roof rises 6 inches per foot of run.

You can determine the pitch of any roof without leaving the ground. Use a carpenter's folding rule in

Fig. 8-15 *Pitch relations.*

Fig. 8-16 *Using a carpenter's rule to find the roof pitch.* (Bird and Son.)

the following manner: Form a triangle with the rule. Stand across the street or road from the building. Hold the rule at arm's length. Align the roof slope with the sides of the rule. Be sure that the base of the triangle is held horizontal. It will appear within the triangle as shown in Fig. 8-16. Take a reading on the base section of the rule. Note the reading point shown in Fig. 8-17. Locate in the top line, headed "Rule Reading," in Fig. 8-18 the point nearest your reading. Below this point is the pitch and the rise per foot of run. Here the reading on the rule is 22. Under 22 in Fig. 8-18, the pitch is designated as a rise of 8 inches per foot of horizontal run.

READING POINT

Fig. 8-17 *Reading the carpenter's rule to find the point needed for the pitch figures.* (Bird and Son.)

Horizontal Area

Figure 8-14 shows a typical dwelling. The roof has valleys, dormers, and variable-height ridges. Below the perspective, the total ground area is covered by the roof. All measurements needed can be made from the ground. Or they can be made within the attic space of the house. No climbing on the roof is needed.

Computation of Roof Areas

Make all measurements. Draw a roof plan. Determine the pitches of the various elements of the roof. Use a carpenter's rule. The horizontal areas now can be worked out quickly.

Include in the estimate only the areas having the same pitch. The rise of the main roof is 9 inches per foot. That of the ell and dormers is 6 inches per foot. The horizontal area under the 8-inch-slope roof will be

$$26 \times 30 = \quad 780 \text{ square feet}$$
$$19 \times 30 = \quad \underline{570} \text{ square feet}$$
$$\text{Total} \quad 1350 \text{ square feet}$$

Less

$$8 \times 5 = 40 \text{ (triangular area under ell roof)}$$
$$4 \times 4 = \underline{16} \text{ (chimney)}$$
$$56 \text{ square feet}$$

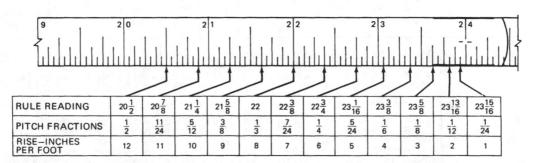

RULE READING	$20\frac{1}{2}$	$20\frac{7}{8}$	$21\frac{1}{4}$	$21\frac{5}{8}$	22	$22\frac{3}{8}$	$22\frac{3}{4}$	$23\frac{1}{16}$	$23\frac{3}{8}$	$23\frac{5}{8}$	$23\frac{13}{16}$	$23\frac{15}{16}$
PITCH FRACTIONS	$\frac{1}{2}$	$\frac{11}{24}$	$\frac{5}{12}$	$\frac{3}{8}$	$\frac{1}{3}$	$\frac{7}{24}$	$\frac{1}{4}$	$\frac{5}{24}$	$\frac{1}{6}$	$\frac{1}{8}$	$\frac{1}{12}$	$\frac{1}{24}$
RISE—INCHES PER FOOT	12	11	10	9	8	7	6	5	4	3	2	1

Fig. 8-18 *Reading point converted to pitch.* (Bird and Son.)

1350 − 56 = 1294 square feet total

The area under the 6-inch-rise roof will be

$$20 \times 30 = 600 \text{ square feet}$$
$$\underline{8 \times \ \ 5 = \ \ \underline{40}} \text{ (triangular area projecting over the}$$
$$640 \text{ main house)}$$

Duplications

Sometimes one element of a roof projects over another. Add duplicated areas to the total horizontal area. If the eaves in Fig. 8-14 project only 4 inches, there will be

1. A duplication of 2 (7 × ⅓ = 4⅔ square feet under the eaves of the main house). This is where the rake of the ell section is overhung.

2. A duplication under the dormer eaves of 2 (5 × ⅓ = 3⅓ square feet).

3. A duplication of 9½ × ⅓ = 3¹⁄₁₆ square feet under the eaves of the main house. This is where the rake of the ell section is overhung.

The total is 11⅛ or 12 square feet. Item 2 should be added to the area of the 6-inch-pitch roof. Items 1 and 3 should be added to the 9-inch-pitch roof. The new totals will be 640 + 4 = 644 square feet for the 6-inch pitch and 1,294 + 8 = 1,302 square feet for the 9-inch pitch.

Converting Horizontal to Slope Areas

Now convert horizontal areas to slope areas. Do this with the aid of the *conversion table* (Table 8-1). Hori-

TABLE 8-1 *Conversion Table*

Rise, Inches per Foot of Horizontal Run	1	2	3	4	5	6	7	8	9	10	11	12
Pitch, Fractions	1/24	1/12	1/8	1/6	5/24	1/4	7/24	1/3	3/8	5/12	11/24	1/2
Conversion Factor	1.004	1.014	1.031	1.054	1.083	1.118	1.157	1.202	1.250	1.302	1.356	1.414
Horizontal, area in square feet or length in feet												
1	1.0	1.0	1.0	1.1	1.1	1.1	1.2	1.2	1.3	1.3	1.4	1.4
2	2.0	2.0	2.1	2.1	2.2	2.2	3.2	2.4	2.5	2.6	2.7	2.8
3	3.0	3.0	3.1	3.2	3.2	3.2	3.5	3.6	3.8	3.9	4.1	4.2
4	4.0	4.1	4.1	4.2	4.3	4.5	4.6	4.8	5.0	5.2	5.4	5.7
5	5.0	5.1	5.2	5.3	5.4	5.6	5.8	6.0	6.3	6.5	6.8	7.1
6	6.0	6.1	6.2	6.3	6.5	6.7	6.9	7.2	7.5	7.8	8.1	8.5
7	7.0	7.1	7.2	7.4	7.6	7.8	8.1	8.4	8.8	9.1	9.5	9.9
8	8.0	8.1	8.3	8.4	8.7	8.9	9.3	9.6	10.0	10.4	10.8	11.3
9	9.0	9.1	9.3	9.5	9.7	10.1	10.4	10.8	11.3	11.7	12.2	12.7
10	10.0	10.1	10.3	10.5	10.8	11.2	11.6	12.0	12.5	13.0	13.6	14.1
20	20.1	20.3	20.6	21.1	21.7	22.4	23.1	24.0	25.0	26.0	27.1	28.3
30	30.1	30.4	31.0	31.6	32.5	33.5	34.7	36.1	37.5	39.1	40.7	42.4
40	40.2	40.6	41.2	42.2	43.3	44.7	46.3	48.1	50.0	52.1	54.2	56.6
50	50.2	50.7	51.6	52.7	54.2	55.9	57.8	60.1	62.5	65.1	67.8	70.7
60	60.2	60.8	61.9	63.2	65.0	67.1	69.4	72.1	75.0	78.1	81.4	84.8
70	70.3	71.0	72.2	73.8	75.8	78.3	81.0	84.1	87.5	91.1	94.9	99.0
80	80.3	81.1	82.5	84.3	86.6	89.4	92.6	96.2	100.0	104.2	108.5	113.1
90	90.4	91.3	92.8	94.9	97.5	100.6	104.1	108.2	112.5	117.2	122.0	127.3
100	100.4	101.4	103.1	105.4	108.3	111.8	115.7	120.2	125.0	130.2	135.6	141.4
200	200.8	202.8	206.2	210.8	216.6	223.6	231.4	240.4	250.0	260.4	271.2	282.8
300	301.2	304.2	309.3	316.2	324.9	335.4	347.1	360.6	375.0	390.6	406.8	424.2
400	401.6	405.6	412.4	421.6	433.2	447.2	462.8	480.8	500.0	520.8	542.4	565.6
500	502.0	507.0	515.5	527.0	541.5	559.0	578.5	601.0	625.0	651.0	678.0	707.0
600	602.4	608.4	618.6	632.4	649.8	670.8	694.2	721.2	750.0	781.2	813.6	848.4
700	702.8	709.8	721.7	737.8	758.1	782.6	809.9	841.4	875.0	911.4	949.2	989.8
800	803.2	811.2	824.8	843.2	864.4	894.4	925.6	961.6	1000.0	1041.6	1084.8	1131.2
900	903.6	912.6	927.9	948.6	974.7	1006.2	1041.3	1081.8	1125.0	1171.8	1220.4	1272.6
1000	1004.0	1014.0	1031.0	1054.0	1083.0	1118.0	1157.0	1202.0	1250.0	1302.0	1356.0	1414.0

zontal areas are given in the first column. Corresponding slope areas are given in columns 2 through 12.

The total area under the 9-inch rise is 1,302 square feet. Under the column headed "9" (for 9-inch rise) on the conversion table is found:

	Horizontal Area		Slope Area
Opposite	1000	is	1250.0
Opposite	300	is	375.0
Opposite	00	is	00.0
Opposite	2	is	2.5
Totals	1302		1627.5

The total area under the 6-inch rise is 644 square feet.

	Horizontal Area		Slope Area
Opposite	600	is	670.8
Opposite	40	is	44.7
Opposite	4	is	4.5
Totals	644		720.0

The total area for both pitches is 1,627.5 + 720 = 2,347.5 square feet.

Now add a percentage for waste. This should be 10 percent. This brings the total area of roofing required to 2,582 square feet. Divide 2,582 by 100 and get 25.82 or, rounded, 26 squares.

One point about this method should be emphasized. The method is possible because of one fact: Over any given horizontal area, at a given pitch, a roof always will contain the same number of square feet regardless of its design. A shed, a gable, and a hip roof, with or without dormers, each will require exactly the same square footage of roofing—that is, if each is placed over the same horizontal area with the same pitch.

Accessories

Quantities of starter strips, edging strips, ridge shingles, and valley strips all depend on linear measurements. These measurements are taken along the eaves, rake ridge, and valley. The eaves and ridge are horizontal. The rakes and valleys run on a slope. Quantities for the horizontal elements can be taken off the roof plan. True lengths of rakes and valleys must be taken from conversion tables.

LENGTH OF RAKE

Determine the length of the rake of the roof. Measure the horizontal distance over which it extends. In this case, the rakes on the ends of the main house span distances of 26 and 19 feet. More rake footage is 26 + 19 + 13 + 3½ ± 61½ feet.

Refer to Table 8-1 under the 9-inch-rise column. Opposite the figures in column 1, find the length of the rake:

	Horizontal Run	Length of Rake
	60	75.00
	1	1.3
	0.5	0.6
Totals	61.5	76.9 (actual length of the rake)

Apply the same method to the rake of the ell. This will indicate its length, including the dormer, to be 39.1 inches. Add these amounts to the total length of the eaves. The figure obtained can be used for an estimate of the amount of edging needed.

Hips and Valleys

Hip-and-valley lengths can be determined. Use the run off the common rafter. Then refer to the hip-and-valley table (Table 8-2).

Common rafter run is one-half the horizontal distance that the roof spans. This determines the length of a valley. The run of the common rafter should be taken at the lower end of the valley.

Figure 8-14 shows the portion of the ell roof that projects over the main roof. It has a span of 16 feet at the lower end of the valley. Therefore, the common rafter at this point has a run of 8 feet, 0 inches.

There are two valleys at this roof intersection. The total run of the common rafter is 16 feet, 0 inches (refer to Table 8-2). Opposite the figures in the column headed "Horizontal," find the linear feet of the valleys. Then check the column under the pitch involved.

One of the intersecting roofs has a rise of 6 inches. The other has a rise of 9 inches. Length for each rise must be found. The average of the two then is taken. This gives a close approximation of the true length of the valley. Thus

Horizontal	6-inch rise	9-inch rise
10	15	16
6	9	9.6
16	24	25.6

24 + 25.6 = 49.6

49.6 ÷ 2 = 24.8 length of valleys

TABLE 8-2 *Hip-and-Valley Conversions*

Rise, Inches per Foot of Horizontal Run	4	5	6	7	8	9	10	11	12	14	16	18
Pitch — Degrees	18°26'	22°37'	26°34'	30°16'	33°41'	36°52'	39°48'	42°31'	45°	49°24'	53°8'	56°19'
Pitch — Fractions	1/6	5/24	1/4	7/24	1/3	3/8	5/12	11/24	1/2	7/12	2/3	3/4
Conversion Factor	1.452	1.474	1.500	1.524	1.564	1.600	1.642	1.684	1.732	1.814	1.944	2.062
Horizontal Length in Feet												
1	1.5	1.5	1.5	1.5	1.6	1.6	1.6	1.7	1.7	1.8	1.9	2.1
2	2.9	2.9	3.0	3.0	3.1	3.2	3.3	3.4	3.5	3.6	3.9	4.1
3	4.4	4.4	4.5	4.6	4.7	4.8	4.9	5.1	5.2	5.4	5.8	6.2
4	5.8	5.9	6.0	6.1	6.3	6.4	6.6	6.7	6.9	7.3	7.8	8.2
5	7.3	7.4	7.5	7.6	7.8	8.0	8.2	8.4	8.7	9.1	9.7	10.3
6	8.7	8.8	9.0	9.1	9.4	9.6	9.9	10.1	10.4	10.9	11.7	12.4
7	10.2	10.3	10.5	10.7	10.9	11.2	11.5	11.8	12.1	12.7	13.6	14.4
8	11.6	11.8	12.0	12.2	12.5	12.8	13.1	13.5	13.9	14.5	15.6	16.5
9	13.1	13.3	13.5	13.7	14.1	14.4	14.8	15.2	15.6	16.3	17.5	18.6
10	14.5	14.7	15.0	15.2	15.6	16.0	16.4	16.8	17.3	18.1	19.4	20.6
20	29.0	29.5	30.0	30.5	31.3	32.0	32.8	33.7	34.6	36.3	38.9	41.2
30	43.6	44.2	45.0	45.7	46.9	48.0	49.3	50.5	52.0	54.4	58.3	61.9
40	58.1	59.0	60.0	61.0	62.6	64.0	65.7	67.4	69.3	72.6	77.8	82.5
50	72.6	73.7	75.0	76.2	78.2	80.0	82.1	84.2	86.6	90.7	97.2	103.1
60	87.1	88.4	90.0	91.4	93.8	96.0	98.5	101.0	103.9	108.8	116.6	123.7
70	101.6	103.2	105.0	106.7	109.5	112.0	114.9	117.9	121.2	127.0	136.1	144.3
80	116.2	117.9	120.9	121.9	125.1	128.0	131.4	134.7	138.6	145.1	155.5	165.0
90	130.7	132.7	135.0	137.2	140.8	144.0	147.8	151.6	155.9	163.3	175.0	185.6
100	145.2	147.4	150.0	152.4	156.4	160.0	164.2	168.4	173.2	181.4	194.4	206.2

Dormer Valleys

The run of the common rafter at the dormer is 2.5 feet. Check Table 8-2. It is found that

Horizontal	6-inch rise
2.0	3.0
0.5	0.75
2.5	3.75 (length of valley)

Two such valleys will total 7.5 feet.

The total length of valley will be 24.8 + 7.5 = 32.3 feet. Use these figures to estimate the flashing material required.

ROOFING TOOLS

Most roofing tools are already in the carpenter's toolbox. All the tools needed for roofing are shown in Fig. 8-19. Here's the list:

Roof brackets. Can be used to clamp onto ladder.

Ladders. A pair of sturdy ladders with ladder jacks are needed. Shingles are placed on the roof using a hoist on the delivery truck. These ladders come in handy for side roofing.

Staging. These are planks for the ladder jacks. They hold the roofer or shingles. They are very useful on mansard roof jobs.

Apron. The carpenter's apron is very necessary. It holds the nails and hammer. Other small tools can fit into it. It saves time in many ways. It keeps needed tools handy.

Hammer. The hammer is a necessary device for roofing. It should be a balanced hammer for less wrist fatigue.

Chalk and line. This combination is needed to draw guidelines. Shingles need alignment. The chalk marks are needed to make sure that the shingles line up.

Tin snips. Heavy-duty tin snips are needed for trimming flashing. They also can be used for trimming shingles.

Kerosene. A cleaner is needed to remove tar from tools. Asphalt from shingles can be removed from tools with kerosene.

Tape measure. A roofer has to make many measurements. This is a necessary tool.

Utility knife. A general-purpose knife is needed for close trimming of shingles.

Putty knife. This is used to spread roofing cement.

Carpenter's rule. This makes measurements and also serves to determine the pitch of a roof (see Fig. 8-16).

Stapler. Some new-construction roofing can use a stapler. This device replaces the hammer and nails.

Fig. 8-19 *Carpenter's tools needed for roofing: A. planking support; B. ladder with planking; C. claw hammer; D. carpenter's apron; E. chalk and cord; F. snips; G. tape measure; H. kerosene; I. stapler; J. carpenter's folding rule; K. utility knife; L. putty knife.* (Bird and Son.)

SAFETY

Working on a roof can be dangerous. Here are a few helpful hints. They may save you broken bones or pulled muscles.

1. Wear sneakers or rubber-soled shoes.
2. Secure ladders and staging firmly.
3. Stay off wet roofs.
4. Keep away from power lines.
5. Don't let debris accumulate underfoot.
6. Use roofing brackets and planks if the roof slopes 4 inches or more for every 12 inches of horizontal run.

APPEARANCE

How the finished job looks is important. Here are a few precautions to improve roof appearance:

1. Avoid shingling in extremely hot weather. Soft asphalt shingles are easily marred by shoes and tools.

2. Avoid shingling when the temperature is below 40°F. Cold shingles are stiff and may crack.

3. Measure carefully, and snap the chalk line frequently. Roof surfaces aren't always square. You'll want to know about problems to come so that you can correct them.

4. Start at the rear of the structure. If you've never shingled before, this will give you a chance to gain experience before you reach the front.

APPLYING AN ASPHALT ROOF

Asphalt roofing products will serve well when they are applied correctly. Certain fundamentals must be considered. These have to do with the deck, flashing, and application of materials.

Roof Problems

A number of roof problems are caused by defects in the deck. A nonrigid deck may affect the lay of the roofing. Poorly seasoned deck lumber may warp. This can cause cocking of the shingle tabs. It also can cause wrinkling and buckling of roll roofing.

Improper ventilation can have an effect similar to that of green lumber. The attic area should be ventilated. This area is located directly under the roof deck. It should be free of moisture. In cold weather, be sure that the interiors are well ventilated. This applies when plaster is used in the building. A positive ventilation of air is required through the building during roofing. This usually can be provided by opening one or two windows. Windows in the basement or on the first floor can be opened. This can create a positive draft through the house. Open windows at opposite ends of the building. This also will create a flow of air. The moving air has a tendency to dry out the roof deck. This helps to eliminate excess moisture. Condensation under the roofing can cause problems.

Deck construction Wood decks should be made from well-seasoned tongue-and-groove lumber that is more than 6 inches wide. Wider sheathing boards are more likely to swell or shrink, producing a buckling of the roof material. Sheathing should be tightly matched. It should be secured to the rafter with at least two 8d nails. One should be driven through the edge of the board. The other should be driven through the board face. Boards containing too many resinous areas should be rejected. Boards with loose knots should not be used. Do not use badly warped boards.

Figure 8-20 shows how a wood roof deck is constructed. In most cases today, 4- × 8-foot sheets of plywood are used as sheathing. The plywood goes over the rafters. C–D grade plywood is used.

Underlayment Apply one layer of No. 15 asphalt-saturated felt over the deck as an underlayment if the deck has a pitch of 4 inches per foot or greater. The felt should be laid horizontally (see Fig. 8-20). Do not use No. 30 asphalt felt. Also do not use any tar-saturated felt. Laminated waterproof papers should not be used either. Do not use any vapor barrier type of material. Lay each course of felt over the lower course. Lap the courses 4 inches. Overlap should be at least 2 inches where ends join. Lap the felt 6 inches from both sides over all hips and ridges.

1 X 6 MATCHED SHEATHING
END JOINTS STAGGERED
TWO 8d NAILS AT EACH RAFTER
ATTIC SPACE MUST BE WELL VENTILATED
2"
4"
NO. 15 ASPHALT-SATURATED FELT
METAL PATCH OVER DEFECT

Fig. 8-20 *Features of a good wood roof deck.* (Bird and Son.)

Apply underlayment as specified for low-slope roofs, where the roof slope is less than 4 inches per foot and not below 2 inches. Check the maker's suggestions (Fig. 8-21). Felt underlayment performs three functions:

1. It ensures a dry roof for shingles. This avoids buckling and distortion of shingles. Buckling may be caused by shingles being placed over wet roof boards.

2. Felt underlay prevents the entrance of wind-driven rain onto the wood deck. This may happen when shingles are lifted up.

3. Underlayment prevents any direct contact between shingles and resinous areas. Resins may cause chemical reactions. These could damage the shingles.

Plywood decks Plywood decks should meet the Underwriters' Laboratories standards. Standards are set according to grade and thickness. Design the eaves, rake, and ridge to prevent problems. Openings through the deck should be made in such a way that the plywood will not be exposed to the weather (Fig. 8-22).

Fig. 8-21 *A. Eaves flashing for a low-slope roof. B. Placing of sheathing and drip edge. C. Placement of the underlayment for a shingle roof.* (Bird and Son.)

Fig. 8-22 *Preparing the roof deck for shingling.* (Bird and Son.)

(A)

(B)

Fig. 8-23 *A. Flashing patterns and in place around a chimney. B. Flashing around a soil pipe.* (Bird and Son.)

Nonwood deck materials Nonwood materials are sometimes used in decks. Such things as fiberboard, gypsum, concrete plank, and tile are nonwoods used for decks. These materials have their own standards. Check with the manufacturer for suggestions.

Flashings Roofs often are complicated by intersections with other roofs. Some adjoining walls have projections through the deck. Chimneys and soil stacks create leakage problems. Special attention must be paid to protecting against the weather here. Such precautions are commonly called *flashing*. Careful attention to flashing is critical. It helps to provide good roof performance (Fig. 8-23).

Valleys Valleys exist where two sloping roofs meet at an angle. This causes water runoff toward and along the joint. Drainage concentrates at the joint. This makes the joint an easy place for water to enter. Smooth, unobstructed drainage must be provided. The valley should have enough capacity to carry away the collected water.

There are three types of valleys: open, woven, and closed-cut (Fig. 8-24).

Each type of valley calls for its own treatment. Figure 8-25 shows felt being applied to a valley. A 36-inch-wide strip of No. 15 asphalt-saturated felt is centered in the valley. It is secured with nails. They hold it in place until shingles are applied. Courses of felt are cut to overlay the valley strip. The lap should not be less than 6 inches. Eave flashing then is applied.

PUTTING DOWN SHINGLES

Before you put the shingles down, you need an underlayment (Fig. 8-26). Covering the underlayment is a sheet of saturated felt or tar paper. Table 8-3 lists characteristics of typical asphalt rolls. It is best not to put down the first shingle until you know what is avail-

able. Study Table 8-4 to check the characteristics and sizes of typical asphalt shingles. Remember that the number symbol (#) means pound, as in 285# = 285 pounds per square of shingles. A square covers an area of 100 square feet.

Nails

Nails used in applying asphalt roofings are large headed and sharp pointed. Some are hot-galvanized steel. Others are made of aluminum. They may be barbed or otherwise deformed on the shanks. Figure 8-27 shows three types of asphalt nails.

Roofing nails should be long enough to penetrate through the roofing material. They should go at least ¾ inch into the wood deck. This requires that they be of the lengths indicated in Table 8-5.

Number of nails The number of nails required for each shingle type is given by its maker. Manufacturer's recommendations come with each bundle. Use 2-inch

EACH STRIP TO EXTEND AT LEAST 12" BEYOND CENTER OF VALLEY

EXTRA NAIL IN END OF STRIP

(A)

(B)

36" ROLL ROOFING 50# OR HEAVIER

6" MIN.

EACH STRIP TO EXTEND AT LEAST 12" BEYOND CENTER OF VALLEY

PLASTIC ASPHALT CEMENT

EXTRA NAIL IN END OF STRIP

(C)

NO. 15 FELT

12" LAP

18" STRIP FACE DOWN

36" STRIP FACE UP

PLASTIC ASPHALT CEMENT

PLASTIC ASPHALT CEMENT

12" LAP

CHALK LINES

Fig. 8-24 *A. Woven valley roof. B. Closed-cut valley. C. Preparing an open valley.* (Bird and Son.)

centers in applying roll roofing. This means that 252 nails are needed per square. If 3-inch centers are used, then 168 nails are needed per square.

Fasteners for Nonwood Materials

Gypsum products, concrete plank and tile, fiberboard, or unusual materials require special fasteners. This

Fig. 8-25 *Felt underlay centered in the valley before valley linings are applied.* (Bird and Son.)

Fig. 8-26 *A. Underlayment and drip edge. B. Eaves flashing for a roof. C. Underlayment, eaves flashing, and metal drip edge.* (Bird and Son.)

type of deck varies with its manufacturer. In such cases, follow the manufacturer's suggestions.

Shingle Selection

A number of types of shingles are available. They may be used for almost any type of roof. Various colors are used to harmonize with buildings. Some are made for various weather conditions. White and light colors are used to reflect the sun's rays. Pastel shingles are used to achieve a high degree of reflectivity. They still permit color blending with siding and trim. Fire and wind resistance should be considered. Simplicity of application makes asphalt roofings the most popular in new housing. Asphalt roofings are also rated high for reroofing.

Farm buildings   There is no one kind of asphalt roofing for every job. Building types are numerous. The style of roof on a farmhouse may affect the choice of roofing on other buildings. They probably should have the same color roofing. This would make the group harmonize. A poultry laying house or machine storage shed near the house calls for a roof like the farmhouse. An inexpensive roll roofing might be used for an isolated building.

The main idea is to select the right product for the building. The main reason for selecting a roofing mater-

TABLE 8-3 *Typical Asphalt Rolls*

1	2		3	4		5		6	7
Product	Approximate Shipping Weight		Squares Per Package	Length, Feet	Width, Inches	Side or End Lap, Inches	Top Lap, Inches	Exposure, Inches	Underwriters' Listing
	Per Roll	Per Square							
Mineral surface roll double coverage	75# to 90#	75# to 90#	1	36 38	36 38	6	2 4	34 32	C
	Available in some areas in 9/10 or 3/4 square rolls.								
Mineral surface roll	55# to 70#	55# to 70#	1/2	36	36	6	19	17	C
Coated roll	50# to 65#	50# to 65#	1	36	36	6	2	34	None
Saturated felt	60# 60# 60#	15# 20# 30#	4 3 2	144 108 72	36 36 36	4 to 6	2	34	None

ial is protection of the contents of a building. The second reason for selecting a roofing is low maintenance cost.

Staples Staples may be used as an alternative to nails. This is only for new buildings. Staples must be zinc coated. They should be no less than 16 gauge. A semiflattened elliptical cross section is preferred. They should be long enough to penetrate ¾ inch into the wood deck. They must be driven with *pneumatic* (air-driven) staplers. The staple crown must bear tightly against the shingle. However, it must not cut the shingle surface. Use four staples per shingle (Fig. 8-28). The crown of the staple must be parallel to the tab edge. Position it as shown in the figure. Figure 8-29 shows how shingles are nailed. Figure 8-30 shows how the shingles are overlapped.

Cements

Six types of asphalt coatings and cements include

1. Plastic asphalt cements
2. Lap cements

Fig. 8-27 *A. Screw-threaded nail. B. Annular threaded nail. C. Asphalt shingle nail—smooth.*

TABLE 8-4 *Typical Asphalt Shingles*

Product*	Configuration	Per Square			Size		Exposure, Inches	Underwriters' Listing
		Approximate Shipping Weight	Shingles	Bundles	Width, Inches	Length Inches		
Wood appearance strip shingle more than one thickness per strip Laminated or job-applied	Various edge, surface texture, and application treatments	285# to 390#	67 to 90	4 or 5	11 1/2 to 15	36 or 40	4 to 6	A or C, many wind-resistant
Wood appearance strip shingle single thickness per strip	Various edge, surface texture, and application treatments	Various, 250# to 350#	78 to 90	3 or 4	12 or 12 1/4	36 or 40	4 to 5 1/8	A or C, many wind-resistant
Self-sealing strip shingle	Conventional three-tab	205# to 240#	78 or 80	3	12 or 12 1/4	36	5 or 5 1/8	A or C, all wind-resistant
	Two- or four-tab	Various, 215# to 325#	78 or 80	3 or 4	12 or 12 1/4	36	5 or 5 1/8	
Self-sealing strip shingle No cutout	Various edge and texture treatments	Various, 215# to 290#	78 to 81	3 or 4	12 or 12 1/4	36 or 36 1/4	5	A or C, all wind-resistant
Individual lock-down Basic design	Several design variations	180# to 250#	72 to 120	3 or 4	18 to 22 1/4	20 to 22 1/2		C, many wind-resistant

*Other types available from manufacturers in certain areas of the country. Consult your regional Asphalt Roofing Manufacturers Association manufacturer.

TABLE 8-5 *Recommended Nail Lengths*

Purpose	Nail Length, Inches
Roll roofing on new deck	1
Strip or individual shingles—new deck	1¼
Reroofing over old asphalt roofing	1¼ to 1½
Reroofing over old wood shingles	1¾

Fig. 8-28 *Nailing or stapling a strip asphalt shingle.* (Certain-Teed.)

Fig. 8-29 *Nailing points on a strip shingle.* (Bird and Son.)

Fig. 8-30 *Overlap of shingles.* (Bird and Son.)

3. Quick-setting asphalt adhesives
4. Asphalt water emulsions
5. Roof coatings
6. Asphalt primers

Methods of softening   The materials are flammable. Cement should be applied to a dry, clean surface. It should be troweled or brushed vigorously to remove air bubbles. The material should flow freely. It should be forced into all cracks and openings. An emulsion may be applied to damp or wet surfaces. It should not be applied in an exposed location. It should not be rained on for at least 24 hours. Emulsions are water soluble.

Uses   Plastic asphalt cements are used for flashing cements. They are so processed that they will not flow at summer temperatures. They are elastic after setting. This compensates for normal expansion and contraction of a roof deck. They will not become brittle at low temperatures.

Lap cements come in various thicknesses. Follow the manufacturer's suggestions. Lap cement is not as thick as plastic cement. It is used to make a watertight bond. The bond is between lapping elements of roll roofing. It should be spread over the entire lapped area. Nails used to secure the roofing should pass through the cement. The shank of the nail should be sealed where it penetrates the deck material.

Seal down the free tabs of strip shingles with quick-setting asphalt adhesive. It also can be used for sealing laps of roll roofing.

Quick-setting asphalt adhesive is about the same thickness as plastic asphalt cement. However, it is very adhesive. It is mixed with a solvent that evaporates quickly. This permits the cement to set rapidly.

Roof coatings are used in spray or brush thickness. They are used to coat the entire roof. They can be used to resurface old built-up roofs. Old roll roofing or metal roofs also can be coated.

Asphalt water emulsions are a special type of roof coating made with asphalt and sometimes mixed with other materials. Because they are emulsified with water, they can freeze. Be sure to store them in a warm location. They should not be rained on for at least 24 hours.

Masonry primer is very fluid. Apply it with a brush or by spray. It must be thin enough to penetrate rapidly into the surface pores of masonry. It should not leave a continuous surface film. Thin the primer, if necessary, by following the instructions on the can.

Asphalt primer is used to prepare the masonry surface. It should bond well with other asphalt products. These are found on built-up roofs. Other products are plastic asphalt cement and asphalt coatings.

Starter Course

Putting down the shingles isn't too hard—that is, if you have the roof deck in place. It should be covered by the proper underlayment. The eaves should be properly prepared. Refer back to Fig. 8-26.

Starting at the rake   Use only the upper portion of the asphalt shingle. Cut off the tabs. Position it with the adhesive dots toward the eaves. The starter course should overhang the eaves and rake edges by ¼ inch. Nail it in a line 3 to 4 inches above the eaves (Fig. 8-31).

Fig. 8-31 *Starting asphalt shingles at the rake.* (Bird and Son.)

Fig. 8-32 *Starting asphalt shingles at the center.* (Bird and Son.)

Start the first course with a full strip. Overhang the drip edges at the eaves and rake by ¼ inch. Nail the strip in place. Drive the nails straight. The heads should be flush with the surface of the shingle.

Snap a chalk line along the top edge of the shingle. The line should be parallel with the eaves. Snap several others parallel with the first. Make them 10 inches apart. Use the lines to check alignment at every other course. Snap lines parallel with the rake at the shingle cutouts. Use the lines to check cutout alignment.

Start the second course with a full strip less 6 inches. This means that half a tab is missing. Overhang the cut edge at the rake. Nail the shingle in place. Start the third course with a full strip less a full tab. Start the fourth course with half a strip. Continue to reduce the length of the first shingle in each course by an additional 6 inches. The sixth course starts with a 6-inch strip.

Return to the eaves. Apply full shingles across the roof, finishing each course. Dormer, chimney, or vent pipe instructions are another matter. They will be found later in this chapter.

For best color distribution, lay at least four strips in each row. Do this before repeating the pattern up the roof.

Start the seventh course with a full shingle. Repeat the process of shortening. Each successive course of shingles is shortened by an additional 6 inches. This continues to the twelfth course. Return to the seventh course. Apply full shingles across the roof.

Starting at the Center (Hip Roof)

Snap a vertical chalk line at the center of the roof (Fig. 8-32). Put down starter strips along the eaves. Do this in each direction from the chalk line. Go slightly over the centerlines of the hips. Overhang the eaves by ¼ inch.

Align the butt edge of a full shingle with the bottom edge of the starter strip. Also align it with its center tab centered on the chalk line.

Snap a chalk line along the top of the shingle parallel with the eaves. Snap several others parallel with the first. They should be 10 inches apart. Use the lines to check the alignment of alternate courses. Finish the first course with full shingles. Extend the shingles partway over the hips. Finish the remaining courses with full shingles.

Valleys

There are three types of valleys. One is the open type. Here, the saturated felt can be seen after the shingles are applied. Another type is the woven valley. This one has the shingles woven. There is no obvious valley line. The other is the closed-cut valley. This one has a straight line where the roofs intersect.

Open valleys Use mineral-surfaced-material roll roofing for this valley. Match or contrast the color with

Fig. 8-33 *Use of roll roofing for open valley flashing.* (Bird and Son.)

that of the roof covering. The open-valley method is shown in Fig. 8-33. The felt underlay is centered in the valley before shingles are applied (Fig. 8-34).

Center an 18-inch-wide layer of mineral-surfaced roll roofing in the valley. The surfaced side goes down. Cut the lower edge to conform to and be flush with the eave flashing strip. The ends of the upper segments

Fig. 8-34 *Felt underlay centered in the valley before valley linings are applied.* (Bird and Son.)

overlap the lower segments in a splice. The ends are secured with plastic asphalt cement (see Fig. 8-33). Use only enough nails, 1 inch in from each edge, to hold the strip smoothly. Press the roofing firmly in place into the valley as you nail. Place another 36-inch-wide strip on top of the first strip. This is placed surfaced side up. Center it in the valley. Nail it in the same manner as the underlying 18-inch strip.

Do this before the roofing is applied: Snap two chalk lines the full length along the valley, one line on each side of the valley. They should be 6 inches apart at the ridge or 3 inches when measured from the center of the valley. The marks diverge at the rate of ⅛ inch per foot as they approach the eaves. A valley of 8 feet in length will be 7 inches wide at the eaves. One 16 feet long will be 8 inches wide at the eaves. The chalk line serves as a guide in trimming the last unit to fit the valley. This ensures a clean, sharp edge. The upper corner of each end shingle is clipped (see Fig. 8-34). This keeps water from getting in between the courses. The roofing material is cemented to the valley lining. Use plastic asphalt cement.

Woven and closed-cut valleys Some roofers prefer woven or closed-cut valleys. These are limited to strip-type shingles. Individual shingles cannot be used. Nails may be required at or near the center of the valley lining. Avoid placing a nail in an overlapped shingle too close to the center of the valley. It sometimes may be necessary to cut a strip short. This is done if it otherwise would end near the center. Continue from this cut end over the valley with a full strip. This method increases the coverage of the shingles throughout the length of the valley. This adds to the weather resistance of the roofs at these points.

Woven valleys There are two methods of weaving shingles (Fig. 8-35). They can be applied on both roof areas at the same time. This means that you weave each course, in turn, over the valley. Or you may cover each roof area first. Do this to a point about 3 feet from the center of the valley. Then weave the valley shingles in later.

In the first method, lay the first course. Place it along the eaves of one roof area up to and over the valley. Extend it along the adjoining roof area. Do this for a distance of at least 12 inches. Then lay the first course along the eaves of the intersecting roof area. Extend it over the valley. It goes on top of the previously applied shingle. The next courses go on alternately. Lay along one roof area and then along the other. Weave the valley shingles over each other (see Fig. 8-35). Make sure that the shingles are pressed

Fig. 8-35 *Weaving each course in turn to make a woven valley.*
(Bird and Son.)

tightly into the valley. Nail them in the normal manner. No nails are located closer than 6 inches to the valley centerline. Two nails are located at the end of each terminal strip (see Fig. 8-35).

Closed-cut valleys For a closed-cut valley, lay the first course of shingles along the eaves of one roof area up to and over the valley. Extend it along the adjoining roof section. The distance is at least 12 inches. Follow the same procedure when applying the next courses of shingles (Fig. 8-36). Make sure that the shingles are pressed tightly into the valley. Nail in the normal manner, except that no nail is to be located closer than 6 inches to the valley centerline. Two nails are located at the end of each terminal strip (Fig. 8-37).

Apply the first course of shingles. Do this along the eaves of the intersecting roofs. Extend it over previously applied shingles. Trim a minimum of 2 inches up from the centerline of the valley. Clip the upper corner of each end shingle. This prevents water from getting between courses. Embed the end in a 3-inch-wide strip of plastic asphalt cement. Other courses are applied and completed (see Fig. 8-37).

An open valley for a dormer roof A special treatment is needed where an open valley occurs at a joint between a dormer roof and the main roof through which it projects (Fig. 8-38). First apply the underlay. Main roof shingles are applied to a point just above the lower end of the valley. The course last applied is fitted. It is fitted close to and flashed against the wall of the dormer. The wall is under the projecting edge of the dormer eave. The first strip of valley lining then is applied. Do this the same way as for an open valley.

Fig. 8-36 *Worker installing shingles. Note that the first strip is at the bottom. Also note the felt strip in the valley.*

Fig. 8-37 *Closed-cut valley.* (Bird and Son.)

Fig. 8-38 *An open valley for a dormer roof. Shingles have been laid on the main roof up to lower end of the valley.* (Bird and Son.)

Fig. 8-39 *Application of valley lining for a dormer roof.* (Bird and Son.)

Fig. 8-40 *Dormer valley completed.* (Bird and Son.)

The bottom end is cut so that it extends ¼ inch below the edge of the dormer deck. The lower edge of the section lies on the main deck. It projects at least 2 inches below the joining roofs. Cut the second or upper strip on the dormer side. It should match the lower end of the underlying strip. Cut the side that lies on the main deck. It should overlap the nearest course of shingles. This overlap is the same as the normal lap of one shingle over another. It depends on the type of shingle being applied. In this case, it extends to the top of the cutouts. This is a 12-inch-wide three-tab square-butt strip shingle.

The lower end of the lining then is shaped (Fig. 8-39). It forms a small canopy over the joint between the two decks. Apply shingles over the valley lining. The end shingle in each course is cut. It should conform to the guidelines. Embed the ends in a 3-inch-wide strip of plastic asphalt cement. Valley construction is completed in the usual manner (Fig. 8-40).

Flashing Against a Vertical Wall

Step flashing is used when the rake of a roof abuts a vertical wall. It is best to protect the joint by using metal flashing shingles. They are applied over the end of each course of shingles.

The flashing shingles are rectangular in shape. They are from 5 to 6 inches long. They are 2 inches wider than the exposed face of the roofing shingles. When used with strip shingles laid 5 inches to the weather, they are 6 to 7 inches long. They are bent so as to extend 2 inches out over the roof deck. The remainder goes up the wall surface. Each flashing shingle is placed just uproof from the exposed edge of the single that overlaps it. It is secured to the wall sheathing with one nail in the top corner (Fig. 8-41). The metal is 7 inches wide. The roof shingles are laid 5 inches to the weather. Each element of flashing will lap the next by 2 inches.

Fig. 8-41 Use of metal flashing shingles to protect the joint between a sloping roof and a vertical wall. (Bird and Son.)

Fig. 8-42 Cricket or saddle built behind the chimney. (Bird and Son.)

The finished siding is brought down over the flashing to serve as a cap flashing. However, it is held far away from the shingles. This allows the ends of the boards to be painted. Paint excludes dampness and prevents rot.

Chimneys

Chimneys are usually built on a separate foundation. This avoids stresses and distortions owing to uneven settling. A chimney is subject to differential settling from a house. Flashing at the point where the chimney comes through the roof calls for something that will allow movement without damage to the water seal. It is necessary to use base flashings. They should be secured to the roof deck.

The counter, or cap, flashings are secured to the masonry. Figures 8-42 through 8-46 show how roll roofing is used for base flashing. Metal is used for cap flashing.

Apply shingles over the roofing felt up to the front face of the chimney. Do this before any flashings are placed. Make a saddle or cricket (see Fig. 8-42). This sits between the back face of the chimney and the roof deck. The cricket keeps snow and ice from piling up. It also deflects downflowing water around the chimney.

Apply a coat of asphalt primer to the brick work. This seals the surface. This is where plastic cement will later be applied. Cut the base flashing for the

Fig. 8-43 Base flashings cut and applied. (Bird and Son.)

Fig. 8-44 *Flashing over the cricket in the rear of the chimney.*
(Bird and Son.)

LENGTH OF CRICKET
ALONG CHIMNEY FACE

6" CUT

FOLD

8" CUT

LENGTH OF
SADDLE RIDGE

(A)

12"

(B)

Fig. 8-45 *Flashing patterns.* (Bird and Son.)

front. Cut according to the pattern shown in Fig. 8-43 (pattern *A*). This one is applied first. The lower section is laid over the shingles in a bed of plastic asphalt cement. Secure the upper vertical section against the masonry with the same cement. Nails also can be used here, driven into the mortar joints. Bend the triangular ends of the upper section around the corners of the chimney. Cement in place.

Cut the side base flashings next. Use pattern *B* of Fig. 8-43. Bend them to shape, and apply them as shown. Embed them in plastic asphalt cement. Turn

METAL CAP
FLASHING

3" SIDE LAP

Fig. 8-46 *Metal cap flashing applied to cover the base flashing.*
(Bird and Son.)

the triangular ends of the upper section around the chimney corners. Cement them in place over the front base flashing.

Figure 8-44 shows the cutting and fitting of base flashings over the cricket. The cricket consists of two triangular pieces of board. The board is cut to form a ridge. The ridge extends from the centerline of the chimney back to the roof deck. The boards are nailed to the wood deck. They are also nailed to one another along the ridge. This is done before felt underlayment is applied. Cut the base flashing (see Fig. 8-45A). Bend it to cover the entire cricket. Extend it laterally to cover part of the side base flashing. Cut a second rectangular piece of roofing (see Fig. 8-45B). Make a cutout on one side to conform to the rear angle of the cricket. Set it tightly in plastic asphalt cement. Center it over that part of the cricket flashing extending up to the deck. This piece provides added protection where the ridge of the cricket meets the deck. Cut a second similar rectangular piece of flashing. Cut a V from one side. It should conform to the pitch of the cricket. Place it over the cricket ridge and against the chimney. Embed it in plastic asphalt cement.

Use plastic asphalt cement generously. Use it to cement all standing portions of the base flashing to the brickwork.

Cap flashings are shown in Fig. 8-46. They are made of 16-ounce or heavier sheet copper. You also can make the caps of 24-gauge galvanized steel. If steel is used, it should be painted on both sides.

Fig. 8-47 *Method of securing cap flashing to the chimney.* (Bird and Son.)

Fig. 8-48 *Alternative base flashing method.* (Bird and Son.)

Brickwork is secured to the cap flashing in Figs. 8-46 and 8-47. These drawings show a good method. Rake the mortar joint to a depth of 1½ inches. Insert the bent back edge of the flashing into the cleared space between the bricks. It is under slight spring tension. It cannot be dislodged easily. Refill the joint with portland cement mortar. Or you can use plastic asphalt cement. This flashing is bent down to cover the base flashing. The cap lies snugly against the masonry.

The front unit of the cap flashing is one continuous piece. On the sides and rear, the sections are of similar size. They are cut to conform to the locations of brick joints. The pitch of the roof is also needed. The side units lap each other (see Fig. 8-46). This lap is at least 3 inches. Figure 8-48 shows another way to flash a sloping roof abutting a vertical masonry wall. This is known as the *step-flashing method.* Place a rectangular piece of material measuring 8 × 22 inches over the end tab of each course of shingles. Hold the lower edge slightly back of the exposed edge of the covering shingle. Bend it up against the masonry. Secure it with suitable plastic asphalt cement. Drive nails through the lower edge of the flashing into the roof deck. Cover the nails with plastic asphalt cement. Repeat the operation for each course. Flashing units should be wide enough to lap each other by at least 3 inches. The upper one overlays the lower one each time.

Asphalt roofing can be used for step flashing. It simply replaces the base flashing already shown. Metal cap flashings must be applied in the usual manner. The metal cap completes a satisfactory job.

Soil Stacks

Most building roofs have pipes or ventilators through them. Most are circular in section. They call for special flashing methods. Asphalt products may be used successfully for this purpose. Figures 8-49 through 8-54 show a step-by-step method of flashing for soil pipes. A soil pipe is used as a vent for plumbing. It is

Fig. 8-49 *Roofing is first applied up to the soil pipe and fitted around it.* (Bird and Son.)

Fig. 8-50 *First step in marking an opening for flashing. (Bird and Son.)*

Fig. 8-51 *Second step in marking an opening for flashing. (Bird and Son.)*

Fig. 8-52 *Cut an oval in the flange. (Bird and Son.)*

Fig. 8-53 *Cement the collar molded around the pipe. (Bird and Son.)*

Fig. 8-54 *Shingling completed past and above the pipe. (Bird and Son.)*

made of cast iron or copper. The pipe gets its name from being buried in the soil as a sanitary sewer pipe.

An alternative procedure for soil stacks can be used. Obtain noncorrodible metal pipes. They should have adjustable flanges. These flanges can be applied as flashing to fit any roof pitch.

STRIP SHINGLES

Prepare the deck properly before starting to apply strip shingles. First, place an underlayment down.

Deck Preparation

Metal drip edge Use a metal drip edge made of noncorrodible, nonstaining metal. Place it along the eaves and rakes (Figs. 8-55 and 8-56). The drip edge is designed to allow water runoff to drip free into a gutter. It should extend back from the edge of the deck not

Fig. 8-55 *Application of the metal drip edge at eaves directly onto the deck.* (Bird and Son.)

Fig. 8-56 *Application of the metal drip edge at the rakes over the underlayment.* (Bird and Son.)

more than 3 inches. Secure it with nails spaced 8 to 10 inches apart. Place the nails along the edge. Drip edges of other materials may be used. They should be of approved types.

Underlayment   Place a layer of No. 15 asphalt-saturated felt down. Cover the entire deck as shown in Fig. 8-20.

Chalk lines   On small roofs, strip shingles may be laid from either rake. On roofs 30 feet or more long, it is better to start at the center and then work both ways from a vertical line. This ensures better vertical alignment. It also provides for meeting and matching above a dormer or chimney. Chalk lines are used to control shingle alignment.

Eaves flashing   Eaves flashing is required in cold climates (January daily average temperatures of 25°F or less call for eaves flashing). In cold climates, there is a possibility of ice forming along the eaves. If this happens, it causes trouble. Flashing should be used if there is doubt. Ice jams and water backup should be avoided. They can cause leakage into the ceiling below.

There are two flashing methods to prevent leakage. The methods depend on the slope of the roof. Po-

tentially severe icing conditions are another factor in choice of method.

Normal slope is 4 inches per foot or over. Install a course of 90-pound mineral-surfaced roll roofing. Or apply a course of smooth roll roofing. It should not be less than 50 pounds. Install it to overhang the underlay and metal drip edge from ¼ to ⅜ inch. It should extend up to the roof. Cover a point at least 12 inches inside the building's interior wall line. For a 36-inch eave overhang, the horizontal lap joint must be cemented. It should be located on the roof deck extending beyond the building's exterior wall line (Fig. 8-57).

Fig. 8-57 *Eave flashing strip for normal-sloped roof. This means 4 inches or more per foot.* (Bird and Son.)

First and Succeeding Courses

Start the first course with a full shingle. Succeeding courses are started with full or cut strips. It depends on the style of shingles being applied. There are three major variations for square-butt strip shingles:

1. Cutouts break at joints on the thirds (Fig. 8-58).
2. Cutouts break at joints on the halves (Fig. 8-59).
3. Random spacing (Fig. 8-60).

Random spacing can be done by removing different amounts from the rake tab of succeeding courses. The amounts are removed according to the following scheme:

1. The width of any rake tab should be at least 3 inches.
2. Cutout centerlines should be located at least 3 inches laterally from other cutout centerlines. This means both the course above and the course below.

Fig. 8-58 *Applying three-tab square-butt strips so that cutouts break the joints at thirds.* (Bird and Son.)

Fig. 8-59 *Applying three-tab square-butt strips so that cutouts are centered over the tabs in the course below.* (Bird and Son.)

3. The rake-tab widths should not repeat closely enough to cause the eye to follow a cutout alignment.

Ribbon Courses

Use a ribbon course to strengthen horizontal roof lines. It adds a massive appearance that some people prefer (Fig. 8-61).

One method involves special starting procedures. This is repeated every fifth course. Some people prefer this method:

1. Cut 4 inches off the top of a 12-inch-wide strip shingle. This will give you an unbroken strip 4 × 36 inches. You also get a strip 8 × 36 inches. Both strips contain the cutouts.

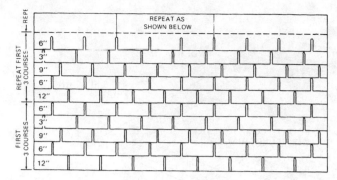

Fig. 8-60 *Random spacing of three-tab square-butt strips.* (Bird and Son.)

Fig. 8-61 *Ribbon courses—side view.* (Bird and Son.)

Fig. 8-62 *Laying the ribbon courses.* (Bird and Son.)

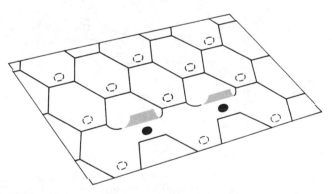

Fig. 8-63 *Location of tab cement under square-butt tabs.* (Bird and Son.)

Fig. 8-64 *Location of tab cement under hex tabs.* (Bird and Son.)

2. Lay the 4- × 36-inch strip along the eave.

3. Cover this with the 8- × 36-inch strip. The bottom of the cutout is laid down to the eave.

4. Lay the first course of full (12- × 36-inch) shingles. It goes over layers *B* and *C*. The bottom of the cutout is laid down to the eave (Fig. 8-62).

Cutouts should be offset. This is done according to thirds, halves, or random spacing.

Wind Protection

High winds call for specially designed shingles. Cement the free tabs for protection against high winds (Figs. 8-63 and 8-64).

With a putty knife or caulking gun, apply a spot of quick-setting cement on the underlying shingle. The cement should be about the size of a half-dollar. Press the free tab against the spot of cement. Do not squeeze the cement beyond the edge of the tab. Don't skip or miss any shingle tabs. Don't bend tabs back farther than needed.

Two- and Three-Tab Hex Strips

Nail two- and three-tab hex strips with four nails per strip. Locate the nails in a horizontal line 5¼ inches above the exposed butt edge. Figure 8-65 shows how the two-tab strip is applied. Use one nail 1 inch back from each end of the strip. One nail is applied ¾ inch back from each angle of the cutouts. Three-tab shingles require one nail 1 inch back from each end. One nail is centered above each cutout (Fig. 8-66).

Hips and Ridges

Use hip and ridge shingles to finish hips and ridges. They are furnished by shingle manufacturers. You can cut them from 12- × 36-inch square-butt strips. They should be at least 9 × 12 inches. One method of applying them is shown in Fig. 8-67.

Bend each shingle lengthwise down the center. This gives equal exposure on each side of the hip or ridge. Begin at the bottom of a hip. Or you can begin at one end of a ridge. Apply shingles over the hip or ridge. Expose them by 5 inches. Note the direction of

Fig. 8-65 *Application of two-tab hex strips.* (Bird and Son.)

Fig. 8-66 *Application of three-tab hex strips.* (Bird and Son.)

Fig. 8-67 *Hip-and-ridge shingles applied with hex strips.* (Bird and Son.)

the prevailing winds. This is important when you are placing ridge shingles. Secure each shingle with one nail on each side. The nail should be 5½ inches back from the exposed end. It should be 1 inch up from the end. Never use metal ridge roll with asphalt roofing products. Corrosion may discolor the roof.

STEEP-SLOPE AND MANSARD ROOFS

New rooflines have caused changes recently. The mansard roof requires some variations in shingle application (Fig. 8-68). Excessive slopes give reduced results with factory self-sealing adhesives. This becomes obvious in colder or shaded areas.

Fig. 8-68 *Mansard roof.*

Maximum slope for normal shingle application is 60 degrees, or 21 inches per foot. Shingles on a steeper slope should be secured to the roof deck with roofing nails. See the manufacturer's suggestions. Nails are placed 5⅝ inches above the butt. They should not be in or above the self-sealing strip.

Quick-setting asphalt adhesives should be applied as spots. The spots should be about the size of a quarter. They should be applied under each shingle tab. Do this immediately on installation. Ventilation is needed to keep moisture-laden air from being trapped behind sheathing.

INTERLOCKING SHINGLES

Interlocking shingles are designed for windy areas. They have high resistance to high winds. They have an integral locking device. They can be classified into five groups (Fig. 8-69).

These shingles generally do not require the use of adhesives. They may require a restricted use of cement. This is needed along the rakes and eaves. This is where the locking device may have to be removed.

Interlocking shingles can be used for both new and old buildings. Roof pitch is not too critical with these shingles. The designs can be classified into types 1 through 5. Type 5 is a strip shingle with two tabs per strip.

Figure 8-70 shows how interlocking shingles work. Nail placement is suggested by the manufacturer. Proper nail placement is important for good results. Figure 8-71 shows how the drip edge should be placed. The drip edge is designed to allow water runoff to drip free into gutters. The drip edge should not extend more than 3 inches back from the edge of the deck. Secure it with appropriate nails. Space the nails 8 to 10 inches apart along the inner edge.

Use the manufacturer's suggestions for starter-course placement. Chalk lines will be very useful. Because it is very short, this type of shingle needs chalk lines for alignment.

Fig. 8-69 *Locking devices used in interlocking shingles.* (Bird and Son.)

TYPE 1: OUTSIDE LOCK

TYPE 2: INSIDE LOCK

TYPE 3: INSIDE AND OUTSIDE LOCK

TYPE 4: SIDE LOCK

TYPE 5: BOTTOM LOCK

Fig. 8-70 *Methods of locking shingles types 1 through 5. In each case, only the locking device is shown.* (Bird and Son.)

AT EAVES

AT RAKE

Fig. 8-71 *Placement of the metal drip edge for interlocking shingles.* (Bird and Son.)

Hips and Ridges

Hips and ridges require shingles made for that purpose. However, you can cut them from standard shingles. They should be 9 × 12 inches. Either 90-pound mineral-surfaced roofing or shingles can be used (Fig. 8-72).

In cold weather, warm the shingles before you use them. This keeps them from cracking when they are bent to lock. Do not use metal hip or ridge materials. Metal may become corroded and discolor the roof.

Time is an important element in any job. You can speed up the roofing job when you don't have to cut

Fig. 8-72 *Hip-and-ridge application of interlocking shingles.* (Bird and Son.)

and fit the hips and ridges with shingles you are using to cover the roof. You can speed things up and become more efficient by using precut fiberglass high-profile hip and ridge caps. High profile means that the cap sticks up higher than usual and gives a more "finished" quality to the roof. The high profile adds texture and shadow line to improve the appearance and enhance the beauty of any roof. Such caps come in boxes with 48 pieces that cover 30 linear feet of finished hip or ridge when applied with 8 inches to the weather (Fig. 8-73).

Fig. 8-73 *Back nail each shingle with two nails 9.5 inches from the exposed end. (Ridglass.)*

The ridge and hip caps are backnailed with two nails 9.5 inches from the exposed end (Fig. 8-74). Then continue to apply the caps with an 8-inch exposure. All nails should be covered by the succeeding shingle by at least 1 inch (Fig. 8-75). In high-wind areas, seal down each installed ridge unit with elastomer adhesive or face nail each shingle with two nails, one on each side, as shown in Fig. 8-75. In general, use 11- or 12-gauge galvanized roofing nails with $\frac{7}{16}$-inch heads that are long enough to penetrate the roof deck by $\frac{3}{4}$ inch. When applying Ridglass caps in cold weather (under 50°F), unpack the carton on the roof and allow the shingles to warm up before application.

These caps are made of fiberglass, not paper. They are SBS-modified asphalt with no granule loss. There is no cutting, folding, or fabrication on the roof, and matching colors are available for all manufacturers' shingles.

Fig. 8-74 *Continue to apply Ridglass with 9-inch exposure. All nails should be covered by the succeeding shingle by at least 1 inch. (Ridglass.)*

Fig. 8-75 *Seal down each installed ridge unit with elastomeric adhesive, and face nail each shingle with two nails. (Ridglass.)*

ROLL ROOFING

Roofing in rolls in some instances is very economical. Farm buildings and sheds are usually covered with inexpensive roll roofing. It is easier to apply and cheaper than strip shingles.

Do not apply roll roofing when the temperature is below 45°F. If it is necessary to handle the material at lower temperatures, warm it before unrolling it. Warming avoids cracking the coating.

The sheet should be cut into 12- and 18-foot lengths. Spread them in a pile on a smooth surface until they flatten.

Windy Locations

Roll roofings are recommended for use in windy locations. Apply them according to the maker's suggestions. Use the pattern edge and blind nail. This means that the nails cannot be seen after the roofing is applied. Blind nailing can be used with 18-inch-wide mineral-surfaced or 65-pound smooth roofing. Both can be used in windy areas. The 19-inch selvage double-coverage roll roofing is also suited for windy places.

Use concealed nailing rather than exposed nailing to apply roll roofing. This ensures maximum life in service. Use only lap cement or quick-setting cement. It should be cement the maker of the roofing suggests. Cements should be stored in a warm place until you are ready to use them. Place the unopened container in hot water to warm. Never heat asphalt cements directly over a fire. Use 11- or 12-gauge hot-dipped galvanized nails. Nails should have large heads. This means at least $\frac{3}{8}$-inch-diameter heads. The shanks should be $\frac{7}{8}$ to 1 inch long. Use nails that are long enough to penetrate the wood below.

Exposed Nails— Parallel to the Rake

Exposed nailing, parallel to the eaves, is shown in Fig. 8-76. Figure 8-77 also shows the exposed-nail method. It is parallel to the rake in this case. The overhang is $\frac{1}{4}$

to ⅜ inch over the rake. End laps are 6 inches wide and cemented down. Stagger the nails in rows, 1 inch apart. Space the nails on 4-inch centers in each row. Stagger all end laps. Do not have an end lap in one course over or adjacent to an end lap in the preceding course.

Hips and Ridges

For the method used to place a cap over the hips and ridge, see Fig. 8-78. Butt and nail sheets of roofing as they come up on either side of a hip or ridge. Cut strips of roll roofing 12 inches wide. Bend them lengthwise through their centers. Snap a chalk line guide parallel to the hip or ridge. It should be 5½ inches down on each side of the deck.

Cement a 2-inch-wide band on each side of the hip or ridge. The lower edge should be even with the chalk line. Lay the bent strip over the hip or ridge. Embed it in asphalt lap cement.

Secure the strip with two rows of nails. One row is placed on each side of the hip or ridge. The rows should be ¾ inch above the edges of the strip. Nails are spaced on 2-inch centers. Be sure that the nails penetrate the cemented portion. This seals the nail hole with some of the asphalt.

WOOD SHINGLES

Wood shingles are the oldest method of shingling. In early U.S. history, pine and other trees were used for shingles. Then the western United States discovered the cedar shingle. It is resistant to water and rot. If cared for properly, it will last at least 50 years. Application of this type of roofing material calls for some different methods.

Fig. 8-76 *Application of roll roofing by the exposed-nail method (parallel to the eaves).* (Bird and Son.)

Fig. 8-77 *Application of roll roofing by the exposed-nail method (parallel to the rake).* (Bird and Son.)

NAIL EACH EDGE—NAILS
SPACED 2" APART

6"

2" BAND OF LAP CEMENT
UNDER EACH EDGE

Fig. 8-78 *Hip-and-ridge application of roll roofing.* (Bird and Son.)

Sizing Up the Job

You need to know a few things before ordering these shingles. First, find the pitch of your roof (Fig. 8-79). Simply measure how many inches it rises for every foot it runs.

Remember that a square contains four bundles. It will cover 100 square feet of roof area (Fig. 8-80).

THIS IS CALLED "PITCH"

THIS IS CALLED "RISE"

THIS IS CALLED "RUN"

Fig. 8-79 *Figuring the pitch of a roof.* (Red Cedar Shingle & Hand-Split Shake Bureau.)

10'

10'

Fig. 8-80 *A square of shingles contains four bundles.* (Red Cedar Shingle & Hand-Split Shake Bureau.)

Roof Exposure

Exposure refers to the area of the shingle that comes into contact with the weather (Fig. 8-81). Exposure depends on roof pitch. A good shingle job is never less than three layers thick. See Table 8-6 for important information about shingles.

There are three lengths of shingles: 16, 18, and 24 inches. If the roof pitch is 4 inches in 12 inches or steeper (three-ply roof):

- For 16-inch shingles, allow a 5-inch exposure.
- For 18-inch shingles, allow a 5½-inch exposure.
- For 24-inch shingles, allow a 7½-inch exposure.

If the roof pitch is less than 4 inches in 12 inches but not below 3 inches in 12 inches (four-ply roof):

- For 16-inch shingles, allow a 3¾-inch exposure.
- For 18-inch shingles, allow a 4¼-inch exposure.
- For 24-inch shingles, allow a 5¾-inch exposure.

If the roof pitch is less than 3 inches in 12 inches, cedar shingles are not recommended. These exposures are for No. 1 grade shingles. In applying No. 3 shingles, make sure that you check with the manufacturer.

Estimating Shingles Needed

Determine the ground area of your house. Include the eaves and cornice overhang. Do this in square feet. If the roof pitch found previously:

- Rises 3 in 12, add 3 percent to the square foot total.

Fig. 8-81 *Shingle exposure to the weather.* (Red Cedar Shingle & Hand-Split Shake Bureau.)

- Rises 4 in 12, add 5½ percent to the square foot total.
- Rises 5 in 12, add 8½ percent to the square foot total.
- Rises 6 in 12, add 12 percent to the square foot total.
- Rises 8 in 12, add 20 percent to the square foot total.
- Rises 12 in 12, add 42 percent to the square foot total.
- Rises 15 in 12, add 60 percent to the square foot total.
- Rises 18 in 12, add 80 percent to the square foot total.

Divide the number you have found by 100. The answer is the number of shingle squares you should order to cover your roof if the pitch is 4 inches in 12 inches or steeper. If the roof is of lesser pitch, allow

TABLE 8-6 *Summary of Sizes, Packing, and Coverage of Wood Shingles*

Shake Type, Length, and Thickness, Inches	No. of Courses per Bundle	No. of Bundles per Square	Approximate Coverage (in Square Feet) of One Square, When Shakes are Applied with ½-Inch Spacing, at Following Weather Exposures (in Inches):								
			5½	6½	7	7½	8½	10	11½	14	16
18 × ½ medium resawn	9/9[a]	5[b]	55[c]	65	70	75[d]	85[e]	100[f]			
18 × ¾ heavy resawn	9/9[a]	5[b]	55[c]	65	70	75[d]	85[e]	100[f]			
24 × ⅜ handsplit	9/9[a]	5		65	70	75[g]	85	100[h]	115[i]		
24 × ½ medium resawn	9/9[a]	5		65	70	75[c]	85	100[j]	115[i]		
24 × ¾ heavy resawn	9/9[a]	5		65	70	75[c]	85	100[j]	115[i]		
24 × ½ to ⅝ tapersplit	9/9[a]	5		65	70	75[c]	85	100[j]	115[i]		
18 × ⅜ true-edge straight-split	14[k] straight	4								100	112[l]
18 × ⅜ straight-split	19[k] straight	5	65[c]	75	80	90[j]	100[i]				
24 × ⅜ straight-split	16[k] straight	5		65	70	75[c]	85	100[j]	115[i]		
15 starter-finish course	9/9[a]	5	Use supplementary with shakes applied with not over 10-inch weather exposure.								

[a]Packed in 18-inch-wide frames.
[b]Five bundles will cover 100 square feet of roof area when used as starter-finish course at 10-inch weather exposure; six bundles will cover 100 square feet wall area when used at 8½-inch weather exposure; seven bundles will cover 100 square feet roof area when used at 7½-inch weather exposure. [m]
[c]Maximum recommended weather exposure for three-ply roof construction.
[d]Maximum recommended weather exposure for two-ply roof construction; seven bundles will cover 100 square feet of roof area when applied at 7½-inch weather exposure. [m]
[e]Maximum recommended weather exposure for sidewall construction; six bundles will cover 100 square feet when applied at 8½-inch weather exposure. [m]
[f]Maximum recommended weather exposure for starter-finish course application; five bundles will cover 100 square feet when applied at 10-inch weather exposure. [m]
[g]Maximum recommended weather exposure for application on roof pitches between 4 in 12 in 8 in 12.
[h]Maximum recommended weather exposure for application on roof pitches of 8 in 12 and steeper.
[i]Maximum recommended weather exposure for single-coursed wall construction.
[j]Maximum recommended weather exposure for two-ply roof construction.
[k]Packed in 20-inch-wide frames.
[l]Maximum recommended weather exposure for double-coursed wall construction.
[m]All coverage based on ½-inch spacing between shakes.

one-third more shingles to compensate for reduced exposure. Also add 1 square for every 100 linear feet of hips and valleys.

Tools of the Trade

A shingler's hatchet speeds the work (Fig. 8-82). Sneakers or similar traction shoes make the job safer. A straight board keeps your rows straight and true.

Fig. 8-82 *Tools of the trade. (Red Cedar Shingle & Hand-Split Shake Bureau.)*

APPLYING THE SHINGLE ROOF

Begin with a double thickness of shingles at the bottom edge of the roof (Fig. 8-83). Let the shingles protrude over the edge to ensure proper spillage into the eave trough or gutter. See *A* in Fig. 8-83.

Fig. 8-83 *Applying a double thickness of shingles at the bottom edge of the roof. (Red Cedar Shingle & Hand-Split Shake Bureau.)*

Figure 8-84 shows how the nails are placed so that the next row above will cover the nails by not more than 1 inch. Use the board as shown in Fig. 8-85. Use it as a straightedge to line up rows of shingles. Tack the board temporarily in place as a guide. It makes the work faster and the results look professional.

Figure 8-86 shows the locations of the nails. They should be placed no farther than ¾ inch from the edge of the shingle.

Fig. 8-84 *Covering nails in the previous course. (Red Cedar Shingle & Hand-Split Shake Bureau.)*

Fig. 8-85 *Using a straightedge to keep the ends lined up. (Red Cedar Shingle & Hand-Split Shake Bureau.)*

Fig. 8-86 *Placement of nails in a single shingle. (Red Cedar Shingle & Hand-Split Shake Bureau.)*

Figure 8-87 shows how the shingles are spaced ¼ inch apart to allow for expansion. Other simple rules are also shown in Fig. 8-88.

Valleys and Flashings

Extend the valley sheets beneath the shingles. They should extend 10 inches on either side of the valley

Fig. 8-87 *Spacing of shingle between courses.* (Red Cedar Shingle & Hand-Split Shake Bureau.)

¼" BETWEEN SHINGLES

DO NOT LINE UP
TWO JOINTS SEPARATED
BY ONLY ONE COURSE

SIDE LAP OF 1½" BETWEEN
JOINTS IN SUCCESSIVE
COURSES

SPACING
VARIES
WITH PITCH

Fig. 8-88 *Spacing for valleys.* (Red Cedar Shingle & Hand-Split Shake.)

center. This is the case if the roof pitch is less than 12 inches in 12 inches. For steeper roofs, the valley sheets should extend at least 7 inches (see Fig. 8-88).

Most roof leaks occur at points where water is channeled for running off the roof. Or they occur where the roof abuts a vertical wall or chimney. At these points, use pointed metal valleys and flashings to assist the shingles in keeping the roof sound and dry. Suppliers will provide further information on which of the various metals to use. Figure 8-89 shows flashing installed around a chimney.

Shingling at Roof Junctures

Apply the final course of shingles at the top of the wall. Install metal flashing (26-gauge galvanized iron 8 inches wide). Cover the top 4 inches of the roof slope. Bend the flashing carefully; avoid fracturing or

Fig. 8-89 *Flashing around a chimney.* (Red Cedar Shingle & Hand-Split Shake Bureau.)

DOUBLE STARTER COURSE

METAL FLASHING

FINAL SHINGLE COURSE

SHEATHING

Fig. 8-90 *Convex roof juncture shingling.* (Red Cedar Shingle & Hand-Split Shake Bureau.)

breaking it. Make sure that the flashing covers the nails that hold the final course. Apply a double starter course at the eave. Allow for a 1½-inch overhang of the wall surface. Complete the roof in the normal manner. Figure 8-90 shows a convex juncture.

For a concave juncture, apply the final course of shingles as shown in Fig. 8-91. Install the metal flashing to cover the last 4 inches of roof slope and bottom 4 inches of wall surface. Make sure that the flashing covers the nails that hold the final course. Apply a dou-

DOUBLE STARTER COURSE

METAL FLASHING

SHEATHING

Fig. 8-91 *Concave roof junction shingling.* (Red Cedar Shingle & Hand-Split Shake Bureau.)

ble starter course at the bottom of the wall surface. Complete the shingling in the normal manner.

Before applying the final course of shingles, install 12-inch-wide flashing. This is to cover the top 8 inches of the roof. Bend the remaining 4 inches to cover the top portion of the wall. Figure 8-92 shows how to treat apex junctures. Complete the roof shingling to cover the flashing. Allow the shingle tips to extend beyond the juncture. Complete the wall shingling. Trim the last courses to fit snugly under the protecting roof shingles. Apply a molding strip to cover the top-most portion of the wall. Trim the roof shingles even with the outer surface of the molding. Apply a conventional shingle ridge across the top edge of the roof. This is done in a single strip without matching pairs.

Fig. 8-92 *Apex roof juncture shingling. (Red Cedar Shingle & Hand-Split Shake Bureau.)*

Applying Shingles to Hips and Ridges

The alternative overlap-type hip and ridge can be built by selecting uniform-width shingles. Lace them as shown in Fig. 8-93. Time can be saved if factory-assembled hip-and-ridge units are used. These are shown in Fig. 8-94.

Nails for Wooden Shingles

Rust-resistant nails are very important. Zinc-coated or aluminum nails can be used. Don't skimp on nail quality (Fig. 8-95).

CHAPTER 8 STUDY QUESTIONS

1. How many courses of shingles are used with wood shingles?

Fig. 8-93 *Applying overlap-type roofing ridges. (Red Cedar Shingle & Hand-Split Shake Bureau.)*

Fig. 8-94 *Applying a factory-assembled hip or ridge unit. (Red Cedar Shingle & Hand-Split Shake Bureau.)*

2. Sketch a gable roof shape on a simple building.
3. What is the main purpose of a roof?
4. How does soffit ventilation affect ice dams?
5. What is a valley?
6. What are eave troughs used for?
7. How are plywood and board sheathing different when used on a roof to provide a deck?
8. What are three types of shingles that can be used on a roof?
9. How many shingles are there in a square?
10. What is meant by shingle exposure?
11. What is meant by underlayment?

FOR NEW ROOF CONSTRUCTION			OVER-ROOFING CONSTRUCTION		DOUBLE COURSING
3d	3d	4d	5d	6d	5d
FOR 16" AND 18" SHINGLES		FOR 24" SHINGLES	FOR 16" AND 18" SHINGLES	FOR 24" SHINGLES	FOR ALL SHINGLES
1¼" LONG	1¼" LONG #14½ GAGE	1½" LONG #14 GAGE	1¾" LONG #14 GAGE	2" LONG #13 GAGE	1¾" LONG #14 GAGE
APPROX. 376 NAILS TO LB	APPROX. 515 NAILS TO LB	APPROX. 382 NAILS TO LB	APPROX. 310 NAILS TO LB	APPROX. 220 NAILS TO LB	APPROX. 380 NAILS TO LB

Fig. 8-95 *Nails used for wood shingles.*

12. What is the difference between pitch and slope?

13. What is a woven valley?

14. How does a woven valley differ from a closed-cut valley?

15. What is staging?

16. How long should roofing nails be?

17. What is galvanized iron used for when roofing a house?

18. What are asphalt-water emulsions?

19. Why is plastic asphalt cement used in roofing?

20. Why can't one type of asphalt roofing be used for every job?

9
CHAPTER

Windows and Doors

OTH WINDOWS AND DOORS PLAY AN IMPORTANT part in any type of house or building. They allow the air to circulate. They also permit passage into and out of the structure. Doors open to allow traffic in a planned manner. Windows are closed or open in design. There may be open and closed combinations, too. The design of a window or door is dictated by the building's use.

Buildings such as those in Fig. 9-1 use the window as part of the design. The shape of a window may add to or detract from the design. The designer must be able to determine which is the right window for a building. The designer also must be able to choose a door that is architecturally compatible. Doors and windows come in many designs. However, they are limited in their function. This means that some standards are set for the design of both. Most residential doors, for instance, are 6 feet, 8 inches high. If you are taller than that, you have to duck to pass through. Since most people are shorter, it is a safe height to use for doors.

Windows should be placed so that they will allow some view from either a standing or a sitting position. In most instances, the window top and the door top are even. This way they look better from the outside.

In this chapter you will learn how carpenters install windows and doors. You will learn how windows and doors are used to enhance a design. You also will learn the various sizes and shapes of windows and the different sizes and hinging arrangements of doors. Locks will be presented so that you can learn how to install them.

Details for the placement of windows and doors are given. Things you will learn to do include

- Prepare a window for installation
- Shim a window if necessary
- Secure a window in its opening
- Level and check for proper operation of a window
- Prepare a door for installation
- Shim a door if necessary
- Secure a door in its opening
- Level and check for proper operation

SEQUENCE

A carpenter should install a window in this order:

1. Check for proper window opening.
2. Uncrate the prehung window.
3. Remove the braces, if called for by the manufacturer.
4. Place builder's paper or felt between the window and the sheathing.
5. Place the window in the opening, and check for level and plumb.
6. Attach the window at the corner, or place one nail in the casing or flange (depending on window design).
7. Check for proper operation of the window.
8. Secure the window in its opening.

A carpenter should install a door in this order:

1. Check the opening for correct measurements.
2. Uncrate the prehung door. If the door is not prehung, place the molding and trim in place first. Attach the door hinges by cutting the gains and screwing in the hinges.
3. Check for plumb and level.
4. Temporarily secure the door with shims and nails.

Fig. 9-1 *Windows can add to or detract from a piece of architecture.* (Western Wood Products.)

5. Check for proper operation.

6. Secure the door permanently.

7. Install the lock and its associated hardware.

TYPES OF WINDOWS

There are many types of windows. Each hinges or swings in a different direction. They may be classified as

1. Horizontal sliding windows

2. Awning picture windows

3. Double-hung windows

4. Casement windows

Most windows are made in factories today, ready to be placed into a rough opening when they arrive at the site. There are standards for windows. For instance, Commercial Standard 190 is shown in Fig. 9-2. The window requires a number of features, which are pointed out in the drawing. It must be weatherstripped to prevent air infiltration in excess of 0.75 cubic feet per minute per perimeter foot. This is under a 25 mi/h wind.

ONE 24" DIAGONAL FASTENED TO PRESQUARED UNIT

FIVE PINS PER HORIZONTAL CASING ON 2'8" WIDE WINDOW

ONE HORIZONTAL BRACE TO MAINTAIN EXACT SASH OPENING WIDTH

MITERED CORNERS CROSS-NAILED

FOUR 6-PENNY COATED NAILS

DADOED

INSIDE VIEW

SEVEN PINS PER VERTICAL CASING ON 4'6" WINDOW

BOTH SASH OPERATIVE & EQUIPPED WITH BALANCES

GLASS "B" GRADE

SASH MORTISE AND TENON CONSTRUCTION

JAMBS 3/4" THICK, MINIMUM

GLASS BEDDED AND FACE-PUTTIED

FOUR 6-PENNY COATED NAILS

SILL BEVELED 14°, MINIMUM 15/16" THICKNESS

DADOED

FRONT VIEW

SIDE VIEW

ONE 8-PENNY COATED NAIL

1. Weather-stripped to prevent air infiltration in excess of 0.75 cubic feet per minute per perimeter foot, under 25 MPH wind pressure.
2. No more than two species of wood per unit.
3. Chemically treated in accordance with NWMA minimum standards.
4. Finger-jointing permitted. See Commercial Standard 190, Par. 3.1.7.
5. Sash manufactured under Commercial Standard 163.
6. Ease of operation. See Commercial Standard 190. Par. 3.1.5.

Fig. 9-2 *Standard features of a window required to meet Commercial Standard 190.* (C. Arnold & Sons.)

No more than two species of wood can be used in a unit. The wood used is ponderosa pine or a similar type of pine. Spruce, cedar, redwood, and cypress also may be used in the window frame, sill, and sash.

The wood has to be chemically treated in accordance with National Window Manufacturers Association (NWMA) minimum standards. Finger jointing is permitted. The sash has to be manufactured under Commercial Standard 163. Ease of operation is also spelled out in the written standard. Note how braces are specified to hold the window square and equally distant at all points. These braces are removed once the window has been set in place.

Horizontal sliding window   This type of window is fitted with a vinyl coating. The wood is not exposed at any point, which means less maintenance in the way of painting or glazing. The window is trimmed in vinyl so that it can be nailed in place on the framing around the rough opening. Figure 9-3 shows the window and details of its operation.

Figure 9-4 shows how the rough opening is made for a window. In this case, the framing is on a 24-inch on-center (O.C.) spacing. The large timber over the opening has to be large enough to support the roof without a stud where the window is placed. This prevents the window from buckling, which would stop the window from sliding. The cripple studs under the window opening are continuations of the studs that would be there normally. They are placed there to properly support the opening and to remove any weight from the window frame.

Sliding windows are available in a number of sizes. Figure 9-5 shows the possibilities. To find the overall unit dimension for a window that has a non-supporting mullion, add the sum of the unit dimensions and subtract 2 inches. The mullion is the vertical bar between windows in a frame that holds two or more windows. The overall rough-opening dimension is equal to ¾ inch less than the overall unit dimension.

Double-hung window   This window gets its name from the two windows that slide past one another. In this case, they slide vertically (Fig. 9-6). This is the most common type used today. The window shown is coated with plastic (vinyl) and can be easily attached to a stud through holes already drilled into the plastic around the frame. This plastic is called the *flashing* or the *flange*.

The double-hung window can be installed rather easily (Fig. 9-7). The distance between the side jambs is checked to make sure that they are even. Once the window is in the opening, place a shim where necessary (Fig. 9-8). Note the placement of the nails here.

(A)

(B)

Fig. 9-3   *A. Horizontal sliding window. B. Details.* (Andersen.)

Notice that this window does not have a vinyl flange around it. This is why the nails are placed as shown in Fig. 9-8. Figure 9-9 shows how shims are placed under the raised jamb legs and at the center of the long sills

WINDOW ON MODULE

6' WALL SECTION

STUDS	24 L.F.
JACKS	14 L.F.
CRIPPLES	3 L.F.
	41 L.F.

24" 24" 24" 24"

WINDOW OFF MODULE

6' WALL SECTION

STUDS	32 L.F.
JACKS	14 L.F.
CRIPPLES	6 L.F.
	52 L.F.

(23% MORE FRAMING REQUIRED)

24" 24" 24" 24" 24"

L.F. = LINEAR FEET

Fig. 9-4 *Window openings in a house frame. Window on module and off module with 24-inch centers.* (American Plywood Association.)

UNIT DIMENSION	3-8	4-8	5-8
ROUGH OPENING	3-7 1/4	4-7 1/4	5-7 1/4
SASH OPENING	3-4 1/8	4-4 1/8	5-4 1/8
GLASS SIZE	15 5/8	21 5/8	27 5/8

NOTE: Glass sizes shown are unobstructed.

Fig. 9-5 *Sliding-window sizes.* (Andersen.)

of a double window. Figure 9-10 shows a sash out of alignment (see the arrow). The sashes will not be parallel if the unit is out of square.

Once the window is in place and seated properly, you can pack insulation between the jambs and the trimmer studs (Fig. 9-11). Figure 9-12 shows how 1¾-inch galvanized nails are placed through the vinyl anchoring flange. This flange then is covered by the outside wall covering. The nails are not exposed to the weather.

Double-hung windows can be bought in a number of sizes. Figure 9-13 shows some of the sizes. Figure 9-14 shows the windows all in place. The upper-story windows will be butted by siding. The downstairs windows are sitting back inside the brick. The upper-story windows will have part of the trim sticking out past the exterior siding.

Once the windows are in place and the house is completed, the last step is to put the window dividers in place. They snap into the small holes in the sides of the window. The design may vary (Fig. 9-15). The plastic grill patterns can be changed to meet the needs of the architectural style of the house. They can be removed by snapping them out. In this way, it is easier to

Fig. 9-6 *Double-hung window. (Andersen.)*

Fig. 9-7 *Measure the distance in at least three places to make sure that the window is square. (Andersen.)*

Fig. 9-8 *Shim the window where necessary. Side jambs are nailed through the shims. (Andersen.)*

Fig. 9-9 *Shims raise the jamb legs and keep the window square. (Andersen.)*

clean the window panes. Figure 9-16 shows a house with the diamond light pattern installed in windows that swing out.

Casement window A casement window is hinged so that it swings outward. The whole window opens, allowing for more ventilation (Fig. 9-17). This particular type has a vinyl flange for nailing it to the frame opening. It can be used as a single window or in groups. This type of window can be weatherproofed more easily if it opens outward instead of inward.

Plastic muntins can be added to give a varied effect. They can be put into the windows as shown in

Fig. 9-10 *If the unit is not square, the sash rails will not be parallel. This can be spotted by eye.* (Andersen.)

Fig. 9-11 *Loose insulation batting can be placed around the window to prevent drafts.* (Andersen.)

Fig. 9-16. The diamond light muntins divide the glass space into small diamonds that resemble individual panes of glass.

The crank is installed so that the window opens outward with a twist of the handle. Figure 9-18

Fig. 9-12 *Installing the plastic flange around a prehung window with 1½-inch galvanized nails.* (Andersen.)

shows some of the ways this type of window may be operated.

Screens are mounted on the inside. Storm windows are mounted on the outside, as in Fig. 9-19. In most cases, however, a thermopane is used for insulation purposes. A thermopane is a double sheet of glass welded together with an airspace between. This is then set into the sash and mounted as one piece of glass (Fig. 9-20).

Multiple units are available. They may be movable or stationary. They may have one stationary part in the middle and two movable parts on the ends. Various combinations are available, as shown in Fig. 9-21.

Awning picture window This type of window has a large glass area. It also has a bottom panel that swings outward. A crank operates the bottom section. As it swings out, it has a tendency to form an awning effect—thus the name (Fig. 9-22). A number of sizes and combinations are available in this type of window (Fig. 9-23). The fixed sash with an awning sash is also available in multiple units. Glass sizes are given in Fig. 9-23. If you need to find the overall basic unit dimen-

Fig. 9-13 Various sizes of double-hung windows. (Andersen.)

Fig. 9-14 Double-hung windows in masonry (bottom floor) and set for conventional sliding (top floor).

HORIZONTAL BAR   DIVIDED LIGHT   DIAMOND LIGHT

SNAP-IN UNIT

Fig. 9-15 Plastic dividers make possible different windowpane treatments. (Andersen.)

sion, add the basic unit to 2⅞ inches. The rough opening is the sum of the basic units plus ½ inch.

Figure 9-24 shows the inswinging hopper type and the outswinging awning type of casement window. The awning sash type, shown in Fig. 9-22, has a bottom sash that swings outward. You have to specify which type of opening you want when you order. Specifying bottom or top hinged is the quickest way to order.

Figure 9-25 shows the various sizes of hopper- and awning-type casements available. They can be stacked vertically. If this is done, the overall unit dimension for stacked units is the sum of the basic units plus ¾ inch for two units high, plus ¼ inch for three units high, and less 1¼ inches for four units high. To find the overall basic unit width of multiple units, add the basic unit dimensions plus 2⅞ inches to the total. To find the rough opening width, add the basic unit width plus ½ inch.

Figure 9-26 shows how a number of units may be stacked vertically. All units in this case open for maximum ventilation.

PREPARING THE ROUGH OPENING FOR A WINDOW

It is important that you consult the window manufacturer's specifications before you make the rough opening. Installation techniques, materials, and building codes vary according to area. Contact the window dealer for specific recommendations.

The same rough opening preparation procedures are used for wood and Perma-Shield windows. Fig-

Fig. 9-16 *Diamond light pattern installed with plastic dividers.*

ures 9-27 and 9-28 show the primed wood window and the Perma-Shield window made by Andersen. These will be the windows discussed here. The instructions for installation will show how a manufactured window is installed.

Figure 9-29 shows how the wood casement window operates. Note the operator and its location. Fig-

ure 9-30 gives the details of the Perma-Shield Narroline window.

Figure 9-31 shows some of the possible window arrangements available from a window manufacturer. There is a window for almost any use. Select the window needed, and follow the instructions or similar steps for your window.

Brick veneer with a frame backup wall is similar in construction to the frame walls in the following illustrations. When the opening must be enlarged, make certain that the proper size header is used. Contact the dealer for the proper header size. To install a smaller size window, frame the opening as in a new installation.

Fig. 9-17 *Casement window.* (Andersen.)

Fig. 9-18 *Methods of operating casement windows: A. standard push bar; B. lever lock; C. rotary gear.* (Andersen.)

Fig. 9-19 *Double glass insulation. A ¹³⁄₁₆-inch airspace is placed between panes.* (Pella.)

Fig. 9-20 *Welded-glass or thermopane windows.*

Steps in Preparing the Rough Opening

In some remodeling jobs, the following must be done:

1. Lay out the window opening width between regular studs to equal the window rough opening width plus the thickness of two regular studs (Fig. 9-32). Normally, in new construction, the rough opening is already there, so all you have to do is install the window in it.

Fig. 9-21 *Various sizes of casement windows.* (Andersen.)

Fig. 9-22 *Picture window with awning bottom.* (Andersen.)

BASIC UNIT	⊢2-10⊣	⊢3-10⊣
SASH OPENING	⊢2-7⊣	⊢3-7⊣
GLASS	⊢28¾⊣	⊢40¾⊣

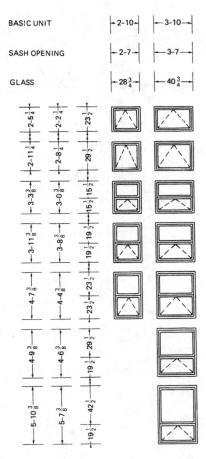

BASIC UNIT	⊢2-7⅝⊣	⊢3-5⊣	⊢4-1⊣
SASH OPENING	⊢2-4⅝⊣	⊢3-2⊣	⊢3-10⊣
GLASS	⊢26½⊣	⊢35¾⊣	⊢43¾⊣

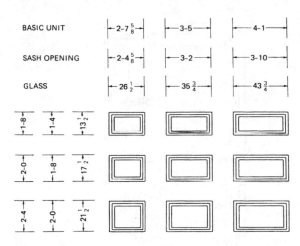

Fig. 9-25 *Various sizes of outswinging and inswinging windows. (Andersen.)*

Fig. 9-23 *Various sizes of the fixed sashes with awning-sash windows. (Andersen.)*

IN SWINGING HOPPER

OUT SWINGING AWNING

Fig. 9-24 *Inswinging hopper and outswinging awning types of windows. These are casement windows.*

Fig. 9-26 *Vertical stacking of outswinging windows.*

ANDERSEN WOOD CASEMENT WINDOW

Fig. 9-27 *Primed-wood window. (Andersen.)*

2. Cut two pieces of window header material to equal the rough opening of the window plus the thickness of two jack or trimmer studs. Nail the two header members together using an adequate spacer so that the header thickness equals the width of the jack or trimmer stud (Fig. 9-33).

Fig. 9-28 *Perma-Shield window.* (Andersen.)

Fig. 9-30 *Details of the Perma-Shield window.* (Andersen.)

Fig. 9-29 *Details of the primed-wood casement window.* (Andersen.)

Fig. 9-31 *Various types of manufactured windows ready for quick installation.* (Andersen.)

3. Cut the jack or trimmer studs to fit under the header for support. Nail the jack or trimmer studs to the regular studs (Fig. 9-34).

4. Position the header at the desired height between the regular studs. Nail through the regular studs into the header to hold the header in place until the next step is completed (Fig. 9-35).

5. Measure the rough opening height from the bottom of the header to the top of the rough sill. Cut 2- × 4-inch cripples and the rough sill to the proper length (Fig. 9-36). The rough sill length is equal to the rough opening width of the window.

Fig. 9-32 *How to locate a window rough opening.* (Andersen.)

Fig. 9-33 *Making the header outside the window opening.* (Andersen.)

HEADER

TRIMMER STUDS

Fig. 9-34 *Placement of the jack studs.* (Andersen.)

HEADER

REGULAR STUDS

Fig. 9-35 *Placing the header where it belongs.* (Andersen.)

Fig. 9-36 *Assembling the cripples for easy placement.* (Andersen.)

Assemble the cripples by nailing the rough sill to the ends of the cripples.

6. Fit the rough sill and cripples between the jack studs (Fig. 9-37). Toenail the cripples to the bottom plate and the rough sill to the jack studs at the sides. See the round insert in Fig. 9-37.

7. Apply the exterior sheathing (e.g., fiberboard, plywood, etc.) flush with the rough sill, header, and jack or trimmer stud framing members (Fig. 9-38).

WINDOW ROUGH OPENING

TOENAIL

Fig. 9-37 *Placing the rough sill and cripples between the jack studs. Note the inset showing the toenailing.* (Andersen.)

JACK STUD

ROUGH SILL

Fig. 9-38 *Applying the exterior sheathing.* (Andersen.)

INSTALLING A WOOD WINDOW

Installation of a wood window is slightly different from that of a Perma-Shield window. However, there are many similarities. The following steps will show you how the windows are installed in the rough opening you just made from the preceding instructions:

1. Set the window in the opening from the outside with the exterior window casing overlapping the exterior sheathing. Locate the unit on the rough sill, and center it between the side framing members (jack studs). Use 3½-inch casing nails, and partially secure one corner through the head casing (Fig. 9-39). Drive the nail at a slight upward angle through the head casing into the header (Fig. 9-40).

2. Level the window across the casing, and nail it through the opposite corner with a 3½-inch cas-

(A)

(B)

Fig. 9-41 A. Level the window across the casing, and nail through the opposite corner. B. Location of the shim that holds the window level. (Andersen.)

Fig. 9-39 Placing the window in the rough opening. (Andersen.)

Fig. 9-40 Use 3½-inch nails to partially secure one corner through the head casing. (Andersen.)

ing nail. It may be necessary to shim the window under the side jambs at the sill to level it. This is done from the interior (Fig. 9-41).

3. Plumb (check the vertical of) the side jamb on the exterior window casing, and drive a nail into the lower corner (Fig. 9-42). Complete the installation by nailing through the exterior casing with 3½-inch nails. Space the nails about 10 inches apart.

4. Before you finally nail in the window, make sure that you check the sash to see that it operates easily.

5. Apply a flashing with the rigid portion over the head casing (Fig. 9-43). Secure this flashing with 1-inch nails through the flexible vinyl into the sheathing. Do not nail into the head casing.

6. Caulk around the perimeter of the exterior casing after the exterior siding or brick is applied.

Masonry or brick veneer wall window   This type of window can be installed in masonry wall construction. Fasten the wood buck to the masonry wall, and nail the window to the wood buck using the procedures just shown for frame-wall construction.

Fig. 9-42 *Check for plumb, and nail the lower corner.*

(A)

Fig. 9-43 *Apply the flashing with the rigid part over the head casing.* (Andersen.)

(B)

Fig. 9-44 *Details of a window installation in a masonry or brick veneer wall: A. wood window; B. Perma-Shield window.* (Andersen.)

Figure 9-44A shows a wood window installed in a masonry wall. Figure 9-44B shows a Perma-Shield window installed with metal jam clips in a masonry wall with brick veneer. The metal jam clips and auxiliary casing are available when specified.

Keep in mind that when brick veneer is used as an exterior finish, adequate clearance must be left for caulking between the window sill and the masonry. This will prevent damage and bowing of the sill. The bowing is caused by settling of the structural member. Shrinkage also will cause damage. Shrinkage takes place as the roughed-in lumber dries after enclosure and the heat is turned on in the house.

Installing Windows by Nailing the Flange to the Sheathing

A simple procedure is used to place windows into the rough openings left in the framing of the house for such purposes (Fig. 9-45). Most windows come with a flange that can be nailed to the sheathing or window framing (Fig. 9-46). This eliminates cold air seepage in the winter and some noise generated by brisk winds. Figure 9-47 illustrates details of how the flange or flashing is covered by the exterior wall covering.

In most instances, it is possible for the window unit to be handed out the opening to an outside carpenter who can nail it in place with little effort while the inside carpenter holds the unit in place or levels it with shims. Figure 9-48 shows a house with windows installed using the methods just described.

Nail

Nail

Nail

Fig. 9-45   *A double-hung window mounted by nailing the flange to the sheathing and structural frame of the house.*

SKYLIGHTS

Having windows in walls is not enough today. Houses also have windows in the ceiling. This new demand has been popular where more light is needed and an air of openness is desired. The skylight seems to be the answer to the demands of today's lifestyles. Bathrooms and kitchens are the rooms most often fitted with skylights.

Skylights can be installed when the house is built, or they can be added later. In the examples here, the second approach is selected because it really encompasses both methods and can be easily adapted for original construction. Describing how to do it in original construction, however, would not necessarily serve those who want to install skylights after the house is built.

Four basic types of skylights are shown here, ranging from flush-mount to venting types. Figure 9-49 shows the flush-mount type. The low-profile flush-mount skylight provides the most economical solution to skylight installation. The flush-mount skylight in-

cludes two heavy-gauge domes that are formed to provide a built-in deck-mounting flange. This flange reduces installation error and allows fast and easy roof attachment when used with the mounting clips prepackaged in the carton. The frameless/seamless feature eliminates potential leakage and provides airtight reliability. The one in the figure is designed specifically for residential use. It is designed to be installed on a pitched roof of 20 degrees or more. It is available in four roof opening sizes.

Figure 9-50 is a curb-mount model that is ideal for locations where water, leaves, or snow collect on a roof, making an elevated skylight desirable. It is designed to use a wood curb put in place by the installer. This type can be installed on either a flat or pitched roof. It also is available in four roof opening sizes.

Figure 9-51 shows a self-curbing type of skylight, which eliminates on-roof curb construction. This feature allows simple, time-saving installation while eliminating the potential for leakage around a wood curb. The premanufactured curbing is fully insulated and includes an extrawide deck-mounting flange with

Fig. 9-46 *Using a lightweight plastic "glass block" window that fits as a unit and can be mounted by nailing the flange to the frame.*

INSTALLATION
FLANGE/FLASHING

HEAD

Fig. 9-47 *Details of how the flange or flashing is covered by the exterior wall covering.*

Fig. 9-48 *Windows help to make a house attractive.*

Fig. 9-49 *Flush-mount skylight.* (Novi.)

Fig. 9-50 *Curb-mount skylight.* (Novi.)

predrilled holes. This unit is easily installed by placing it directly over the roof opening and fastening it through the flange. It also has built-in condensation channels that get rid of unwanted moisture for maintenance-free operation. It, too, can be installed on a flat or pitched roof and comes in four roof opening sizes.

If you want something in the venting type, Fig. 9-52 has it. A built-in crank system allows operation with a hand crank or optional extension pole. The venting system is chain driven for better control and

Fig. 9-51 *Self-mount skylight.* (Novi.)

(A)

provides airtight closure when required. It can be installed on a flat or pitched roof and comes in two roof opening sizes. This skylight not only provides light but also gives you a way to allow for hot air to escape in the summer.

Installing a Skylight

Determine the roof opening location for the skylight on an inside ceiling or attic surface (Fig. 9-53). Roof openings should be positioned between rafters whenever possible to keep rafter framing to a minimum. Take a look at Fig. 9-54 for framing diagrams for both 16- and 24-inch O.C. rafter spacing.

In a room with a cathedral ceiling, use a drill and wire probe to determine rafter spacing. When working in an attic, position the roof opening so that it is relative to the proposed opening (Fig. 9-53). Make sure that the installation area does not have plumbing or electrical wires inside the opening area.

Preparing the Roof Opening

From inside the house, square the finished roof opening dimensions for the skylight between the roof rafters. If a roof rafter does not cross the proposed opening, mark the four corner points of the roof opening (Fig. 9-55).

When a rafter must be cut away from the opening, measure 1½ inch beyond the finished roof opening dimensions on the upper and lower sides of the proposed opening, and mark the four corner points (see Fig. 9-55B). This extra measurement allows for lumber that will frame the opening.

(B)

Fig. 9-52 *A. Venting self-mount skylight.* (Novi.) *B. Using a crank pole to open the skylight.* (Velux-America.)

Cutting the Roof Opening

Drive a nail up through the roof at each designated corner point. Go to the roof, and remove the shingles covering the proposed opening, as indicated by the nails. Remove the shingles 12 to 14 inches beyond the pro-

Fig. 9-53 *Measuring the opening for the skylight.*

Fig. 9-54 *Framing diagrams. (Novi.)*

MODEL	KEY	OVERALL SIZE	FINISHED ROOF OPENING DIMENSION
16 x 32	A	24" x 40"	14" x 30"
24 x 24	B	32" x 32"	22" x 22"
24 x 48	C	32" x 56"	22" x 46"
32 x 32	D	40" x 40"	30" x 30"

posed opening at the top and on both sides, leaving the bottom row of shingles in place. If roof felt underlies the shingles, it is not necessary to remove it from the installation area.

Draw connecting lines between the nails, and cut a hole in the roof using a saber saw or circular saw set to a depth of about 1 inch (Fig. 9-56).

The rafter sections that remain in the opening must be cut away perpendicular to the roof-deck surface with a handsaw. Temporary rafter supports should be installed to maintain structural alignment of the rafters.

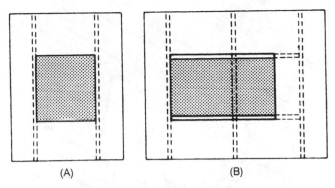

Fig. 9-55 *A. Roof rafter does not cross the proposed opening. B. Rafter must be cut to make room for the opening. (Novi.)*

Fig. 9-56 *Cutting the hole in the roof. (Novi.)*

Framing the Roof Opening

Frame the roof opening at the top and bottom by cutting two sections of lumber to fit between the existing rafters under the roof deck. Refer to the framing diagrams in Fig. 9-54.

Lumber used in framing should be the same size as the existing roof rafter lumber. When working with a hole such as that shown in Fig. 9-55, position each header under the roof deck to align with the top and bottom edges of the roof opening, and secure it with nails.

If a rafter has been removed from the opening, as shown in Fig. 9-55B, secure each header between the rafters and into the cut rafter ends with nails (Fig. 9-57). Nail a 1½-inch-wide sheathing patch over the top of each header to make the opening level with the roof.

Install a side header where the roof opening does not align with an existing roof rafter. The finished rafter frame should align with the finished roof opening dimension.

Fig. 9-57 *Framing the roof opening.* (Novi.)

Mounting the Skylight

Working on the roof, apply a layer of roof mastic (¼ inch deep) around the outer edge of the roof opening, 3 to 4 inches wide. Keep the mastic 1 inch away from the edge of the roof opening to prevent oozing (Fig. 9-58).

For installations made on a pitched roof, you have to rotate the skylight until the runoff diverter strip is positioned at the top side of the roof openings, as shown in Fig. 9-59.

Set the skylight into the mastic to align with the roof opening. Be sure that the lower skylight flange overlaps the bottom row of shingles by at least 1 inch. Secure the skylight to the roof with screws using the predrilled flange holes.

Fig. 9-59 *Making sure that the diverter strip is located properly.* (Novi.)

Sealing the Installation

Select a flashing material that will seal three sides of the installation area. You may want to use metal, aluminum, or asphalt. If you use asphalt, select a minimum 30-pound rolled asphalt. Begin by cutting three sheets 5 inches longer than each exterior deck flange located at the top and both sides of the skylight. The width of each sheet should be cut to measure 8 to 10 inches (see Fig. 9-59).

Apply a layer of roof mastic to cover the exterior deck flange and roof deck at the top and on both sides of the skylight. Spread the mastic to completely cover the deck flange and 8 to 10 inches of the roof deck around three sides of the installation (Fig. 9-60).

Working with either side of the skylight, center the asphalt over the deck flange, and press it into the mastic. Asphalt should extend 2½ inches over the flange at

Fig. 9-58 *Spreading mastic around the opening.* (Novi.)

Fig. 9-60 *Applying mastic to cover the exterior deck flange.* (Novi.)

RUNOFF DIVERTER
POSITION AT TOP OF ROOF OPENING
FOR PITCHED ROOF

SHINGLES

MASTIC

ASPHALT

MASTIC

MASTIC

LUMBER
INSULATOR

FOAM INSULATOR

ROOF DECK
EXTERIOR DECK FLANGE
PRE-DRILLED FLANGE HOLE

Fig. 9-61 *Applying asphalt or tar paper or roofing paper.* (Novi.)

Fig. 9-62 *Replacing the shingles.* (Novi.)

the top and bottom for adequate coverage. Continue to apply asphalt to the opposite side and then to the top. Asphalt at the top must overlap the asphalt on both sides (Fig. 9-61).

Replacing the Shingles

After the asphalt has been set into place, apply a layer of mastic to completely cover the asphalt around the skylight. Starting at the bottom, replace each row of shingles. Trim the shingles around the skylight where necessary (Fig. 9-62).

Preparing the Ceiling Opening

Drop a plumb line from each inside corner of the framed roof opening to the ceiling, and mark four corner points (Fig. 9-63). Remove the insulation between the joists 4 to 6 inches beyond the proposed ceiling opening.

If you want a ceiling opening that is larger than the roof opening or if an angled shaft is desired, some

Fig. 9-63 *Dropping a plumb line to find the ceiling opening.* (Novi.)

more measuring will have to be done at this time. To angle the base of the skylight shaft beyond a parallel ceiling opening, pull the plumb line taut to the floor at the desired angle, and mark each point. Using a tape measure and carpenter's square, determine the exact size and location of the proposed ceiling opening.

Tap a nail through the ceiling at each corner point. Find the locator nails from the room above, and draw connecting lines between each point. Cut through the ceiling along the lines, and remove the section (Fig. 9-64).

Fig. 9-65 *Framing the light shaft.* (Novi.)

Fig. 9-64 *Removing the ceiling section.* (Novi.)

Framing the Ceiling Opening

The ceiling opening may be framed using procedures that apply to the framing of any roof opening. Using the same dimensional lumber as existing ceiling joists, cut the headers to fit between the joists, and secure them in place with nails. The finished inside frame should align with the ceiling opening.

Constructing the Light Shaft

The light shaft is constructed with bevel-cut 2- × 4-inch lumber hung vertically from the corners of the roof frame to the corners of the ceiling frame. Right angles are formed using two 2 × 4s at each corner of the proposed shaft to provide a nailing surface for the shaft liner. Additional 2 × 4 nailers are spaced to reinforce the shaft frame (Fig. 9-65).

Cut the drywall or plywood to the size of the inside shaft walls, and nail it into place. The shaft may be insulated from the attic for maximum efficiency (Fig. 9-66). The interior light shaft surfaces may be finished to match the room decor. Place trim around the opening in the ceiling, and finish it to match the room decor.

Fig. 9-66 *Finishing or closing up the light shaft.* (Novi.)

OPERATION AND MAINTENANCE OF SKYLIGHTS

Condensation

Drops of condensation may appear on the inner dome surfaces with sudden temperature changes or during periods of high humidity. These droplets are condensed moisture and do not indicate a water leak from outside moisture. Condensation will evaporate as conditions of temperature and humidity normalize.

Figure 9-67 shows how light shaft installations can be used to present the light from the skylight to various parts of the room below. Figures 9-68 and 9-69 show how the original installations are made in houses under construction. The details and basic sizes are

SUGGESTED LIGHT SHAFT INSTALLATIONS

Where a roof window is installed above a flat ceiling, a light shaft will be needed. Typical installations are shown below. Flaring the shaft will give broader light distribution. Shaft construction by others.

TUNNEL LIGHT SHAFT
(90° TO CEILING ON ALL FOUR SIDES)

ANGLED LIGHT SHAFT
(90°TO PORCH PITCH)

TUNNEL FLARE LIGHT SHAFT
(SHAFT FLARED AT HEAD & 90° TO CEILING
AT SILL)

WIDE ANGLE LIGHT SHAFT
(FLARED ON ALL FOUR SIDES)

Fig. 9-67 *Suggested light shaft installations.* (Andersen.)

given, along with the roof pitch/slope chart. These will help you to plan the installation from the start.

Care and Maintenance

If the dome is made of plastic, the outer dome surface may be polished with paste wax for added protection from outdoor conditions. If it is made of glass, you may want to wash it before installation and then touch up the finger marks after it is in place. Roofing mastic can be removed with rubbing alcohol or lighter fluid. Avoid petroleum-based or abrasive cleaners, especially on the clear plastic domes. Roof inspection should be conducted every 2 years to determine potential loosening of screws, cracked mastic, and other weather-related problems that may result from the normal exposure to outdoor conditions.

Tube-Type Skylights

The newer tube-type skylights can be installed during construction or after. They are designed to provide maximum light throughput from a relatively small unit. They are right for areas where a larger standard skylight may not be practical (Fig. 9-70). The tube shaft can be designed to reflect 95 percent of available sunlight. The low-profile ceiling diffuser spreads natural light evenly through interior space. Early-morning and late-afternoon light can be captured and used by the dome to provide good illumination even during winter months in northern locations (Fig. 9-71).

The tube-type skylight comes in kit form with everything needed, including illustrated instructions,

for the do-it-yourselfer. It installs in a few hours with basic hand tools. No framing, dry-walling, mudding, or painting is required. It is available in both 10- and 14-inch diameters and therefore fits easily between standard 16- and 24-inch O.C. rafters (Fig. 9-72).

Most people are concerned with skylights because they have heard of them leaking, especially during the winter with snow pileup and melting. The skylight shown here has a one-piece roof flashing that eliminates leaks. Flashing is specific to the roof type and ensures a perfect fit. The 14-inch skylight spreads light up to 300 square feet. There is also an electric light kit available that makes the skylight into a standard light fixture at night and during dark periods of the day. It is designed to work from a wall switch and is an Underwrtiers' Laboratories–approved installation (Fig. 9-73).

Installation   To start, locate the diffuser position on the ceiling. Check the attic for any obstructions or wiring. Locate the position on the roof for flashing and dome. If the skylight is being installed in new construction, you can make sure that plumbing and electrical subcontractors take the skylight into consideration during the construction phase. Measure and cut an opening in the roof. Loosen shingles, and install the flashing. (In new construction, it may be best to install the flashing before the shingles are in place.) Insert the adjustable tube, and attach the dome (Fig. 9-74). Measure and cut an opening in the ceiling. Install the ceiling trim ring, and attach the diffuser. In the attic, assemble, adjust, and install the tubular components. In colder climates, it is necessary to insulate the tube shaft.

VERTICAL DETAIL VENT UNIT

Optional Water Deflector

9" Flashing

Apply Caulking

Insulation

Sash Frame Screen

Double Pane Tempered High Performance Insulating Glass

Operator Handle

Vapor Barrier

HEAD

6" Flashing

1/2"

1-3/4"

Pine Extension Jambs By Others

3/8"

UNIT DIMENSION HEIGHT

ROUGH OPENING HEIGHT

3/8"

SILL

1/2"

scale 1½" = 1'0"

34°-8/12 Roof pitch shown. For complete specifications see installation instructions.

BASIC SIZES

UNIT DIM.	1'-9¼"	2'-5⅛"	3'-5⅝"
RGH. OPG. †	1'-10"	2'-5⅞"	3'-6"
GLASS VENT.*	15⅟₁₆"	23"	35³⁄₁₆"

2'-9¼"   2'-10"   27⅞"

RW2133V

*Unobstructed glass size shown in inches.

3'-8⅛"   3'-8¾"   37¾"

RW2144V   RW2944V   RW4144V

4'-9¼"   4'-9¼"   50¾"

RW2957V   RW4157V

HORIZONTAL DETAIL VENT UNIT

Double Pane Tempered High Performance Insulating Glass

6" Step Flashing

3-11/32"

Apply Caulking

Insulation

Pine Extension Jambs By Others

Vapor Barrier

JAMB

3/8"

UNIT DIMENSION WIDTH

3/8"

JAMB

ROUGH OPENING WIDTH

INCLINE CURB FLASHING ROUGH OPENINGS

†When installing units with incline curb flashing use these rough openings.

Unit	Width Dim. A	Height Dim. B
2133	21-5/8"	34-3/4"
2144	21-5/8"	45-1/2"
2944	29-1/2"	45-1/2"
2957	29-1/2"	58-3/4"
4144	41-3/4"	45-1/2"
4157	41-3/4"	58-3/4"

Header Set Back

Roof Thickness	Dim. C
6-1/2"	13/16"
8-1/2"	1-1/8"
10-1/2"	1-1/2"
12-1/2"	1-13/16"

ROOF PITCH/SLOPE CHART

Roof Pitch	Roof Slope
2/12	9° 26'
3/12	14°
4/12	18° 26'
5/12	22° 37'
6/12	26° 34'
7/12	30° 15'
8/12	33° 41'
9/12	36° 52'
10/12	39° 48'
11/12	42° 30'
12/12	45°
14/12	49° 24'
20/12	59°
40/12	70°
68/12	80°

VERTICAL DETAIL VENT UNIT

Incline Curb Flashing

Batt Insulation

Header

Vapor Barrier   Roof Thickness

HEAD

SILL

1/2" Drywall

Blocking

ROUGH OPENING HEIGHT

Header Set-Back Determined by Roof Thickness

Incline curb flashing is recommended for roof installations less than 18½° (4/12 pitch) to 9° (2/12 pitch) minimum.

3/12 roof pitch (14°) shown. For complete specifications see installation instructions.

Fig. 9-68 *Roof window vent unit in place.* (Andersen.)

VERTICAL DETAIL STATIONARY UNIT

Double Pane Tempered High Performance Insulating Glass

9" Flashing

Frame

Apply Caulking

Insulation

Vapor Barrier

HEAD

6" Flashing

Pine Extension Jambs By Others

UNIT DIMENSION HEIGHT

ROUGH OPENING HEIGHT

3/8"

3/8"

SILL

34°-8/12 Roof pitch shown. For complete specifications see installation instructions.

BASIC SIZES

UNIT DIM.	1'-9 1/4"	2'-5 1/8"	3'-5 5/16"
RGH. OPG.	1'-10"	2'-5 7/8"	3'-6"
GLASS STAT*	18 5/16	26 3/16	38 7/16

2'-9 1/2" 2'-10 1/4" 30 9/16"
RW2133S

3'-8 3/16" 3'-8 15/16" 41 3/16"
RW2144S   RW2944S   RW4144S

4'-9 1/16" 4'-9 13/16" 54 3/16"
RW2957S   RW4157S

scale 1½" = 1'0"

*Unobstructed glass sizes shown in inches.

HORIZONTAL DETAIL STATIONARY UNIT

Double Pane Tempered High Performance Insulating Glass

Step Flashing

Caulk

Insulation

Drywall Return (by others)

JAMB

UNIT DIMENSION WIDTH

ROUGH OPENING WIDTH

JAMB

Fig. 9-69 *Roof window stationary unit in place.* (Andersen.)

TERMS USED IN WINDOW INSTALLATION

Now is a good time to review the terms associated with installation of a window. This will make it possible for you to understand the terminology when you work with a crew installing windows.

Plumb The act of checking the vertical line of a window when installing it in a rough opening.

Level The act of checking the horizontal line of a window when installing it in a rough opening.

Regular stud A vertical frame member that runs from the bottom plate on the floor to the top plate at the ceiling. In normal construction, this is a 2 × 4 approximately 8 feet long.

Jack or trimmer stud A vertical frame member that forms the window rough opening at the sides and supports the header. It runs from the bottom plate at the floor to the underside of the header.

Header A horizontal framing member located over the window rough opening supported by the jack studs. Depending on the span, headers usually are double 2 × 6s, 2 × 8s, or 2 × 10s in frame-wall construction or steel I-beams in heavier masonry construction.

Rough sill A horizontal framing member, usually a single 2 × 4, located across the bottom of the window rough opening. The window unit rests on the rough sill.

Cripples Short vertical framing members spaced approximately 16 inches O.C. and located be-

Fig. 9-70 *Skylight installation.* *(ODL, Inc.)*

(A)

(B)

Fig. 9-71 *A. The dome above the roof line. B. The dome reflects the sunlight coming from any angle throughout the day in any season.* *(ODL, Inc.)*

low the rough sill across the width of the rough opening. Also used between the header and the top plate depending on the size of the headers required.

Shim An angled wood member—wedge shaped—used as a filler at the jamb and sill. (Wood shingle makes a good shim.)

Wood buck A structural wood member secured to a masonry opening to provide an installation frame for a window unit.

Fig. 9-72 *Exploded view of the skylight. (ODL, Inc.)*

- Solar Lens® Dome
- Flashing (Asphalt Type)
- 15" Adjustable Tube with Reflective Lining
- 20" Extension Tube with Reflective Lining
- 15" Adjustable Tube with Reflective Lining
- Tube Ring Seal
- Ceiling Trim Ring
- Low-Profile Diffuser

Fig. 9-73 *Conversion of the skylight to a light fixture. (ODL, Inc.)*

Fig. 9-74 *Installation of the skylight. (ODL, Inc.)*

PREHUNG DOORS

Types of Doors

Exterior doors are made in many sizes and shapes (Fig. 9-75). They may be solid with a glass window. They may have an X shape at the bottom, in which case they are referred to as a *cross-buck door*. These doors are made in a factory and crated and shipped to the site. There they are unpacked and placed in the proper openings. There is little to do with them other than level them and nail them in place. The hardware is already mounted on the door.

Fig. 9-75 *Various door designs. (National Woodwork Manufacturers.)*

Figure 9-76 shows a door that was prehung and shipped to the site. Note how it sticks out from the sheathing so that the siding can be applied and butted to the side jamb. Doors are chosen for their contribution to the architecture of the building. They must harmonize with the design of the house. Figure 9-77 shows a door that adds to the design of the house in which it is installed. The door facing or trim adds to the column effect of the porch.

Flush doors Flush doors are made of plywood or some facing over a solid core. The core may be made of a variety of materials. In some instances, where the door is used inside, the inside of the door is nothing

Fig. 9-76 *A prehung exterior door with three panels of glass. Note how the trim sticks out sufficiently for the siding to butt against it.*

Fig. 9-77 *The proper door can do much to improve the looks of a house.*

more than a mesh or strips (Fig. 9-78). Wood is usually preferred over metal for exterior doors of homes. Wood is nature's own insulator. While metal readily conducts heat and cold, wood does not. Wood is 400 times more effective an insulator than steel and 1,800 times more effective than aluminum.

Stock wood flush doors come in a wide variety of sizes, designs, and shapes. Standard wood door frames will accommodate wood combination doors, storm doors, and screen doors without additional framing expense.

Panel doors   This type of door has solid vertical members, rails, and panels. Many types are available (Figs. 9-79 and 9-80). The amount of wood and glass varies. Many people want a glass section in their front door. The four most popular types are clear, diamond

obscure, circle obscure, and amber Flemish. Figure 9-80 shows some of the decorative variations in doors. Note how the type of door can improve the architecture. The main entrance may be highlighted with sidelights on one or both sides of the door (Fig. 9-81). These are 12 or 14 inches wide. The panels of glass are varied to meet different requirements.

Sliding doors   Sliding doors are just what the name suggests. They usually have tempered or safety glass. They can slide to the left or to the right. You have to specify a right- or left-sliding door when ordering. Most have insulating construction. They have two panes of glass with an airspace in between (Fig. 9-82). Figure 9-83 shows the sizes available. Also note the arrow, which indicates the direction in which the door slides.

French doors   This is usually two or more doors grouped to open outward onto a patio or veranda. They have glass panes from top to bottom. They may be made of metal or wood. Later in this chapter you will see two- and three-window groupings mounted step by step.

INSTALLING AN EXTERIOR DOOR

The door frame has to be installed in an opening in the house frame before the door can be hung. Figure 9-84 shows the parts of the door frame. Note the way it goes together. Figure 9-85 shows the frame in position with a wire on the right. It has been pulled through for installation of the doorbell push button. Note the spacing of the hinges. The door in the background is a six-panel type that is already hung. The siding has not yet been butted against the door casing. It will be placed as close as possible and then caulked to prevent moisture from damaging the wood over a period of time.

Figure 9-86 shows the general information you need to identify the parts of a door. It is important that the door be fitted so that there is a uniform ⅛-inch clearance all around to allow for free swing. Allow ½ to ¾ inch at the bottom for a better fit with carpet.

There are nine steps to installing an exterior door:

1. Trim the height of the door (see Fig. 9-86B). Many doors are made with extralong stiles or horns. Before proceeding, cut the horns off the *top* of the door, even with the top of the top rail B. When cutting, start with the saw at the outside edge to avoid splintering the edges of the door.

2. On the inside of the room, place the door into the opening upside down and against the door jamb

Fig. 9-78 *Various types of materials are used to fill the interior space in flush doors.* (National Woodwork Manufacturers.)

Fig. 9-79 *Add-on panels and lights give designers options with flush doors.* (General Products.)

(see Fig. 9-86C). Keep the hinge stile tight against the hinge jamb A1 of the door frame. This edge must be kept straight to ensure that the hinges will be parallel when installed. Place two ¼-inch blocks under the door. These will raise the door to allow for ⅛-inch clearance at both the top and bottom when cut. Mark the door at C, the top of the door frame opening, for cutting.

3. After cutting the door to the proper height, place the door into the opening again—this time right side up, with ⅛-inch blocks under the door for clearance (see Fig. 9-86D). With the door held tightly against the hinge jamb of the door frame, have someone mark a pencil line along the lock

CHATEAU 8-PANEL

WILLIAMSBURG 6-PANEL

COLONIAL 9-LIGHT

CROSS BUCK 9-LIGHT

Fig. 9-80 *The four most popular door styles. They have clear, diamond obscure, circle obscure, and amber Flemish safety glass.* (General Products.)

Fig. 9-81 *Sidelights are designed for fast installation as integrated units in wood or plastic models. This insulated safety glass comes in 12- or 14-inch widths. (General Products.)*

stile of the door from outside the door opening (line 1 to 1A), holding a ⅛-inch block between the pencil and the door frame. This will automatically allow for the necessary ⅛-inch clearance needed.

4. Trim the width of the door (see Fig. 9-86E). If the amount of wood to be removed from the door (line 1 to 1A in Fig. 9-86D) is more than ¼ inch, it will be necessary to trim both edges of the door. Trim one-half the width of the wood to be removed from each edge of the door. Use a smooth or jack plane.

5. Bevel the lock stile of the door (see Fig. 9-86F). The lock stile of the door should be planed to about a 3-degree angle so that it will clear the door frame when the door is swung shut.

6. Install the hinges (see Fig. 9-86G). Be sure that markings for hinges are uniform for all hinges

Fig. 9-82 *Sliding door.*

Fig. 9-83 *Various sizes of sliding doors. (Andersen.)*

Fig. 9-84 *Assembling a door frame (left to right)*

Fig. 9-85 *Prehung door in place. Note the wires hanging out on the left side. They indicate where a doorbell pushbutton will go.*

used. The mortises (cutouts) for hinges should be of uniform depth (thickness of the hinge). Measure 7 inches down from the top of the door and 7⅛ inches down from the underside of the door frame top. Mark the locations of the top edge of the upper hinge. Placement of hinges will be unnecessary if the door is prehung.

7. The bottom hinge is 9 inches up from the door bottom (9⅛ inches up from the threshold) (see Fig. 9-86H). The middle hinge is centered in the door height. Attach the hinge leafs to the door and door frame. Hang the door in the opening. If the mortises are cut properly but the hinges still bind, the frame jamb may be distorted or bowed. It may be necessary to place a thin shim under one edge of the frame hinge leaf to align it parallel and relieve the binding.

8. Install the lockset. Because of the variety of styles of locksets, there is no one way to install them. This subject will be discussed in detail later in this chapter. Each lockset comes with a complete set of instructions. The best advice is to follow these instructions.

9. Finish the door. Care must be taken to paint or seal all door surfaces. Top, bottom, edges, and

DOOR TERMINOLOGY

A. HINGE STILE
   (WHERE HINGES ARE TO BE
   APPLIED)

B. LOCK STILE
   (WHERE LOCK IS TO BE
   INSTALLED)

C. BOTTOM RAIL

D. TOP RAIL

(A)

(B)

(C)

(D)

DOOR FRAME

BEVEL

(F)

LOCK EDGE OF DOOR BEVELED

(E)

FITTING BUTT HINGE TO DOOR

MARKING LOCATION

MARKING DEPTH

SCORING

COMPLETED MORTISE

(G)

(H)

Fig. 9-86 *A. Terms used with doors. (Grossman Lumber.) B. Trim the door height. C. Mark the door to fit the frame. D. Mark the other end of the door. E. Trim the width of the door. F. Bevel the lock stile of the door. G. Install the hinges. H. Check the height of the hinges.*

faces should be sealed and painted. Weather and moisture can hurt a door and decrease its performance. A properly treated door will give many years of satisfactory service.

Hanging a Two-Door System

In some cases, a two-door system is chosen for the main entrance (Fig. 9-87). In other cases, the two-door system may be French doors (Fig. 9-88). This type of door system comes in pieces and has to be assembled. The details for hanging of a two-door system are shown in Fig. 9-89.

Figure 9-90 shows how the active and inactive doors are identified first. In most cases, both doors do not open. This is especially the case when the entrance door is involved. In the case of French doors, both doors can open.

Fig. 9-87 *Double doors installed. The one on the left is the active door.*

Fig. 9-88 *Double door ready for assembly.* (General Products.)

Handing Instructions

With the assembly kit furnished, one of the first things you will want to check is the "hand" of the door. Hand of doors is always determined by the *outside*. Inswinging doors are more common (Fig. 9-91). The right-hand symbol is *RH*, and it means that the door swings on hinges that are mounted on the right. If the door swings out, it is a right-hand reverse door, and the symbol is *RHR*. It is still hinged on the right looking at it from the outside.

The left-hand symbol is *LH*. This means that on an inswinging door the hinges are on the left. This is most convenient for persons who are right-handed. If the door is outswinging, the left-hand reverse symbol is *LHR*. Doors usually swing into a wall so that they can rest against the wall. They usually swing back only 90 degrees. The traffic pattern also determines the way a door swings. Doors swing out in most commercial, industrial, and school buildings. This lets a person open the door outward so that it will be easy to leave the building if there is a fire. Safety is the prime consideration in this case.

Figure 9-92 shows how energy conservation has entered the picture. The figure shows how the top and bottom of the door are fitted to make sure that air does not leak through the door. The door may be made of metal. Because metal conducts heat and cold, the door is insulated (Fig. 9-93).

Metal Doors

Metal doors also may be used for residential houses. In some cases they are used to replace old or poorly fitting wooden doors. Figure 9-94 shows how doors with metal frames are designed for ease of installation. Figure 9-95 shows how the metal frame is attached to concrete, wood, or concrete blocks.

One of the advantages of metal doors is their fire resistance. Frames for 2-foot, 8-inch, and 3-foot, 0-inch doors carry a 1½-hour label. The frames for 3-foot, 6-inch, and double doors are not labeled as to fire resistance.

In the case of metal frames, the frame has to be installed before the walls are constructed. The frame requires a rough opening 4½ inches wider and 2¼ inches higher than nominal. The stock frame is usually 5⅓ inches wide.

INSTALLING FOLDING DOORS

Folding doors are used to cover closets with any number of interesting patterns. They may be flush and plain or mirrored. They may have two or four panels.

Step 1   Attach Astragal to Inactive Door

Remove plastic filler plates from deadbolt and lock locations. Remove appropriate metal knock-out at deadbolt location.

Place door on edge, place astragal with *notch to door top* as shown. Compress outer flange against face of door, install bolt retainer spring at top and bottom of astragal, and secure astragal with five (5) self-tapping screws using power driver or drill.

Place two nylon screw bosses under the strike route and secure strike with 2 No. 8 screws provided.

Strike has tab that can be adjusted to assure proper closing while hanging door.

Place the bolt assembly in top and bottom of astragal. Adjust the bolt retainer spring to proper position and secure bolt retainer with Allen wrench provided.

Snap in 2 ea. channel closures in bolt recess above and below strike location. (proper lengths provided).

Tape one pile pad to interior face of inactive door for installation on astragal after door is installed in opening.

Step 2   Attach Sweep

Place inactive door on prehang table, interior face up. Pick up sweep with flush bolt hole, place door bottom spacer on latch end of sweep as shown.

Place sweep on bottom of door, flush latch end with latch edge of astragal.

Compress tightly against door bottom and drive self drilling screws into door skin at extreme bottom of slot. Tighten screws moderately to hold sweep in up position.

Check operation of bolt assembly.
Turn door over (interior face down)
Place active door on table (interior face up) and install sweep in same manner as first door.
Turn door over and proceed with frame prehanging. (Step 3)

Step 3   Attach Hinge Jambs

Place hinge jambs beside edge of doors as shown and attach hinges to doors with No. 10 machine screws.
Apply caulking tape to jambs at threshold locations. Make sure it follows contour of vinyl threshold part.

Fig. 9-89 *Details for hanging a two-door system.*

The sizes and door widths vary to suit the particular application (Table 9-1).

Figure 9-96 shows the openings and details of fitting a metal bifold door. Figure 9-97 shows the details of the four panels to be installed and different panel styles available. Note the names given to the parts so that you can follow the installation instructions.

1. Carefully center the top track lengthwise in the finished opening. This should suit both flush and recessed mountings. Attach the track with No. 10 × 1¼-inch screws in the provided holes. The bifold track is assembled for four-panel door installation. The track may be separated at the center without the use of tools when a two-panel door is

Step 4 — Installing Header and Threshold

Stick ⅛" PAK-WIK spacers—2 ea. on header jambs, and 3 ea. on the astragal side at locations marked "X", to maintain ⅛" clearance between doors and jambs. Place header jamb against header stop block. Line up doors with header jamb and press doors firmly against it. Raise jambs up, make sure they are flush with header jamb at top corners. Drive 3 ea. 2¼" long staples in each corner of frame.

Assemble vinyl and aluminum threshold parts. Place threshold in frame and secure with #10 x 1½" screws through pre-drilled holes. Back edge to be flush with frame. (Make sure pile is firmly in contact with threshold and weatherstripping.) If not, remove and reposition.

Step 5 — Attaching Brickmold

NOTE: If door is to be outswing, proceed with bracing shown in Step 6. Then turn unit upside down and install bolt strike as in Step 6. Proceed with brickmold in Step 5. NOTE: Outswing frame requires ⅝" longer header brickmold than inswing. Place brickmold gage at top and bottom of jambs. Position miter joints of jamb and header brickmold. Align fit of mitered corners and properly space reveal. Tack each corner, nail header brickmold, move gage down jambs and nail brickmold. Use 6 ea. No. 10 x 2½" finishing nails as shown, drive 2 ea. No. 10 x 2½" finishing nails in two corners as shown.

Step 6 — Installing Flush Bolt Strike and Bracing Frame

Turn unit upside down and replace on table. Place pencil mark at flush bolt locations on header and threshold. Open inactive door, place thin dab of putty at bolt pencil marks on header and threshold. Close inactive door. Move bolts to mark the putty. Open inactive door. Center punch top and bottom bolt locations with nail. Remove putty. Drill ⅝" hole in header and threshold. Place bolt strike on header and align with drilled hole. Install 2 ea. No. 6 x 1" screws provided. Close inactive door and secure bolts. Close active door. Tack corner braces as shown. Tack strip of wood across frame, approximately 12" above threshold as shown. Use 8d coated box nails. On outswing unit, cut bracing to fit between brickmold.

Fig. 9-89 (Continued)

installed (Fig. 9-98). Knobs, screws, and rubber stops are packaged for two-panel installation.

2. Place the bottom track, either round edge or square edge, toward the room. Plumb the groove with the top track (Fig. 9-99). Screw the track to the floor with ½-inch screws, or fasten it to a clean floor with 3M double-coated tape (No. 4432).

3. On all two-panel sections, lower the bottom pivot rod until it projects ½ inch below the edge of the door. Make it 1¼ inches if carpet is under the door.

4. Attach the doorknobs.

5. Lift one door set. Insert the bottom pivot rod (threaded) into the bottom pivot bracket. Pull down the top spring-loaded pivot rod. Insert it

64 17/32"-5'-4" DOOR
72 17/32"-6'-0" DOOR

4 1/2"
5 1/8"

INACTIVE DOOR

ACTIVE DOOR

Fig. 9-90 *Active and inactive door designation.* (General Products.)

INSWING

OUTSWING

RIGHT HAND
SYMBOL: RH

LEFT HAND
SYMBOL: LH

RIGHT HAND
REVERSE
SYMBOL: RHR

LEFT HAND
REVERSE
SYMBOL: LHR

Fig. 9-91 *Handing chart.* (General Products.)

into the pivot bracket in the top rack. Insert the top and bottom nylon glide rod tips into the track (Fig. 9-100).

6. Install the second door set the same way.

7. Insert the rubber stop in the center of the top and bottom tracks. Make sure that the stop seats firmly in the track. For a two-panel installation, cut the rubber stop to the proper length.

8. Because of their design, the bifold doors are rigid enough to operate smoothly without a full bottom track. This permits better carpeting in the closet. Saw off a 4-inch section of the bottom track. Place this on the floor. Use a plumb bob to pivot the bottom points with the top pivot (Fig. 9-101). Fasten the section to the floor with two ½-inch screws. Remove the bottom glide rods from the doors. Single-

track installation is not recommended for 8-foot, 0-inch-high doors; 7-foot, 0-inch-wide four-panel doors; or 3-foot, 6-inch-wide two-panel doors.

Final adjustments To raise or lower the doors to the desired height, turn the threaded bottom pivot rod with a screwdriver. Make sure that the doors are even and level across the top. Tighten the locknut.

Doors should close snugly against the rubber stop. For horizontal (lateral) alignment, loosen the screw holding the top or bottom pivot brackets in the track. Adjust the door in or out. Retighten the screw.

Keep all glides and track free from paint and debris. The aluminum track is already lubricated to ensure smooth operation. Occasionally repeat the lubrication with silicon spray, paraffin, or soap. This keeps door operating free and easy.

FBI INSWING
WITH FTAS SWEEP

4 1/2"

6' 9 3/4"
ROUGH
OPENING

6' 10 1/2"
BRICK
OPENING

6' 9 5/16"
FRAME
SIZE

SHIM

SUBFLOOR   FINISH FLOOR

FBO OUTSWING
WITH FTAS SWEEP

4 1/2"

6'10 1/2"
BRICK
OPENING

6' 9 3/4"
ROUGH
OPENING

6' 9 5/16"
FRAME
SIZE

SHIM

FINISH
FLOOR   SUBFLOOR

FBEI ENERGY THRESHOLD
WITH FTAS SWEEP

6 1/2"

6' 9 3/4"
ROUGH
OPENING

6' 10 1/2"
BRICK
OPENING

6' 9 5/16"
FRAME
SIZE

SHIM

SUBFLOOR   FINISH FLOOR

Fig. 9-92 *Finished dimensions on a double-hung door.* (General Products)

"U" CHANNELS

"U" channels on door edges add to rigid construction and provide thermal barrier to prevent heat transfer. No-wood design.

CHOICE OF JAMBS

Extra-wide jamb (6½") is designed for the new Energy Construction method in home building.

MAGNETIC WEATHERSTRIPS

Magnetic weatherstrips on head and strike jambs "reach out" for positive sealing like a refrigerator door.

COMPRESSION WEATHERSTRIP

Double-compression vinyl weatherstrip on hinge jamb completes the all-around weather-stopping action.

DEEP-EMBOSSED ON BOTH SIDES

CORNER PADS

Flexible pads at both lower corners seal out wind and water.

ADJUSTABLE DOOR SWEEP

Bottom sweep can be adjusted for snug fit against the threshold to seal against air and water leakage.

THRESHOLD

Aluminum threshold was designed for the new Energy Construction. Two other frost-break options: aluminum outswing threshold; aluminum-vinyl threshold for standard construction.

FOAM-FILLED INSULATING SANDWICH

Thick density (2.8 lb/ft²) of special foamed-in-place polyurethane uses natural insulation of tiny trapped gas cells.

FROST-BREAK THERMAL BARRIER

Fig. 9-93 *Insulated metal door for commercial, industrial, or residential use. Note the energy-saving features.* (General Products.)

These instructions are for a particular make of door. However, most manufacturers' instructions are basically the same. There are some minor adjustments you will have to make for each manufacturer. Make sure that you follow the manufacturer's recommendations.

DOOR AND WINDOW TRIM
Interior Door Trim

Most inside or interior doors have two hinges. They usually come in a complete package. Once they are set in place, the casing has to be applied. Figure 9-102 shows the location of the casing around an interior door. Note how the jamb is installed. The stop is attached with nails and has a bevel cut at the bottom of the door. It prevents the door from swinging forward more than it should.

Figure 9-103 shows the two of the most commonly used moldings applied to the trim of a door. These two are colonial and ranch casing moldings. These are the names you use when ordering them. They are ordered from a lumberyard or mill.

CORNER DETAIL
ASSEMBLE WITH 4 TEK SCREWS

CROSS SECTION

1 1/8" x 2 3/4"
MORTISE,
NO LIP,
FILLER PLATE
INSTALLED

PREPPED FOR
2 3/4" T STRIKE
WITH LIP

79 1/2"

32 1/16"
36 1/16"
42 1/16"

T ANCHOR
FOR MASONRY
(3 PER JAMB)

WOOD STUD ANCHOR
(3 PER JAMB)

FIXED FLOOR
ANCHOR

Fig. 9-94 *Putting together a metal frame for a door.* (General Products.)

5 3/4"

4 3/4"

3. 3 5/8" CONCRETE
BLOCK AND PLASTER

Fig. 9-95 *Typical installations with metal door frames.* (General Products.

TABLE 9-1 *Finished Opening Sizes for Bifold Doors*

Door Width	Number of Panels	Door opening, Inches*		Actual Door Width, Inches
		6'8"	8'0"	
1'6"	2	18½ × 80¾	18½ × 95¼	17⁷/₁₆
2'0"	2	24½ × 80¾	24½ × 95¼	23⁷/₁₆
2'6"	2	30½ × 80¾	30½ × 95¼	29⁷/₁₆
3'0"	2	36½ × 80¾	36½ × 95¼	35⁷/₁₆
3'0"	4	36½ × 80¾	36 × 95¼	35
3'6"	2	42½ × 80¾	42½ × 95¼	41⁷/₁₆
4'0"	4	48 × 80¾	48 × 95¼	47
5'0"	4	60 × 80¾	60 × 95¼	59
6'0"	4	72 × 80¾	72 × 95¼	71
7'0"	4	84 × 80¾	84 × 95¼	83

*Finished opening width shown provides ¹/₂ inch clearance each side of door. Finished opening width may be reduced by ¹/₂ inch provided finished opening is square and plumb. This will require cutting track. Finished opening heights shown provide ³/₈ inch clearance between door and track—top and bottom. (This makes ⁷/₈ inch between door and floor.) Doors can be raised to have 1¹/₈ inch clearance door to floor without increasing opening height.

hold the two pieces securely. Also notice the other nail locations. Why do you need to drill the nail hole for the toenailed side?

Installation of the strike plate in the side jamb is shown in Fig. 9-105. It has to be routed or drilled out. This allows the door-locking mechanism to move into the hole. Figure 9-106 shows how the strike plate is mounted onto the door jamb.

Window Trim

Windows have to be trimmed. This completes the installation job (Fig. 9-107). There are a couple of ways to trim a window. Take a look at Fig. 9-108, and note the difference. Shown is a trimmed window with casing at the bottom instead of a stool and apron. This is a quicker and simpler method of finishing a window. There is no need for a stool to overlap the apron or casing in some instances. This is the choice of the architect or the owner of the home. There are problems with

1. 3 1/4" STEEL STUD
3/8" SHEET ROCK AND PLASTER

2. 3 1/2" WOOD STUD
1/2" DRYWALL

4. 5 5/8" CONCRETE
BLOCK WALL

5. 7 5/8" CONCRETE
BLOCK WALL

SIDE TRIM
INSTALLATION

JAMB

1/8''

STACK TRACK HIGH
CARPET ADAPTER

1/2''

1/8'' MIN.

3/8'' NOMINAL

FULL 1 1/4''

80 3/4'' OR 95 1/4'' FINISHED OPENING

79'' ACTUAL DOOR – 6' 8''

93 1/2'' ACTUAL DOOR – 8' 0''

NUT AND
LOCKWASHER

3/8'' NOMINAL
5/8'' MAX.

DOOR
BOTTOM

1/2''

FLOOR

Normal clearance is 7/8'' to 1 1/8'' between
floor and bottom of door in standard 80 3/4''
finished opening.

Up to 1 3/4'' clearance under door may
be obtained by increasing finished
opening height. Do not increase to
more than 81 3/4'' without using
"Stack Track."

Clearance from 1 3/4'' to 2 1/4'' may be
obtained by using "Stack Track."
Finished opening height should not exceed
82 1/4''

1/2''

Floor   "Stack Track"

Fig. 9-96   *Installation of the bifold door.* (General Products.)

the apron and stool method. The apron and stool will pull away from the inside casing. This leaves a gap of up to ¼ inch. It can become unsightly in time.

Figure 9-109 shows some of the moldings that can be used in trimming windows, doors, or panels. These moldings are available in prepared lengths of 8 and 12 feet. Generally speaking, the simpler the molding design, the easier it is to clean. Many depressions or designs in a piece of wood can allow it to pick up dust. Some are very difficult to clean.

INSTALLING LOCKS

There are seven simple steps to installing a lock in a door. Figure 9-110 shows them in order.

In some cases, you might want to reverse the lock. This may be the case when you change the lock from one door to another. The hand of the door might be different. Figure 9-111 shows you how simple it is to change the hand of a lock. In some cases, you may have bought the lock without noticing how it should fit. In this way, you are able to make it fit in either direction.

There are a number of locks available. Figure 9-112 shows how 18 different locksets can be replaced by National Lock's locksets or lever sets. Figure 9-113 shows some of the designs available for strikes. The strike is always supplied with the lockset. Figure 9-114 shows the latch bolts. They may have round or square corners, and they may or may not have dead-lock capability.

Entrance handle locks   Most homes have an elaborate front door handle. Figure 9-115 shows two of the types of escutcheons used to decorate a doorknob.

Door handles also become something of a decorative item. They come in a number of styles. Each lock manufacturer offers a complete line. Figure 9-116 shows two such handles. These handles are usually cast brass.

A number of lockset designs are available for entrance and interior doors. They may lock then require a nail or pin to be opened, or they may require a key (Fig. 9-117).

Fig. 9-97 *A. Details of the metal bifold door.* (General Products.) *B. Different designs for bifold doors.*

FLUSH DOORS

CLASSIC COLONIAL

PANEL LOUVER

FULL LOUVER

Fig. 9-98 *Vinyl connectors let you snap apart four-panel track instantly to install two-panel bifolds.* (General Products.)

Fig. 9-99 *Using a plumb bob to make sure that the tracks line up.* (General Products.)

Fig. 9-100 *Pivoting the bottom pin in the track.* (General Products.)

Fig. 9-101 *You can remove most of the bottom track if you don't want it on the floor. This lets carpeting run straight through to the closet wall.* (General Products.)

Fig. 9-102 *Trim details for a door frame.*

FRAMING STUDS
PLASTER GROUND
SHINGLE WEDGE
NAIL
STOP
JAMB
CASING
NAIL
3/16″
BEVEL CUT
6d FINISHING NAIL
4d FINISHING NAIL

COLONIAL   RANCH

Fig. 9-103 *Two popular types of molding used for trim around a door.*

NAIL
NAIL
CASING
NAIL (DRILL)
NAIL
MITER JOINT
NAIL
BUTT JOINT

Fig. 9-104 *Two methods of joining trim over a door.*

BORE HOLE
ROUT OR CHISEL PLATE AREA

FRAMING STUDS
SIDEJAMB
STOP
NAILS
CASING
STRIKE PLATE (ROUT)

Fig. 9-105 *Installing a strike plate for the lockset.*

CLEARANCE HOLE FOR LATCH BOLT

Fig. 9-106 *Installing the strike plate on the door jamb.*

Auxiliary locks   Auxiliary locks are those placed on exterior doors to prevent burglaries. They are called *deadlocks.* They usually have a 1-inch bolt that projects past the door. It fits into the door jamb (Fig. 9-118). Figure 9-119 shows three of the various deadlock designs used. The key side, of course, goes on the outside of the door.

A standard type of lock is exploded for you in Fig. 9-120. Note the names of the parts. These locks have keys that fit into a cylinder. The keys lock or unlock the latch bolt. Exposed brass, bronze, or aluminum parts are buffed or brushed. They are protected with a coat of lacquer. Aluminum is brushed and anodized.

Construction keying   There are a couple of methods used for keying locksets. One of them makes it possible for a construction supervisor to get into a number of buildings with one key. Once the building is occupied, the lock is converted. The lock's key and no other then will operate it (Fig. 9-121).

Fig. 9-107 *Window trim. Note the apron and stool.*

Fig. 9-108 *Window trim. Note the absence of the apron.*

STOPS	
WP 444	11/16″ × 3-1/2″
WP 412	11/16″ × 3-1/2″
WP 432	9/16″ × 3-1/2″
WP 433	9/16″ × 3-1/4″
WP 452	11/16″ × 2-1/2″
WP 472	9/16″ × 2-1/2″
WP 620	9/16″ × 4-1/4″
WP 622	9/16″ × 3-1/2″
WP 623	9/16″ × 3-1/4″
WP 662	9/16″ × 3-1/2″
WP 663	9/16″ × 3-1/4″
WP 664	9/16″ × 3″
WP 712	9/16″ × 3-1/2″
WP 713	9/16″ × 3-1/4″
WP 714	9/16″ × 3″

STOPS	
WP 816	7/16″ × 1-3/8″
WP 818	7/16″ × 1-1/8″
WP 820	7/16″ × 7/8″
WP 846	7/16″ × 1-3/8″
WP 848	7/16″ × 1-1/8″
WP 850	7/16″ × 7/8″
WP 876	7/16″ × 1-3/8″
WP 878	7/16″ × 1-1/8″
WP 880	7/16″ × 7/8″
WP 906	7/16″ × 1-3/8″
WP 908	7/16″ × 1-1/8″
WP 910	7/16″ × 7/8″
WP 936	7/16″ × 1-3/8″
WP 938	7/16″ × 1-1/8″
WP 940	7/16″ × 7/8″

MULLION CASING

WP 978	3/8″ × 1-3/4″
WP 983	3/8″ × 1-3/4″

Fig. 9-109 *Different designs of molding used as trim for windows, doors, and other parts of the house.*

Builders use a short four-pin tumbler key to operate the lock during the construction period. A nylon wafer is inserted in the keyway at the factory. This blocks operation of the fifth and sixth tumblers. Accidental deactivation of the builder's key is unlikely. This is so because a conscious effort to apply a 10- to 15-pound force is required to dislodge the nylon wafer the first time the five-tumbler key is used. The owner's keys are packed in specially marked sealed envelopes.

When construction is completed, the unit is ready for occupancy. The homeowner inserts the regular five-pin tumbler key to move the nylon wafer. This makes the fifth tumbler operative. It automatically de-activates the four-pin tumbler arrangement, making the builder's key useless. Now the locksets can be operated only by the owner's keys.

Other lockmakers have different methods for this key operation (Fig. 9-122). Figure 9-123 shows another method of key operation of locks. In this case, the whole cylinder of the new lock is removed. The construction worker inserts a different cylinder. This cylinder works with the master key. Once the job is finished, the original cylinder is reinserted. The construction worker's key no longer operates the lock. Only the homeowner can operate it.

STORM DOORS AND WINDOWS

In most windows, thermopane is used. It consists of two pieces of glass welded together with an airspace in between (see Fig. 9-20). Some windows have the space evacuated so that a vacuum exists inside. This cuts down on the transfer of heat from the inside of a heated building to the cold outside or the reverse dur-

3. Position lock and latch by slot engagement.

4. Fasten face plate with 2 wood screws.

ADJUSTABLE LOCKSETS
FACTORY PRESET LOCKSETS

1. Mark hole centers with template. Drill for bolt and lock mechanism.

2. Mortise for face plate and strike plate.

3. Insert latch in 7/8" hole and fasten face plate with 2 wood screws.

4. Insert lockset into 1 7/8" hole so lock case "slot-engages" the latch.

5. Put on clamp plate by turning to engage keyway. Tighten screws.

6. Snap on rose. Depress spring retainer and apply knob.

7. Fasten strike plate with 2 wood screws.

Fig. 9-110 *Seven simple steps to install a lockset.* (National Lock.)

Apply lock to door with key to outside as shown above. (See installation instruction sheet.) Turn turnbutton or pushbutton on inside of door to locked position.

Insert key in lock. Turn key 30° either left or right. Do not retract latch bolt.

Keeping key in 30° position, remove lock and knob. First depress knob retainer pin. Then pull on key and knob together. Do not pull the knob out ahead of the key.

Be sure plug is in locked position. Then remove key approximately halfway out of plug. Turn entire plug, key, and cylinder in knob so bitting of key is up.

Apply knob on cam by engaging tab on knob with slot in tube. Keep key in vertical position. Push knob on tube while rotating key and plug gently. You will feel the proper engagement of locating tab on lock plug with slot in tube.

Depress knob retainer pin and push on knob as far as possible.

Push key all the way in keyway. Turn key to 30° position and push knob until retainer pin engages slot in knob shank.

Fig. 9-111 *Reversing the hand of a lock.* (National Lock.)

ing the summer. In some cases, as shown in Chapter 13, another piece of glass or plastic is added. It fits over the twin panes of glass (Fig. 9-124).

Metal transfers heat faster than wood. Wood is 400 times more effective than steel as an insulator, and it is 1,800 times more effective than aluminum.

Storm doors come in a wide variety of shapes and designs. They usually have a combination of screen

and glass. The glass is removed in the summer, and a screen wire panel is inserted in its place. In this way, the storm door serves year round (Fig. 9-125). Storm doors are delivered prehung and ready for installation. All that has to be done to install them is to level the door and add screws in the holes around the edges. A door closer is added to make sure that the door closes after use. In some cases, a spring adjustment device is

EASY LOCKSET INSTALLATION

1. Prepare door; drill for bolt and lock mechanism.

2. Insert bolt and lock mechanism.

3. Engage bolt and lock mechanism; fasten face plate.

4. Put on clamp plate; tighten screws.

5. "Snap-On" rose.

6. Apply knob on spindle by depressing spring retainer.

7. Mortise for latch bolt; fasten strike with screws.

EASY LOCKSET REPLACEMENT

With just a screwdriver, the following 18 residential lockset brands can be replaced by National Lock locksets or lever sets.

ARROW COMET CORBIN DONNER ELGIN HARLOC KWIKSET LOCKWOOD MEDALIST NATIONAL RUSSWIN SARGENT SCHLAGE TROJAN TROY WEISER WESLOCK YALE

Fig. 9-112 *Replacement of a lockset.* (National Lock.)

Fig. 9-113 *Strike designs.* (National Lock.)

Fig. 9-114 *Latch bolt designs.* (National Lock.)

Fig. 9-115 *Some of the many escutcheons for locks.* (Weiser Lock.)

added to the top so that the door is protected from wind gusts. Most storm doors are made of metal. However, they are available in wood or plastic.

Standard sizes are for openings from 35¾ to 36⅜ inches wide and from 79¾ to 81¼ inches high. There is an extender Z bar available for openings up to 37⅛ inches wide. According to the company's testing lab, a plastic (polypropylene) door has 45 percent more heat retention than an aluminum door (Fig. 9-126).

Fig. 9-118 *A deadlock projects out from the door and fits into the door jamb to make a secure door.* (Weiser Lock.)

Fig. 9-116 *Entrance-handle locks.* (Weiser Lock.)

Fig. 9-117 *Locksets for interior and exterior doors. (Weiser Lock.)*

Fig. 9-119 *Auxiliary locks. Chain and bolt and two types of deadbolts.* (Weiser Lock.)

TELESCOPING PRECISION
ALIGNMENT BETWEEN THE
TWO KNOB ASSEMBLIES

ACCESSIBLE ATTACHMENT SCREWS
(RECESSED AND REINFORCED
UNDER HEAD)

STEEL
REINFORCED
ROSES

LARGE SIZE,
PIN TUMBLER
CYLINDER
ASSEMBLY AND
BRASS PINS

STAINLESS STEEL
MECHANISM SPRINGS

BIND-FREE
FLOATING HUB

½" LATCH BOLT
PROJECTION

POST INDEXING
ALIGNMENT
WITH LATCH

SOLID BRASS, BRONZE, ALUMINUM,
OR STAINLESS STEEL BAR SHANKS.
BRASS BUSHINGS

ANTI-FRICTION
NONCORROSIVE
DELRIN BEARING

★NO CYLINDER REVERSING NECESSARY
★STEEL OR BRASS MECHANISM THROUGHOUT
★AUTOMATIC ADJUSTMENT TO DOOR THICKNESS OF 1 3/8" to 1¾"

Fig. 9-120 *Exploded view of a lockset.* (Weiser Lock.)

DURING CONSTRUCTION

AFTER CONSTRUCTION

Fig. 9-121 *Construction keying of locksets.* (National Lock.)

KEY BLOCK

PROJECT KEY

Lock cylinder is operated by the special "project key." The last two pins in the cylinder are held inoperative by the key block.

KEY BLOCK

REMOVAL TOOL

The special "project key" is canceled out by removal of the key block. A key block removal tool is furnished with the master keys for the locks. Simply push the removal tool into the keyway. Upon withdrawal, the key block will come out of the keyway. Thereafter, the "project key" no longer will operate the lock cylinder.

REGULAR CHANGE KEY OR MASTER KEY

Fig. 9-122 *Another method of construction keying.* (Weiser Lock.)

INSTALLING A SLIDING DOOR

Sliding doors are a common addition to a house today. The doors slide open so that the patio can be reached easily. It takes some special precautions to make sure that sliding doors will operate correctly. Most of these doors are made by a manufacturer such as Andersen.

KEY IN 60° POSITION

PULL TAILPIECE

REMOVE REGULAR KEY CYLINDER

KEY IN 60° POSITION

PULL TAILPIECE

INSERT CONSTRUCTION CYLINDER

Fig. 9-123 *Some types of construction cylinders have to be removed.* (Weiser Lock.)

Fig. 9-124 *Examples of thermopane windows.* (Andersen.)

They require a minimum of effort on the part of the carpenter. However, some very special steps are required. This portion of the chapter will deal primarily with the primed-wood type of sliding door and the Andersen Perma-Shield type of sliding door.

The primed-wood sliding door (Fig. 9-127) does not have a flange around it for quick installation. It requires some special attention. You will get an idea of how it fits into the rough opening from Fig. 9-128.

The Perma-Shield sliding door is slightly different from the primed-wood type (Fig. 9-129). Figure 9-130

Fig. 9-125 *Storm door.* (EMCO.)

shows you the details of the Perma-Shield door so that you can see the differences between the two types.

Installation of both types of doors requires a rough opening in the frame structure of the house or building. The rough opening is constructed the same way for both types of doors.

Preparation of the Rough Opening

Installation techniques, materials, and building codes vary from area to area. Contact your local material supplier for specific recommendations for your area.

The same rough opening procedures can be used for both doors. There are, however, variations to note in sliding-door installation procedures. These will be looked at more fully as we go along here.

If you need to enlarge the opening, make sure that you use a proper-sized header. Header size usually is

RAIN CAP ABOVE THE DOOR DIVERTS WATER AWAY FROM THE DOOR OPENING.

MOLDED OF RUGGED STRUCTURAL POLYPROPYLENE

TEMPERED SAFETY GLASS WINDOW

SOLID BRASS PERSONALIZED NAMEPLATE

SECURITY KEY LOCK FOR EXTRA PROTECTION (DOOR CAN BE LOCKED FROM BOTH INSIDE AND OUT).

55 POUNDS, 1 INCH THICK

FIBERGLAS SCREEN

WOOD-GRAINED TEXTURE LOOKS AND FEELS LIKE REAL WOOD.

ADJUSTABLE TRIPLE DOOR SWEEP ENSURES TIGHT SEAL AT THE BOTTOM OF THE DOOR.

NO MITERED CORNERS TO LOOSEN OR SEPARATE

HEAVY-DUTY DOOR CLOSER ALWAYS RETURNS DOOR TO THE SEALED, CLOSED POSITION, WHILE THE SAFETY CHAIN PROTECTS HINGE AGAINST WIND DAMAGE.

FLEXIBLE WEATHER SEAL ON THE TOP AND SIDES OF THE DOOR HELPS KEEP WEATHER OUT AND COMFORT IN.

FULL LENGTH CONTINUOUS HINGE ADDS STRENGTH AND EASE OF OPERATION. KEEPS THE DOOR PROPERLY ALIGNED FOR A TIGHTER SEAL AND BETTER INSULATION.

PAINT RESISTS SCRATCHING, FADING, AND CHIPPING. (MAY BE REPAINTED WITH ANY GOOD QUALITY EXTERIOR LATEX PAINT.)

SELF-STORING SCREEN AND WINDOW

TO AID IN ENERGY CONSERVATION, THE DOOR INCORPORATES AN INSULATING AIR BARRIER — ONE INCH THICK.

SOLID STEEL SUPPORT CHANNELS ON ALL FOUR SIDES.

COLONIAL CROSS-BUCK DESIGN ON BOTH THE INTERIOR AND EXTERIOR SIDES.

Mr. & Mrs. John Jones

Fig. 9-126 *Energy-saving plastic storm door.* (EMCO.)

Fig. 9-127 *Primed-wood sliding door.* (Andersen.)

Fig. 9-129 *Perma-Shield sliding door.* (Andersen.)

Fig. 9-128 *Details of the primed-wood sliding door.* (Andersen.)

Fig. 9-130 *Details of the Perma-Shield type of sliding door.* (Andersen.)

given by the manufacturer of the door, or you can obtain it from your local supplier.

Preparation of the rough opening should follow these steps:

1. Lay out the sliding-door opening width between the regular studs to equal the sliding-door rough opening width plus the thickness of two regular studs (Fig. 9-131).

2. Cut two pieces of header material to equal the rough opening width of the sliding door plus the thickness of two trimmer studs. Nail the two

Installing a Sliding Door   289

Fig. 9-131 *Layout of a rough opening for a sliding door.*
(Andersen.)

Fig. 9-133 *Placement of the header in the rough opening.*
(Andersen.)

Fig. 9-132 *Header for a sliding-door opening.* (Andersen.)

Fig. 9-134 *Placement of jack or trimmer studs in the rough opening.* (Andersen.)

header members together using an adequate spacer so that the header thickness equals the width of the trimmer stud (Fig. 9-132).

3. Position the header at the proper height between the regular studs. Nail through the regular studs into the header to hold the header in place until the next step is completed (Fig. 9-133).

4. Cut the jack or trimmer studs to fit under the header. This will support the header. Nail the jack or trimmer studs to the regular studs (Fig. 9-134).

5. Apply the exterior sheathing (e.g., fiberboard, plywood) flush with the header and jack-stud members (Fig. 9-135).

Installation of a Wood Sliding Door

Keep in mind that all these illustrations are as viewed from the outside. Be sure that the subfloor is level and the rough opening is plumb and square before installing the sliding-door frame. If you follow these steps closely, the door should be installed properly.

Fig. 9-135 *Application of exterior sheathing over the header.*
(Andersen.)

1. Run caulking compound across the opening to provide a tight seal between the door sill and the floor. Remove the shipping skids from the sill of

the frame if the sliding door has been shipped set up. Follow the instructions included in the package if the frame is not set up (Fig. 9-136).

2. Position the frame in the opening from the outside (Fig. 9-137). Apply pressure to the sill to properly distribute the caulking compound. The sill must be level. Check carefully, and shim if necessary.

3. After leveling the sill, secure it to the floor by nailing along the inside edge of the sill with 8d coated nails spaced approximately 12 inches apart (Fig. 9-138).

4. The jamb must be plumb and straight. Temporarily secure it in the opening with 10d casing nails through each side casing into the frame members. Using a straightedge, check the jambs for bow and shim. Shim solidly (five per jamb) between side jambs and jack studs.

Fig. 9-138 *Securing the sill to the floor by nailing.*

5. Complete the exterior nailing of the unit in the opening by nailing through the side and head casings into the frame members with 10d casing nails (Fig. 9-139).

6. Position the flashing on the head casing, and secure it by nailing through the vertical leg. The vertical center brace now may be removed from the frame. Be sure to remove and save the head and sill brackets (Fig. 9-140).

7. Apply the treated wood sill support under the protruding metal sill facing (Fig. 9-141). Install it tight to the underside of the metal sill with 10d casing nails.

8. Position the stationary door panel in the outer run. Be sure that the bottom rail is straight with the sill. Force the door into the run of the side jamb with a 2 × 4 wedge (Fig. 9-142). Check the position by aligning the screw holes of the door bracket with the holes in the sill and head jamb.

Fig. 9-136 *Running a bead of caulking to seal the sill and door for a sliding door.*

Fig. 9-137 *Leveling the sill.*

Fig. 9-139 *Exterior nailing of the unit.* (Andersen.)

Fig. 9-140 *Securing the flashing on the head casing by nailing.*

Fig. 9-141 *Applying the sill support under the metal sill facing.*

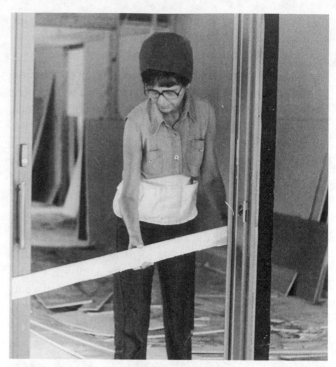

Fig. 9-142 *Positioning the stationary door panel in the outer run by using a 2 × 4 wedge.*

Fig. 9-143 *Securing the bottom bracket with a screw.* (Andersen.)

Fig. 9-144 *Securing the top bracket with a screw.* (Andersen.)

Repeat this procedure for stationary panels of a triple door (if one is used here). Before the left-hand stationary panel is installed in a triple door, be sure to remove the screen bumper on the sill. Keep in mind, however, that only a double door is shown here.

9. Note the mortise in the bottom rail for a bracket. Secure the bracket with 1-inch No. 8 flathead screws through the predrilled holes (Fig. 9-143). Align the bracket with the predrilled holes in the head jamb, and secure it with No. 8 one-inch flathead screws (Fig. 9-144). Repeat the procedure for the stationary panels of a triple door. The head

stop is now removed if the unit has been shipped set up.

10. Apply security screws. Apply the two 1½-inch No. 8 flathead painted-head screws through the parting stop into the stationary door top rail (Fig. 9-145). Repeat for the stationary panels of a triple door.

11. Place the operating door on the rib of the metal sill facing, and tip the door in at the top (Fig. 9-146). Position the head stop, and apply with 1⁹⁄₁₆-inch No. 7 screws (Fig. 9-147).

12. Check the door operation. If the door sticks or binds or is not square with the frame, locate the two adjustment sockets on the outside of the bot-

Fig. 9-147 *Positioning the head stop.* (Andersen.)

Fig. 9-145 *Applying the security screws.* (Andersen.)

Fig. 9-148 *Adjusting the door for square.* (Andersen.)

tom rail (Fig. 9-148). Simply remove the caps, insert the screwdriver, and turn to raise or lower the door. Replace the caps firmly.

13. If it is necessary to adjust the throw of the latch on two-panel doors, turn the adjusting screw to move the latch in or out (Fig. 9-149). The lock may be adjusted on triple doors by loosening the screw to move the lock plate (Fig. 9-150).

Fig. 9-146 *Placing the operating door on the rib of the metal sill facing.* (Andersen.)

Fig. 9-149 *The throw of the door is adjusted by this screw.* (Andersen.)

Fig. 9-150 *Lock adjustment on a triple door.* (Andersen.)

Fig. 9-152 *Perma-Shield door installed with metal wall plugs or extender plugs in a masonry wall with a brick veneer.* (Andersen.)

Masonry or Brick-Veneer Wall Installation of a Sliding Door

Sliding doors can be installed in masonry wall construction. Fasten a wood buck to the masonry wall, and nail the sliding door to the wood buck using the procedures shown for frame wall construction.

Figure 9-151 shows a wood sliding door installed in a masonry wall. Figure 9-152 shows a Perma-Shield door installed with metal wall plugs or extender plugs in a masonry wall with brick veneer. Metal wall plugs or extender plugs and auxiliary casing can be specified when the door is ordered.

Keep in mind that when brick veneer is used as an exterior finish, adequate clearance must be left for caulking between the frame and the masonry. This will prevent damage and bowing caused by shrinkage and settling of the structural lumber.

Installation of a Perma-Shield Sliding Door

The Perma-Shield type of door is installed in the same way as has just been described, with some exceptions. The exceptions include the following:

1. Note that wide vinyl flanges that provide flashing are used at the head and side jambs (Fig. 9-153A). Locate the side member flush with the bottom of the sill with an offset leg pointing toward the inside of the frame. Tap with the hammer using a wood block to firmly seat the flashing in the groove (Fig. 9-153B). Apply the head member similarly. Overlap the side flange on the outside.

2. After securing the frame to the floor with nails through the sill, apply clamps to draw the flanges tightly against the sheathing (Fig. 9-154).

3. Temporarily secure the door in the opening with 10d casing nails through each side casing into the frame members. Using a straightedge, check the jambs for bow and shim. The jamb must be plumb and straight. Shim solidly using five shims per jamb between the side jambs and the jack studs.

4. Side members on the Perma-Shield sliding doors have predrilled holes to receive 2½-inch No. 10 screws (Fig. 9-155). Shim at all screw holes between the door frame and the studs. Drill pilot holes into the studs. Secure the door frame to the studs with screws.

5. The head jamb also has predrilled holes to receive the 2½-inch No. 10 screws. Shim at all screw holes between the door frame and header. Drill the pilot holes into the header. Insert the screws

Fig. 9-151 *Wood sliding door installed in masonry wall.* (Andersen.)

VINYL
FLASHING

(A)

(B)

Fig. 9-153 *A. Location of the vinyl flashing after it has been applied. B. Using a wooden block to apply the vinyl flashing.* (Andersen.)

Fig. 9-154 *Clamps draw the flanges tightly against the sheathing.* (Andersen.)

Fig. 9-155 *Securing the door frame to the studs with screws.* (Andersen.)

Fig. 9-156 *Placing screws in predrilled holes in the door header.* (Andersen.)

Fig. 9-157 *Finished installation of a Perma-Shield sliding door.*

through the predrilled holes, and draw up tightly. Do not bow the head jamb (Fig. 9-156).

6. Figure 9-157 shows a Perma-Shield door completely installed. Interior surfaces of the panel and frame should be primed before or immediately after installation for protection.

INSTALLING A GARAGE DOOR

Garage doors are available in metal and wood. Wood doors are available in a high-grade hemlock/fir-frame construction and recessed hardboard or raised redwood panels. Rough-sawn flush wood doors are also available. They come ready to be primed, painted, or stained to match the house finish. Steel doors come with a primer and need to be painted with a second coat to match the owner's preference in color or finish.

Figures 9-158 and 9-159 show the types of springs used to aid in raising a garage door. Torsion springs usually are used for heavy doors for two-car garages.

A garage door with extension springs usually is used in single-car garages.

A word of caution: To avoid installation problems that could result in personal injury or property damage, use only the track specified and supplied with the door unit. Some large doors weigh as much as 400 pounds when the spring tension is released. A single door weighs up to 200 pounds, and two people should work on it to prevent damage.

Fig. 9-158 *A garage door with torsion springs.* *(Clopay.)*

One of the primary concerns about installing a door is headroom. Headroom is the space needed above the top of the door for the door, the overhead tracks, and the springs. Measure to check that there are no obstructions in your garage within that space. The normal space requirement is shown in Table 9-2. The backroom distance is measured from the back of the door into the garage and should be at least 18 inches more than the height of the garage door. A minimum sideroom of 3.75 inches (5.5-inch EZ-Set Spring)

Fig. 9-159 *A garage door with extension springs.* (Clopay.)

TABLE 9-2 *Headroom Requirement Chart*

SPRING TYPE	TRACK RADIUS	HEADROOM REQUIRED
EZ-Set Spring™/ Extension	12"	10"
EZ-Set Spring™/ Extension	15"	12"
Torsion	12"	12"
Torsion	15"	14"

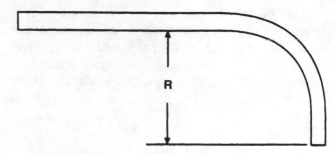

Fig. 9-161 *Track radius measurement.* (Clopay.)

should be available on each side of the door on the interior wall surface to allow for attachment of the vertical track assembly (Fig. 9-160).

Track radius is another important concern in installation of the track. The radius of the track can be determined by measuring the dimension *R* in Fig. 9-161. If the dimension *R* measures 11 to 12 inches, then you have a 12-inch-radius track. If *R* equals 14 to 15 inches, then you have a 15-inch-radius track (Fig. 9-161). About 3 inches of additional headroom height at the center plus additional backroom is needed to install an automatic garage door opener.

Check the door opener instructions to be sure. If the headroom is low, there are a few options to compensate. For instance, a double-track low-headroom conversion kit reduces the headroom requirement to 4.5 inches on EZ-Set Spring and extension springs,

(9.5 inches on front mount torsion springs). Instructions are provided with the track. This option is designed to modify the standard track. Instructions are provided in the kit. Another way to reduce the headroom requirement is to use the Quick-Turn bracket (Fig. 9-162). The Quick-Turn bracket cannot be used in conjunction with any other low-headroom option. This is used in place of the existing top roller. Instructions are included with the kit.

The next step is to prepare the opening. Figure 9-163 shows the rough opening for the door and the necessary additions. An option is stop molding featuring a built-in weather seal, as shown in Fig. 9-164.

Next, prepare for installing the door sections. Spread the hardware on the floor in groups so that you can easily find the parts. Assemble as the directions require. One thing to keep in mind is that if the door is going to be equipped with an automatic garage door opener, make sure that the door is always unlocked when the operator is being used. This will avoid damage to the door. Instructions come with the door in the form of a pamphlet with detailed line drawings.

Assemble and install the track. Follow the directions for the specific door being used. Pay special attention and use adequate-length screws to fasten the

Fig. 9-160 *Required headroom.* (Clopay.)

Fig. 9-162 *Quick-Turn bracket for low headroom.* (Clopay.)

Fig. 9-163 *Preparing the opening.* (Clopay.)

Fig. 9-164 *Door-stop molding.* (Clopay.)

rear track hangers into the trusses. A door may fall and cause serious injury if not secured properly.

Attaching springs Lifting cables and springs can be dangerous when installed incorrectly. It is very important to follow the instructions closely when installing the springs. Garage door springs can cause serious injury and property damage if they break under tension and are not secured with safety cables.

Keep your head well below the track when the spring is under tension or being tensioned because springs are dangerous when they are fully or partially wound. The first time the door is opened, make sure that the door doesn't fall. This might happen if the tracks are not aligned correctly or the rear track hangers are not strong enough. Proceed slowly and carefully, and follow the directions provided by the manufacturer. Both springs should be adjusted equally for proper operation.

The EZ-Set Spring by Clopay can be adjusted by using a ⅜-inch drill with the ¼-inch hex driver provided in the door kit. Make sure that the ¼-inch hex driver is inserted completely into the worm drive. The spring is tensioned by operating the drill in the clockwise direction (Fig. 9-165).

Extension springs are used on single doors and have two springs, as in Fig. 9-166. The door with less weight uses a single spring on each side of the door. The heavier door may use a double spring on each side. If your door was supplied with four extension springs, take notice of the color coding on the ends of the springs. If there are two color codes, be sure to use one of each on either side so that the spring tension is equal on both sides.

Fig. 9-165 *Adjusting the spring tension with a hand drill.* (Clopay.)

Fig. 9-166 *Single- and double-extension springs.* (Clopay.)

Torsion-spring installation Torsion springs can be very dangerous if they are installed improperly or mishandled. Do not attempt to install them alone unless you have the right tools and reasonable mechanical aptitude or experience and you follow instructions very carefully. It is important to firmly and securely attach

Winding Chart	
Door Height	**Spring Turns**
6'0"	6.75
6'3"	7.00
6'6"	7.25
6'9"	7.50
7'0"	7.75
7'3"	8.00
7'6"	8.25
7'9"	8.50
8'0"	8.75

Number of turns on the spring can be counted by drawing a chalk line on them. Doors with two springs are wound the same number of turns on each spring.

Double Springs (Shown)
Red Winding Cone Spring - Mount on left side
Black Winding Cone Spring - Mount on Right Side

Fig. 9-167 *Torsion-spring installation.* (Clopay.)

the torsion-spring assembly to the frame of the garage (Fig. 9-167).

Attaching an automatic opener   When installing an automatic garage door opener, make sure to follow the manufacturer's installation and safety instructions carefully. Remove the pull-down rope and unlock or remove the lock. If attaching an operator bracket to the wood anchor pad, make sure that the wooden anchor pad is free of cracks and splits and is firmly attached to the wall. Always drill pilot holes before attaching lag screws.

To avoid damage to the door, you must reinforce the top section of the door to provide a mounting point for the opener to be attached. Failure to reinforce the door, as illustrated in Fig. 9-168, will void the manufacturer's warranty. Note: All reinforcing angles are to be attached with ⅝-inch No. 14 sheet-metal screws at the reinforcement backup plate locations.

Cleaning and painting   Before painting the door, it must be free of dirt, oil, chalk, wax, and mildew. The prepainted surfaces can be cleaned of dirt, oil, chalk, and mildew with a diluted solution of trisodium phos-

phate. Trisodium phosphate is available over the counter at most stores under the name of Soilax, in many laundry detergents without fabric softener additives, and in some general-purpose cleaners. Check the label for trisodium phosphate content. The recommended concentration is ⅓ cup of powder to 1.5 to 2 gallons of water. After washing the door, always rinse it well with clear water and allow it to dry.

The steel door can be painted with a high-quality flat latex exterior-grade paint. Because all paints are not created equal, the following test needs to be performed. Paint should be applied on a small area of the door (following the instructions on the paint container), allowed to dry, and evaluated prior to painting the entire door. Paint defects to look for are blistering and peeling. An additional test is to apply a strip of masking tape over the painted area and peel it back, checking to see that the paint adheres to the door and not to the tape.

After satisfactorily testing a paint, follow the directions on the container, and apply it to the door. Be sure to allow adequate drying time should you decide to apply a second coat.

Fig. 9-168 *Mounting the backup plate for door opener. (Clopay.)*

Window frames and inserts can be painted with a high-quality latex paint. The plastic should be lightly sanded first to remove any surface gloss.

ENERGY FACTORS

There are some coatings for windows and door glass that reflect heat. A thin film is placed over the glass area. This thin film reflects up to 46 percent of the solar energy. Only 23 percent is admitted. This means that 77 percent of the solar energy is reflected or turned away. Figure 9-169 shows the extent of this ability to absorb and reflect energy.

A thin vapor coating of aluminum prevents solar radiation from passing through glass. It does this by reflecting it back to the exterior. The temperature of the glass is not raised significantly. The coating minimizes undesirable secondary radiation from the glass. Visible light is reduced. However, the level of illumination remains acceptable. During the winter, this coating reflects long-wavelength radiation and keeps heat in the room.

The film is easily applied to existing windows (see Fig. 9-169). Step 1 calls for spraying the entire window with cleaner. Scrape every square inch of the window with a razor blade. This removes the paint, varnish, and excess putty. Wipe the window clean and dry. Step 2 calls for respraying the window with cleaner. Squeegee the entire window. Use vertical strokes from top to bottom of the window. Use paper towels to dry the edges and corners.

Step 3 calls for clear water to be sprayed onto the glass area. Remove the separator film (Fig. 9-170). Test for the tacky side of film by folding it over and touching it to itself. The tacky side will stick slightly. Position the tacky side of the film against the top of the glass. Smooth the film by carefully pulling on the edges, or gently press the palms of your hands against the film. Slide the sheet of film easily to remove large wrinkles.

In step 4, spray the film, which is now on the glass, with water. Standing at the center, use a squeegee in gentle, short vertical and horizontal strokes to flatten the film. Slowly work out the wrinkles. Work out the

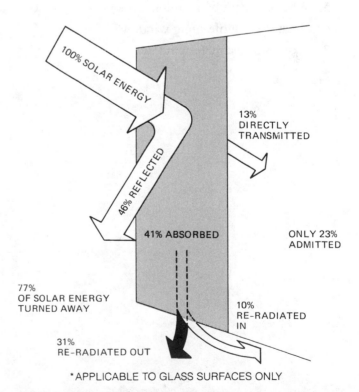

Fig. 9-169 *Energy-saving film for windows. (Kelly-Stewart.)*

1   2   3   4

Fig. 9-170 *Step-by-step application of energy-saving film.* (Kelly-Stewart.)

bubbles, and remove excess water from under the film. Be careful not to crease or wrinkle the film. Trim the edges with a razor blade. Wipe the excess water from the edges. The film will appear hazy for about 2 days until the excess water evaporates.

This is shown here to illustrate the possibilities of making your home an energy saver. Films and other devices will be forthcoming as we look forward to conserving energy.

CHAPTER 9 STUDY QUESTIONS

1. What purpose do windows serve?
2. What purpose does a door serve?
3. What is the height of a door?
4. Where are windows made today?
5. What is meant by a prehung door?
6. What is a double-hung window?
7. What is a casement window?
8. What is an awning-type window?
9. What is a hopper-type window?
10. What is a flush door?
11. Why does a door need a 3-degree angle on the lock stile?
12. What determines the hand of a door?
13. What does LHR mean?
14. What is a construction key?
15. How are the locks in a house changed after the house is completed?
16. How are storm doors installed?
17. Where are storm windows installed?
18. What type of garage door uses extension springs?
19. Why is headroom important in garage door installation?
20. How much can a double garage door weigh?

10
CHAPTER

Exterior Walls

HREE METHODS ARE USED TO FINISH EXTERIOR walls. The first is by covering the wall with a wood or wood-product material called *siding*. Another is to cover the area with a siding made of aluminum, plastic, or coated steel. The other is to cover the wall with a masonry material such as brick, stone, or stucco. The carpenter will install the exterior siding and sometimes will prepare the exterior wall for the masonry materials. However, people employed in the *trowel trades* (i.e., bricklayers and stone masons) install the masonry materials.

Siding is applied over the wall sheathing and adds protection against weather, strengthens the walls, and also gives the wall its final appearance or beauty. Siding may be made of many materials. Often, more than one type of material is used. Wood and brick could be combined for a different look. Other materials also may be combined.

This chapter will aid you in learning these new skills and how to

- Prepare the wall for an exterior finish
- Estimate the amount of siding needed
- Select the proper nails for the procedure
- Erect scaffolds
- Install flashing and water tables to help waterproof the wall
- Finish the roof edges

- Install exterior siding
- Trim out windows and doors

WALLS

Once the windows and doors are installed, it is time for putting on the exterior finish. The sheathing should be on the walls. Then, to finish the exterior, three things are needed. First, the cornices and rakes around the roof are enclosed. Next, the siding is applied. Finally, the finish trim for windows and doors is installed.

Cornice

The *cornice* is the area beneath the roof overhang. This area is usually enclosed or boxed in. Figure 10-1 shows a typical cornice. In many areas, the cornice is boxed in as part of the roofing job. The cornice is often painted before siding is installed. This is particularly likely when brick or stone is used.

Several types of siding may be installed by carpenters. Most types of siding are made of wood, plywood, or wood fiber. Some types of siding are made of plastic and metal. These are usually stamped or embossed to resemble wood siding.

Siding is the most visible part of the house because it is on the outside. Beauty and appearance are important factors. However, siding is also selected for other reasons. The builder may want a siding that can

Fig. 10-1 *Typical box cornice framing. (Forest Products Laboratory.)*

be installed quickly and easily. This means that it costs less to install. The owner may want a siding that requires little maintenance. Both will want a siding that will not rot, warp, crack, or craze. Whether insects will attack the siding is another factor to be considered.

TYPES OF SIDING

Siding can consist of plywood, wood boards, wood fiber, various compositions, metal, or plastics (Fig. 10-2).

However, the shape is the main factor that determines how siding is applied. Wood fiber is made in two main shapes, "boards" and sheets or panels, which often look like boards. However, the sheets are put up as whole sheets. Fiber panels also can be made to resemble other types of siding.

Plywood panels also can be made to look like other types of siding. Figure 10-2B shows a house with plywood siding that looks like individual boards. Ply-

(A)

(B)

(C)

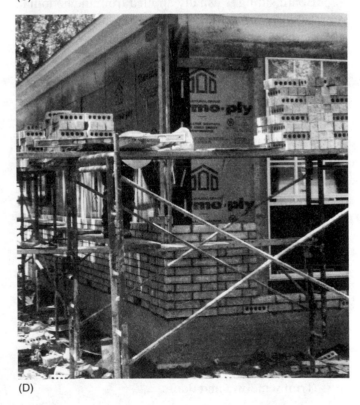

(D)

Fig. 10-2 *A. Cedar boards are used both horizontally and diagonally for this siding.* (Potlatch.) *B. Plywood siding has many looks and styles. This looks like boards.* (American Plywood Association.) *C. This siding combines rough brick, smooth stucco, and boards. D. Here, brick siding is applied over frame construction.*

wood also can be made in "board" strips. These are applied like boards.

Shingles and shakes also may be used for siding. Shingles are made from wood or asbestos mineral compounds. Actually, both types of shingle siding are applied in much the same manner. The same shingles made of cedar that are used on the roof can be used as siding. Cedar is usually good for exposure to the weather for a period of 40 to 50 years. Many people prefer the "weathered" look of cedar shingles. However, there are some disadvantages (e.g., they burn quickly in a fire).

SEQUENCE

The sequence is determined by

- The type of siding
- The type of roof
- The type of sheathing (if any) used

How high the building is also affects the sequence. To work on high places, carpenters must erect scaffolds.

Board siding is usually applied from the bottom to top. However, if scaffolds have to be nailed to the wall, the siding on the bottom could be damaged. In such cases, the sequence can be changed. The scaffolds can be put up, and the siding can be put on the top first. In this way, the siding is not damaged by scaffold nails or bolts. Many scaffolds can stand alone. They don't need to be nailed to the wall.

The common sequence for finishing an exterior wall is

1. Prepare for the job. Make sure that the windows are installed, the vapor barrier is in place, the nails are selected, and the amount of siding is estimated.
2. Erect necessary scaffolds.
3. Install the flashing and water tables.
4. Finish the roof edges.
5. Install siding on the upper gable ends and on the upper stories.
6. Install siding on the sides.
7. Finish the corners.
8. Trim windows and doors.

JOB PREPARATION

Before siding is nailed to the wall, several things should be done. Windows and doors should be installed properly. Rough openings should have been moisture shielded. Then any spaces between window units and

Fig. 10-3 *A window unit blocked for brick siding.*

the wall frame are blocked in. An airspace should be present. Some types of sheathing are also good moisture barriers. However, other types are not. These must have a moisture barrier installed. Figure 10-3 shows a properly blocked window in a wall that is ready for siding.

Vapor Barrier

A good reason for installing a vapor barrier is to protect the outside walls from vapor originating inside the house. Paint problems on the siding of the house can arise from too much moisture escaping from inside the house. The main purpose of the barrier is to prevent water vapor from entering the enclosed wall space, where condensation might occur and cause rot, mold, and odor problems. Make the warm side of the wall as vapor-tight as possible. The asphalt- or tar-saturated felt papers are not vaporproof. However, they may be used on sheathing for their water repellency.

Vapor barriers are available on insulation. This means that the barrier side of the insulation should be installed toward the inside of the house. In some instances, the entire inside walls and ceiling are covered with a plastic (polyethylene) film to ensure a good vapor barrier. Heated air contains moisture. It travels from inside the building to the outside. This means that it will blister and bubble paint on the siding. Damp air also can increase the susceptibility to rot and damage.

Tar-saturated paper is used to cover the sheathing. This improves the siding's ability to shed water and at the same time improve the energy efficiency of the house by preventing air leakage through the walls. Reflective energy barriers are also used (Fig. 10-4). They help to keep energy on one side of a wall. They can cause the inside of the house to be warmer in the winter and cooler in the summer.

To put up an air barrier, start at the bottom. Nail the top part in place. Most air-barrier materials will be applied in strips. The strips should shed water to the outside (Fig. 10-5). To do this, the bottom strip is installed first. Each added strip is overlapped. Moisture barriers also should be lapped into window openings. See Chapter 9 for this procedure.

Plastic film, metal foil, and builder's felt (also called *tar paper*) are vapor barriers. Builder's felt and some plastic films are used most frequently for these purposes.

Nail Selection

Methods of nailing can vary according to need or application. Some siding may be put up by any of several methods. Other siding should be put up with special fasteners. The important thing is that the nail must penetrate into something solid to hold. For example, nails driven into fiber sheathing will not hold. Siding over this type of sheathing must be nailed at the studs. If the studs are placed 16 inches on center (O.C.), the nails are driven at 16-inch intervals. Likewise, splices in the siding should be made over studs. Also, the nail must be long enough to penetrate into the stud.

Fig. 10-5 *Barrier is applied from the bottom to the top. The overlap helps shed moisture.*

If plywood or hardboard sheathing is used, then nails may be driven at any location. The plywood will hold the siding even with a short nail.

Strips of wood are also used to make a nail base. These strips are placed over the sheathing. Then they are nailed to the studs through the sheathing. Nail strips are used for several types of shingled siding (Fig. 10-6).

Fig. 10-4 *Foil surfaces reflect energy.*

Fig. 10-6 *Nail strips can be used over sheathing.* (Forest Products Laboratory.)

The nail should enter the nail base for at least ½ inch. A nonstructural sheathing such as fiber sheathing cannot be the nail base. With fiber or foam sheathing, the nail must be longer. The nail must go through siding and the sheathing and ½ inch into a stud.

Also, the type of nail should be considered. When natural wood is exposed, a finishing or casing-head nail would look better. The head of the nail can be set beneath the surface of the wood. The head will not be seen. In this way, the nail will not detract from the appearance. However, if the wood is to be painted, a common or a box-head nail may be used. Coated nails are preferred for composition and mineral siding. These nails are coated with zinc to keep them from rusting. This makes them more weather resistant.

For some siding, the nail is driven at an angle. This means that it must be longer than the straight-line distance. This type of nailing is frequently done in grooved and edged siding. The nails are driven in the grooves and edges. In this way, the siding is put on so that the nails are not exposed to the weather or the eye of the viewer.

Estimating the Siding Needed

Keep in mind that a "one by six" (1- × 6-inch) board is not 1 inch thick and 6 inches wide. However, a 1 × 6 board is actually ¾ inch thick by 5½ inches wide. Also, most board siding overlaps. Rabbet, bevel, and drop sidings all overlap. When overlapped, each board would not expose 5½ inches. Each board would expose only about 5 inches.

A number of things must be known to estimate the amount of siding needed. First, you must determine the height and width of the wall. Then find the type of siding and the sizes of the windows and doors. Consider the following example:

Siding:	1- × 8-inch bevel siding
Overlap:	1½ inches
Wall height:	8 feet
Wall length:	40 feet
Windows and doors:	Two windows, each 2 × 4 feet
	One door, 8 × 3 feet

First, find the total area to be covered. To do this, multiply the length of the wall times the height:

$$\text{Area} = 40 \times 8$$
$$= 320 \text{ square feet}$$

Next, subtract the area of the doors and windows from the area of the wall. To do this,

$$\text{Area of windows} = \text{width} \times \text{length}$$
$$= 2 \times 4$$
$$= 8 \text{ square feet} \times \text{number of windows}$$
$$= 16 \text{ square feet}$$
$$\text{Door area} = \text{width} \times \text{length}$$
$$= 3 \times 8$$
$$= 24 \text{ square feet}$$
$$\text{Total opening area} = 16 \text{ square feet}$$
$$+ 24 \text{ square feet}$$
$$\text{Total} \quad 40 \text{ square feet}$$

To get enough siding to cover this area, the percentage of overlap must be considered. Also, there is always some waste in cutting boards so that they join at the proper place. When slopes and corners are involved, there is more waste. The amount of overlap for 8-inch siding lapped 1¼ inches is approximately 17 percent. However, it is best to add another 15 percent to this for waste. Thus about 32 percent would be added to the total requirements. Thus to side the wall would require 320 – 40 = 280 square feet of siding. However, enough siding should be ordered to cover 280 square feet plus 32 percent (90 board feet). A total of 370 board feet should be ordered. Table 10-1 shows allowances for different types of siding.

TABLE 10-1 *Allowances That Must Be Added to Estimate Siding Needs*

Type	Size, Inches	Amount of Lap, Inches	Allowance, Percent (Add)
Bevel siding	1 × 4	¾	45
	1 × 6	1	35
	1 × 8	1¼	32
	1 × 10	1½	30
	1 × 12	1½	25
Drop siding (shiplap)	1 × 4		30
	1 × 6		20
	1 × 8		17
Drop siding (matched)	1 × 4		25
	1 × 6		17
	1 × 8		15

Use the same procedure to estimate the area for gable ends. Find the length and the height of the gable area. Then multiply the two dimensions for the total. However, since slope is involved, as shown in Fig. 10-7, only one-half of this figure is required. The allowance for waste and overlap is based on this halved figure.

The amount of board siding required for an entire building may be estimated. First, the total area of all the walls is found. Then the areas of all the gable sec-

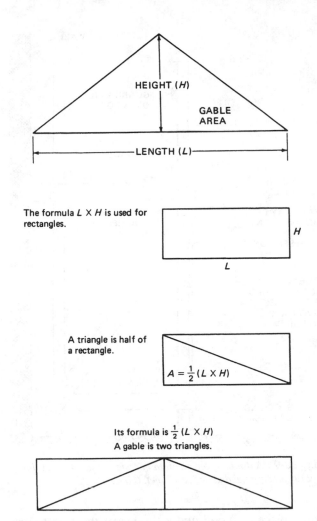

HEIGHT (H)

GABLE AREA

LENGTH (L)

The formula L × H is used for rectangles.

H

L

A triangle is half of a rectangle.

$A = \frac{1}{2}(L \times H)$

Its formula is $\frac{1}{2}(L \times H)$
A gable is two triangles.

Fig. 10-7 *Estimating gable areas.*

tions are found. These areas are added together. The allowances are made on the total figure.

Ordering paneled siding Paneled external plywood siding is sold in sheets. The standard sheet size is 4 feet wide by 8 feet long. The standard height of the wall is 8 feet. Thus a panel fully covers a 4-foot length of wall. To estimate the amount needed, find the length of the wall. Then divide by 4, the width of a panel. For example, a wall is 40 feet long. The number of panels is 40 divided by 4. Thus it would take 10 panels to side this wall with plywood paneling. If, however, the length of the wall is 43 feet, then 11 panels would be required. Only whole panels may be ordered. No allowances for windows and doors are made. Sections of panels are cut out for windows and doors.

Estimating shingle coverage To find the amount of shingles needed, the actual wall area should be found. The areas of the doors and windows should be deducted from the total area. Shingles vary in length and width. The size of the shingle must be considered when estimating. Also, shingles come in packages called *bundles*. Generally, four bundles are approximately 1 square (100 square feet) of coverage. This means that four bundles will cover about 100 square feet of surface area. However, the amount of shingle exposed determines the actual coverage. If only 4 inches of a 16-inch-long shingle is exposed, the coverage will be much less. Also, if more of the shingle is exposed, a greater area can be covered. For walls, more of the shingle can be exposed than for roofs. Table 10-2 shows the coverage of 1 square of shingles when different lengths are exposed.

SCAFFOLDING

When siding has to be applied high off the ground, the carpenter must work up high. In order to reach all these areas so high off the ground, a scaffold is necessary. Figure 10-8 shows a typical pump-jack scaffold. Scaffolds are also sometimes called *stages*.

Scaffolding allows you access to high places quickly and efficiently. It can be movable, or it can be

TABLE 10-2 *Coverage of Wood Shingles for Varying Exposures*

Length and Thickness[b]	Approximate Coverage of Four Bundles or One Carton, [a]Square Feet									
	Weather Exposure, Inches									
	5½	6	7	7½	8	8½	9	10	11	11½
Random-width and dimension 16″ × 5/2″	110	120	140	150[c]	160	170	180	200	220	230
18″ × 5/2¼″	100	109	127	136	145½	154½[c]	163½	181½	200	209
24″ × 4/2″	—	80	93	100	106½	113	120	133	146½	153[c]

[a] Nearly all manufacturers pack four bundles to cover 100 square feet when used at maximum exposures for roof construction; rebutted-and-rejoined and machine-grooved shingles typically are packed one carton to cover 100 square feet when used at maximum exposure for double-coursed sidewalls.
[b] Sum of the thickness, e.g. 5/2″ means 5 butts equal 2″.
[c] Maximum exposure recommended for single-coursed sidewalls.

Fig. 10-8 *Pump-jack scaffold for light, low work such as brick-laying, painting, and the application of siding.*

Fig. 10-9 *A double-pole scaffold can stand free of the building. Guard rails are needed above 10 feet, 0 inches.*

held in place with permanent brackets or nails to boards. A number of features should be examined in the interest of safety. State and federal regulations apply to all scaffolding, so the manufacturer is aware of the limitations of a specific arrangement. Follow the manufacturer's recommendations for a safe tower or support platform.

A scaffold must be strong enough to hold up the weight of everything on it. This includes the people, tools, and building materials. Several types of scaffolds can be used. The type used depends on the number of workers involved and the weight of the materials. Also, how high a scaffold will be is a factor. How long the scaffold will remain up also must be considered.

Job-Built Scaffolds

Scaffolds are often built from job lumber. The maximum distance above the ground should be less than 18 feet. Three main types of job scaffolds can be built. Supports for the platform should be no more than 10 feet apart, and less is desirable.

Double-pole scaffold This style of scaffold is shown in Fig. 10-9. The poles form each support section. They should be made from 2- × 4-inch pieces of lumber. A 1 × 6 board is nailed near the bottom of both poles for a bottom brace. Three 12d nails should be used in each end of the 1- × 6-inch board. The main support for the working platform is called a *ledger*. It is nailed to the poles at the desired height. It should be made of 2- × 4-inch or 2- × 6-inch lumber. It is nailed with three 16d common nails in each end of the ledger. Sway braces should be nailed between each end pole, as shown in the figure. The sway braces can be made of 1-inch lumber. They are nailed as needed.

After two or more of the pole sections have been formed on the ground, they can be erected. Place a

small piece of wood under the leg of the pole to provide a bearing surface. This keeps the end of the pole from sinking into the soft earth. The sections should be held in place by the carpenters. Then another carpenter nails braces between each pole section. These are nailed diagonally, as are sway braces. They may be made of 1- × 6-inch lumber. They are nailed with three 14d common nails at each end.

When the ledgers are more than 8 feet above the ground, a guard rail should be used. The guard rail is nailed to the pole about 36 inches above the platform. Note that the guard rail is nailed to the inside of the pole. In this way, a person can lean against the guard rail without pushing it loose.

The double-pole type of scaffold can be freestanding. That is, it need not be attached to the building. However, carpenters often nail a short board between the scaffold and the wall. This holds the scaffold safely. If this is done, the board should be nailed near the top of the scaffold.

Single-pole scaffolds Single-pole scaffolds are similar to double-pole scaffolds. However, the building forms one of the "poles." Blocks are nailed to the building. They form a nail base for the brace and

WALL

GUARD RAIL
ON INSIDE

2 × 4 OR 2 × 6 LEDGER

2 × 10 OR
LARGER

USE 2 × 4 BLOCK
UNDER LEDGER
FOR HEAVY LOADS

CLEATS NAILED
TO STUDS OR
SOLID WALL
PIECES

1 × 6 SWAY BRACES
ACROSS EACH
SECTION

WALL

BRACE

BLOCKING OR
PADS

Fig. 10-10 *A single-pole scaffold must be attached to the building. Guard rails are needed above 10 feet, 0 inches.*

ledgers. Figure 10-10 shows a single-pole scaffold. Sway braces are not needed on the pole sections. However, sway braces should be used between each section of the scaffold.

Wall brackets   Wall brackets are often made on the job (Fig. 10-11). Wall brackets are used most often when the distance above the ground is not very great. They are quicker and easier to build than other scaffold types. Of course, they are also less expensive.

On the ends, the wall brackets are nailed to the outside corners of the wall. On intersections, nail blocks are used, similar to the single-pole scaffold blocks. Wall brackets must be nailed to solid wall members. It is best to use 20d common nails to fasten them in place. Figure 10-12 shows another type of wall bracket. This type is sturdier than the other.

Factory Scaffolds

Many builders use factory-made scaffolds. These have several advantages for a builder. They are quick and easy to erect. They are strong and durable and can be taken down and reused easily.

No lumber is used. Also, no cutting and nailing are needed to erect the scaffold. Thus no lumber is ruined

TOP PLATES

TEMPORARY
BRACE

LAP
TOP PLATES
AND NAIL

WINDOW
HEADER

STUD

LET-IN
CORNER BRACE

SHEATHING

2 × 10 OR
2 × 12

WINDOW SILL

SOLE PLATE

NAIL
TO
LEDGER

SPIKE TO
AT LEAST
2 STUDS

Fig. 10-11 *A typical low-level wall bracket.*

HEADER
JOIST

SUBFLOOR

FOUNDATION WALL

STRINGER JOIST

SPACER BLOCK

ANCHORED SILL PLATE

Fig. 10-12 *A job-made wall-bracket scaffold. It is dangerous to work on this kind of scaffold in high places without safety equipment.*

Fig. 10-13 *Metal double-pole scaffolds are used widely.* (Beaver-Advance.)

or made unusable for the building. The metal scaffold parts are easily stored and carried. They are not affected by weather and will not rot. There are several different types of scaffolds.

Double-pole scaffold sections Double-pole scaffolds feature a welded-steel frame. They include sway braces, base plates, and leveling jacks (Figs. 10-13 and 10-14). Two or more sections can be used to gain greater distance above the ground. Special pieces provide sway bracing and leveling. Guard rails and other scaffold features also may be included.

Wall brackets Wall brackets, as in Figs. 10-15 through 10-17, are common. No nail blocks are needed on the walls for these metal brackets. These brackets are nailed or bolted directly to the wall. Be sure that they are nailed to a stud or another structural member. Use 16d or 20d common nails. After the nails are driven in place, be sure to check the heads. If the nail heads are damaged, remove them, and renail the bracket with new nails. Remember that it is the nail head that holds the bracket to the wall. A dam-

Fig. 10-14 *Metal sections can be combined in several ways to get the right height and length.*

EASY TO ERECT

SIMPLE INSTALLATION

SIMPLE REMOVAL

EASY STORAGE

SHIPPING/STORAGE CONFIGURATION

Fig. 10-15 *Metal wall brackets are quick to put up and take down.* (Richmond Screw Anchor.)

Fig. 10-16 *Wall brackets can be used on concrete forms.*

Fig. 10-17 *Guard rails may be added through an end bracket.* (Richmond Screw Anchor.)

aged nail head may break. A break could let the entire scaffold fall.

Trestle jack   Trestle jacks are used for low platforms. They are used both inside and outside. They can be moved very easily. Trestle jacks are shown in Fig. 10-18. Two trestle jacks should be used at each end of the section. A ledger, made of 2- × 4-inch lumber, is used to connect the two trestle jacks. Platform boards then are placed across the two ledgers. Platform boards always should be at least 2 inches thick.

As you can see, it takes four trestle jacks for a single section. Trestle jacks can be used on uneven areas, but they provide for a platform height of only about 24 inches. However, this is ideal for interior use.

Fig. 10-18 *Two trestle jacks can form a base for scaffold planks.* (Patent Scaffolding Co.)

Ladder jacks   Ladder jacks hang a platform from a ladder. They are most suitable for repair jobs and for light work where only one carpenter is on the job. Two types of jacks are used. The type shown in Fig. 10-19 puts the platform on the outside of the ladder. The type shown in Fig. 10-20 places the platform below or on the inside of the ladder.

The aluminum ladder jack comes in handy when you need to put a plank between two ladders to serve as a platform. The aluminum ladder jack has an adjustable arm with a positive lock. Adjustments are every inch, so it can be used on either side of the ladder. It can be installed on ladders with roof hooks, enabling the user to work on the roof or the face of a dormer. It accommodates forms up to 18 inches wide but folds to 6 inches for deep storage.

Ladder Use

Using ladders safely is an important skill for a carpenter. A ladder to be erected is first laid on the ground. The bottom end of the ladder should be near the building. The top end is raised and held overhead. The carpenter gets directly beneath the ladder. The hands are

Fig. 10-19 *Outside ladder jacks.*

(A)

(B)

(C)

Fig. 10-20 *A, B. Inside ladder jacks. C. Raising a ladder can be a two-man job. In fact, a heavier ladder should have two men on it. However, if you have a single ladder that's not too heavy, you can place the end of the ladder against the house or some obstruction and walk it up one rung at a time.*

then moved from rung to rung. As the top end of the ladder is raised, the carpenter walks toward the building. Thus the ladder is raised higher and higher with every step. Care is taken to watch where the top of the ladder is going. The top of the ladder is guided to its proper position. Then the ladder is leaned firmly against some part of the building. The base of the ladder is made secure on solid ground or concrete. Both ends of the bottom should be on a firm base.

Both wooden and aluminum ladders are used commonly by carpenters. Aluminum ladders are light and easy to manage. However, they have a tendency to sway more than wooden ladders. They are very strong and safe when used properly. Wooden ladders do not sway or move as much, but they are much heavier and harder to handle.

Wooden ladders are sometimes made by carpenters. A ladder is made from clear, straight lumber. It should have been well seasoned or treated. The sides are called *rails*, and the steps are called *rungs*. Joints are cut into the rails for the rungs. Boards should never just be nailed between two rails.

Ladder Safety

1. Ladder condition should be checked before use.
2. The ladder should be clean. Grease, oil, or paint on rails or rungs should be removed.
3. Fittings and pulleys on extension ladders should be tight. Frayed or worn ropes and lines should be replaced.
4. The bottom ends of the ladder must rest firmly and securely on a solid footing.
5. Ladders should be kept straight and vertical. Never climb a ladder that is leaning sideways.
6. The bottom of the ladder should be one-fourth the height from the wall. For example, if the height is 12 feet, the bottom should be 3 feet from the wall.

Scaffold Safety

1. Scaffolds should be checked carefully before each use.
2. Follow the specifications, as stated by the manufacturer.
3. State codes and local safety rules also should be followed. Pads should be under poles.
4. Flimsy steps on scaffold platforms or ladders should never be used. Height should be increased only with scaffold of sound construction.
5. Planking for platforms should be heavy enough to carry the load and span.
6. Platform boards should hang over the ledger at least 6 inches. In this way, when boards overlap, the total overlap should be at least 12 inches.
7. Guard rails and toe boards should be used.
8. Scaffolds should never be put up near power lines without proper safety precautions. The electric service company can be consulted for advice when a procedure is not known.
9. Remove all materials and equipment before a platform or a scaffold is moved.

ROOF EDGES

The part of the roof that overhangs the walls is called the *eave*. If eaves are enclosed, they are also called a *cornice* (Fig. 10-21). Usually, the edges of roofs are finished when the roof is sheathed. However, building sequences vary from place to place. Two methods are commonly used to finish the eaves. These are the open method and the closed method. Several versions of the closed method exist.

Open Eaves

A board across the ends of the rafters ties them together and is usually placed when the roof is sheathed (Fig. 10-22). This board, called the *fascia*, helps to brace and strengthen the rafters. Fascia should be joined or spliced as shown in Fig. 10-23. However, the fascia is not needed structurally. Some types of open-eave construction do not use fascia and leave the overhanging rafters exposed.

Open eaves expose the area where the rafters and joists rest on the top plate of a wall. This area should be sealed either by a board or by the siding (Fig. 10-24). Sealing this area helps to prevent air currents from entering the wall. It also keeps insects and small animals from entering the attic area. However, to completely seal this restricts the airflow. This airflow is important to keep the roof members dry. It also prevents rotting caused by moisture. Airflow also can cool a building. To allow airflow, vents should be installed. These vents are backed with screen or hardware cloth to prevent animals and insects from entering.

Enclosed Cornices

Enclosed eaves have a neater appearance. These eaves do not overhang. Enclosed eaves are called *cornices*. Some houses do not have eaves or cornices. These are called *close cornices* (Fig. 10-25). Notice how the

Fig. 10-21 *Narrow box cornice. A closed over-hang is called a* cornice. (Forest Products Laboratory.)

Fig. 10-22 *Open overhangs are called* eaves.

Fig. 10-23 *Splicing the fascia: A. No fascia backer is used; B. a fascia backer is used.*

sheathing from the wall extends up to cap off the end of the roof sheathing board or plywood.

Several methods are used to enclose cornices. Two of the most common methods are the standard-slope cornice and the flat cornice. Both types seal the cornice area with panels called *soffits*. The panels usually are made of plywood or metal. Note that both types have some type of ventilation. Special cornice vents are used. However, when plywood is used for the soffits, an opening is often left, as shown in Fig. 10-26.

Fig. 10-24 *Open eaves should be sealed and vented properly.*

Fig. 10-26 *A strip opening may be used for ventilation instead of cornice vents.* (Forest Products Laboratory.)

This provides a continuous vent strip. This strip is covered with some type of grill or screen.

Standard-slope cornice The standard-slope cornice is the simplest and quickest to make. It is shown in Fig. 10-27A. The soffit panel is nailed directly to the underside of the rafters. If ¼-inch paneling is used, a 6d box nail is appropriate. Rust-resistant nails are recommended. Casing nails, when used, should be set and covered before painting. Panels should join on a rafter. In this way, each end of each panel has firm support. One edge of the panel is butted against the fascia. The edge next to the wall is closed with a piece of trim called a *frieze*.

Standard flat cornice The standard flat cornice has a soffit that is flat or horizontal with the ground. It is not sloped with the angle of the roof. A nail base must be built for the soffit. Special short joists are con-

Fig. 10-25 *With close cornices, the roof does not project over the walls.* (Forest Products Laboratory.)

RAFTER

ROOF SHEATHING

NAILING
BLOCK

WOOD SHINGLES

FRIEZE BOARD
AND MOLDING

SIDING

SHINGLE MOLDING

FASCIA

NAILING HEADER
(FASCIA BACKER)

WALL
SHEATHING

VENTILATOR

SOFFIT
(NAILED TO RAFTER)

(A)

RAFTER

ROOF SHEATHING

SHINGLE
STARTING
COURSE
(DOUBLE FOR
WOOD SHINGLES)

PLATE

FRIEZE BOARD
AND MOLDING

SIDING

SHINGLE MOLDING

FASCIA

NAILING HEADER
(FASCIA BACKER)

VENTILATOR

LOOKOUT

SOFFIT

WALL
SHEATHING

(B)

Fig. 10-27 *A. Standard-slope cornice. B. Standard flat cornice.* (Forest Products Laboratory.)

structed. These are called *lookouts* (see Fig. 10-27B). The flat cornice should be vented, as are other types. Either continuous strips or stock vents can be used.

Lookouts need a nail base on both the wall and the roof inasmuch as they are nailed to the tail of the rafter. On the wall, they can be nailed to the top plate or to a stud. More often than not, neither the top plate nor the stud can be used. This means that the ledger is nailed to the studs through the sheathing (Fig. 10-28). Either 16d or 20d common nails should be used to nail the ledger to the wall. The bottom of the ledger should be level with the bottom of the rafters.

Next, determine the correct length needed for the lookouts. Cut the lookouts to length. Keep both ends square. Drive nails into one end of the lookout. The nails are driven until the tips just barely show through the lookout. Then they can be held against the rafter and driven down completely. Butt the other end of the lookout against the ledger. Then toenail the lookout on each side using 8d or 12d common nails.

Cut soffit panels to size. Cornice vents should be cut or spaced next. Cornice vents can be attached to

Fig. 10-28 *Lookouts are nailed first to the rafter tail. Then they are toenailed to a solid part of the wall.*

soffits before mounting. After the soffits are ready, they are nailed in place. Use, for ¼-inch panels, a 6d common nail or box nail.

Soffit panels should be joined on a solid nail base. The ends should join at the center of a lookout. One edge is butted against the fascia. The inner edge is sealed with a frieze board or a molding strip.

Many builders use prefabricated soffit panels. Sometimes fascia is grooved for one edge of the soffit panel. When prefabricated soffits are used, a vent is usually built into the panel.

Closed rakes The rake is that part of the roof hanging over the end of a gable (Fig. 10-29). When a cornice is closed, the rake also should be closed. However, most rakes are not vented. This is so because they are not connected with the attic space, as are the cornices.

Usually, it is common practice to add the gable siding before the rake soffit is added. Then a 2- × 4-inch nailing block is nailed to the end rafter (Fig. 10-30). Use 16d common nails. A fly rafter is added and supported by the fascia and the roof sheathing. Once the soffit is nailed to the bottom of the nail block and the fly rafter, a frieze board or bed molding is added to finish the soffit.

Solid rake On roofs that do not extend over the ends of the gables, a fascia is nailed directly onto the last rafter. The roof then is finished over the fascia. This becomes what is called a *solid rake* (Fig. 10-31).

Siding the Gable Ends

Cornices are painted before any siding is installed in many instances. In other cases, the gable siding is installed first, and then both the cornice and gables are painted. After these are done, the siding is installed. This is common when two different types of siding are used. For example, wood siding is put on the gables and painted; then a brick siding is laid (Figs. 10-32 and 10-33).

There are several ways that gable walls may be sided. Gable siding may be the same as that on the rest of the walls. In this case, the gable and the walls are treated in the same way; however, gable walls are often covered with different siding. Brick exterior walls topped by wooden gable walls are common. Different types of wood and fiber sidings also may be combined (Figs. 10-34 and 10-35). Different textures, colors, and directions are used to make a pleasing contrast.

How gable siding is applied is important. Gable siding is applied in such a way that water is shed (cascades) properly from gable to wall. For brick or stone,

(A)

Fig. 10-30 *A nailing block is nailed to the end rafter for the soffit.* (Forest Products Laboratory.)

(B)

Fig. 10-31 *Detail for a solid rake.* (Forest Products Laboratory.)

Fig. 10-29 *Closed rakes: A. narrow cornice with boxed return; B. narrow box cornice and closed rake; C. wide overhang at cornice and rake.* (Forest Products Laboratory.)

the gable must be framed to overhang the rest of the wall (Fig. 10-36). This overhang should be sufficient to provide an allowance for the thickness of the stone or brick. Wood, plywood, or fiber siding uses special drainage joints. Also, metal strips and separate wooden moldings are used.

Framed overhangs   Gable ends are framed to overhang a wall, especially if bricks are used for the siding below (see Fig. 10-36). Remember that this must be done to allow for the thickness of a brick wall.

Several methods are used, and one of the most common is to extend the top plate as in Fig. 10-37A. This puts the regular end rafter just over the wall for the overhang. A short (2-foot) piece is nailed to the top plate. The rafter bird's mouth is cut 1½ inches deeper than the others. A solid header is added on the bottom, as shown in Fig. 10-37B. The gable is framed in the usual manner. Gable siding is done before wall siding (Figs. 10-38 and 10-39).

Fig. 10-32 *Here, the gable siding is added and painted before brick siding is added.*

Fig. 10-33 *A wall and gable prepared for a combination of diagonal siding and brick veneer. Note that the wood has been painted before the brick has been applied.*

Drainage joints Drainage joints are important and are made in several ways. They are important because water must be shed from gable to wall properly. Otherwise, rain water or snow would run inside the siding and damage the wall. Drainage joints are used wherever two or more pieces are used, one above the other. Figure 10-40 shows the most common drainage joints.

GABLE (BOARDS AND BATTENS)

CLEARANCE

FLASHING

DRIP CAP

SIDING

Fig. 10-34 *Different types of siding can be used for the gable and the wall.* (Forest Products Laboratory.)

Fig. 10-35 *Smooth gable panels contrast with the brick siding.*

Fig. 10-36 *This gable is framed to overhang the wall.*

(A)

(B)

Fig. 10-37 *A. Extending the top plate so that the gable siding overhangs a brick or masonry wall. B. With a solid header added on the bottom.*

Fig. 10-38 *Gable siding is done before wall siding.*

Fig. 10-39 *Installing gable siding.*

In one method, molded wooden strips are used. These strips are called *drip caps*.

Notice how a plywood panel has a rabbet joint cut on the ends. These are overlapped to stop water from running to the inside. Metal flashing is also used. This flashing may be used alone or with drip caps.

At the bottom of panel siding, a special type of drip cap is used. It is called a *water table*. It does the same thing as a drip cap but has a slightly different shape. The water table also may be used with or without metal flashing.

INSTALLING SIDING

Shapes of siding vary. Boards, panels, shakes, and shingles are the most common. These may be made of many different materials. However, many of the techniques for installing different types of siding are simi-

Fig. 10-40 *Drainage joints between gable and siding.* (Boise-Cascade.)

Fig. 10-41 *Board-and-batten vertical siding.* (Forest Products Laboratory.)

Fig. 10-42 *Batten-and-board vertical siding.* (Forest Products Laboratory.)

lar. Shape determines how the siding is installed. For example, wood or asbestos shingles are installed in about the same way. Boards made of any material are installed alike. Panels made of plywood, hardboard, or fiber are installed in a similar manner. Special methods are used for siding made from vinyl or metal.

Board Siding

The three major types of board siding are plain boards, drop siding, and beveled siding. Each type is put up in a different manner.

Board siding may be of wood, plywood, or composition material. This material is a type of fiberboard made from wood fibers. Generally, plain boards and drop siding are made from real wood. Composition siding is usually made in the plain beveled shape.

Plain boards Plain boards are applied vertically. This means that the length runs up and down. There are no grooves or special edges to make the boards fit together. Three major board patterns are used.

The *board and batten* pattern is shown in Fig. 10-41. In this style, the board is next to the wall. The board is usually not nailed in place except sometimes at the top end. It is sometimes nailed at the top to hold it in place. A narrow opening is left between the boards. This opening is covered with a board called a *batten*. The batten serves as the weather seal. The nails are driven through the batten but not the boards, as shown in Fig. 10-41.

The *batten and board* style is the second pattern. The batten is nailed next to the wall. A typical nailing pattern is shown in Fig. 10-42. The wide board is then fastened to the outside as shown.

Another style is the Santa Rosa style, shown in Fig. 10-43. All the boards are about the same width in this style. A typical nail pattern is shown in the figure. The inner board can be thinner than the outer board.

Keep in mind that plain board siding can be used only vertically. This is so because it will not shed water well if it is used flat or horizontally. The surface appearance of the boards may vary. The boards may be rough or smooth. For the rough effect, the boards are taken directly from the saw mill. In order to make a smooth board finish, the rough-sawn boards must be surfaced.

Drop siding Drop siding differs from plain boards. Drop siding has a special groove or edge cut into it.

NOTE: Nail for first board should be 8d or 9d. Nail for second board should be 12d.

Fig. 10-43 *Santa Rosa or board-and-board style vertical siding.* (Forest Products Laboratory.)

Fig. 10-44 *Shapes and nailing for drop siding. (Forest Products Laboratory.)*

DROP
OR
RABBETED

2 NAILS FOR
WIDTHS 8" AND OVER
AND WHEN USED
WITHOUT SHEATHING

BLIND NAIL
(FINISHING NAIL)

PANELING

FOR WIDTHS GREATER
THAN 6" USE EXTRA FACE
NAIL OR 2 FACE NAILS

DROP
(PATTERN 106)

DROP
(PATTERN 124)

HORIZONTAL APPLICATION

PANELING
(WC 130)

PANELING
(WC 140)

HORIZONTAL OR VERTICAL APPLICATION

Fig. 10-45 *Major types of beveled siding. (Forest Products Laboratory.)*

TYPE

BEVEL

"ANZAC"
(BEVEL)

DOLLY
VARDEN

This edge lets each board fit into the next board and makes the boards fit together. This arrangement resists moisture and weather. Figure 10-44 shows some common types of drop siding. Nailing patterns for each type are also shown.

Beveled siding   *Beveled* siding is made with boards that are thicker at the bottom. Figure 10-45 shows the major types of beveled siding. Common beveled siding is also called *lap* siding, and the nailing pattern for it is shown in Fig. 10-46.

The minimum amount of overlap for lap siding is about 1 inch. For 10-inch widths, which are common, about 1½ inches of lap is suggested. Most lap siding is made of wood fibers. However, in many areas, wood siding is still used. When wood siding is used, the standard nailing pattern shown in Fig. 10-46 should be used. Some authorities suggest nailing through both boards. However, this is not recommended. All wood tends to expand and contract, and few pieces do so evenly. Nailing the two boards together causes them to bend and bow. When this occurs, air and moisture can easily enter.

BEVEL SIDING

NAIL TO STUD OR
WOOD SHEATHING
(TO CLEAR TOP OF
LOWER SIDING COURSE)

Fig. 10-46 *Nailing pattern for lap siding.* (Forest Products Laboratory.)

Another type of beveled siding is called *rabbeted beveled siding*. This is also called *Dolly Varden siding*. Dolly Varden siding has a groove cut into the lower end of the board. This groove is called a *rabbet*. When installed from the bottom up, each successive board should be rested firmly on the one beneath it. The nails then are driven into the tops of the boards.

Siding Layout

In all types of board siding, two or three factors are important to remember. Boards should be spaced to lay even with windows and doors. One board should fit against the bottom of a window. Another should rest on the top of a window. In this way, there is no cutting for an opening. Cutting also should be avoided for vertical siding. Adjust the spacing so that a board rests against the window on either side.

Before siding is installed, guide marks are laid out on the wall. These are used to align the boards properly. To do this, a pattern board, called a *story pole*, is used just as in wall framing. The procedure for laying out horizontal lap siding will be given. The procedures for other types of siding are similar. This procedure may be adapted accordingly.

Procedure Choose a straight 1-inch board. Cut it to be exactly the height of the area to be sided. Many siding styles are allowed to overlap the foundation 1 to 2 inches. Find the width of the siding and the overlap to be used. Subtract the overlap from the siding width. This determines the spacing between the bottom ends of the siding. The bottom end of lap siding is called the *butt* of the board. For example, siding 10 inches wide is used. The overlap is 1½ inches. Thus the distance between the board butts is 8½ inches. On the story pole, lay out spaces 8½ inches apart.

Place the story pole beside a window. Check to see if the lines indicating the spacing between the boards line up even with the top and bottom of the window. The amount of overlap over the foundation wall may be varied slightly. The amount of lap on the siding also may be changed. Small changes will make the butts even with the tops of the windows. They also will make the tops even with the window bottoms (Fig. 10-47). Story-pole markings should be changed to show adjustments. New marks then are made on the pole.

Use the story pole to make marks on the foundation. Also make marks around the walls at appropriate places. Siding marks should be made at the edges of windows, corners, and doors (Fig. 10-48). If the wall is long, marks also may be made at intervals along the wall. In some cases, a chalk line may be used to snap guide lines. A chalk line is often used to snap a line on the foundation. This shows where the bottom board is nailed.

Nailing

Normally, siding is installed from the bottom to the top, with the first board overlapping the foundation at least 1 inch. The first board is tacked in place and checked for level. After leveling, the first board then is nailed firm. Usually bottom boards will be placed for the whole wall first. Then the other boards are put up, bottom to top. The level of the siding is checked after each few boards.

Siding also can be installed from the top down. This is done when scaffolding is used. The layout is the same. The lines should be used to guide the butts of the boards. However, the first board is nailed at the top. Two sets of nails are used on the top board. The first set is nailed near the top of the board. These may be nailed firm. Then the bottom nail is driven 1½ inches from the butt of the board. This nail is not nailed down firm. About ¾ inch should be left sticking out. The butt of the board is pried up and away from the wall. A nail bar or the claw of a hammer is used. The board is left this way. Then the next board is pushed against the nails of the first board. The second board is checked for level. Hold the level on the bottom of the board, and check the second level. When the board is level, it is nailed at a place 1½ inches from the butt. Again, the nail in the second board is left out. About ¾ inch should be left sticking out. The board is nailed down at the butt for its length. Then the first board is nailed down firm. The second board then is pried up for the third board. This process is repeated until the siding has been applied.

FLASHING

NO SHEATHING
PAPER REQUIRED
FOR PLYWOOD
OR FIBERBOARD
SHEATHING

SIDING TO CLEAR
DRIP CAP

BUILDING PAPER
AROUND AND
UNDER FRAME

SIDING FLUSH
WITH BOTTOM
OF SILL

USE SHEATHING
PAPER OVER
LUMBER
SHEATHING

BUTT JOINTS
MADE OVER STUD

8″ MINIMUM CLEARANCE

STARTING STRIP

Fig. 10-47 *Beveled siding application. Note that pieces are even with top and bottom of window opening.* (Forest Products Laboratory.)

Corner Finishing

There are three ways to finish corners for siding. The most common ways are shown in Fig. 10-49. Corner boards can be used for all types of siding. In one style of corner board, the siding is butted next to the boards. In this way, the ends of the siding fit snugly, and the corner is the same thickness as the siding (see Fig. 10-49C).

Figure 10-49B shows special metal corner strips. These are separate pieces for each width of board. The carpenter must be careful to select and use the right-size corner piece for the board siding being used.

In other methods, the siding is butted at the corners. The corner boards then are nailed over the siding (Fig. 10-50). This type of corner board is best used for plywood or panel siding. However, it is also common for board siding.

Another common method uses metal corners for lap siding. Figure 10-49B shows metal corner installation. These corners have small tabs at the bottom. They fit around the butt of each board. At the top is a small tab for a nail. The corners are installed after the siding

is up. The bottom tab is put on first. Tabs then are put on other boards from bottom to top. The corners can be put on last. This is so because the boards will easily spread apart at the bottoms. A slight spread will expose the tab for nailing.

Another method of finishing corners is to miter the boards. Generally, this method is used on more expensive homes. It provides a very neat and finished appearance. However, it is not as weatherproof. In addition, it requires more time and is thus more expensive. To miter the corners, a miter box should be used. Lap siding fits on the wall at an angle. It must be held at this angle when cut. A small strip of wood as wide as the siding is used for a brace. It is put at the base of the miter box (Fig. 10-51). This positions the siding at the proper angle for cutting. If this is not handy, an estimate may be made. It is generally accurate enough. This method is shown in Fig. 10-52. A distance equal to the thickness of the siding is laid off at the top. The cut then is made on the line as shown.

Inside corners are treated in two ways. Both metal flashing and wooden strips are used. The most common method uses wooden strips (Fig. 10-53). A ¾-

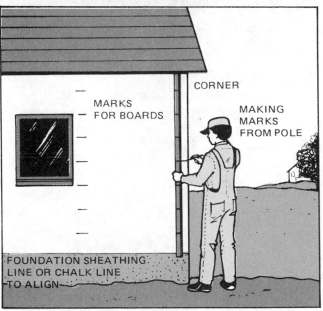

Fig. 10-48 *Marks are made to keep siding aligned on long walls.*

inch-square strip is nailed in the corner. Then each board of the siding is butted against the corner strip. After the siding is applied, the corner is caulked. This makes the corner weathertight.

Metal flashing for inside corners is similar. The metal has been bent to resemble a wooden square. The strips are nailed in the corner as shown in Fig. 10-54. As before, each board is butted against the metal strip.

PANEL SIDING

Panel siding is now used widely. It has several advantages for the builder. It provides a wider range of appearance. Panels may look like flat panels, boards, or battens and boards. Panels also can be grooved, and they can look like shingles. The surface textures range from rough lumber to smooth, flat board panels. Panels can show diagonal lines, vertical lines, or horizontal lines. Also, panels may be used to give the appearance of stucco or other textures.

Panel siding is easy to lay out. There is little waste, and the installation is generally much faster. It is often stapled in place rather than nailed (Fig. 10-55). As a rule, corners are made faster and easier with panel siding.

Panel siding can be made from a variety of materials. Plywood is commonly used. Hardboard and various other fiber composition materials are also used. A number of finishes also are available on panel siding. These are prefinished panels. Panels may be prefinished with paint, chemicals, metal, or vinyl plastics.

The standard panel size is 4 feet wide by 8 feet long. The most common panel thickness is ⅜ inch. However, other panel sizes are also used. Lengths up to 14 feet are also available. Thickness ranges from ⁵⁄₁₆ to ¾ inch in ¹⁄₁₆-inch intervals.

The edges of panel siding may be flat. They also may be grooved in a variety of ways. Grooved edges form tight seams between plywood panels. Seams may be covered by battens. They also may be covered by special edge pieces. Outside corners may be treated in much the same manner as corners of regular board siding. Corners may be lapped and covered with corner boards. They may be covered with metal. They also may be mitered. Inside corners may be butted together. The edge of one panel is butted against the solid face of the first.

As a rule, panels are applied from the bottom up. However, they may be applied from top to bottom.

NAILS AND NAILING

Correct nails and nailing practices are essential in the proper application of wood siding. As a rule, siding and box nails are used for face nailing, and casing nails are used for blind nailing. Nails should be corrosion resistant and preferably rustproof. Avoid using staples. Stainless steel nails are the best choice. High-tensile-strength aluminum nails are economical and corrosion resistant, and they will not discolor or cause deterioration of wood siding. However, aluminum nails will react with galvanized metal, causing corrosion. Do not

STRIP MUST BE
THICKNESS OF SIDING

Fig. 10-51 *A strip must be used to cant the siding for mitering.*

MARK A
LINE

LINE

45°

Fig. 10-52 *To cut miters without a miter box, make the top distance B equal to the thickness A. Then cut back at about 45 degrees.*

NAIL

METAL INSIDE
CORNER

Fig. 10-54 *Inside corners may be finished with metal strips.*

NOTE:
Nails in field spaced 16″ O.C. (horizontally).
Nails on both sides of joints should be 1/2″
from joint edge and 3/4″ up from butt.

VAPOR BARRIER
ON WARM SIDE
OF WALL

STUDS
16″ O.C.
MAX.

1 1/8″ X 1 1/8″
WOOD STRIP

CAULK

LEAVE 1/32″ GAP AT
JOINT OR USE JOINT
COVERS (SECT. 10)

1/2″

3/4″

16″

3/8″ X 1 1/2″
CANT STRIP

6″ MINIMUM

CRAWL SPACE

METAL CORNER

Fig. 10-53 *Inside corners may be finished with square wooden strips.*
(Boise-Cascade.)

Fig. 10-55 *Panel siding is often stapled in place.*

Nail Shanks

Nails with smooth shanks will loosen as the siding expands and contracts under the extremes of seasonal changes in temperature and humidity. Ring- or spiral-threaded nail shanks will increase holding power. Both types of shanks are readily available.

Nail Points

The most commonly used nail points include blunt, diamond, and needle, as shown in Fig. 10-56. Blunt points reduce splitting, whereas diamond points are used most often. Needle points should be avoided because they tend to cause splitting.

Fig. 10-56 *Nail types.* (Western Wood Products Association.)

use aluminum nails on galvanized flashing (nor galvanized nails on aluminum flashing). The hot-dipped galvanized nail is the least expensive, but it might cause discoloration if precautions are not taken.

In some cases, the use of hot-dipped galvanized nails along with clear finishes on western red cedar has resulted in stains around the nails. While this occurrence seems to be limited to the northeastern and north central regions of the country, the combination of hot-dipped galvanized nails and clear finishes on western red cedar is not recommended.

Plastic hammer-head covers can be used when driving hot-dipped galvanized nails. This will reduce the potential for chipping and the subsequent potential for corrosion.

Use of staples or electroplated nails on siding often results in black iron stains that can be permanent. Copper nails are not suitable for western red cedar because cedar's natural extractives will react with the copper, causing the nails to corrode and resulting in stains on the siding.

High-quality nails for solid-wood siding are a wise investment. The discoloration, streaking, or staining that can occur with inappropriate nails ruins the appearance of the project and is very difficult to remove.

Recommended penetration into studs or blocking or into a combination of wood sheathing and these members is 1.5 inches. Penetration is 1.25 inches with ring-shank nails.

Vertical siding, when applied over wood-bed sheathing, should be nailed to horizontal blocking or other wood framing members not exceeding 35 inches O.C. when face nailed or 32 inches O.C. when blind nailed. Vertical siding, when installed without sheathing, should be nailed to wood framing or blocking members at 24 inches O.C.. Some building codes require 24 inches O.C. with or without sheathing. Check your local code for the requirements. Horizontal and diagonal siding should be nailed to studs at a maximum of 24 inches O.C. when applied over wood-based solid sheathing and a maximum of 16 inches O.C. on center when applied without sheathing.

The siding pattern will determine the exact nail size, placement, and number of nails required. As a rule, each piece of siding is nailed independently of its neighboring pieces. Do not nail through two overlapping pieces of siding with the same nail because this practice will restrict the natural movement of the siding and might cause unnecessary problems. Nail joints into the studs or blocking members—nailing into sheathing alone is not adequate.

Be careful in driving nails. Hand nailing is preferred over pneumatic nailing because there is less control of placement and driving force with pneumatic nailers. Nails should be snug but not overdriven. Nails that are overdriven can distort the wood and cause excessive splitting. Predrilling near the ends helps to reduce any splitting that can occur with the thinner patterns. Some modern siding patterns with their recommended nailing procedures are shown in Fig. 10-57.

SHINGLE AND SHAKE SIDING

Shingles and shakes are sometimes used for exterior siding and are very similar in appearance. However, shakes have been split from a log. They have a rougher surface texture. Shingles have been sawn and are smoother in appearance. The procedure is the same for either shingles or shakes.

Shingles are made from many different materials. Shakes are made only from wood. The standard lengths for wood shakes and shingles are 16, 18, and 24 inches.

Shingles

Shingles may be made of

- Wood
- Flat composition
- Mineral fiber composition

The last is a combination of fiber and portland cement. Shingles for roofs are lapped about two-thirds. About one-third of the shingle is exposed. When the shingles are laid, or coursed, there are three layers on the roof. However, when shingles are used for siding, the length exposed is greater. Slightly more than half is exposed. This makes a two-layer thickness instead of three as on roofs. For mineral sidings, sometimes less lapping is used.

Wooden shingles are made in random widths. Shingles will vary from 3 to 14 inches in width. The better grades will have more wide shingles than narrow shingles.

Nailing

All types of shingles may be nailed in either of two ways. In the first method, shown in Fig. 10-58, nailing strips are used. Note that a moisture barrier is almost always used directly under shingles. Builder's felt (tar paper) is the most common moisture barrier. Shingles should be spaced like lap siding. The butt line of the shingles should be even with the tops of wall openings. Top lines should be even with the bottoms. For utility buildings such as garages, no sheathing is needed. A moisture barrier can be put over the studs. Nailing strips then are nailed to the studs. However, for most residential buildings, a separate sheathing is suggested. Strips are often nailed over the sheathing. This is done over sheathing that is not a good nail base. Shingles may be nailed directly to board or plywood sheathing. This is shown in Fig. 10-59.

The bottom shingle course is always nailed in place first. Part of the bottom course can be laid for large wall sections. The course may reach for only a part of the wall. Then that part of the wall can be shingled. The shingled part will be a triangle from the bottom corner upward. As a rule, two layers of shingles are used on the bottom course. The first layer is nailed in place, and a second layer is nailed over it. The edges of the second shingle should not line up with the edges of the first. This makes the siding much more weather resistant. This layout is shown in Fig. 10-59.

Another technique for shingles is *double coursing*. This means that two thicknesses are applied. The first layer is often done with a cheaper grade of shingle. Again, each course is completed before the one above is begun. The edges should alternate for best weather-resistive qualities. Double coursing is shown in Fig. 10-60. The outside shingle covers the bottom of the inside shingle. This gives a dramatic and contrasting effect.

Shakes

A shake is similar to a shingle, but shakes may be made into panels. Figure 10-61 shows this type of siding. Shake panels are real wooden shakes glued to a plywood base. The strips are easier and quicker to nail than individual shakes. Also, shakes can be spaced more easily. A more consistent spacing is also attained. The individual panels may be finished at the factory. Color, spacing, appearance, and texture can be factory-made. Panels also may be combined with insulation and weatherproofing. The panels are applied in the same manner as shingles.

Corners

Corners are finished the same as for other types of siding. Three corner types are used: corner boards, metal corners, and mitered corners. Again, mitering generally is used for more expensive buildings.

SIDING PATTERNS	NOMINAL SIZES*	NAILING	
	Thickness & Width	6" & Narrower	8" & Wider
TRIM **BOARD-ON-BOARD** **BOARD-AND-BATTEN** Boards are surfaced smooth, rough or saw-textured. Rustic ranch-style appearance. Provide horizontal nailing members. Do not nail through overlapping pieces. Vertical applications only.	1 x 2 1 x 4 1 x 6 1 x 8 1 x 10 1 x 12 1¼ x 6 1¼ x 8 1¼ x 10 1¼ x 12	Board and Batten ½" Recommend ½" overlap. One siding or box nail per bearing.	Board and Batten / Board on Board Increase overlap proportionately. Use two siding or box nails, 3-4" apart.
BEVEL OR BUNGALOW Bungalow ("Colonial") is slightly thicker than Bevel. Either can be used with the smooth or saw-faced surface exposed. Patterns provide a traditional-style appearance. Recommend a 1" overlap. Do not nail through overlapping pieces. Horizontal applications only. Cedar Bevel is also available in ⅞ x 10,12.	½ x 4 ½ x 5 ½ x 6 ⅝ x 8 ⅝ x 10 ¾ x 6 ¾ x 8 ¾ x 10	Plain Recommend 1" overlap. One siding or box nail per bearing, just above the 1" overlap.	Plain Recommend 1" overlap. One siding or box nail per bearing, just above the 1" overlap.
DOLLY VARDEN Dolly Varden is thicker than bevel and has a rabbeted edge. Surfaced smooth or saw textured. Provides traditional-style appearance. Allows for ½" overlap, including an approximate ⅛" gap. Do not nail through overlapping pieces. Horizontal applications only. Cedar Dolly Varden is also available ⅞ x 10,12.	Standard Dolly Varden ¾ x 6 ¾ x 8 ¾ x 10 Thick Dolly Varden 1 x 6 1 x 8 1 x 10 1 x 12	Rabbeted Edge Allows for ½" overlap. One siding or box nail per bearing, 1" up from bottom edge.	Rabbeted Edge ← approximate ⅛" gap for dry material 8" and wider ½" = full depth of rabbet Allows for ½" overlap. One siding or box nail per bearing, 1" up from bottom edge.
DROP Drop siding is available in 13 patterns, in smooth, rough and saw textured surfaces. Some are T&G, others shiplapped. Refer to "Standard Patterns" for dimensional pattern profiles. A variety of looks can be achieved with the different patterns. Do not nail through overlapping pieces. Horizontal or vertical applications. Tongue edge up in horizontal applications.	¾ x 6 ¾ x 8 ¾ x 10	T&G Pattern / Shiplap Patterns Use casing nails to blind nail T&G patterns, one nail per bearing. Use siding or box nails to face nail shiplap patterns, one inch up from bottom edge.	T&G Pattern / Shiplap Patterns ← approximate ⅛" gap for dry material 8" and wider ½" = full depth of rabbet Use two siding or box nails, 3-4" apart to face nail, 1" up from bottom edge.

Fig. 10-57 *Siding patterns, nominal sizes, and recommended nailing.* (Western Wood Products Association.)

SIDING PATTERNS	NOMINAL SIZES*	NAILING	
	Thickness & Width	6" & Narrower	8" & Wider
TONGUE & GROOVE Tongue & groove siding is available in a variety of patterns. T&G lends itself to different effects aesthetically. Refer to WWPA "Standard Patterns" (G-16) for pattern profiles. Sizes given here are for Plain Tongue & Groove. Do not nail through overlapping pieces. Vertical or horizontal applications. Tongue edge up in horizontal applications.	1 x 4 1 x 6 1 x 8 1 x 10 Note: T&G patterns may be ordered with ¼, ⅜ or 7/16" tongues. For wider widths, specify the longer tongue and pattern.	Plain Use one casing nail per bearing to blind nail.	Plain Use two siding or box nails 3-4" apart to face nail.
CHANNEL RUSTIC Channel Rustic has ½" overlap (including an approximate ⅛" gap) and a 1" to 1¼" channel when installed. The profile allows for maximum dimensional change without adversely affecting appearance in climates of highly variable moisture levels between seasons. Available smooth, rough or saw textured. Do not nail through overlapping pieces. Horizontal or vertical applications.	¾ x 6 ¾ x 8 ¾ x 10	Use one siding or box nail to face nail once per bearing, 1" up from bottom edge.	approximate ⅛" gap for dry material 8" and wider ½" = full depth of rabbet Use two siding or box nails 3-4" apart per bearing.
LOG CABIN Log Cabin siding is 1½" thick at the thickest point. Ideally suited to informal buildings in rustic settings. The pattern may be milled from appearance grades (Commons) or dimension grades (2x material). Allows for ½" overlap, including an approximately ⅛" gap. Do not nail through overlapping pieces. Horizontal or vertical applications.	1½ x 6 1½ x 8 1½ x 10 1½ x 12	Use siding or box nail to face nail once per bearing, 1½" up from bottom edge.	approximate ⅛" gap for dry material 8" and wider ½" = full depth of rabbet Use two siding or box nails, 3-4" apart, per bearing to face nail.

SIDING INSTALLATION TIPS

Do not nail through overlapping pieces. Use stainless steel, high tensile strength aluminum or hot-dipped galvanized nails with ring or spiral-threaded shanks. Use casing nails to blind nail; siding or box nails to face nail.

Horizontal applications only for Bevel, Bungalow and Dolly Varden.

Vertical applications only for Board-and-Board or Board-and-Batten; bevel cut ends of pieces and install so water is directed to outside.

Horizontal or vertical applications for Tongue & Groove, Channel Rustic, Log Cabin or Drop patterns. Tongue edge up in horizontal applications of Drop and T&G patterns

Read the section on Nail Penetration & Spacing to determine nail size.

Read the sections on Moisture Content and Prefinishing before installing siding.

Fig. 10-57 (Continued)

WOOD
SHINGLES

FIBERBOARD OR
GYPSUMBOARD

FOUNDATION
WALL

GYPSUM BOARD

1" × 2" NAILING STRIPS

Fig. 10-58 *Nail strips are used over studs and sheathing. Shingles then are nailed to the strips. At least two nails are used for each shingle.*

STUD

NO. 1 SHINGLE

BUILDING PAPER

SUBFLOOR

$\frac{3}{4}$"

2"

5 d

NOTE: Use
2 nails for shingles
up to 8" wide and
3 nails for shingles
over 8" wide.
Use threaded nails
for plywood
sheathing.

NO. 3 OR
UNDERCOURSE
SHINGLE

SILL

SHEATHING

Fig. 10-60 *Double-coursed shingle siding.* (Forest Products Laboratory.)

NAIL ABOVE
BUTT OF
NEXT COURSE

CLOSED
JOINT

$\frac{3}{4}$"

1"

SPACED
JOINT

MAXIMUM FOR
16" SHINGLE

USE BUILDING PAPER
OVER BOARD SHEATHING

$7\frac{1}{2}$"

DOUBLE STARTING COURSE

NOTE: Use 2 nails for shingles
up to 8" wide and 3 nails for
wider shingles. Use threaded
nails for plywood sheathing.

Fig. 10-59 *Single-course shingle nailed directly to solid sheathing.* (Forest Products Laboratory.)

Fig. 10-61 *Shakes are often made into long panels.* (Shakertown.)

Fig. 10-62 *Shingle and shake corners may be woven or lapped.*

Corners may also may be woven. Shingles may be woven as in Fig. 10-62. This is done on corners and edges of door and window openings.

PREPARATION FOR OTHER WALL FINISHES

There are other common methods for finishing frame walls. These include stucco, brick, and stone. As a rule, the carpenter does not put up these wall finishes. However, the carpenter sometimes prepares the sheathing for these coverings. The preparation depends on how well the carpenter understands the process.

Stucco Finish

Stucco is used widely in the South and Southwest. It is durable and less expensive than brick or stone. Like brick or stone, it is fireproof. It may be put over almost any type of wall. It can be prepared in several colors. Several textures also can be applied to give different appearances.

Wall preparation   A vapor barrier is needed for the wall. The vapor barrier can be made of builder's felt or plastic film. The sheathing also may be used for the vapor barrier. It is a good idea to apply an extra vapor barrier over most types of walls. This includes insulating sheathing, plywood, foam, and gypsum. The vapor barrier is applied from the bottom up. Each top layer overlaps the bottom layer by a minimum of 2 inches. Then a wire mesh is nailed over the wall (Fig. 10-63). Staples generally are used rather than nails. A solid nail base is essential. The mesh should be stapled at 18- to 24-inch intervals. The intervals should be in all directions—across, up, and down. The wire mesh may be of lightweight "chicken wire." However, the mesh should be of heavier material for large walls.

The mesh actually supports the weight of the stucco. The staples only hold the mesh upright, close to the wall.

Apply the stucco   Usually two or three coats are applied. The first coat is not the finish color. The first coat is applied in a very rough manner. It is applied carefully, but it may not be even in thickness or appearance. It is called the *scratch coat*. Its purpose is to provide a layer that sticks to the wire. The coat is spread over the mesh with a trowel (Fig. 10-64).

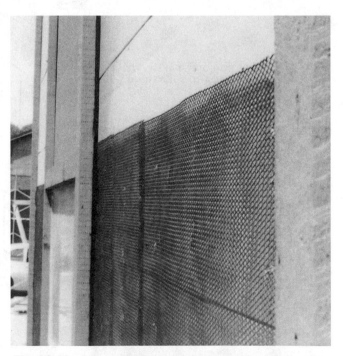

Fig. 10-63 *A wire mesh is nailed to the wall. Stucco then will be applied over the mesh.*

Fig. 10-64 *Stucco is spread over the mesh with a trowel.*

This scratch coat should be rough. It may be scratched or marked to provide a rough surface. The rough surface is needed so that the next coat will stick. A trowel with ridges may be used. The ridges make grooves in the coat. These grooves may run in any direction. However, most should run horizontally. Another way to get a rough surface is to use a pointed tool. After the stucco sets, grooves are scratched in it with the tool.

One or more coats may be put up before the finish coat is applied. These coats are called *brown coats*. They are scratched so that the next coat sticks better. The last coat is called the *finish coat*. It is usually white or tan in color. However, dyes may be added to give other colors. The colors are usually light in shade.

Brick and Stone Coverings

Brick and stone are used to cover frame walls. The brick or stone is not part of the load-bearing wall. This means that it does not hold up any roof weight. The covering is called a *brick* or *stone veneer*. It adds beauty and weatherproofing. Such walls also increase the resistance of the building to fire.

The preparation for either brick or stone is similar. First, a moisture barrier is used. If a standard wall sheathing has been used, no additional barrier is needed. However, it is not bad practice to add a separate moisture barrier.

The carpenter then may be asked to nail *ties* to the wall. These are small metal pieces, shown in Fig. 10-65, that are bent down and embedded in the mortar. After the mortar has set, these ties form a solid connection. They hold the brick wall to the frame wall. They also help to keep the space between the walls even. Small holes usually are left at the bottom. These are called *weep holes*. Moisture can soak through brick and stone. Moisture also collects at the bottom from condensation. The small weep holes allow the moisture to drain out. By draining the moisture, damage to the wood members is avoided. The joints are trimmed after the brick is laid (Fig. 10-66).

Fig. 10-65 *Ties help to hold brick or stone to the wall. They are nailed to studs and embedded in the mortar.*

Fig. 10-66 *Joints are trimmed after brick is laid.*

ALUMINUM SIDING

Aluminum siding is used widely on houses. Aluminum is used for new siding, or it can be applied over an old wall.

A number of vertical and horizontal styles are used. Probably the most common type looks like lap siding. However, even this type may come as individual "boards" or as panels two or three "boards" wide (Fig. 10-67). These types of siding may be hollow backed or insulated (Fig. 10-68).

Aluminum siding also may look like shingles or shakes, as in Fig. 10-69. For all types of aluminum siding, a number of surface textures and colors are available.

Aluminum siding is put up using a special system. Note in Fig. 10-70 that the top edge has holes in it. All nails are driven in these holes. The bottoms or edges of the pieces interlock. In this way, the tops are nailed, and the bottoms interlock. Each edge then is attached to a nailed portion for solid support.

Fig. 10-68 *Aluminum siding may be hollow backed or insulated.* (ALCOA.)

Fig. 10-67 *The most common type of aluminum siding locks like lap siding: A. single-board lap siding; B. two-board panel lap siding; C. double-board drop siding; D. vertical board siding.*

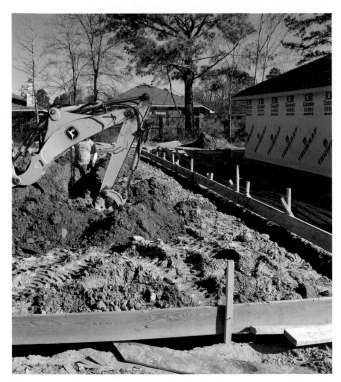

Using a backhoe to dig trenches for the footings and plumbing of a slab foundation.

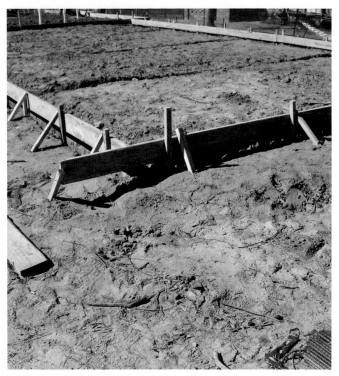

Wooden forms shored up with wooden stakes prior to pouring a concrete slab. Note three trenches that will form footings to support the internal load of the home. Two post-tensioned cables and possibly some plumbing lines will be placed in these trenches before the concrete is poured.

Squaring up the footings and cleaning the trenches so the plumbing drain lines can be sloped properly. Notice the string that is supported by batter boards to guide the workers when placing the wooden forms for the concrete slab.

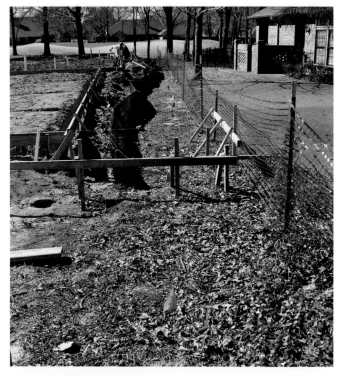

Base for slab with excavating, leveling, and plumbing drains in place. Notice the batter boards by the fence that were used to hold string for a guide to place the concrete slab forms.

Black plastic covers the foundation to prevent moisture from seeping up underneath the floor of the home. Note the black "chairs" that are to be installed under the cable before concrete is poured. In this way, the cables will become totally immersed in concrete and not break out of the sides when being tensioned a few days after the concrete has hardened.

Close-up view of plastic covering with slab post-tensioning cables in place. Notice the black support chairs have yet to be installed under the cables.

A close-up view of a trench and corner fireplace location.

Installation of rebar to reinforce slab at critical weight-bearing points in the slab.

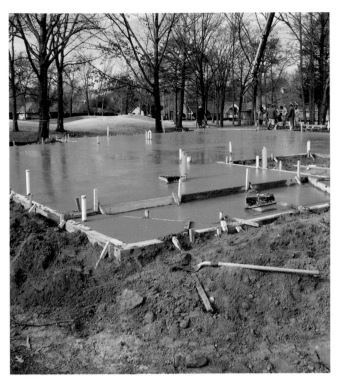

Concrete slab poured and leveled.

Finishing the concrete slab by hand.

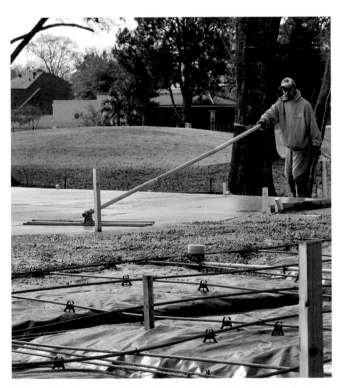

Using a float to smooth the slab surface. Notice the black chairs that are supporting the red post tensioning cables.

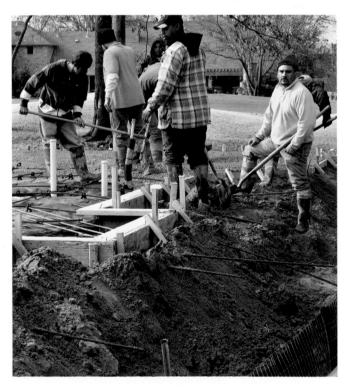

Filling in dirt around open areas under the wooden forms to avoid blowouts, and waiting for truck to bring last batch of redi-mixed concrete.

Spreading and leveling the poured concrete to make a slab.

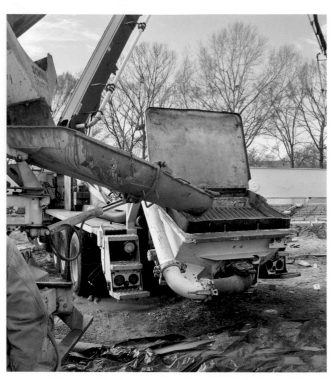

Filling the hopper of the concrete pump truck so it can deliver the concrete to the right spot.

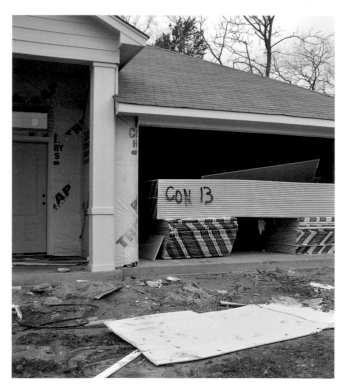

Storing drywall in the garage to prevent damage from the elements.

Drywall installed on the ceiling for the garage. Notice the laminated header for the garage door opening.

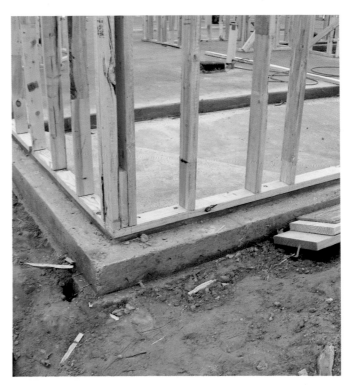

Note the excess length of cable protruding from the slab. This cable will be attached to a portable hydraulic jack that will then stretch the cable to just over 30,000 psi. Wedges are pushed into the tapered hole, thereby locking the cable to the anchor. The cable is then cut and the exposed anchor is grouted over.

Corner slab detail. Note the level of the garage floor and that of the living area are different. Also, take note that the exposed concrete area of the slab indicates that room has been allowed for placement of the brick veneer type for an exterior wall finish.

Notice the bracing for the wall to support it until the remaining walls are finished.

The studs are usually delivered already cut to length.

Note the soleplate and top plate. Also notice the double plate that is lapped over the top plates to attach walls together from the top.

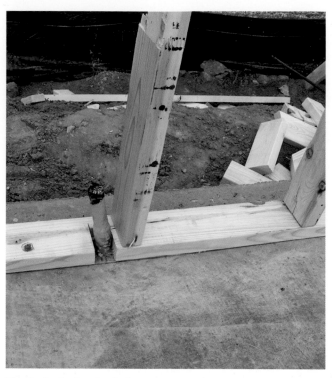

Notched soleplate for drain or water service.

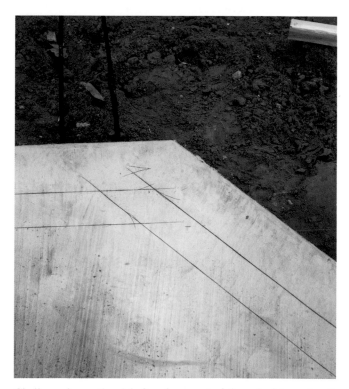

Chalk marks on the slab for placement of the soleplate and wall erection guidance.

Erecting walls in progress. Notice the saw horses that hold studs while they are cut to length with a circular saw.

Notched soleplate for drain and larger pipe for the toilet.

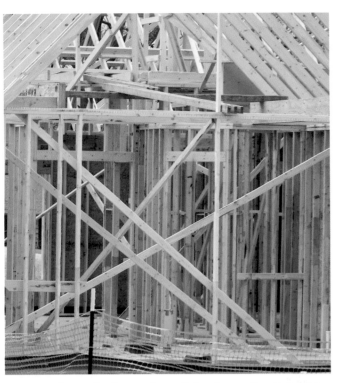

Temporary braces for the exterior wall framing until wall sheathing is applied.

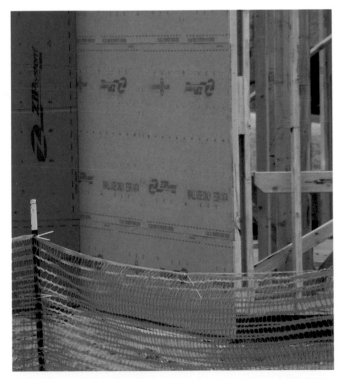

Green exterior sheathing from Zip Systems is 7/16-inch OSB that is treated with a water-resistant green coating. Special tape is rolled over the panel joints and no building wrap, such as Tyvek, is needed.

Braces for exterior framing of the wall. Notice the miter saw that is used to cut the lumber to the correct angle, especially for roof rafters.

Switch box with wires in place. Note extra length of the Romex for wiring the switch. Also note the "S" on the stud indicating where to place the switch box. The horizontal line on the stud locates the bottom of the switch box.

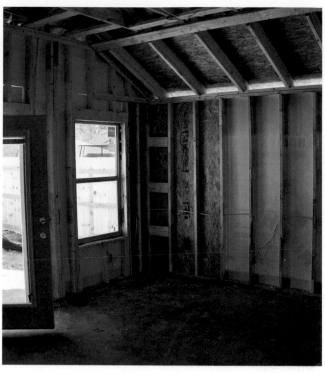

Notice that plywood sheathing is only placed in the corners for strength and how the light shines through the pink house wrap. This is acceptable since the home will have brick covering these sides.

Note the shower drain. The black plastic box was placed before the concrete was poured to allow room for the drain. This allows room to attach the drain to the shower floor pan.

Notice how the switch box extends past the wall stud the thickness of a sheet of drywall that will be installed once the electrical inspector approves the work.

Plumbing installation for the clothes washer.

Plumbing and electrical for the kitchen sink, dishwasher, and garbage disposal. Notice how the inverted "U" is both a drain and vent for the kitchen sink.

Water line with plastic box and brass valve is installed for refrigerators with ice makers.

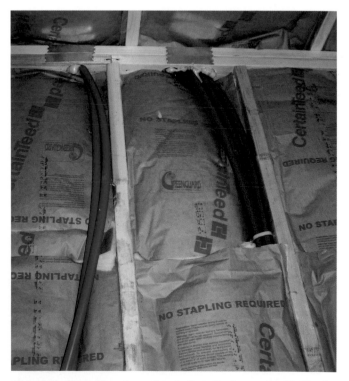

Blue pipes for cold water and red pipes for hot water. Notice the metal plates protect the pipes from sheetrock screws.

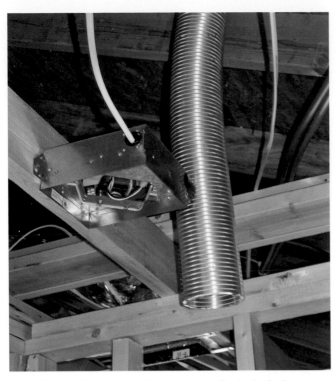

Metal box for forced ventilation or exhaust fan in the bathroom.

Flexible aluminum exhaust duct fitted to fan housing.

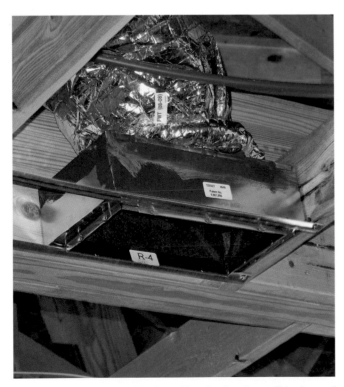

A return air duct mounted in the ceiling for the air conditioning and heating system.

Opening in the garage ceiling for a pull-down attic stairs unit.

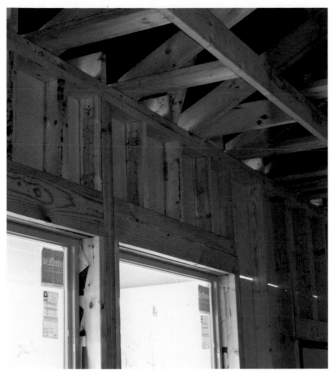

Notice headers and cripple studs above doorways and windows.

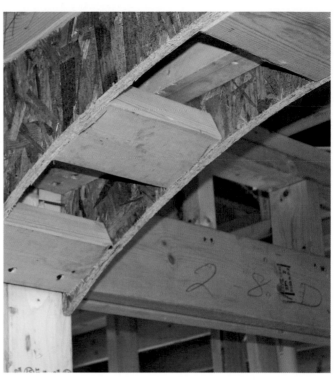

Detail of archway buildup for drywall in arch between living room and dining room.

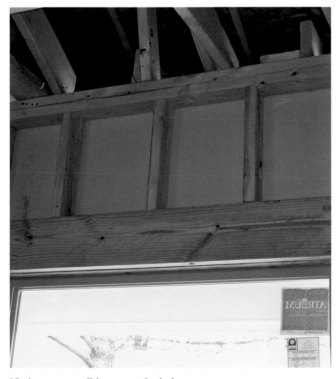

Notice gray caulking around windows.

Notice the wood has been treated with a blue solution that repels termites and is water resistant. Typically, it is sprayed to a height of at least 12 inches above the slab.

Ductwork entering a central plenum box for the HVAC system.

Look at the metal framing that is attached to ceiling joists for the installation of a ceiling return air vent.

Notice the notches cut in the rafters to allow for the additional framing that will support the drywall used for the vaulted ceiling.

Man wearing stilts to aid in insulating ceiling. The man is also wearing a dust mask to prevent inhalation of fiberglass insulation fibers.

Plumbing arrangement for clothes washing machine. Red for hot water, blue for cold water.

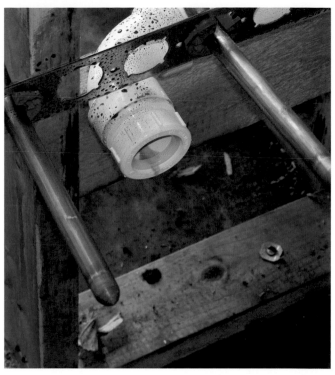

Drain and copper pipes for hot and cold water in place for lavatory sink installation. Copper ends are sealed to prevent contamination and will later be cut open to install valves and flexible lines for water fixtures.

Framing for tub enclosure.

Installation of mixer valve for bathtub and/or shower.

Central control station for all the hot and cold water lines throughout the house. Notice that each separate water line can be turned on or off at this station.

Terminating the red and blue water line tubing with copper pipes to fit to fixtures.

Electrician pulling wire.

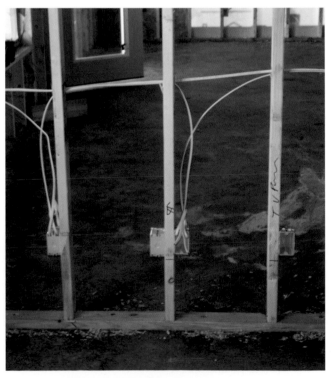

Note the installation of the outlets for two adjacent rooms.

Using stilts for installing drywall on ceiling joists. Note insulation in the background is paper faced.

Stilts are useful for using a cordless Rotozip to cut around electrical boxes after drywall is installed over them. The Rotozip is also being used to cut around doorway and window openings. It is much quicker and easier to install drywall over everything and then trim it quickly with the Rotozip tool.

Attaching shingles with a pneumatic nailer.

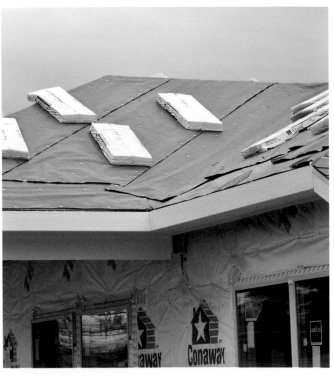

Bundles of shingles on the builders felt (tar paper) ready for installation.

Two dormers jutting out of the top of the roof to bring light into the attic or second story.

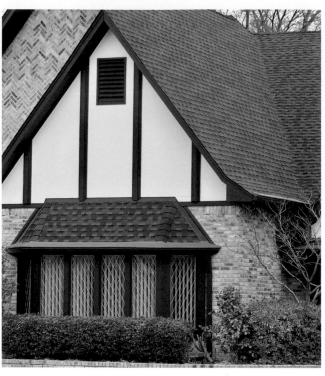

A stucco-and-beam English Tudor exterior appearance with a protruding bay window.

A Southern Colonial–style porch with large, two-story-high columns, wide front entry door, balcony with wrought ironwork, and shutters. This can be used for "street appeal" or enhancing the aesthetics of a finished home.

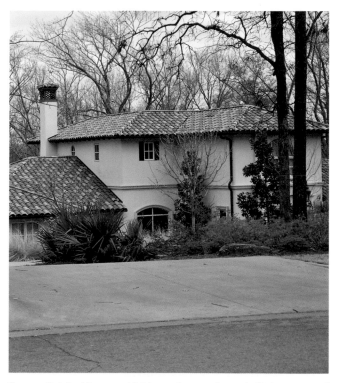

Stucco-finished house with hip roofs over the main living area and the garage. Note how the arched window and tile roof add to the architect's emphasis of the Spanish or Mediterranean appearance.

The authors would like to thank Conaway Homes of Tyler, Texas, for their assistance in producing these color photographs and excellent technical information. Photos by Mark R. Miller.

Fig. 10-69 Aluminum "shake" siding. (ALCOA.)

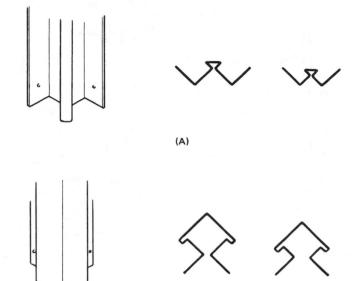

(A)

(B)

Fig. 10-72 Corner strips for aluminum siding: A. inside; B. outside. (ALCOA.)

Fig. 10-70 Panels are nailed at the top and interlock at the bottom. (ALCOA.)

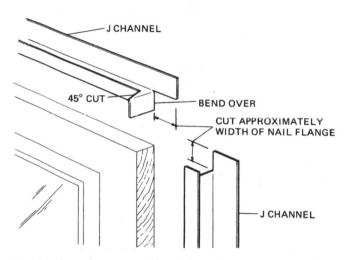

J CHANNEL

45° CUT

BEND OVER

CUT APPROXIMATELY WIDTH OF NAIL FLANGE

J CHANNEL

Fig. 10-73 Special pieces are used around windows and gables. (ALCOA.)

Fig. 10-71 A starter strip is nailed at the bottom of the wall. The lower edge of the first board can interlock with it for support. (ALCOA.)

To start, a special starter strip is nailed at the bottom (Fig. 10-71). Then corner strips, as in Fig. 10-72, are added. Special shapes are placed around windows and other features, as in Fig. 10-73. For these, the manufacturer's directions should be carefully followed.

Next, the first "board" is nailed in place. Note that it is placed at the bottom of the wall. The bottom of the first piece is interlocked with the starter strip and nailed. It is best to start at the rear of the house and work toward the front. In this way, overlaps do not show as much. For the same reason, factory-cut ends should overlap ends cut on the site (Fig. 10-74).

Backer strips (Fig. 10-75) should be used at each overlap. These provide strength at the overlaps. Also, overlaps should be spaced evenly (Fig. 10-76). Grouping the overlaps together gives a poor appearance.

Allowance must be made for heat expansion. Changes in temperature can cause the pieces to move. To allow movement, the nails are not driven up tightly. It is best to check the instructions that come with the siding.

Vertical Aluminum Siding

Most procedures for vertical siding are the same. Strips are put up at corners, windows, and eaves. How-

Fig. 10-74 *Siding is started at the back.* (ALCOA.)

Fig. 10-75 *Backer strips support the ends at overlaps.* (ALCOA.)

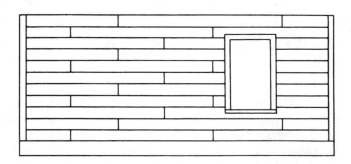

Fig. 10-76 *Overlaps should be equally spaced for best appearance.* (ALCOA.)

ever, the starter strip is put near the center. A plumb line is dropped from the gable peak. A line then is located half the width of a panel to one side. The starter strip is nailed to this line. Panels then are installed from the center to each side (Fig. 10-77).

Fig. 10-77 *Vertical aluminum siding is started at the center.* (ALCOA.)

SOLID VINYL SIDING

Solid vinyl siding is also used widely. It is available in both vertical and horizontal applications. A wide variety of colors and textures is also available.

As with aluminum siding, vinyl siding is nailed on one edge. The other edge interlocks with a nailed edge for support (Fig. 10-78). Special pieces again are needed for joining windows, doors, gables, and so forth. As with aluminum siding, the finish is a permanent part of the siding. No painting will ever be needed. The siding may be cleaned with a garden hose.

Fig. 10-78 *Solid vinyl siding.*

CHAPTER 10 STUDY QUESTIONS

1. What are the methods for enclosing eaves?
2. What sequence is used for finishing an exterior wall?
3. What can happen to painted wooden walls that have no vapor barrier?
4. How far should siding nails penetrate into the nail base?
5. How much siding is needed for

   Wall size   = 8 feet high × 30 feet long

   Openings   = one, 3 × 4 feet

   Overlap   = 1½ inches

   Siding width   = 10 inches
6. Why is some siding started from the top?
7. What is the usual starting point for siding?
8. When are guard rails required on scaffolds?
9. What is a wall bracket?
10. What are the advantages of steel double-pole scaffolds?
11. How thick should platform planks be?
12. List six rules for ladder safety.
13. List six rules for scaffold safety.
14. Why are soffit vents used?
15. What types of board siding are used?
16. Why are boards not nailed together on lap siding?
17. Why is siding spaced even with openings?
18. What is the difference between shingles and shakes?
19. Why are shingles double-coursed on walls?
20. What is a square in shingling?
21. How are gable overhangs framed?
22. How is wooden gable and wall siding joined?
23. What holds brick veneer to a wall?
24. How are corners treated for wood-panel siding?
25. What should be done to prepare a wall for stucco?
26. When can hot-dipped galvanized nails be used with western red cedar?
27. Should vertical siding, when installed without sheathing, be nailed to wood framing or blocking members every 24 inches O.C.?
28. What does a blunt-point nail do to wood?
29. Where can you use a needle-point nail?
30. Why is hand nailing preferred over pneumatic nailing?

11
CHAPTER

House
Wiring

ELECTRICITY IS AN ABSOLUTE NECESSITY FOR MODERN living. So many devices in the home rely on electricity to function and render the style and comfort expected of a new house that it would be almost impossible to keep a house operating as a home without electricity. This electricity can be generated by fossil-fuel-powered generators, by atomic-powered generators, or by the use of falling water to drive generators. It is also possible to generate power by engine-driven generators. In Alaska, for instance, long distribution lines are impractical due to ice conditions that damage lines during storms. Also in Alaska, engine-driven generators are common.

Once electricity is generated in sufficient amounts for consumption in large quantities, the second necessary step is to get the energy to the consumer. Herein lies a distribution problem: that of stringing and maintaining long lines.

Figure 11-1 illustrates the process of getting power from the generating plant along high-voltage transmission lines to voltage step-down substations and then to office buildings, stores, schools, apartments, and large factories. Further reductions in voltage are necessary to reduce the power to the proper voltages (120–240 volts) for home use.

LOCAL DISTRIBUTION

Underground systems for distributing electrical energy are in the middle of a revolution that began about 30 years ago. A prime objective has been to reduce the cost of these systems so that they could be used in areas previously served only by overhead systems. Aluminum, long known as an economical electrical conductor, has contributed to this revolution.

Actually, underground distributing is as old as the electrical power industry itself. In 1882, Thomas A. Edison built the first central power station on Pearl Street in New York City. He installed feeders and mains under the city streets to distribute power generated in the station to its customers. These were pipe-type cables with a pair of copper conductors that were separated from each other and from the enclosing pipe by specially designed washers. The pipe was then filled with an insulating compound. This compound consisted of asphaltum boiled in oxidized linseed oil, with paraffin and a little beeswax added.

The development of lead-sheathed insulated cables that could be pulled into previously installed ducts greatly simplified construction of underground distribution systems. However, the high costs of such systems remained. As a result, for many years, underground distribution was limited largely to the central portions of cities. Here load densities are high, and the congestion that would result from overhead systems is highly undesirable.

Today, costs of underground distribution have been reduced to levels that are economically practical for service to light- and medium-density load areas. Underground systems now are being installed in most residential areas. Home buyers in these areas generally agree that the improved appearance and enhanced

Fig. 11-1 *Generation and distribution of electricity.*

LINE-VOLTAGE SYMBOLS		LOW-VOLTAGE SYMBOLS	
Symbol	Description		
———————	2-Cond. 120V Wire or Cable	— - — - —	Remote-Control Low-Voltage Wire
—///—	3-Cond. 120V Wire or Cable	T	Low-Voltage Transformer
◯	Ceiling Receptacle	—▶—	Rectifier for Remote Control
◯	Floodlight	B$_R$	Box for Relays and Motor Master Controls
◻—◯	Valance Light	R	Remote-Control Relay
—◯$_C$	Clock Receptacle	R$_P$	Remote-Control Pilot-Light Relay
◯$_L$	Keyless Lampholder	P$_{11}$	Separate Pilot Light, R.C. Plate
◯$_L$ PS	Pull-Chain Lampholder	P$_{10}$	Separate Pilot Light, Inter. Plate
—⊖	Double Receptacle, Split-Wired	MS	Master-Selector Switch
—⊖	Grounding Receptacle	MM$_R$	Motor Master Control for ON
—⊕ WP	Weatherproof Grounding Receptacle	MM$_B$	Motor Master Control for OFF
—⊕ R	Range Receptacle	S$_M$	Switch for Motor Master
—◉ CD	Clothes Dryer Receptacle	S$_{F6}$	R.C. Flush Switch
S$_L$	Lighted-Handle Mercury Switch	S$_{F7}$	R.C. Locator-Light Switch
S$_P$	Push-Button, Pilot Switch	S$_{F8}$	R.C. Pilot-Light Switch
S$_D$	Closet Door Switch	S$_{K6}$	R.C. Key Switch
		S$_{K7}$	R.C. Locator Light Key Switch
		S$_{K8}$	R.C. Pilot-Light Key Switch
		S$_{T6}$	R.C. Trigger Switch
		S$_{T7}$	R.C. Locator Light Trigger Switch
		S$_{T8}$	R.C. Pilot-Light Trigger Switch
		S$_{T4}$	Interchangeable Trigger Switch, Brown
		S$_{T5}$	Interchangeable Trigger Switch, Ivory
		—⊖ RO	Remote-Control Receptacle for Extension Switch

Fig. 11-1 *Generation and distribution of electricity.*

value of their property more than justify the added cost of underground service. However, in some places the transformers are located in the front of the house, and this negates any advantage to the appearance value of underground feeds. The transformers in the front yard are difficult to landscape, and if they are landscaped, it is hard to gain access to them in the case of an emergency. Many communities either place the service completely underground with buried transformers or distribute the power from the rear of the house. Exposed upright telephone-line terminals and cable television (CATV) boxes also detract from the beauty of a residential area when they, too, are located in the front yard.

FARM ELECTRICITY

In 1935, President Franklin D. Roosevelt created the Rural Electrification Administration as an emergency relief program. In May 1936, the Congress of the United States passed the Rural Electrification Act. This established the REA as a leading agency in the federal government. It had the responsibility of developing a program for rural electrification. The act authorized

Fig. 11-2 *A farm meter pole. Single-phase, three-wire, 120/240-volt service for loads exceeding 40 kilowatts of demand. Some farmers purchase power from utility companies other than the REA.*

WEATHERHEAD

REDUCER

SERVICE BRACKET AND
MOUNTING STRAP
FURNISHED BY COMPANY
INSTALLED BY CUSTOMER

BACKBRACE

BOND RISER PIPE
TO SERVICE NEUTRAL

ALTERNATE PIPE OR
CONDUIT LOCATION
(IN EXTERIOR
BUILDING WALL
ONLY.)

ALTERNATE
SERVICE
EQUIPMENT

ALTERNATE
SERVICE
ENTRANCE

INSULATING
BUSHING

MAXIMUM LENGTH OF
UNGUARDED SERVICE
ENTRANCE CONDUCTORS
WITHIN WALL SHALL
BE 12"

SERVICE
ENTRANCE

SERVICE
EQUIPMENT

LEAVE SERVICE CONDUCTORS 24"LONG
FOR CONNECTION TO SERVICE DROP BY
COMPANY

COMPANY'S TRIPLEX SERVICE DROP
1000 LBS. STRAIN

9"

2' - 0" MAX.
WITHOUT
BACK BRACE

THE RISER SHALL BE CAPABLE
OF WITHSTANDING A HORIZONTAL
PULL OF 1000 LBS. AT THE
SERVICE DROP ATTACHMENT.
PROVIDE BACK BRACE WHERE
NECESSARY.

APPROVED VENT PIPE FLANGE

3/8"U BOLT AS CLOSE TO
ROOF AS POSSIBLE

2" OR 2-1/2" GALVANIZED STEEL
PIPE OR RIGID CONDUIT

WHERE PIPE IS USED, SERVICE
ENTRANCE CABLE IS REQUIRED
WHERE CONDUIT IS USED
SEPARATE CONDUCTORS MAY
BE USED.

3/8"U BOLT RISER SUPPORT

WEATHERTIGHT FITTING

METER SOCKET TROUGH
FURNISHED BY COMPANY
INSTALLED BY CUSTOMER

PROVIDE VERTICAL SURFACE FOR
MOUNTING METER SOCKET TROUGH

4'
to
6 1/2'

NOTE BELOW

3' CLEAR SPACE TO
PROPERTY LINE

FINAL GRADE

WHERE THE BUILDING IS TOO LOW TO OBTAIN PROPER CLEARANCES IT IS
RECOMMENDED THAT CUSTOMER INSTALL AN UNDERGROUND SERVICE LATERAL.

Fig. 11-3 *Service entrance riser support on a low building.*

and empowered the REA to make self-liquidating loans to companies, cooperatives, municipalities, and public power districts. These loans were to finance the construction and operation of generating plants and transmission and distribution lines (Fig. 11-2).

Whether on the farm or in the suburbs, an electrician is usually required to place the meter socket trough. It is usually supplied by the power company and installed at the expense of the customer. Most meters are of the plug-in type. Once the service has been

connected and the meter socket and house are wired properly, the meter is put into the socket, completing the circuit for power into the house (Fig. 11-3).

SAFETY AROUND ELECTRICITY

Most carpenters use extension cords to obtain power for their power tools. The proper size of cord makes a difference in the performance and life of the power equipment. A proper extension cord can make a difference in the safe operation of a piece of equipment. Make sure that it has the proper capacity to handle the current needed (Table 11-1).

TABLE 11-1 *Flexible Cord Ampacities**

Size	Type S, SO, ST	Type SJ, SJO, SJT
AWG	Amperes	
18	7, 10*	7, 10*
16	10, 13*	10, 13*
14	15, 18*	
12	20	
10	25	
8	35	
6	45	
4	60	

**Table 400.5. Reprinted by permission from NFPA 70-1981, National Electric Code, © 1980, National Fire Protection Association, Boston, Massachusetts.
*Where third conductor is used for equipment grounding only and does not carry load current.

Wire size and insulation of extension cords must be considered to ensure proper operation with a piece of equipment. Table 11-2 shows what the letters on a cord body mean. This table aids in determining what type of service is recommended for a specific cord.

Length of the extension cord is important in the sense that it must have a correct size wire to allow voltage to reach the consuming device. For instance, a 25-foot cord with No. 18 wire size is good for 2 amperes. If the distance is increased to 50 feet, you are still safe with No. 18 and 2 amperes. However, for a distance of 200 feet from the source, the size of the wire must be increased to No. 16. This is necessary to carry the 2 amperes without dropping the voltage along the cord and therefore producing a low voltage at the consuming device at the end of the line (Table 11-3).

A visual inspection of the cord and harness should be made before each day's use. Check for loose or missing screws in the end and for loose or broken

TABLE 11-2 *Types of Flexible Cords*

Trade name	Type letter	Size AWG	No. of conductors	Insulation braid on	Braid on each conductor	Outer covering	Use
Junior Hard Service Cord	SJ SJO SJT	18, 14	2, 3 or 4	Rubber Thermoplastic or Rubber	None	Rubber Oil Resist Compound Thermoplastic	Pendant or Portable Damp Places Hard Usage
Hard Service Cord	S SO ST	6 18, to 10 incl.	2 or more	Rubber Thermoplastic or Rubber	None	Rubber Oil Resist Compound Thermoplastic	Pendant or Portable Damp Places Hard Usage

TABLE 11-3 *Extension Cord Sizes for Portable Electric Tools*

THIS TABLE FOR 115-VOLT TOOLS						
Full-load ampere rating of tool	0 to 2.0 A	2.1 to 3.4 A	3.5 to 5.0 A	5.1 to 7.0 A	7.1 to 12.0 A	12.1 to 16.0 A
Length of Cord	Wire size (AWG)					
25 ft	18	18	18	16	14	14
50 ft	18	18	18	16	14	12
75 ft	18	18	16	14	12	10
100 ft	18	16	14	12	10	8
200 ft	16	14	12	10	8	6
300 ft	14	12	10	8	6	4
400 ft	12	10	8	6	4	4
500 ft	12	10	8	6	4	2
600 ft	10	8	6	4	2	2
800 ft	10	8	6	4	2	1
1000 ft	8	6	4	2	1	0

NOTE—If voltage is already low at the source (outlet), have voltage increased to standard, or use a much larger cable than listed in order to prevent any further loss in voltage.

Fig. 11-4 *A. After an accident with construction equipment, the cord should be checked for damage. B. Check the extension cord before it is placed in service each day. C. Check the wire clamp on the plug for proper fit.*

blades in the plug end. Check the continuity of each conductor with an ohmmeter at least once a week to make sure that there are no broken conductors. Check the cord when it is first used or taken from the box. Check again after it has been repaired or after an accident such as shown in Fig. 11-4. If the plug has been repaired, check to see if the clamp on the wire is securely attached.

GROUNDED CONDUCTORS

A grounded conductor has a white-colored jacket in a two- or three-wire cable (neutral wire). It is terminated to the white- or silver-colored terminal in a plug cap or connector, and it is terminated at the neutral bar in the distribution box.

An electrical fault might allow the hot line to come into contact with the metal housing of electrical equipment (in a typical two-wire system) or some other ungrounded conductor. In such a case, any person who touches the equipment or conductor will be shocked. The person completes the circuit from the hot line to the ground, and current passes through the body. Because a human body is not a good conductor, the current is not high enough to blow the fuse. It continues to pass through the body as long as the body remains in contact with the equipment (Fig. 11-5).

A *grounding conductor*, or *equipment ground*, is a wire attached to the housing or other conductive parts of electrical equipment that are not normally energized to carry current from the equipment to the ground. Thus, if a person touches a part that is accidentally energized, there will be no shock because the grounding line furnishes a much lower-resistance path to the ground (Fig. 11-6). Moreover, the high current passing through the wire conductor blows the fuse and stops the current. In normal operation, a ground conductor does not carry current.

The grounding conductor in a three-wire conductor cable has a green jacket. It is always terminated at the green-colored hex-head screw on the cap or connector. It uses either a green-colored conductor or a metallic conductor as its path to ground. In Canada, this conductor is referred to as the *earthing conductor*. This term is somewhat more descriptive and helpful in distinguishing between grounding conductors and neutral wires or grounded conductors.

HOUSE SERVICE

For an electrical system to operate properly, it is necessary to design the system so that the proper amount of power is available. This requires the house to be wired so that wire of the proper size serves each plug in the house.

Ground-Fault Circuit Interrupter (GFCI)

Some dangerous situations have been minimized by using ground-fault circuit interrupters (GFCIs) (Fig. 11-7A). Since 1975, the *National Electrical Code* (NEC) has required installation of GFCIs in outdoor outlets, outlets near kitchen sinks, and bathroom outlets in new construction, but most homes built before 1975 have no GFCI protection.

Retrofit GFCIs can protect one outlet or an entire circuit with multiple outlets. They can be installed in older homes to reduce the danger of electric shock. One of the simplest ways to achieve this protection in outdoor outlets or other outlets in which shock dangers are high is to use a plug-in type of GFCI.

Two kinds of plug-ins are available. One has contact prongs attached to the housing, and it is simply plugged into a grounded outlet. The device to be used is plugged into a receptacle in the housing of the GFCI.

THIS SYSTEM IS DANGEROUS

DON'T YOU ACT AS A GROUNDING CONDUCTOR
– MAKE SURE EQUIPMENT IS GROUNDED THROUGH THE CORD!

White-jacketed system grounds cannot conduct electricity from short circuits to the ground. Thus, they do not prevent the housing of faulty equipment from becoming charged. Therefore, a person who contacts the charged housing becomes the conductor in a short circuit to the ground. (National Safety Council)

(A)

Fig. 11-5 *A. Dangerous system. You become part of the circuit to ground and can be fatally injured. B. Safe system uses the green wire to protect the power tool operator from shock.*

THIS SYSTEM IS SAFE

LET THE GREEN GROUNDING CONDUCTOR TAKE THE CHARGE – NOT YOU!

Equipment stays at ground potential in spite of the short circuit, if the circuit has a grounding conductor. Internal electrical faults cause current to short-circuit harmlessly to equipment ground in systems with green-jacketed grounding conductors. (National Safety Council)

(B)

Fig. 11-6 *A pigtail adapter can cause problems if it is not properly attached to the screw.*

Fig. 11-7 *B. This 15-amp GFCI is used to protect 120-volt receptacles in hotels and motels and homes where each bathroom is required to have GFCI protection. This is a Class A GFCI that trips when it senses 4 to 6 mA of ground-fault leakage current. It is made in five colors to color-coordinate with bathroom interiors. Kitchens and outdoor outlets on homes also must be GFCI protected according to the NEC. (It takes over 7 mA for a human to feel the jolt of electricity.)*

Another GFCI, more suitable for outdoor use, has a heavy-duty housing attached to a short extension cord. The extension cord type of GFCI is easily plugged into an outdoor outlet without its getting tangled with the outlet's lid or cover.

Keep in mind that GFCIs are not foolproof. They do, however, switch off the current in less than 0.025 second if a leakage is detected. They can detect as lit-

Fig. 11-7 *A. Three types of GFCIs.*

Fig. 11-7 *C. This GFCI provides protection for swimming pools, hot tubs, and industrial controls. It is designed for installation within 10 feet of a swimming pool. The reset button is red and will pop out when the circuit has been tripped and power turned off. Swimming pool wiring installed before the 1965 code may have wiring that can cause a GFCI to trip continuously.*

tle as 2 milliamperes, although most operate at a threshold of 5 milliamperes—well below the level that would affect a person.

The GFCI should be tested about once a month to make sure that it is working. Plug in something that consumes power, and then push the test button. The red button should pop out, and the power should be off. Then push in the red button to reset the device. This should turn the power on again. Some GFCIs that are built so that they fit outlet boxes have two buttons—black and red. These buttons usually are located side by side as in Fig. 11-7B, or they can be one above the other and a little longer than those in Fig. 11-7C.

Most power is distributed locally within a neighborhood by overhead wires (Fig. 11-8). The wires usually are located in the rear of the house. In some communities, the wires are buried and come into the house near the basement or foundation wall. Power is brought from the pole or transformer into the rear of the house by one black, one red, and one white (uninsulated) wire (Fig. 11-9). Once the cable is connected to the house, it is brought down to the meter by way of a sheathed cable with three wires. These are red, white, and black. The white wire is usually uninsulated and twisted as shown in Fig. 11-10. Figure 11-11 shows how the power is fed from the transformer to the distribution box.

Service Entrance

All services should be three-wire. The capacity of service-entrance conductors and the rating of service equipment should not be less than that shown in Table 11-4. This is a quick way to determine the type of service required. There are more accurate ways, which are explained later.

Fig. 11-9 *A transformer on a pole located at the rear of a house.*

Fig. 11-8 *Stringing overhead lines for local service in a residential neighborhood.*

Twisted bare conductor

Insulated conductors

Bare conductor helically

(A)

(B)

Fig. 11-10 *A. Note how the twisted bare conductor is made from the outside coating of the other two wires. B. Type SE cable. This cable has three wires: one red, one black, and an uninsulated wire that forms a protection for the other two. The stranded uninsulated wire is twisted at the end to make a connection, as in part A.*

3 WIRES FROM TRANSFORMER

3 WIRE CABLE

KILOWATT HOUR METER

DISTRIBUTION BOX AND MAIN DISCONNECT INSIDE HOUSE

(A)

(B)

Fig. 11-11 *A. Power from the transformer is fed through the wires and down the side of the house to the meter. Then it goes through the wall at a lower level to a distribution box in the basement. B. Underground service feeds the meter on a slab-type house. The distribution box for the house is to the left of the meter housing.*

These capacities are sufficient to provide power for lighting, portable appliances, equipment for which individual appliance circuits are required, electric space heating of individual rooms, and air conditioning. A larger service might be required for larger houses. It also might be required if a central furnace or central hydronic boiler is used for electric space heating.

Many major appliances in the kitchen require individual equipment circuits. Thus, where practical, electric service equipment should be located near or on a kitchen wall. This will minimize installation and wire costs. In addition, such a location is often convenient to the laundry. This minimizes circuit runs to laundry appliances.

Planning

The first step in wiring a house is planning. This means that you should make accurate plans and know exactly how much material you need. Once you have learned about the boxes, wire, and tools, you can take a close look at the codes.

The local power company can advise you regarding the kind of service to use for your house. In most instances, the local power company will take the power right up to the house. The rest of the wiring is your responsibility.

The electrician works on a house at two of the construction phases. The electrician is needed to put in the wiring while the house is still in the roughed-in stage. This means that a new house will have the wiring installed before the drywall goes onto the studs. Once

(C)

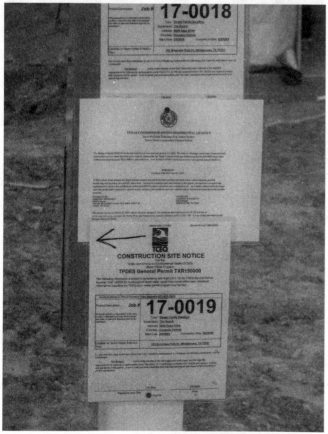

(D)

Fig. 11-11 *C. Temporary power connection for use while house is under construction. D. Construction site permit.*

TABLE 11-4A *Minimum Service Capacities for Various House Sizes*

Floor area		Minimum service capacity
m²	sq ft	
Up to 93	Up to 1000	125 amperes
93–186	1001–2000	150 amperes
186–279	2001–3000	200 amperes

the drywall workers and carpenters have finished their work, the electrician needs to return. At this time, the electrician installs the receptacles, receptacle faceplates, circuit breakers, chandeliers, and lights. Switches are connected, and circuits tested for proper operation. In some cases, the electrician has to install the furnace, the electric heating, or the electric range. Various states of construction make it obvious that certain wiring should be done in an area before it is closed up. This is where planning is very important. It is much easier to wire a house while it is still unplastered than when it is walled in and ready for occupancy.

Permits

Before starting installation, check with your local power company to determine what permits are necessary. Most permits specify that you will have someone inspect the installation after you have finished wiring it. This inspection will determine if the installation is safe and free from possible fire hazards.

Local Regulations

In some communities, local regulations supersede the NEC. Make sure that you know what these local regulations are before starting the installation. Make sure that the materials you use are approved by the local power company. If not, you might have trouble getting service once the job is complete.

Some regulations apply as a minimum. Those provided as guides by the NEC relative to the stringing of feeder wires are similar to those shown in Fig. 11-12. Here, a roof with a rise greater than 4 inches in 1 foot would be difficult to walk on easily. Therefore, one of the exceptions to the code will allow a clearance of 3 feet between the roof and an overhead power line if the line does not have over 300 volts. Figure 11-12 shows such an installation.

The electrician's main concern is the actual service drop and its entry into the house. The location of the entrance is important. The NEC does have something to say about where the service can be located (Fig. 11-13). Note that the service is required to be at least 3 feet from a window.

TABLE 11-4B *Conductor Capacities for Household Equipment*

Item	Conductor capacity
Range (Up to 21-kW rating) *or*	50 A-3 W-115/230 V
⌈Built-In Oven	30 A-3 W-115/230 V
⌊Built-In Surface Units	30 A-3 W-115/230 V
Combination Washer-Dryer *or*	40 A-3 W-115/230 V
Electric Clothes Dryer	30 A-3 W-115/230 V
Fossil-Fuel-Fired Heating Equipment (if installed)	15 A or 20 A-2 W-115 V
Dishwasher and Waste Disposer (if necessary plumbing is installed)	20 A-3 W-115/230 V
Water Heater (if installed)	Consult Local Utility

The possibility of rot caused by water can be minimized by placing the service head as shown in Fig. 11-14. The installation of this service head is usually the responsibility of the residential electrician.

SERVICE FROM HEAD TO BOX

The service wires are not connected to the power company's lines until the inside of the house is wired properly and inspected. Figure 11-15 shows how the service is connected and the meter socket is inserted in the line. The overhead connections can be made if the meter is left out of the socket. This causes the rest of the service to be inoperative until the meter is inserted properly in the socket by the power company.

Notice how the neutral wire is also connected to the cold water pipe inside the house (Figs. 11-16 and 11-17). In some cases, the house does not have a cold water system furnished by a local community water source. If a well is used for water, the neutral must be grounded as shown in Fig. 11-18.

A conduit might be used for the installation of the service into the house. Here, three wires are used. One is black, one is white, and the other is red. The size of the wires will vary according to the amount of current needed.

Conduit is limited as to the size of wire it can handle. For instance, a 0.75-inch conduit can take three No. 8 wires. A 1.25-inch conduit can take three No. 2 wires, three No. 3 wires, three No. 4 wires, or three No.

Fig. 11-12 *Clearance required for a 4/12 pitched roof in the path of a power line.*

Fig. 11-13 *Minimum clearances for windows and doors for the installation of a service drop.*

6 wires. A 1.5-inch conduit can take three No. 1 wires. A 2-inch conduit will handle three No. 1/0 wires, three No. 2/0 wires, or three No. 3/0 wires. Therefore you can see the importance of knowing the requirements first. If you know the amount of current needed, you can determine the size of the wire. Then you can figure out which size conduit to use to hold the wires safely.

Wires from the utility company's pole to your building are called the *service drop*. These are usually furnished by the company. These wires must be high enough to provide proper clearance above grade. They

also must not come within 3 feet of doors, windows, fire escapes, or any opening (see Fig. 11-12). The structure to which the service drop wires are fastened must be sturdy enough to withstand the pull of ice and wind.

Installation of the Service

Figure 11-19 shows how the conduit for the service is connected to the head and to the meter socket. Note how it enters the house with a waterproof fitting and then goes unspliced into the distribution panel in the basement.

Circuits are run from the distribution panel to various parts of the house. Some examples of circuits are shown in Fig. 11-19. Notice that there are 22 circuits available. Figure 11-15 shows how a fuse-type box is serviced by the entrance wires.

Conduit is fastened to the wall of the building by clamps placed every 4 feet. The conduit should use an entrance ell to turn the conduit into the house. Such an ell has two threaded openings corresponding to conduit size. Use an adapter to fasten the conduit into the threaded opening at the top of the ell. Into the lower

Fig. 11-14 *Proper and improper ways of installing a service head and service drop.*

Metal strap

3 lead wires must extend out of entrance head at least 36 in. Middle wire is the neutral bare wire and usually connects to center wire in view above.

Watertight connector

Meter socket

Non-watertight connector

Clamped to a cold waterpipe. However, in some areas the ground wire should be connected to a rod driven into the ground at least 8 ft.

Neutral wire
Red wire
Black wire
(240-volt circuit)

MAIN

RANGE

Fig. 11-15 *Connected service using a fuse box and external ground.*

opening, fasten a piece of conduit to run through the side of the house. Use a connector to attach the conduit to the electric service panel.

Inserting Wire into a Conduit

After the conduit is installed, you can push the wires through the top hub of the meter, through the conduit, and out the service head. All three wires must extend at least 36 inches out of the service head. This will allow enough wire for connecting to the power lines. Then the wires are brought down from the meter to the entrance ell. Remove the ell cover, and pull the wires through to the service panel inside the house. Use a white-covered insulated wire for the neutral in the conduit.

Fig. 11-16 *The usual method of grounding city and town systems.*

Fig. 11-17 *The usual method of grounding city and town systems using conduit.*

Fig. 11-18 *The approved REA method of grounding a wire system with a ground rod.*

Distribution Panels

There are many types of distribution panels made for homes. One type has fuses that screw in. These fuses must be replaced when they are blown or open (Fig. 11-20). Another type is the circuit-breaker box. A circuit breaker can be reset if the device is tripped by an overload. This type is gaining in popularity for home use. Many people do not want to be bothered with looking for a new fuse every time one blows. New builds do not use fuses. The circuit breaker can be reset by pushing it to the "off" position and then to the "on" setting. If it trips a second time, the circuit trou-

ble is still present. It should be located and removed before resetting the breaker again (Fig. 11-21).

For a closer look at fuse replacement or resetting a circuit breaker, take a look at Fig. 11-22. This is a 100-ampere fuseless service entrance panel [this size (100 amperes) is obsolete today]. Most houses demand a larger current rating serving the home because of the number of electrical devices consuming power.

Romex Cable

Romex cable is used to carry power from a distribution panel box to the individual outlets within a house. This nonmetallic sheathed cable has plastic insulation covering the wires to insulate it from the environment. Some types of cable can be buried underground. Figures 11-23 and 11-24 show the types used most commonly in house wiring.

Most new homes are wired with 12/2 (No. 12 wire with two conductors). One of the wires is white, and the other is black. The uninsulated conductor in the cable is usually the same size as the insulated black and white wires. In the past, the smallest wire size used in homes was No. 14, with two conductors (written on the cable as "14/2 with ground" or "14/2WG"). Single conductors can be used in conduits as stranded or solid

SERVICE ENTRANCE FITTINGS
FOR HEAVYWALL OR THINWALL

Threaded caps

Clamp type cap

Service entrance cable connector

Service entrance cable fittings

Cable straps

THREADED CAPS FOR HEAVYWALL (RIGID) INSTALLATION

Service mast

Entrance ell for conduit

-or-

Sill plate for non-metallic cable

1
2
3
4
5

15- or 20-ampere general-purpose circuits

6
7

Two 20-ampere kitchen-appliance circuits

8 — 20-ampere laundry appliance circuit
9 — 15-ampere circuit for fuel-fired furnace
10 — 240-volt central air conditioning circuit
12

11 — 20-ampere workshop circuit
13 —15 — 120-240-volt range circuit
14 —16 — 240-volt hot water heater circuit
— 120-240-volt washer-dryer circuit
17 —19 — 20-ampere circuit for dishwasher
21 — 20-ampere circuit for garbage dispenser
22

Service to barn or outbuilding should be through a separate entrance panel connected to power take-off lugs in main entrance panel.

18 20

60-ampere feed to sub-panel

Outdoor light 150 watts

1-2
4-5
3
6

Fig. 11-19 *Service entrance.*

conductors (Fig. 11-25). Figure 11-26 shows how the wire is cut to remove the insulation without scoring the metal part of the wire. Scoring the copper makes it easy to break when bent around a screw for fastening to a switch or receptacle.

Wire Size

The larger the physical size of the wire, the smaller is the number. For example, No. 14 is smaller than No. 12. Number 14 has been used for years to handle the

Fig. 11-20 *A distribution box. This is a four-circuit box with screw-in fuses. It is used for an add-on in some older systems.*

15-ampere home circuit. Today, the NEC calls for No. 12 for a 15-ampere circuit with aluminum or copper-clad aluminum wire. This is a safety factor. Aluminum wire cannot handle as much current as copper wire.

Figure 11-27 shows how size and numbers relate to determining wire diameter. Table 11-4 lists the capacities of wires needed for various household equipment. Table 11-5 shows conductor capacities for other household equipment. You should consider this information whenever you are planning an installation. Check the NEC and local codes for the latest changes and/or requirement.

Fig. 11-21 *Distribution box with circuit breakers.*

TABLE 11-5 *Conductor Capacities for Other Household Equipment*

Item	Conductor capacity
Room Air Conditioners *or*	20 A-2 W-230 V
Central Air-Conditioning Unit *or*	40 A or 50 A-2 W-230 V
Attic Fan	20 A-2 W-115 V (Switched)
Food Freezer	20 A-2 W-115 or 230 V
Water Pump (where used)	20 A-2 W-115 or 230 V
Bathroom Heater	20 A-2 W-115 or 230 V
Workshop or Bench	20 A-3 W-115/230 V

PLANNING THE RIGHT SIZE SERVICE AND CIRCUITS

Most houses have either a 150- or 200-ampere service from the pole in the back of the house to the circuit breaker box in the basement. Number 1/0 or No. 3/0 (type RHW insulation) three-wire service is usually installed. This usually provides sufficient power for household needs. One way to find out how much power is needed is to list the devices that use power in a house, as well as how much power they use. Add up

EXAMPLE OF 100-AMPERE FUSELESS SERVICE ENTRANCE PANEL

100-ampere main breaker (shuts off all power)

30-ampere circuit (240-volt) for dryer, hot-water heater, central air conditioning, etc.

Four 20-ampere circuits for kitchen and small appliances and power tools.

40-ampere circuit (120- to 240-volt) for electric range.

Four 15-ampere circuits for general-purpose lighting, television, vacuum cleaner.

Space for four 120-volt circuits to be added for future loads as needed.

Fig. 11-22 *An example of a fuseless service entrance panel capable of handling a 100-ampere service.*

Fig. 11-23 *Romex cable (12/2) for use in home branch circuits.*

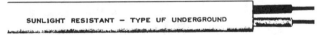

SUNLIGHT RESISTANT — TYPE UF UNDERGROUND

Fig. 11-24 *Romex that can be used for underground wiring.*

Fig. 11-25 *Single wire, stranded, with plastic coating. This can be used in conduits. It also comes in a solid copper wire.*

Wrong way   Right way

Fig. 11-26 *The right and wrong ways to cut insulation from a piece of wire.*

Fig. 11-27 *Sizes of wires. Note the relationship between the size and diameter.*

the requirements. Divide the wattage by 120 to find out how many amperes would be needed at any time.

In a house with an electric range, water heater, high-speed dryer, and central air conditioning, together with lighting and the usual small appliances, there is at least a 150-ampere requirement. If the house also has electric heating, a 200-ampere line should be installed.

150-Ampere Service

A 150-ampere service provides sufficient electrical power for lighting and portable appliances. This would include, for example, a roaster, rotisserie, refrigerator, and clothes iron. If the dryer does not draw more than 8,700 watts, the range no more than 12 kilowatts, and the air conditioner no more than 5,000 watts, then the 150-ampere service is sufficient. Do not add any more than 5,500 watts of appliances, however. Table 11-6 shows how much power is required to operate small devices.

Branch Circuits

General-purpose circuits should supply all lighting and all convenience outlets, except those in the kitchen, dining room (or dining areas of other rooms), and laundry rooms. General-purpose circuits should be provided on the basis of one 20-ampere circuit for not more than every 500 square feet or one 15-ampere circuit for not more than every 375 square feet of floor area. Outlets supplied by these circuits should be divided equally among the circuits (Figs. 11-28 to 11-36).

TABLE 11-6 *Appliances and Their Wattages*

Device	Wattage	Voltage	Current
Small Appliances			
Blender	950	120	7.92
Clothes iron	1100	120	9.17
Electric fryer (small)	1200	120	10.00
Food processor	400	120	3.34
French fryer	900	120	7.50
Hair dryer	1200	120	10.00
Hand drill (⅓ HP)	324	120	2.70
Microwave oven	1680	120	14.00
Mixer	165	120	1.38
Toaster	1200	120	10.00
Vacuum cleaner (1 HP)	746	120	6.22
Waffle grill	1400	120	11.67
Large Appliances			
Clothes dryer	5000	240	20.83
Dishwasher	1200	120	10.00
Laundry circuit	3000	240	12.50
Range	12,000	240	50.00
Space heating	9000	240	37.50
Water heater	2500	240	10.40

Fig. 11-28 *To add a wall switch for control of a ceiling light at the end of the run.*

Fig. 11-29 *To add a new convenience outlet from an existing junction box.*

Fig. 11-30 *To add a wall switch to control a ceiling light in the middle of a run.*

Fig. 11-31 *To add new convenience outlets beyond old outlets.*

Fig. 11-32 *To add a switch and convenience outlet in one box beyond an existing ceiling light.*

Fig. 11-33 *To install two ceiling lights on the same line, one controlled by a switch and the other with a pull chain.*

These requirements for general-purpose branch circuits take into consideration the provision in the current edition of the NEC. Floor area designations are in keeping with present-day usage of such circuits

Fig. 11-34 *To add a switch and convenience outlet beyond an existing ceiling light.*

Fig. 11-35 *To install one ceiling outlet and two new switch outlets from an existing ceiling outlet.*

Fig. 11-36 *Two methods of attaching ground wire to a metal box.*

(Figs. 11-37 through 11-45). For a closer look at the internal wiring of a house, before it is covered with drywall or plaster, see Figs. 11-46 through 11-52. Wiring of switches and receptacles and locations within the circuits are shown in Figs. 11-53 through 11-63. It is recommended that separate branch circuits

Fig. 11-37 *Three-way switches wired to control a lamp from two locations.*

Fig. 11-38 *Another way to connect switches to control a lamp from two locations.*

be provided for both lighting and convenience outlets in living rooms and bedrooms. It is also recommended that the branch circuits servicing convenience outlets in these rooms be three-wire type of equipment with split-wired receptacles.

ELECTRIC SPACE HEATING

The capacity required for electric heating should be determined from Table 11-7. The table shows maximum winter heat loss based on the total square feet of living space in the home. If electric space heating is installed initially, wiring should be as follows:

- For a central furnace, boiler, or heat pump: A three-wire 120/240-volt feeder sized to the installation

- For individual room units: Ceiling cable, or panels, sufficient 15-, 20-, or 30-ampere two-wire

Plaster ears

Silver terminal screws

Brass terminal screws

Green grounding terminal screw

A

SINGLE - POLE

OFF

B

Position one

position two

THREE - WAY

C A

Common terminal (usually darker)

Lighter colored terminals

B

C

position one

C ● → ● A

● B

position two

C ● ● A

● B

Fig. 11-39 *Switching arrangements.*

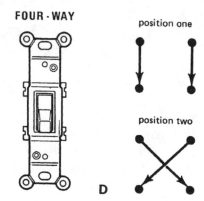

FOUR - WAY

D

position one

position two

TABLE 11-7 *Maximum Heat-Loss Values* (Based on Infiltration Rate of Three-Quarters Air Change per Hour)*

Degree days	Watts		Btuh	
	m^2	sq ft	m^2	sq ft
Over 8000	0.98	10.6	3.3	36
7001 to 8000	0.93	10.0	3.2	34
6001 to 7000	0.88	9.5	3.0	32
4501 to 6000	0.85	9.2	2.9	31
3001 to 4500	0.83	8.9	2.8	30
3000 and under	0.78	8.4	2.7	29

*For new homes. May be exceeded in converting existing homes.

Fig. 11-40 *Two three-way switches and a four-way switch for three-location control of a lamp.*

Fig. 11-41 *Three wires feeding a single receptacle. Switched lower outlet—top outlet hot all the time.*

Fig. 11-42 *Three wires feeding a single receptacle.*

Fig. 11-43 *Breakaway connection between two outlets allows the red and black hot wires to be connected to the same duplex outlet.*

230-volt circuits to supply the heating units in groups or individually.

AIR CONDITIONING

The capacity required for electric air conditioning is determined from the chart of maximum allowable summer heat gain based on the total square feet of living space in the home (Table 11-8). If electric air conditioning is installed initially, wiring should be provided for the following:

• A heat pump, providing both winter heat and summer air conditioning: A three-wire 120- to 240-volt feeder sized to the installation

• Individual room air-conditioning units: Sufficient 15-, 20-, or 30-ampere, 240-volt circuits to supply

Fig. 11-44 *Schematic drawing of two three-way switches controlling a single lamp.*

Fig. 11-45 *Schematic diagram of five locations to control a single lamp. Note there are never more than two three-way switches. Each time another location for control is added, a four-way switch is used.*

TABLE 11-8 *Maximum Allowable Heat Gain (British Thermal Unit Hours)**

Area to be conditioned		Design, dry-bulb temperature**			
m²	sq ft	32°C [90°F.]	35°C [95°F.]	38°C [100°F.]	40°C [105°F.]
70	750	15,750	18,000	19,500	21,000
93	1000	20,500	23,000	24,750	26,500
139	1500	27,000	30,000	31,500	33,000
186	2000	36,000	40,000	42,000	44,000
232	2500	45,000	50,000	52,500	55,000
279	3000	54,000	60,000	63,000	66,000

*Based on FHA minimum property standards.
**Based on dry-bulb temperature of less than 32° C (90° F.); use 32° C (90° F.) values. For designed dry-bulb temperature exceeding 40° C (105° F.); use 40°C (105°F.) values. (To find °C, simply subtract 32 from °F and multiply by 5/9.)

all units and a 20-ampere, 240-volt three-wire outlet in each room, on an outside wall and convenient to a window

In some cases, neither electric space-heating nor electric air-conditioning equipment is installed initially. In such an instance, the service entrance conductors and service equipment should have the capacity required in accordance with the appropriate chart. Spare feeder or circuit equipment should be pro-

(A)

(B)

Fig. 11-46 *A. The cable in this switch box is fed from the top. B. Receptacle box is fed from the top.*

Fig. 11-47 *Location of an outlet where there is no insulation.*

Fig. 11-48 *Three-ganged switch boxes. Note how the wires are stapled to anchor them and how each box has a cable entering and leaving. This will be a three-switch control center.*

Fig. 11-49 *Note that the Romex feeds the box. BX cable is the metallic armored one.*

vided, or provision for these should be made in panel board bus space and capacity.

The plan should allow bus space and capacity for a feeder position that can serve a central electric heating or air-conditioning plant directly or can supply a separate panel board to be installed later. The panel board would control circuits to individual room heating or air-conditioning units. Space heating and air condi-

Fig. 11-50 *Note that the studs are drilled, and the wires are fed to the switch boxes.*

tioning are considered dissimilar and non-coincidental loads. Service and feeder capacity is provided only for the larger load, not for both.

SPACE HEATING AND AIR-CONDITIONING OUTLETS

Because many different systems and types of equipment are available for both space heating and air conditioning, it is impractical to show outlet requirements for those uses in each room. Electrical plans and specifications for the house should indicate the type of system to be supplied and the option of each outlet. Today, there are very few new homes built with individual window units for air conditioners. Most are central air-conditioning units that also serve as part of the heating plant.

ENTRANCE SIGNALS

Entrance push buttons should be installed at each commonly used entrance door and connected to the door chime. They should give a distinctive signal for both front and rear entrances. Electrical supply for entrance signals should be obtained from an adequate bell-ringing or chime transformer.

A voice intercommunications system permits the resident and a caller to converse without opening the door. This is convenient and is an added protection. It might be designed for this purpose alone, or it might be part of an overall intercommunications system for an entire house.

(A)

(B)

Fig. 11-51 *Wires through the siding of a new house, to be attached to the outlet boxes later.*

QUICKWIRE HOLE
easy way to create a loop.

SPEEDWIRE
(Copper Wire Only)

Fig. 11-52 *Duplex receptacles being wired. A speedwire connection means that the wire is pushed into the hole to make contact.*

insert

release

Fig. 11-53 *Note how the wire is released by pressing with a screwdriver blade into the slot.*

Fig. 11-54 *Recessed outlet mounted in a 4-inch square box.*

COMMON FEED ONLY

When plug is inserted into receptacle, pilot light automatically lights up.

Fig. 11-55 Receptacle with pilot light.

WIRED AS PILOT LIGHT

When 3-Way switch is ON, pilot light glows. This circuit requires a ground wire for this operation.

WIRED AS NIGHT LIGHT (Breakoff Tab Removed)

When 3-Way switch is OFF, indicator light glows.

Fig. 11-56 Switch and pilot light.

WIRED AS NIGHT LIGHT

When switch is OFF, indicator light glows.

WIRED AS PILOT LIGHT

When switch is ON, pilot light glows.

Fig. 11-57 Switch and pilot light wired with a three-way switch.

Grounded power outlet controlled from 3-Way switch and 3-Way switch at another location.

COMMON FEED

3-Way switch controls light only. Power outlet for grounded appliances.

Fig. 11-58 Power outlet and switch can be used in a number of configurations for circuit control.

In a smaller home, the door chime is often installed in the kitchen, providing it will be heard throughout the home. If not, the chime should be installed at a more central location, usually in the entrance hallway. In a larger home, a second chime is often needed. This ensures that the signal is heard throughout the living quarters. Entrance signal conductors should be no smaller than the equivalent of a No. 18 AWG copper wire. If two devices are used make sure a larger than usual transformer is selected.

The installation of door bells and chimes is an easy job. The wire used is No. 18. The connections are simple. Usually the wire comes in two- or three-conductor cable. This is taken from the transformer to the switch and the chime, as shown in Fig. 11-64. The other side of the transformer goes to the 120-volt line. Usually the transformer is mounted on the side of the circuit breaker box. The primary wires are fed through a hole in the box. The transformer has a screw that makes it easily secured to the box (see Fig. 11-64).

CABLE TELEVISION

All new housing is made with cable television (CATV) available in at least three locations. More than three locations requires an expensive broadband amplifier and splitter. These cables are pulled into place using the same technique as for Romex wiring. They are terminated in a receptacle box just like the one used for a 120-volt wall outlet. The proper termination for the ca-

COMMON FEED
S. P. switch controls light only. Power outlet for grounded appliances.

Grounded power outlet controlled by S.P. switch.

SEPARATED FEED
(Breakoff Tab Removed)
S. P. switch and grounded power outlet on separate circuits.

Fig. 11-59 *Switch and outlet with a number of arrangements.*

ble is usually placed on the end of the coaxial wire after the wiring job is finished and the plugs and switches are installed. The source of the entry cable varies with the area. Some come in underground, and others might be brought to the house from the street or pole through the air to be anchored near a termination determined by the local cable company. A special tool is used to attach the plug to the cable. The cable company usually has its installer do the job and check it out before energizing the in-house circuitry.

INSTALLING ROMEX

Romex cable of the No. 14 or No. 12 size is usually employed in the wiring of homes, light industries, and businesses. To use the nonmetallic sheathed cable safely in a manner consistent with accepted wiring practices, it is necessary to use switch and outlet boxes for terminations and connections. Splices are not allowed along the run of the cable unless the splice is housed in an appropriate box and totally enclosed with a cover plate.

COMMON FEED
Two S.P. switches on same circuit. Each switch controls an independent light.

SEPARATED FEED
(Breakoff Tab Removed)
Two S.P. switches on separate circuit. Each switch controls an independent light.

Fig. 11-60 *Two switches mounted in one unit with possibilities for controlling two loads in different ways.*

WIRED AS NIGHT LIGHT
When switch is OFF, indicator light glows.

WIRED AS PILOT LIGHT
When switch is ON pilot light glows.

Fig. 11-61 *Switch and nightlight or pilot light with the addition of an external jumper.*

Fig. 11-62 *Baseboard electric heater.*

When insulating material is used to make the boxes and other equipment for wiring a house, it is not necessary to use a clamp or connector to hold the wire in place. Such a box is permitted when the wire or cable is supported within 8 inches of the box (Fig. 11-65). Cable enters the box by way of a knockout. All knockouts should be closed if not used for cable.

Box Volume

The volume of the box is the determining element in the number of conductors that can be allowed in the box. For instance, a No. 14 wire needs 2 cubic inches for each conductor. Therefore, if a two-wire Romex cable of No. 14 conductors is specified, it means that a ground wire (uninsulated) is also included, which will count as a conductor, and the three wires or conductors will require 6 cubic inches of space. A box used for this installation should have at least this amount. A 3- × 2- × 1.5-inch box has only 9.0 cubic inches and is allowed to contain three of either No. 15, No. 12, or No. 10 conductors but only two No. 8 conductors. Boxes manufactured recently have the cubic inch capacity stamped on them. You should check the latest edition of the National Electrical Code Handbook for

Fig. 11-64 *A door bell or chime transformer. Note how the transformer is attached to the entrance panel with screws in the base of its cover. The curled wires attach to the 120-volt source, and the screws between the wires are used to make connections to the stepped-down 16 volts for operation of the chimes.*

the number of wires allowed until you are familiar with the standards for boxes (Table 11-9).

Figures 11-66 through 11-79 show a representative sampling of devices made of insulating material that are acceptable in house wiring. Read the captions for each figure, and identify the characteristics for future use in wiring buildings.

TABLE 11-9 *Volume Required for Each Wire Size**

Size of conductor	Free space within box for each conductor
No. 14	2.00 cubic inches
No. 12	2.25 cubic inches
No. 10	2.50 cubic inches
No. 8	3.00 cubic inches
No. 6	5.00 cubic inches

*Reprinted by permission from NFPA 70-2002, *National Electrical Code*, © 2002 National Fire Protection Association, Boston, Massachusetts.

Fig. 11-63 *Installation of door bells and chimes.*

Fig. 11-65 *Installation of the nonmetallic (NM) wire known as Romex in a single-gang box.*

Fig. 11-66 *An insulated metallic grounding bushing.*

Fig. 11-67 *A male insulating bushing.*

Duplex receptacles come in a variety of shapes and forms. For instance, Fig. 11-77B shows a metal cover plate rated at 10 amperes. This toggle switch comes with a gasket to go underneath to seal the box from rain or snow. The 15-ampere covered receptacles shown here come in both single and duplex types. One thing you should keep in mind is that the receptacles with a cap that stays in place when opened, such as

Fig. 11-68 *A nonmetallic switch box for a house.*

Fig. 11-69 *A nonmetallic switch box. It will hold nine conductors of No. 14 wire, eight conductors of No. 12 wire, or seven conductors of No. 10 wire.*

those shown here, are for use in exterior locations but not in exposed areas. This type can be used on porches, where they are sheltered from the weather. However, those designed for outside use in exposed-to-the-elements locations have a spring-loaded cap that closes automatically when you remove the plug from the socket. Be sure to use the correct type in each location.

Figure 11-80A shows how a connector is used to hold the Romex cable in place. Notice how the grounding clip is attached to the metal box. Also notice that part of the Romex insulation is extended through the connector into the box. Figure 11-80B shows how two or more ground wires are connected (twisted) together and then grounded by a clip attached over the metal edge of the box. Some Romex boxes are designed to be attached to 2 × 4 studs without nails (Figs. 11-81 and 11-82). A switch box can be attached with nails, as shown in Fig. 11-83.

Lighting Fixtures

Lighting fixtures can be installed in several ways. There are hanger supports that thread onto a threaded

Fig. 11-70 *A nonmetallic switch box with nails used to mount it. It will hold seven conductors of No. 14 and six of No. 12 or No. 10 wire.*

Fig. 11-71 *A nonmetallic switch box designed with a bracket to mount to the wood or steel studs with an SST tool. The box will hold 15 No. 14 conductors, 13 No. 12 conductors, or 12 No. 10 conductors.*

Fig. 11-72 *A nonmetallic switch box. It will hold four devices. The bracket will fit between two studs located 16 inches on center. The box will hold 20 No. 14 conductors, 17 No. 12 conductors, or 15 No. 10 conductors. Remember, the ground wire counts as one conductor.*

Fig. 11-73 *A cover for an outlet box.*

Fig. 11-74 *A nonmetallic cover for a toggle switch.*

stud mounted in the box, as in Fig. 11-83. Straps and machine screws are used to mount the fixture in Fig. 11-84. If there is no stud, the metal strap can be used to hold the canopy of the light fixture in place, as in Fig. 11-85.

Glass-enclosed ceiling fixtures are easily attached to the ceiling box by using a threaded stud (Fig. 11-86). Wall lights can be installed using the same method. Make sure that the outlet in the wall fixture is wired for full-time service—not controlled by the switch that turns the light on and off (Fig. 11-87).

Wires and Boxes

The box to be used is determined by its volume and the number of conductors you will be putting into the box.

Fig. 11-75 *A nonmetallic surface box.*

Fig. 11-76 *A nonmetallic conduit box, round, with four 0.5-inch threaded knockouts.*

same size, you have to check the volume of the box and then take a look at Table 11-9, which shows the volume required for each wire. Just add the number of wires of a particular size and multiply by the volume required by that conductor. Then do the same thing for the next size conductor. Once you have figured all the various sizes and their volume requirements, just add them. This will tell you whether or not you have too much for the volume of the box.

Any wire that runs unbroken through a box is counted as one wire. The ground wires can be in non-metallic cable or ground wires that are run in metal or nonmetallic raceways. Wires that come into a box and are spliced are counted as one wire. It can be crimped or twisted in its connection. If a wire that comes into the box is connected to a wiring-device terminal, it is counted as one wire. Cable clamps, hickeys, and fixture studs count as one wire whether the box has one clamp, two clamps, or any combination of clamps, studs, or hickeys.

You can remove unused cable clamps from a box to provide more space in the box. If one clamp is left in the box, you must count the clamp as one connection. If both clamps are removed and you use a box connec-

No matter how many ground wires come into a box, a deduction of only one conductor must be made from the number of wires shown in Table 11-10. This table shows the number of wires in a box. This assumes that all the wires are the same size. If the wires are not the

TABLE 11-10 *Boxes and Maximum Number of Wires*

Box dimension, inches trade size or type	Min. cu. in. cap.	Maximum number of conductors			
		#14	#12	#10	#8
4 × 1¼ Round or Octagonal	12.5	6	5	5	4
4 × 1½ Round or Octagonal	15.5	7	6	6	5
4 × 2⅛ Round or Octagonal	21.5	10	9	8	7
4 × 1¼ Square	18.0	9	8	7	6
4 × 1½ Square	21.0	10	9	8	7
4 × 2⅛ Square	30.3	15	13	12	10
4¹¹⁄₁₆ × 1¼ Square	25.5	12	11	10	8
4¹¹⁄₁₆ × 1½ Square	29.5	14	13	11	9
4¹¹⁄₁₆ × 2⅛ Square	42.0	21	18	16	14
3 × 2 × 1½ Device	7.5	3	3	3	2
3 × 2 × 2 Device	10.0	5	4	4	3
3 × 2 × 2¼ Device	10.5	5	4	4	3
3 × 2 × 2½ Device	12.5	6	5	5	4
3 × 2 × 2¾ Device	14.0	7	6	5	4
3 × 2 × 3½ Device	18.0	9	8	7	6
4 × 2⅛ × 1½ Device	10.3	5	4	4	3
4 × 2⅛ × 1⅞ Device	13.0	6	5	5	4
4 × 2⅛ × 2⅛ Device	14.5	7	6	5	4
3¾ × 2 × 2½ Masonry Box/Gang	14.0	7	6	5	4
3¾ × 2 × 3½ Masonry Box/Gang	21.0	10	9	8	7
**FS-Minimum Internal Depth 1¾ Single Cover/Gang	13.5	6	6	5	4
**FD-Minimum Internal Depth 2⅜ Single Cover/Gang	18.0	9	8	7	6
**FS-Minimum Internal Depth 1¾ Multiple Cover/Gang	18.0	9	8	7	6
**FD-Minimum Internal Depth 2⅜ Multiple Cover/Gang	24.0	12	10	9	8

*Reprinted by permission from NFPA 70-2002, *National Electrical Code*, © 2002 National Fire Protection Association, Boston, Massachusetts.
**FS and FD are designations used for Unilets with conduit.

A duplex receptacle cover (REC),with gasket and stainless steel screws. It may also be obtained with a single receptacle hole. Note the REC on the outside to identify the outlet.

(A)

10-ampere toggle switch

15-ampere covered receptacles. Choice of single or duplex types.

(B)

Fig. 11-77 A. A duplex receptacle cover. B. Metal cover plates for exterior use.

tor, then the clamp is not deducted from the total allowed in the box.

If a jumper is used from the box screw to the receptacle grounding terminal, then the jumper is not counted as a conductor because it does not leave the

A ground continuity tester for use on energized circuits only. It is equipped with a ground and a neutral blade. If the light comes on when it is plugged into a grounding-type receptacle, the ground continuity is complete. If the light does not come on, there is a fault in the circuit. It uses a penlite cell for power.

Fig. 11-78 A ground continuity tester for use on energized circuits only.

A box finder and cable tracer used to locate boxes or cables that are hidden from view. To use, plug it into one of the receptacles on the covered box circuit wires. The transmitter will produce a signal. Tune in a small transistor radio to pick up the transmitted signal. Follow the cable by checking sound build-up due to more wire being folded back in the box. The box will be located at maximum signal strength. The unit is more efficient if the circuit ground wire is disconnected. It uses a 9-volt transistor battery.

Fig. 11-79 A box finder and cable tracer.

box. If a switch or receptacle has a grounding strip on it, this must also be counted as a conductor in terms of the number of conductors used in a box.

One of the factors involved in mounting boxes and electrical equipment is the prevention of fire. Also, in case of fire, it must not be allowed to move from one level to another by way of a raceway or hole. The use of a hexagonal or square box in ceilings where there is sheetrock is not permitted. If hexagonal or square boxes are used, they must have a "mud ring" installed to prevent fire from being allowed to contact wall or ceiling materials that may burn.

In single-gang boxes, the nonmetallic cable does not have to be clamped to the box. It should be secured to a stud within 8 inches of the box, however (see Fig. 11-65). If the box is round or square two- or three-gang, the cable must be clamped to the box (Fig. 11-88).

ELECTRIC RANGES

The largest appliance in the house is usually the electric range for cooking. It is wired with a permanent connection or with a plug and pigtails. The plug and receptacle must be capable of handling at least 50 amperes and provide connections for the three wires used in such circuits. In some instances, a pigtail is used to connect the range to the power source. Figure 11-89 shows the pigtail used in ranges and in some instances for clothes dryers. However, range and dryer pigtails are different because they are capable of supplying either 50 amperes for a range or 30 amperes for a dryer. The plugs are also different (Fig. 11-90). Figure 11-91 shows how a pigtail is installed.

Fig. 11-80 *A. Romex connector attached to a box. B. Note how the ground clip is attached to the metal box.*

Fig. 11-81 *A metal switch box for Romex cable.*

Fig. 11-82 *Beveled-corner switch box for Romex. Note mounting without nails.*

Connecting Ranges Permanently

The NEC specifies the size of wire that can be used for connecting electric ranges. Feeder capacity must be allowed for household cooking appliances according to Table 220-19 of the NEC C. This table applies if the appliance is rated over 1.75 kilowatts.

Section 210-19(b), Exception No. 1, refers to Table 220-19 for the sizing of a branch circuit to supply an electric range, a wall-mounted oven, or a counter-mounted cooking unit. Table 220-19 is also used for sizing the feeder wires that supply more than one electric range or cooking unit.

Figure 11-92 shows that the minimum size for a 12,000-watt range would be 35 amperes, and the wire would have to be No. 8. The type of wire would be TW-40A, THW-45A, XHHW, or THHN-50A. Number 8 wire is the minimum size to be used for any range rated 8.75 kilowatts or more. The overload protection could be either a 40-ampere fuse or a 40-ampere circuit breaker.

Although the two hot legs of the wire must be No. 8, an exception to the NEC says that the neutral conductor can be smaller (see Fig. 11-92). This is so primarily because the heating elements of the range will be connected directly between the two hot wires, and the maximum current will flow through these wires. The devices that operate on 120 volts will be the oven light and the lighting circuits over the top of the range. In some instances, the heating elements for the top portion will be wired to 120 volts at some heat levels.

(A)

Outlet box has a stud in this case. Insert the machine screws in threaded holes of the metal strap shown. Slip the center hole of the strap over the stud in the outlet box. Hold the strap in position by a locknut. Connect wires with wire nuts and slip the canopy over the machine screws; fit flush and secure the fixture with two cap nuts. Don't forget to anchor the uninsulated ground wire to the box with a ground clip or to the box's threaded grounding screw.

Fig. 11-84 *Outlet box has a stud in this case for mounting of light fixture.*

(B)

If there is no stud, insert the machine screws as shown here. Fasten the ears of the outlet box and the strap with screws. Then align the canopy onto the two screws pointing down and cap off with cap screws.

Fig. 11-85 *Use machine screws if there is no stud.*

(C)

Mount large drop fixtures by simply using a screw hanger support onto the threaded stud in the outlet box. Use solderless connectors (wire nuts) to connect the electrical wires and a grounding clip for the extra uninsulated ground wire. Raise the canopy and anchor in position by means of a locknut.

Fig. 11-83 *A. A switch box with nailing holes and hammer driving the nails into the stud. B. A switch box with nails inserted for nailing. C. Mounting of large drop fixture.*

Fig. 11-86 *Glass-enclosed fixtures installed by using a stud.*

Wall brackets or lights are installed by strapping to the ears of the box, then using a nipple and cap to complete. Don't forget the ground wire and the ground clip. If there is an outlet which is "always on," wire according to the insert.

Fig. 11-87 *Wall brackets or lights are installed by strapping to the ears of the box.*

The exception says that the smaller conductor cannot be less than 70 percent of the capacity of the hot wires. In this case, the wire cannot be less than No. 10. The neutral size is figured as follows: 70 percent of 40 amperes is 28 amperes. Thus 28 amperes calls for No. 10 wire. The neutral can be No. 10, but the other two wires must be No. 8.

If the range is over 8.75 kilowatts, then the minimum size of the neutral conductor must not be smaller than No. 10. Hot legs of the range must not be less than No. 8 wire and the neutral less than No. 10 wire.

The maximum demand for a range of 12-kilowatt rating according to the table in the NEC is 8 kilowatts. This 8,000-watt load can be converted to amperes by dividing by 230 volts (the midpoint between 220 and 240 volts). This, then, gives you 35 amperes. Look up the wire for handling 35 amperes, and you find No. 8. Number 8 wire can handle up to 40 amperes.

On modern ranges, the heating elements of the surface units are controlled by five heat switches. The surface unit heating elements will not draw current from the neutral conductor unless the unit switch is in one of the low-heat settings.

Fig. 11-89 *A pigtail and plug.*

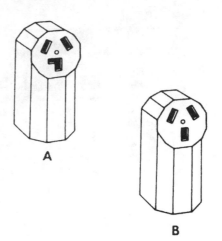

Fig. 11-90 *A. L-type ground for a 250-volt, 30-ampere circuit. B. 250-volt, 50-ampere range receptacle.*

Sizing a Range over 12,000 Watts

If you have a kitchen range rated at 16.6 kilowatts, then the 230-volt supply must be capable of delivering at least 43 amperes. This is determined by the following:

1. Column A of the NEC Table for Demand Loads for Household Ranges, Wall-Mounted Ovens, Counter-Mounted Cooking Units, and Other Household Appliance Ovens 1.75 kW Rating states in note 1 for ranges over 12 kilowatts and up to 27 kilowatts, the maximum demand in column A must be increased by 5 percent for each additional kilowatt (or major fraction thereof) above 12 kilowatts.

Square   Two-Gang

Fig. 11-88 *Boxes must have a clamp or connector if used for wiring.*

Fig. 11-91 *Installation of a pigtail.*

2. This range is rated at 16.6 kilowatts, which is 4.6 kilowatts more than 12 kilowatts. Column A says to use 5 percent of the 8,000 watts, or 400 watts.

3. The maximum demand for this 16.6-kilowatt range must be increased above 8 kilowatts by using the following: 400 watts × 5 (4 kilowatts + 1 for the 0.6 kilowatt more than 16). Therefore, 400 × 5 = 2,000. This is added to the 8,000 watts used as a base demand and produces 10,000 as the maximum demand load.

4. If you want the amount of current drawn at maximum demand, divide the 10,000 watts by 230 volts. The answer is 43.478 amperes. This is rounded to 43 amperes because the fraction is less than 0.5. Look up the wire size for using a 60°C conductor, and you will find that the required Underwriters' Laboratories wire for a branch circuit must have No. 6 TW conductors to handle the 43 amperes

Tap Conductors

Section 210-19(b), Exception 1, of the NEC gives you permission to reduce the size of the neutral conductor for ranges or a three-wire range branch circuit to 70 percent of the current-carrying capacity of the ungrounded conductors. Keep in mind, however, that this does not apply to smaller taps connected to a 50-ampere circuit, where the smaller taps all must be the same size. It does not apply when individual branch circuits supply each wall- or counter-mounted cooking unit, and all circuit conductors are of the same size and less than No. 10.

Exception No. 2 of the section previously mentioned allows tap conductors rated at not less than 20 amperes to be connected to a 50-ampere branch circuit that supplies ranges, wall-mounted ovens, and counter-mounted cooking units. These taps cannot be any longer than necessary for servicing the cooking top or oven (Figs. 11-93 and 11-94).

Figure 11-95 shows two units treated as a single range load. Figure 11-96 shows how to determine branch-circuit load for separate cooking appliances on a single circuit according to Section 210 of the NEC.

Individual branch circuits can be used. There are some advantages to individual circuits. With this arrangement, smaller branch circuits supply each unit with no junction boxes required. Two additional fuses or circuit-breaker poles are required in a panel board, however. Overall labor and material costs are less than those for the 50-ampere circuit shown in Fig. 11-93. There is a disadvantage, though. The smaller circuits will not handle larger units that you may wish to install later (Fig. 11-97).

In some cases, a single 40-ampere circuit might supply the units (Fig. 11-98). The NEC allows 40-

Minimum hot-leg rating = 35 A

Range circuit

Hot

N

Hot

Minimum neutral rating = 28 A

115 V   230 V

Rated at 12 000 W

Fig. 11-92 *A 12-kilowatt range with its minimum ampere rating.*

4-kW, 115/230-volt
oven (17.4 amperes)
PREWIRED

Maximum feeder demand
for both units-8 kW per
column A of Table 220-19
for a single 12-kW range

4 in.
octagonal box

54 in. - $\frac{3}{8}$ in. flexible conduit
w/No. 14 type A wire (Junction
box and leads supplied by
manufacturer)

12/3 NM cable
w/No. 14 bare ground wire
No. 12 ground wire
is attached to
4 in. octagonal box

Note 1: Individual branch circuits supplying
single units are computed at
100 percent demand factor.

Note 2: Equipment grounding
conductors are computed
according to Table 250-95
or Sec. 250-60, whichever
applies. Also Sec. 250-57(a)
permits metallic conduit or
cable armor to be used to
ground fixed equipment.

To 20-ampere
fuse or breaker
in panel

To 40-ampere
fuse or breaker
in panel

8/3 NM
cable

Junction box
on unit
(Neutral
grounded
to unit)

8-kW, 115/230 volt,
four-burner cook top
(34.7 amperes)

NOT PREWIRED

Fig. 11-93 *Cooking units with separate circuits.*

Note: These units are not
prewired types.

6-kW cook top

4-kW
oven

3 foot length of 10/3
(Neutral used to ground
unit)

Neutral grounds
junction boxes

6/3 NM cable, 6/2-8/1 SE cable
6/3 armored cable or 1"C
w/2 No. 6 and 1 No. 8

To 50 ampere
fuse or breaker
in panel

50-ampere circuit
(or 40-ampere circuit)

4 11/16 inch square by
2 1/8 inch deep box
for 6 No. 6 and 3 No. 10

4-inch square by
2 1/8-inch deep box
for 3 No. 6 and 3 No. 10

4 1/2 foot length of 10/3
(Neutral used to ground unit)

Fig. 11-94 *Using one branch circuit for cooking units.*

ampere circuits in place of 50-ampere circuits where
the total nameplate rating of the cook-tops and ovens
is less than 15.5 kilowatts. Because most ranges and
combinations of cook-tops and ovens are less than
15.5 kilowatts, this can be a very popular arrangement.

CLOTHES DRYER

An electric clothes dryer uses 240 volts for the heating
element. It also uses 120 volts for the motor and 120
volts for the light bulb. This means that three wires are
needed for proper operation. Figure 11-99 shows how

Demand for these two units is 8 kW-same as a single range of 12 kW

6-kW 120/240-volt cook top

6-kW 120/240-volt oven

One cook top and two wall ovens with aggregate nameplate ratings not over 15.5 kW may be connected to a 40-ampere branch circuit

Junction box

40-A or 50-A fuse or circuit breaker

No. 8 (40 A) or No. 6 (50 A)

Fig. 11-95 *These two units are treated as one range load.*

6-kW wall ovens

8000 WATTS + (6x400 WATTS) = 10 400 WATTS

All appliances rated 115/230 V and used in kitchen of residence.

6-kW cook top

Branch circuit

Note 4 of Table 220-19 says that the branch-circuit load for a counter-mounted cooking unit and not more than two wall-mounted ovens, all supplied from a single branch circuit and located in the same room, shall be computed by adding the nameplate ratings of the individual appliances and treating this total as a single range.

That means the three appliances shown may be considered to be a single range of 18-kW rating (6 kW + 6 kW + 6 kW).

From Note 1 of Table 220-19, such a range exceeds 12 kW by 6 kW and the 8-kW demand of Column A must be increased by 400 watts (5 percent of 8000 watts) for each of the 6 additional kilowatts above 12 kW.

Fig. 11-96 *Separate cooking units on a single circuit branch.*

6-kW cook top
(26A)

To 30-ampere
fuse or circuit
breaker in panel   10/3 NM cable

4-kW
oven
(17.4A)

To 20-ampere
fuse or circuit
breaker in panel   12/3 w/bare No. 12 ground

Fig. 11-97 *Smaller wire can be used when separate branches are run for each cooking unit.*

4-kW oven
PREWIRED

54-inch length
*³⁄₈ inch flex.

w/3 No. 14
Type A wires

7.5 - kW cook top
PREWIRED

36-inch length
½-inch flexible cable

w/3 No. 12
Type A wires

Removed and flexible cable
is connected to junction box

To 40-A
fuse or circuit
breaker in panel

Junction
box

*4-inch oct. box,
1½ inches deep
(See note)

4 x 2⅛-inch deep
box may be needed

Neutral is bonded to
each junction box
40-A circuit

8/3 NM cable

2-1/8 x 4 inch square box or
1-1/4 x 4-11/16 inch square box
for 6 No. 8 and 3 No. 14

*Furnished with units

Note: Cubic inch capacity of 1½-by 4-inch octagonal
box is 15.5; according to Table 370-6(b), 3 No. 8
and 3 No. 12 would require 15.75 cubic inches.

Fig. 11-98 *In some instances, it is possible for a single 40-ampere circuit to supply both cooking units.*

the green and white wires are treated in the dryer. Because most home dryers draw about 4,200 watts, it is necessary to use a separate 30-ampere circuit with a pullout fuse for disconnecting it from the line. A circuit breaker in the main service panel can also be used to disconnect it from the line. The 30-ampere circuit uses No. 8 wire in most instances, but it should be noted that some high-speed dryers use about 8,500 watts and need a 50-ampere circuit. This would call for a No. 6 wire. There is a difference in the configuration of the receptacle for a 50-ampere device versus a 40-ampere device (see Fig. 11-89). A surface-mounted receptacle with an L-shaped ground is used with dryers drawing up to 30 amperes. A surface-mounted recepta-

cle capable of delivering 50 amperes can also be used, if needed. Note the difference in the slots in the receptacles. The pigtails that fit the two different plugs must be closely examined to make sure that they will fit the L-shaped slot or the straight slots.

MICROWAVE OVENS

Microwave ovens are designed to operate on 120 volts. This means that they can be plugged into the closest convenient wall outlet. This might become an expensive mistake for many. The microwave oven usually pulls around 14 amperes. This means that there is only 1 ampere of spare capacity left for the circuit. If there

Fig. 11-99 *Hookup for an electric dryer.*

is another appliance on the circuit, a circuit breaker or fuse may go. Microwave ovens should have their own circuit. There should be nothing else on the line. They should have a direct connection to the distribution panel. This will also minimize the interference caused by the microwave's high-frequency energy radiation.

OVERHEAD GARAGE DOORS

Most overhead garage doors for residential use are furnished with a 0.33-horsepower, or 248.6-watt, electric motor. This means that it draws about 2.07 amperes when running. It does, however, draw much more when starting, up to 35 amperes. This is what dims the lights when it starts. And every time it turns off, it produces an inductive kickback of up to about 300 or 400 volts. This spike on the line is what reduces the life of the light bulbs in the door opener or anything else on the same circuit in the garage (Fig. 11-100).

Garage Door Opener

The Homelink universal transceiver replaces up to three remote controls (handheld transmitters) that operate devices such as garage door openers, motorized gates, or home lighting. It triggers these devices at the push of a button located on the overhead console (Fig. 11-101). The universal transceiver operates off the car

battery, so no extra batteries are needed. For additional information on Homelink, call 1-800-355-3515, or check the Internet at www.homelink.com.

ELECTRIC WATER HEATERS

Electric water heaters draw current to heat the water. The larger the heater capacity, the higher is the current rating of the heating element. They use 240 volts for heating the water.

A double-element water heater probably is better for larger families because it permits a more constant supply of hot water. Double-element heaters have two thermostats. The single-element type has only one thermostat. The sizes of the elements, the types of thermostats, and the method of wiring for heaters usually are specified by the local power company.

Figure 11-102 shows a typical installation of a hot water heater. Keep in mind that some electric companies offer a special rate for heating water. *Off-peak load* is the term that refers to this special rate. It simply means that the power company furnishes electricity during its low-load time. The time is already known by the electric company, so it can place a meter that measures the use of power for heating water. It also places a time switch on the line so that the use of electricity is controlled or limited to the time when the power company has a very light load or demand for power. If you

(A)

(B)

Fig. 11-100 *A. Wiring of an overhead door opener. B. Connection of a garage door opener to the power line.*

Fig. 11-101 *These three buttons on the overhead console of a Chrysler Concorde operate the garage door, the front gate, or the inside lighting system.* (Homelink–Johnson Controls.)

Fig. 11-102 *One method of installation of an electric hot water heater.*

use power at anytime other than off-peak time, you have to pay at a standard rate instead of the reduced rate.

As far as the NEC is concerned, any fixed-storage water heater with a capacity of 120 gallons or less must be treated as a continuous-duty load. This applies to about 90 percent of the residential water heaters that use electricity for heating water. The *continuous-duty load* means that the ampere rating of the water heater must not exceed 80 percent of the ampere rating of the branch-circuit conductors.

The only case where the water heater current might load the circuit protective device (circuit breaker or fuse) to 100 percent of its rating is where the circuit protective device is listed for continuous operation at 100 percent of its rating. Presently, no standard protective device is rated in this way. Therefore, it is necessary to use the 80 percent rating as a guide to wire sizing.

GARBAGE DISPOSERS

The NEC emphasizes that the receptacle under the sink must be accessible and located to avoid physical damage to the flexible cord used to connect the motor to the power source (Fig. 11-103).

The cord used for this type of installation should be type S, SO, ST, STO, SJO, SJT, SJTO, or SPT-3, which is a three-conductor terminating with a grounding-type plug. The cord must be at least 18 inches and not more than 36 inches long.

The hookup of trash compactors and dishwashers is the same as the hookup for garbage disposers. However, the cord must be from 3 to 4 feet long instead of the 1.5 to 3 feet for the disposer. The ampere rating of an individual branch circuit to furnish power to these appliances must not be less than the marked rating of the appliance.

Fig. 11-103 *Installation of garbage disposers under the sink. Remember, double-insulated disposers do not require a ground.*

Each of the receptacles serving an appliance with a flexible cord must have some way of disconnecting the circuit so that it can be serviced or repaired by removing the plug from the receptacle. Most appliances need service from time to time. Thus it is a good idea to make sure that the repair person is properly protected by removing the plug from the receptacle. It should have an easily accessible receptacle so that the plug can be removed before work begins on the appliance.

The NEC permits electric ranges to be supplied by cord and plug connections to a range receptacle located at the rear base of the range. The rule allows such a plug and receptacle to serve as the means for disconnecting the range if the connection is accessible from the front by removal of a drawer.

In some cases, a gas range is used, and the clock and oven lights are electric. Then the receptacle for the lights and clock must be easily accessible without having to disassemble the range.

AIR CONDITIONERS

Some houses use a central air conditioner for cooling. This requires a unit with the compressor and condenser outside. The evaporator is usually located inside. A duct system carries the cool air throughout the house.

Electrical wiring in and to the units varies with the manufacturer. The extent to which the electrician must be concerned with the fuse and circuit-breaker calculations depends on the manner in which the unit motors are fed. It also depends on the type of distribution system with which the unit is to be connected. A packaged unit is treated as a group of motors. This is different from the approach used for the window units, which plug into an outlet on the wall nearby.

Some equipment will be delivered with the branch-circuit selection current marked. This simplifies the situation. All the required controls and the size of the wire can be judged based on this information.

In sizing the wire for this type of unit, the selection of components for control must be the *rated-load current* marked on the equipment or the compressor. The disconnect for a hermetic motor (used in the compressor of the air conditioner) must be a motor-circuit switch rated in horsepower or a circuit breaker.

If a circuit breaker is used, it must have an ampere rating of not less than 115 percent of the nameplate rated-load current or the branch-circuit selection current. The larger of the two ratings would be taken as a working base for sizing of the wire.

NEWER WIRING SYSTEMS

The current method of wiring a house has not changed for decades. What has been and is usually installed is simple quad wiring supporting plain old telephone service and a low-performance coaxial cable for cable TV service. This cabling provides support for up to two basic analog telephone lines and a limited number of CATV channels. This wiring situation does not allow for any home systems integration and does not include forethought for the future. The characteristics of a home wiring network need to be carefully defined. The products comprising a home network should be specifically designed for residential use. They should be easy to install in a new home construction as well as in existing homes. The system also should be modular and flexible to fit the homeowner's needs. A common in-home network provides great flexibility and lower overall cost than this separate dedicated wiring.

One of the Bell Telephone Companies' newer organizations has come up with a design for the modern home with computers, VCRs, televisions, and fax machines, as well as telephones. This system is especially useful when installed during construction of the house. The system supports interactive voice, high-speed data, multimedia products, and communications services. It is a single-network wiring system that provides instant plug-and-play access to ISDN services, Internet access, CATV programs, video-on-demand, digital satellite signals, and fax/modem, plus controls for security systems and home automation from anywhere in the home.

HIGH-SPEED, HIGH-PERFORMANCE CABLE FOR VOICE AND DATA APPLICATIONS

Today's evolving technologies are requiring line speeds that are fast—up to 155 megabits per second. The uses are diverse—not just voice communications but also fax, e-mail, video, data, and file transfer. Applications include everything from telecommuting to video conferencing to home-based businesses. To help the home keep pace with this changing communications mix, a high-quality coaxial cable must be installed. This high-speed, high-performance cable carries a full range of high-speed communications services up to 100 meters in home distribution systems (Figs. 11-104 and 11-105). Two colors are used for the cables. Black is used for external video applications such as CATV, and white is used for internal video applications such as security cameras.

Installation of Cable

The primary cable installed between the service center and outlets consists of one four-pair Category 5 cable and two RG6 coaxial cables in a single jacket (Fig.

Fig. 11-104 *Twisted-pair cable shown after connectorization.* (Lucent.)

Fig. 11-105 *Coaxial cable.* (Lucent.)

Fig. 11-106 *Hybrid cable for home wiring.* (Lucent.)

11-106). All cable runs are direct from the service center to the outlet. One of the two RG6 coaxial cables in each homerun is external cable and carries output signal that drives TVs and VCRs. It is the electrical combination of the CATV input from the CATV company and all input signals brought to the service center on the other RG6 internal cable. The radio frequency modulated output of VCRs, CD players, and security cameras is sent to the service center on the internal coax cables. Up to 16 internal cable signals are combined together at the service center and then are mixed with the CATV input to form the external signals.

Service Center

The service center (Fig. 11-107) is a centralized distribution point that connects communications devices to a single, uniform structure cabling system in the home. This service center contains distribution devices for voice, data, and both base-band and broad-band video. Up to 31 twisted-pair connections can be made to the service center when it is fully configured. The unshielded twisted-pair distribution can service telephone service, local-area network (LAN) transmission, or computer modems. The service center has many video applications: CATV, satellite dishes, or video cameras for in-home security. They can be connected through the mounting panels in the service center. This family of panels holds video splitters and combiners that service up to 16 dual video outlets when fully configured. A distribution amplifier is also located in the service center. It boosts the CATV feed signal and combines it with in-home sources for distribution throughout the house. It is a necessary component when the home has more than four ports or has cable runs exceeding 150 feet from the service center.

Fig. 11-107 *Service center.* *(Lucent.)*

Figure 11-107 shows a mounting panel, whereas Fig. 11-108 illustrates the distribution amplifier. The amplifier is mounted on a panel, such as that in Fig. 11-109. The amplified video distribution is necessary for excellent picture quality. Figure 11-110 shows the mounting for the service center. It can be surface- or flush-mounted, and the distribution modules are pre-assembled with front-facing fasteners. The enclosure and all components are grounded for safety. Cables can enter the enclosure from the top, bottom, or back for flexible, easy installation. Figure 11-111 shows a crimper and field terminator modulator plugs used with the HomeStar® (Lucent Technologies' registered trademark) wiring system.

Fig. 11-108 *Mounting panel.* *(Lucent.)*

Fig. 11-109 *Distribution amplifier and ac mounting panel.* *(Lucent.)*

Fig. 11-110 *Service center box.* *(Lucent.)*

Fig. 11-111 *Crimper and field terminator for modular plugs used in HomeStar wiring.* *(Lucent.)*

Over 85 percent of American homes have television sets connected to CATV. As of February 17, 2009, all television is digital, and a converter will be needed for analog sets to receive broadcast signals over the air.

With all the demand for DVDs and entertainment centers, electrical wiring in the home will become more complex and more demanding of regular maintenance. Trained electricians will be needed to install and keep the systems operating properly. Keep in mind that the new wiring should be inspected to make sure that it is up to code. Most communities and cities in particular require that all electrical work be inspected and approved as meeting the standards set by the NEC.

CHAPTER 11 STUDY QUESTIONS

1. What are two good electrical conductors?
2. When were farms electrified?
3. Why do we have grounded conductors?
4. What color is the ground wire?
5. When does the electrician work on a house?
6. Who puts in the meter socket?
7. What is Romex cable? Where is it used?
8. How much current does a new house usually require?
9. Who installs the cable television lines?
10. How are glass-enclosed ceiling fixtures attached to the ceiling box?
11. What voltage is needed for electric ovens?
12. How do you figure the size of line needed for a range drawing over 1,200 watts?
13. What is the horsepower rating of most garage door openers?
14. What is meant by the term *code*?
15. What type of wiring is needed for a new house today?
16. What is CATV? Who is responsible for CATV wiring?
17. What voltages are usually available in a residential building?
18. What is another name for a neutral wire?
19. What is meant by service from head to box?
20. What is a conduit?

12
CHAPTER

Plumbing

ATER IS NEEDED FOR LAUNDRY, BATHING, drinking, and removing human waste from the house. A contemporary home uses water for cooking purposes, the sink, the dishwasher, the disposal, and sometimes the refrigerator ice-maker. Water is needed for toilets, lavatories, bathtubs, and showers. Water is also needed for outdoor faucets and lawn sprinklers. Hot water is a required commodity in a modern home.

Providing for the water requirements of the home and for the disposal of human waste calls for a complex system of pipes. Pipes are also needed to carry away waste water from toilets, sinks, washing machines, showers and bathtubs.

In some instances, the term *plumbing* includes pipes used to get hot water from furnaces to various rooms in the house for heating. Other plumbing applications include moving solar-heated water into either the hot water system or a central heating system. In other cases, black pipes are used to carry natural gas (or propane in some locales) to a furnace and hot water heater. This is also considered part of the plumbing system. Steel pipe is recommended for fuel gas.

In this chapter you will learn how to plan and install a plumbing system in a residential building. You will learn how to solder copper pipe, cut and prepare copper tubing, thread iron pipe, and choose fittings for polyvinyl chloride (PVC) pipe as well as iron and steel pipe. You also will learn how to install appliances that use plumbing connections for proper operation and install piping, venting, and drains. Things you can learn to do include

- Design a plumbing system for a house
- Mark off and/or locate plumbing fixtures and devices
- Properly place piping and tubing

- Properly locate water lines, drains, and vents
- How to solder pipe
- How to cut and thread iron or steel pipe
- How to work with PVC piping and tubing
- Identify the tools and fittings used in plumbing

This chapter covers the design and installation of various pipes and drains that carry water to a building and waste water away. Each appliance or fixture is connected by a system in code-approved processes. These processes are available for the do-it-yourselfer in numerous books now on bookstore shelves.

Plumbing for a building is usually done in three stages. These are the installation of

- The main outside supply pipe and main drain
- Inside pipes, vents, and drains
- Water-using fixtures

This chapter's main focus is the first two stages.

First, install the main supply pipe and main drain pipe. This is usually done at the time of the excavation for the footings and the foundation. Water pipes and waste pipes are both called *lines*. They also may be called the *water line* and the *sewage line*.

Both water supply lines and drain lines carry water of some sort. They must be located sufficiently underground to prevent freezing during the winter (see Chapter 3). They are generally put in when the holes and trenches for the footings and foundation are dug. It is cheaper and more efficient to do the digging all at once.

A main supply line typically runs from the street to the building and the water meter (Fig. 12-1). In some areas, the water meter is located in the basement, but the trend today is to locate the water meter near the street. This makes it much easier for the city to read

Fig. 12-1 *Initial supply and drain pipes.*

and maintain the meter, although in some more electronically advanced cities the meters are read via telephone lines connected to the meter and accessible to the water authority's computers.

The city connects the water main to the meter, and the builder connects the meter to the building. The city will not turn the water on until the plumbing has been inspected and approved.

Keep in mind that both the supply and drain lines must be buried below the freeze line. Both pipes carry water and are subject to freezing outside the heated building space.

Water and sewage lines are placed in the area that will become the building. Then they are stubbed out (Fig. 12-2). This means that the pipes are located in areas where they are to be used—the kitchen or bath.

(A)

(B)

(C)

(D)

Fig. 12-2 *Supply lines and drains are stubbed off for later work: A. Main line from street to house; B. plumbing vents and drains as well as soft copper lines encased in a concrete slab; C. vents, drains, and copper lines in a concrete slab; D. vents and drains installed before the slab is poured.*

Some extra pipe is left above the ground or slab level to make full installation a bit easier at a later time. Holes are filled in, and footings, foundations, floors, and such are built over the stubbed pipes.

The second stage of plumbing is done after the floors and wall frames are started. The remaining pipes, drains, and vents are installed as the building is erected. Because the pipes and drains are hidden in the walls and under the floors, they need to be installed before the inside wall board is nailed in place.

Buildings erected on concrete slabs have much of the plumbing placed under the slab. Pipes should be laid under the plastic vapor barrier, and there should be some flexibility left in the pipe. This is so because concrete has a high expansion and contraction rate with temperature changes. Leave some slack in the lines. Wide curves in copper lines allow the pipes to move with expansion and contraction of the slab. Some builders prefer locating all pipes outside the slab for ease in repairs and service. Others locate the pipes in the attic. Attic installations require careful attention to pipe insulation to prevent the pipes from freezing and bursting in winter. This can cause serious water damage to a building.

Plumbing codes are designed to protect residents in structures built for human occupancy. Most state plumbing codes and licensing examinations are designed to establish environmental sanitation and safety through supervision that will ensure properly installed and maintained plumbing systems. Details of plumbing construction vary, but the basic sanitary and safety principles are the same. The desired and required results (to protect the health of people) are similar regardless of locale. These basic principles require that all plumbing in public and private buildings intended for human occupation or use should be installed so as to protect the health, welfare, and safety of the occupants and the public. The following list explains how these requirements can be accomplished. This list puts forth a complete collection of basic principles involved with plumbing system installation:

- Buildings intended for human occupancy or use will be provided with a supply of pure and wholesome water with connections not subjected to the hazards of backflow or back-siphonage and not connected to unsafe water supplies. If there is a public water main available, an individual connection to the public water main shall be made.
- Buildings with plumbing fixtures and devices shall be provided with a supply of water in sufficient volume and pressure to enable them to operate in a satisfactory manner at all times.

- Water heaters and other devices used for purposes of water heating and storing shall be designed and installed to prevent explosion through overheating.
- Where public sanitary sewers are available, buildings intended for human occupancy shall have a connection made to the public sanitary sewer.
- Plumbing fixtures shall be made of materials that are durable, corrosion-resistant, nonabsorbent, and free of concealed fouling surfaces. Rooms in which water closets, urinals, and similar fixtures are installed shall have proper ventilation and adequate lighting.
- It is recommended that family dwelling units adjacent to sanitary sewer lines or having private sewage disposal systems have at least one lavatory, one water closet, a bathtub or shower, and a kitchen-type sink for purposes of sanitation and personal hygiene. Other structures for human occupancy with sanitary public or private sewage-disposal systems should have no less than one water closet and one fixture for hand-washing.
- Building sanitary drainage systems shall be designed, installed, and maintained in a condition so as to conduct waste water and sewage to designated locations from each fixture with a flow that prevents fouling, clogging, and deposits of solids in the piping. Sufficient cleanout shall be installed so that the piping system can be easily cleaned in case of stoppage.
- Plumbing systems shall be maintained in a sanitary condition, and each connection (direct or indirect) to the drainage system shall have a water-seal trap. The system shall be kept in a serviceable condition with adequate spacing of the fixtures. These fixtures should be reasonably accessible for cleaning.
- Drainage pipe shall be designed and installed with a durable material free of water leakage and offensive odors caused by drain sewer air. Installation shall be in accordance with good workmanship practices and use of good-grade materials.
- Plumbing systems shall be designed, installed, and kept in adjustment so as to provide the required quantity of water consistent with adequate performance. There should be no undue noise under normal conditions and use. New systems and/or remodeled systems shall be subjected to tests that will disclose leaks and defects.
- Included in the design shall be every consideration for the preservation of the strength of structural members of the building. Each vent terminal ex-

tending to the outer air shall be designed to minimize clogging and return of foul air to the building.

- Design considerations shall include protection from contamination by sewage backflow of water, food, disposal of sterilized items, and similar materials. Substances that will clog pipes or their joints and interfere with the sewage-disposal process or produce explosive mixtures shall not be allowed in the building sewage drainage system.

- Sewage or other wastes from a plumbing system shall not discharge into subsurface soil or into a water surface unless it has first been treated in an acceptable manner.

SEQUENCE

The plumber should be able to install the plumbing system in the following sequence of operations:

1. Check the plans to see what fixtures and rooms are included in the house design.
2. Select the type of pipes and fixtures to be used.
3. Lay out the location of each fixture and pipe to serve the particular design of the house.
4. Mark the placement of each fixture or device needing plumbing service.
5. Do the trenching necessary for bringing water to the house and devices included in the house.
6. Do the trenching and routing of drain pipes and vents.
7. List the piping and fixtures and other devices needed to complete the job.
8. Acquire the piping and fixtures and have them delivered to the construction site.
9. Install the fixtures and devices.
10. Connect the fixtures and devices with water (hot and cold) pipes as well as drains and vents.
11. Locate holes in the roof for vents.
12. Secure and check the entire system for proper operation.
13. Arrange for local inspection of the plumbing system before allowing carpenters to enclose the interior of the house.

PLUMBING SYSTEMS

Plumbing systems consist of supply lines for bringing water into a building and drains that carry water away from the building. In order for the drains to work effi-

ciently, they are vented to allow air to enter the system. The plumbing system has three major parts:

- A supply
- A drain
- Vents (Fig. 12-3)

Supply Lines

The water line that takes water from the main to the building may have a common size of 1, 1½, or 2 inches in diameter. Once the pipe is inside the building, smaller pipes generally are used. Common sizes for this interior work are 1, ¾, or ½ inch in diameter.

Water lines can be installed through holes in floors, wall studs, and so forth. Most water lines are placed through openings planned for them. It is best to run smaller pipes through holes cut in wall studs or floor joists. Two reasons for this are

1. Any hole or cut into a joist or stud can weaken it.
2. Pipes in these holes might be struck by nails used for hanging wall board. The nails could cause holes in the pipe. The holes would be hard to find until the water is turned on. Then the damage that results could ruin walls and floors.

Supply pipes to most locations are capped with cutoff valves (Fig. 12-4). This capping allows the water to be turned off so that the fixtures can be installed or repairs can be made without turning off the main water supply. Flexible lines, such as the plastic line in Fig. 12-5, can be used to connect the supply valve to the fixture.

Air chambers are installed near each supply outlet to prevent a loud noise from occurring each time the water is turned off. That noise is called *water hammer*. Once the water is turned off, the full force of the moving water (standard city water pressure is 80 pounds per square inch) slams against the valve. This force can make a loud banging noise, and the force can cause the pipes to physically jump or move.

Air in the chamber is momentarily compressed to cushion the force of the water stopping against the valve. Figure 12-6 shows the action and proper location of an air chamber. Remember that each supply outlet must have an air chamber. Most air chambers are made of the same size pipe as the supply line.

Air chambers are usually not enough, and the pipes should be anchored to something solid, such as a stud or joist, and at appropriate intervals. The movement of the water, particularly when it is stopped or started, can cause the pipes to move. The movement

Fig. 12-3 *A typical plumbing system has three parts: (1) supply, (2) drains, and (3) vents.*

can be enough to cause the pipes to bang against the floor or wall. Anchoring them reduces both the movement and the noise.

Drains

Waste waters are divided into two categories:

- The first is called *gray water* and consists of the waste water from sinks, laundries, and bathing units.
- The second type is called *black water* and consists of both liquid and solid wastes from toilets. In most communities, both types of waste water are drained into one sewer system.

Drain systems (often called *DWV*, for drain, waste, and vent) are used to drain away waste water. Pipes

that carry waste water from a fixture (such as a sink) are called *drains*. The drains may empty into a larger pipe to be carried across the building to a main drain. Pipes that carry the water across the building are called *laterals*. Drain pipes that run vertically are called *stacks* or, in some instances, *soil stacks*.

Waste water is not usually under pressure. This means that the drain systems must be angled down to allow gravity to move the water. Typically about ¼ inch of downward slope per foot of horizontal run is sufficient to cause waste water movement. This is usually expressed as 3 inches per 12 feet. This slight slope allows the water to carry the solid waste. If the drain is too steep, the water can run off before the solid wastes are moved. This eventually causes clogs, which stop up the drains. Clogs can force the waste to back up into

FIXTURE SHUTOFF VALVE CONTROLS THE FLOW OF WATER AT EACH FIXTURE; ADDITIONAL VALVES CAN BE INSTALLED TO ISOLATE ONE OR MORE FIXTURES FROM THE WATER SUPPLY SYSTEM FOR REPAIR AND MAINTENANCE. FIXTURE RUNOUT; THE ROUGH-IN DIMENSIONS FOR EACH PLUMBING FIXTURE SHOULD BE VERIFIED WITH THE FIXTURE MANUFACTURER SO THAT THE FIXTURE SUPPLIES CAN BE ACCURATELY INSTALLED DURING THE PROPER PHASE OF CONSTRUCTION. BRANCH SUPPLY LINE.

SUPPLY

Fig. 12-4 *Cutoff valves on the pipes that supply fixtures.*

Fig. 12-5 *Reinforced flexible plastic lines make connecting fixtures easy.*

toilets and other drains, preventing their effective use. These backups can have a terrible odor and can spill over onto a floor, causing considerable damage.

The same slope for these drains must be used from each drain to the main drain line. All drains must slope downward.

Usually, only one main drain connects a building system to the sewer. This means that the several drains

WATER HAMMER IS THE CONCUSSION AND BANGING NOISE THAT CAN OCCUR WHEN A VOLUME OF WATER MOVING IN A PIPE IS SHUT OFF ABRUPTLY. AIR CHAMBERS ARE INSTALLED AT FIXTURE BRANCHES TO PREVENT WATER HAMMER. THE TRAPPED AIR ELASTICALLY COMPRESSES AND EXPANDS TO EQUALIZE THE PRESSURE AND FLOW OF WATER IN THE SYSTEM.

Fig. 12-6 *Typical air chamber that prevents water hammer.*

(A). Direct or "wet" vent drains are connected directly to the stack and vent

(B). Reventing (back venting) two drains connected to a stack. Note the loop connecting both units to the vent.

Fig. 12-7 *Typical multiple drain connections.*

Fig. 12-8 *Plumber's tape holding a lateral drain at the correct slope.*

in the building must be planned so that they can be connected without eventually causing problems. Figure 12-7 shows typical installation factors. Drains that connect sinks, lavatories, tubs, and showers to a system are typically 1¼, 1½, or 2 inches in diameter. Size is determined mainly by available space, building codes, and anticipated drainage flow. The cost of plastic drains is not much different regardless of size.

Laterals are made of larger-diameter pipe. Laterals can be 2, 3, or 4 inches in diameter. Main drains can be larger in comparison and typically are 4 or 6 inches in diameter. Laterals can be suspended using plumber's tape. Plumber's tape is a perforated metal strip that allows the support to be installed at a length that will hold the lateral at the height needed to keep the proper drainage slope (Fig. 12-8).

A vital part of drain systems keeps sewer gases from entering living space. Needless to say, this is very important because sewer gases are very noxious and can be both toxic and explosive.

A simple device called a *trap* is used to block out these gases. Older systems may have an S trap, but most modern systems use a P trap (Fig. 12-9). The name is taken because the trap is shaped like the letter *P*. This keeps a "plug" of standing water between the living space and the sewer. Every time the fixture is used, the water in the trap is replaced, keeping the trap fresh.

Small items, such as rings, can accidentally be dropped into a sink. The trap holds and keeps them from tumbling into the main drain, that is, provided the water

is not allowed to continue to flow through the drain.

Traps can be installed with cleanout plugs (see Fig. 12-9A). They also can be plain, as in Fig. 12-9B. It is not a very difficult job to remove a trap. Both traps allow for the retrieval of lost objects or the removal of solid objects that are impeding the flow of waste water.

Normally, installing traps is not a difficult job. However, a factor called *critical distance* must be considered. Simply expressed, this means that the outlet into the stack must never be completely below the water level of the trap. If it is, a pipe full of water also would siphon out the water in the trap and expose the living space to the odors and gases of the sewer. It is easy to figure the maximum amount of run you can use on a drain. First, determine the diameter of the drain, such as 1.5 inches. Next, divide by 0.25 inch—the amount of slope per foot. This would yield the number 6, or 6 feet. However, 6 feet would be the full 1.5 inches, so the maximum run would be anything less than 6 feet. A good rule of thumb is to use only whole numbers for the distance—which would then be 5 feet.

P Traps and Drains

P traps are used on the drains for every single fixture in the building except water closets (toilets). The reason is that water closets (toilets) have built-in P traps and are usually connected to a fitting called a *closet elbow*.

In many areas, a house trap is also required in the main drain outside the building. Check Fig. 12-1. This house trap protects the entire drain system from the odors and gases of the main sewer.

Drain systems are often made entirely of PVC pipe of appropriate sizes. However, drain pipes also might be made of copper, cast iron, or heat-resistant glass. Glass drains are fragile and are used only when corrosive materials, such as acids, must be drained from the building to an appropriate collection point.

System Vents

Vents allow air to enter a drain system so that water can flow properly. Without access to air, a vacuum could be formed in parts of a drain system, and this would prevent the waste water from flowing. A vent is simply a vertical pipe that rises from a drain up through the roof and allows air to enter the drain, as in Fig. 12-10.

Generally, a vent is needed for every drain. How-ever, this is often impractical and would require three vents for the typical bathroom, three for a kitchen, one for a laundry, and so forth. Vents can be combined as in Fig. 12-11 so that usually there is only one per bathroom, one per kitchen, etc.

Vents usually are made from the same type of pipe as the drain system. It is common practice to use the least expensive material for vents because they do not actually carry water.

Cleanouts

Clogs are fairly common in drain systems, and for this reason, special cleanout plugs are advisable. Cleanouts can be located either inside or outside the building. However, they should be easily accessible (Fig. 12-12). This cleanout location allows a cleanout tool called a *snake* or an auger to be inserted and used to unclog the drain. If no cleanout plug is available and a clog occurs, a hole must be dug to reach the drain. Then an opening must be cut into the drain. Needless to say, cleaning out a drain is a dirty job and can be difficult and time-consuming.

When building a house, it is advisable to install

(A). Cross section shows how the water plug prevents gases entering from drain

Fig. 12-9 Typical P-trap installation: A. cross section shows how the water plug prevents gases entering from the drain; B. typical installation without a cleanout plug.

Fig. 12-10 *Basic drain vent.*

several cleanout plugs. It is even advisable to install a cleanout plug for each drain, as well as the main drain. A typical plug is made by inserting a wye fitting into the line, as shown in Fig. 12-13. Note that the slanted outlet angles into the drain line to make it easier to insert the cleanout tool. The wye must be the same diameter as the drain and must be assembled so as to have the least interference with water flow.

PLANNING

A number of things must be considered before starting to plumb a building. These include who does the work, the type of pipe used for supply and drains, the location of the pipes, the best slope for the drains, and the location of cleanout plugs for drains.

Contractors and Plumbers

Usually, a contractor or builder arranges for professional plumbers, acting as subcontractors, to do the necessary work when it is needed. Developers of large tracts or subdivisions might employ a general contractor to make out all the building schedules. Sometimes, capable individuals who wish to build their own homes either do the work or act as the contractor.

If an individual wishes to do the work, the first step is to check with the local building inspection and permit office (try the local phone book under city or county government) and find out what the rules are. In some areas, work can be done only by professional plumbers who are certified in that area,

Fig. 12-11 *Multiple vent connections.*

Fig. 12-12 *A kitchen cleanout plug located for convenient access.*

Fig. 12-13 *Cleanout plug detail.*

whereas in other areas, the work can be done by anyone but must be checked by a certified plumber. Other areas allow anyone to do the work, provided that it follows the appropriate rules set forth in building or plumbing codes. These codes can be very complex and specify who can do the work, what types and sizes of pipes are to be used, and many other factors.

Some plumbers will provide advice and check the work of individuals and issue the proper certification for a fee. In some areas, inspectors will check the work of individuals when called. It is very important, however, to find out what the rules are before starting.

Pipe Type

Several types of pipe can be used for supply and drain pipes:

1. Copper pipe
2. Galvanized steel pipe
3. Plastic pipe

Certain types may be required for specific things in the local building codes. Other types can be used for various reasons, including durability, cost, ease of working, or special resistance to acids or chemicals.

Several different types of pipe may be used in a single building. One concern when using different types of pipe is whether or not any electrolytic action might result. Electrolytic action happens when the impurities in water (i.e., salt, minerals, alkalis, acids, etc.) react with two or more different metals. If you connect a copper water line directly to a steel water line, the reaction of the water on the two metals (copper and iron) would cause a mineral buildup at the joint that would clog the water line. A simple dielectric union can be used to prevent this. It can be a union with a rubber or plastic washer to separate the two metals, or it can be a simple plastic coupling to separate the two.

Pipe sizes usually are measured by the stream of water that emerges from the pipe under standard city water pressures. However, the actual hole in the pipe is bigger than the "size" of the pipe. This is true even in the metric system, and inch sizes are used in the metric system for water flow planning.

Galvanized water pipe Galvanized water pipe is made of steel and coated with zinc to make it resistant to rust and mineral deposits. It is not used for drains. Steel pipe (black pipe) that is not galvanized never should be used for water but may be used for natural gas.

Galvanized pipe is available in most areas; it is tough, strong, heat resistant, and of moderate cost. Galvanized pipe is durable and can last for decades. Its greatest disadvantage is in the amount of work required to install it across and around the many angles and curves in a typical building. The pieces of pipe must be laboriously assembled by screwing numerous fittings into place. When repairs must be made, in most instances, the whole unit must be taken apart piece by piece.

Working with galvanized pipe means that the various pieces must be screwed together in appropriate lengths and directions. It is important to use a sealing compound (pipe dope) or Teflon tape on the threads before assembling the pipe. Most professionals use Teflon tape more than pipe dope. It is wrapped on the

Fig. 12-14 Using Teflon tape for sealing threads.

SHORT RADIUS 45° ELBOW   SHORT RADIUS 90° ELBOW

Fig. 12-16 Direction is changed with elbows. A 45-degree elbow is depicted on the left and a 90-degree elbow on the right.

end of the threads in the same direction the next piece is to be turned (Fig. 12-14). About three or four wraps is enough, and the tape is simply torn off and the next piece screwed on.

Pipe dope comes in either paste form or in a squeeze tube (similar to toothpaste). It might be a soft, putty-like substance in a stick. In either case, the first three or four threads should be coated. Either dope or tape makes it easier to assemble the pieces and aids in making the joints watertight.

A whole piece of pipe is a *joint* and usually is 21 feet long. Obviously, this is often too long to work with. Shorter pieces are available as *nipples* and are common in lengths of 4, 3, 2, and 1 foot and in inch sizes of 1, 2, 3, 4, 5, 6, and 9. A 4- and a 3-inch nipple can be combined for a 7-inch length. Pieces of pipe or nipples are joined with either a coupling or a union, as shown in Fig. 12-15. A union allows a person to disconnect the water line without taking the whole system apart. However, a union is more expensive than a coupling.

Elbows are used for changing direction. Elbows are made in either 90- or 45-degree turns, as in Fig. 12-16. Size can be changed with a reducing coupling or elbow, as in Fig. 12-17. A single line can be divided into different lines with a T connection (Fig. 12-18). Provision for adding another water line at a later time can be made by using a T and putting a plug into one of the holes. Figure 12-19 shows several typical fittings.

COUPLER   NIPPLE   REDUCER

Fig. 12-17 A reducer coupling joins different pipe sizes.

COUPLING   UNION

Fig. 12-15 Couplings (left) and unions (right) are used to join pipe.

Any length can be cut from a joint, but the ends must be threaded. Most professional plumbers are well equipped to do their own threading. However, threading dies used to do this are expensive. Individuals who wish to do their own plumbing might wish to consider renting the threading dies. It is not cost-effective to rent the threading dies for just a few pieces, but larger jobs are fine. For smaller jobs, it is easier for the individual to build up the right lengths using nipples.

Fig. 12-18 A T coupling is used to split lines.

Special wrenches called *pipe wrenches* are used to assemble and disassemble threaded pipe. The feature that makes these wrenches unique is the teeth in the jaws. These teeth are sharp and will dig into the surface to keep the piece tightly gripped. A professional-quality pipe wrench has jaws that can be replaced as they get dull or wear down.

These wrenches are adjustable so that they can be used on several different sizes of pipes and are available in several lengths to provide more torque to tighten or loosen pipes and fittings. When working on

Fig. 12-19 *Typical pipe fittings. The most common fittings are (left to right): 45-degree elbow, 90-degree elbow, nipple, coupling and union, T with a plug above it, and an end cap. A reducer coupling is at the bottom left.*

threaded pipe systems, two wrenches are usually needed, as in Fig. 12-20. One wrench is used to hold the base, whereas the other is used to turn the piece being worked on.

Copper water pipe Copper pipe is used most often for water lines. It does not corrode easily, and it is easily cut, worked, and repaired. Copper is perhaps the most durable of the pipe materials, but it is also the most expensive. It might cost four or five times that of galvanized pipe. The pieces must be either soldered together or joined using special fittings. Soldering takes

a lot of time and must be done carefully to prevent leaks. The fittings are expensive.

Two types of copper pipe are used commonly. One is called *rigid* and the other is called *soft* copper pipe. The sizes are slightly different because the hole in rigid pipe is measured like that in galvanized pipe, whereas the size of soft pipe is determined by the outside diameter. Because of this, soft copper is not considered pipe but is correctly termed *tube* or *tubing*.

Both types of copper tubing can be joined with the appropriate and easily available fittings. Copper pipe is usually joined by soldering, and tubing is usually

Fig. 12-20 *Proper use of two pipe wrenches.*

Fig. 12-22 Rigid copper pipe fittings (left to right): 45-degree elbow, 90-degree elbow, coupling, T, 90-degree adapter elbow to threaded pipe; threaded plastic adaptor as a dielectric, and galvanized steel nipple.

Fig. 12-21 *Cutting pipe and tubing: A. Cutting galvanized, copper, or plastic pipe with a pipe cutter; B. using a hacksaw and miter box to cut plastic pipe; C. using a hacksaw and a miter box to cut copper pipe.*

joined with either compression or flared fittings. However, solder fittings for tubing can be obtained.

Joining a ¾-inch rigid pipe to a ¾-inch soft pipe does not make much difference in water flow. However, there are advantages to each.

It is very common for soft copper pipe to be used to connect a steel or plastic water line to a sink, refrigerator ice-maker, toilet, or other fixture. This is so because the soft copper can be bent easily to change direction or location (see Fig. 12-4).

Both types of copper pipe can be cut easily to almost any length. Either a tubing cutter or a hacksaw can be used, as in Fig. 12-21. When the pipe or tubing is cut with a saw, the edges must be dressed so that they are smooth both inside and outside.

Rigid copper pipe typically comes in 10- or 20-foot lengths. The most common diameters are ½, ¼, and 1 inch. There is no need to buy nipples because the pipe is so readily cut to any length. However, rigid pipe must be joined by soldering it into flanged fittings (Fig. 12-22).

For best results, it is important to coat both the outside of the pipe and the inside of the fitting with a very thin coat of solder before the pieces are joined. *Tinning* the joint occurs when the end to be soldered is coated with solder. If the tinning is thick or lumpy, you won't be able to slide the pieces together. The excess can be wiped off by using a cloth while the solder is still molten. This wiping with a rag leaves a very thin coating of solder on the pipe.

The handiest method is to use a special soldering compound, which contains powdered solder in a greaselike mixture of flux. Flux is a fluid or paste used to clean the metal when it is being soldered. It is important to use the right type of flux. The flux is simply smeared on both pieces instead of tinning them, and the pieces are then assembled. After assembling either a tinned joint or a compounded joint, the unit must be heated to melt and fuse the solder. More solder can be added, as shown in Fig. 12-23.

Copper pipe also can be soldered without tinning the joints. The solder joints are just coated with flux and assembled. The solder is added to the outside and flows into the joint. This is the easiest method, and it works well.

It does take practice and skill to tin joints for soldering, but it is the best method. When starting a new job, it is a good idea to practice with a few short pieces before moving to the actual pipes. Older systems were soldered using regular solder, which is a mixture of lead and tin. However, lead is a toxic material and

Fig. 12-23 *Soldering procedures. (1) Cut the tubing to length and remove the burrs. (2) Clean the joint area with sandpaper or sand-cloth. (3) Clean inside the fitting. Use sandpaper, sand-cloth, or wire brush. (4) Apply flux to the inside of the fitting. (5) Apply flux to the outside of the tubing. (6) Assemble the fitting onto the tubing. (7) Obtain proper tip for the torch and light it. Adjust the flame for the soldering being done. (8) Apply heat to the joint. (9) When solder can be melted by the heat of the copper (not the torch), simply apply solder so it flows around the joint. (10) Clean the joint of excess solder and cool it quickly with a damp rag.*

could lead to cases of lead poisoning in extreme situations. To prevent lead poisoning, many systems now require a lead-free solder.

Both types of solder are easily obtainable at building-supply stores. Either can be used, but the user should be sure to match the correct flux with the solder used.

Soft copper tubing is sold in rolls of 10, 15, 50, and 100 feet. It is available in diameters from ⅛ through 1 inch in ⅛-inch intervals. It is soft enough to be bent with your hands in any direction. This means that special fittings such as elbows and nipples are not needed. However, it does require either compression or flared fittings (Fig. 12-24) to join it to appliances or

Fig. 12-24 *Compression or flare fittings: A. Compression fitting; B. flare fitting: (1) Slide flare nut onto tube; (2) flare the end of the tube; (3) slide the flare nut over the flare onto the fitting, and tighten.*

other parts of the water system. Compression fittings are easier to install, but they are not as reliable under pressure. Flares on the end of the tubing are for the flare fitting. The flared ends are made easily with a flaring tool, as in Fig. 12-25.

Plastic water pipe Most plastic water pipes are made of PVC. However, care should be taken to make sure that the correct pipe is used for hot and cold water applications. The most common type of PVC is not

Fig. 12-25 *Flaring tool for copper tubing.*

suitable for hot water applications, but a special PVC pipe, CPVC, is made for hot water. CPVC can be used for either hot or cold applications. Cold water PVC is white, whereras CPVC is a tan color. The outside dimensions of CPVC are a little bigger than those of PVC, so the fittings used to join the pipe cannot be mixed up. It would be unwise to use cold water fittings with hot water pipe.

There is not much difference in cost between PVC and CPVC. PVC pipe is the least expensive, most durable, easiest, and quickest type of pipe to install. Its cost is about a tenth of what copper costs and about half of what galvanized pipe costs. It is not very strong nor heat resistant, but it is very light in weight. Like copper, it can be cut easily, which makes the purchase of nipples unnecessary. It can be cut with ordinary wood-cutting tools. PVC is limber and can be bent slightly, but sharp turns require a 45- or 90-degree elbow fitting just like metal pipe (Fig. 12-26). Size can be changed by using reducer couplings, and other couplings can be obtained to connect to threaded systems, as shown in the figure.

PVC pipe is "welded" into the various fittings rather easily. Simply coat the inside of the fitting and the outside of the pipe with a special solvent using the swab in the can, and then push the two pieces together with a slight twisting motion (Fig. 12-27). The solvent is usually available anywhere PVC pipe is sold and comes in various sized containers for big or small jobs. There are separate solvents for PVC and CPVC. How-

Fig. 12-26 *Typical plastic pipe fittings (left to right): 45-degree elbow, 90-degree elbow, nipple, coupling, T, and threaded adaptor.*

ever, there is a solvent that will work with either type of pipe. Some plumbers use a two-stage solvent, where the first application is a cleaning solution. The solvent is applied directly over the first coat, and both are applied in the same manner.

Plastic pipe is easy to work with and can be repaired easily. When trouble occurs, you simply cut out the bad part, and glue in new pieces. Neither the tools nor the processes require any special procedures. Compression unions also can be used easily for either

initial installation or repairs, as in Fig. 12-28. They also can be used to join metal pipe to plastic pipe.

Locating Pipes

The building interior has the various pipes and drains worked between or through joists and studs. This takes a direct route and minimizes the amount of pipe and work needed for installation. However, once the area where the fixture is to be located is reached, care must be taken to locate the pipes or drains so that the fixtures can be installed. As in most systems, locations are fairly standardized for this purpose. This is important so that a fixture, such as a toilet (water closet), can be installed correctly. The pipe outlets must allow the fixture to be the correct distance from the wall and have the water supply available without being very visible and yet remain accessible for repairs. Equally

WELDING PLASTIC PIPE
1. CLEAR INSIDE AND OUTSIDE EDGES OF BURRS
2. SAND GLAZE OFF JOINT AREA
3. COAT PIPE WITH SOLVENT
4. COAT INSIDE OF FITTING WITH SOLVENT
5. PUSH FITTING ONTO PIPE WITH A SLIGHT TWIST

Fig. 12-27 *"Welding" plastic pipe.*

Fig. 12-28 *A plastic compression union; used to join plastic pipes or plastic pipes to metal pipes.*

important is the location of the drain so that the bottom portion can be installed over the drain in a correct manner.

There is more flexibility in the location of supply outlets that will be inside cabinets than for the outlets for tubs and toilets. Typical locating dimensions for various outlets and drains are shown in Fig. 12-29. Note that the hot water line is always installed on the left side.

Once the pipes are installed and either capped off or have cutoff valves installed and all the drains are connected, the system is ready for inspection. If the system passes inspection, then the city will allow the water to be turned on, and the rest of the interior can be completed.

CHAPTER 12 STUDY QUESTIONS

1. In plumbing terms, what is a fixture?
2. What are the stages of plumbing?
3. Why should water supply lines be underground?
4. When does the second stage of plumbing occur?
5. What are the three elements of a plumbing system?
6. Define an air chamber?
7. How is air hammer prevented?
8. What does DWV mean?
9. What is a lateral?
10. What is meant by soil stack?
11. How much slope should a drain system have?
12. What size pipe is used for drains that connect sinks and lavatories, tubs and showers?
13. What is the purpose of a trap?
14. Where do you use a P trap?
15. What purpose does a house trap serve?
16. What is the purpose of a vent?
17. What is a wye?
18. What are the three types of pipe used for plumbing systems?
19. What is the difference between rigid and soft copper?
20. What is CPVC ? What is its color?

Fig. 12-29 *Typical installation dimensions for pipes: A. bathroom sink; B. toilet; C. tub-shower combination.*

13
CHAPTER

Insulation

NSULATION CAN MAKE A DIFFERENCE IN THE AMOUNT of energy used to heat or cool a house. Previously, houses were not insulated because of the low cost of energy. Today, things have changed. It is no longer economical to build without insulating.

According to a study conducted by the U.S. Department of Energy, 70 percent of the energy consumed in a house is used for heating and cooling. Another 20 percent is used to heat water, and 10 percent is used for lighting, cooking, and appliances (Fig. 13-1).

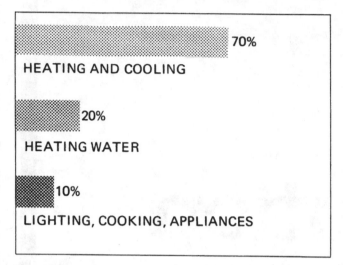

Fig. 13-1 *Consumption of energy in the home.* (U.S. Department of Energy.)

Insulation can help to conserve as much as 30 percent of the energy lost in a home. Note the air leakage test results shown in Fig. 13-2. Wall outlets are a source of 20 percent of this leakage. This can be stopped by using socket sealers (Fig. 13-3). These are

Fig. 13-3 *Socket sealers.* (Manco Tape.)

foam insulation pads that are placed between the faceplate and the socket. Windows and doors have a leakage problem, too. They can be treated in a number of ways to cut down on leakage. The soleplate is an important place to check for proper insulation. To conserve energy, new codes call for leakage seals over corners and soleplates (Fig. 13-4).

Figure 13-4 shows where to insulate by means of the following numbers:

1. In unfinished attic spaces, insulate between and over the floor joists to seal off living spaces below. (Well-insulated attics, crawl spaces, storage areas, and other enclosed cavities should be ventilated to prevent excess moisture buildup.)

2. In finished attic rooms with or without dormers, insulate

   • Between the studs of knee walls

   • Between the studs and rafters of exterior walls and roof

   • Ceilings with cold spaces above

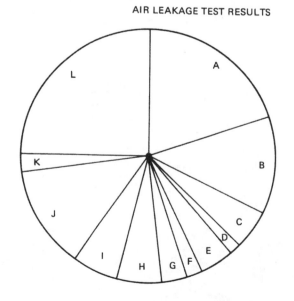

AIR LEAKAGE TEST RESULTS

A – 20% WALL OUTLETS

B – 12% EXTERIOR WINDOWS

C – 5% RECESSED SPOT LIGHTS

D – 1% BATH VENT

E – 5% EXTERIOR DOORS

F – 2% SLIDING GLASS DOOR

G – 3% DRYER VENT

H – 5% FIREPLACE

I – 5% RANGE VENT

J – 14% DUCT SYSTEM

K – 3% OTHER

L – 25% SOLEPLATE

Fig. 13-2 *Air leakage test results.* (Manco Tape.)

Fig. 13-4 *New codes call for leakage seals over corners and soleplates. Examples of where to insulate.*

EARTH

- Into joist space to reduce airflows
- All exterior walls, including
  - Walls between living spaces and unheated garages, shed roofs, or storage areas
  - Foundation walls above ground level
  - Foundation walls in heated basement walls, full wall either interior or exterior
  - Floors above cold spaces, such as vented crawl spaces and unheated garages.
3. Also insulate
   - Any portion of the floor in a room that is cantilevered beyond the exterior wall area
   - Slab floor built directly on the ground (for new construction, slab-on-grade insulation should be installed to the extent required by building codes or greater)
   - Foundation walls of unvented crawl spaces (as an alternative to floor insulation)
   - Band joists
4. Replacement of storm windows and caulk, and sealing around all windows and doors

The ductwork in the heating or cooling system can be insulated with tape. Special tape is made for the job (Fig. 13-5).

Fig. 13-5 *Insulating tape for ductwork.* (Manco Tape.)

There are a number of skills to be learned in this chapter, including

- How to check for insulation requirements
- How to apply insulation
- How to caulk around windows and doors
- How to apply various insulation materials

TYPES OF INSULATION

Batts and blankets are typically made of Fiberglas. They also may be made of cellulose fibers. Batts are best for use by carpenters and do-it-yourselfers. They can be applied easily in attics and between exposed ceiling joists.

Loose-fill insulation comes in bags. It can be poured or blown into the joist spaces in open attics. It can be blown by special machines into closed spaces such as finished walls.

Rigid types of plastic insulation come in boards. Foams can be applied with a spray applicator. Plastic has high insulation quality. Thinner layers of plastic may provide the same protection as thicker amounts of other materials. Thinner sections may provide higher R values. *R values* indicate the relative insulating qualities of a material.

All insulation material should have a vapor barrier. This barrier provides resistance to the passage of water vapor.

Another type of Fiberglas insulation is the tongue-and-groove sheathing that is available for exterior walls.

Batts and blankets (except friction-fit) have an asphalt-impregnated paper vapor barrier. In some cases, they have a foil backing. All other types should have a 2-mil (0.002-inch) plastic sheet installed as a vapor barrier.

HOW MUCH IS ENOUGH?

For years, the recommended amount of insulation was R-19 for ceilings. R-11 was suggested for walls almost everywhere in the country.

Recently, there have been changes in these recommendations. The map in Fig. 13-6 shows how much is needed in various locations. The first number is for the ceiling. The second number is for the wall. Insulate the floor if the basement is not heated. These are the values recommended by Owens-Corning, which manufactures insulation.

In some parts of the country, utilities and cities have separate standards. Be sure to check the local requirements for your area. In most instances, local codes follow Fig. 13-7. At present, it appears that *more is better*. This has not been proven to everyone's satisfaction. More research needs to be done in this area. Different types of insulation are being developed. Foams appear to be the best bet for insulating an older house—that is, if it has no previous wall insulation.

WHERE TO INSULATE

Batts are made in specific thicknesses. Choose the proper R value for the space. Allow the material to expand. Fit it closely at the edges. Don't overlook the space between closely fitted timbers.

Vapor-barrier protection is important. It always should be installed toward the warm side of the area being insulated. Vapor barriers must not be torn or broken. Repair any breaks with a durable tape.

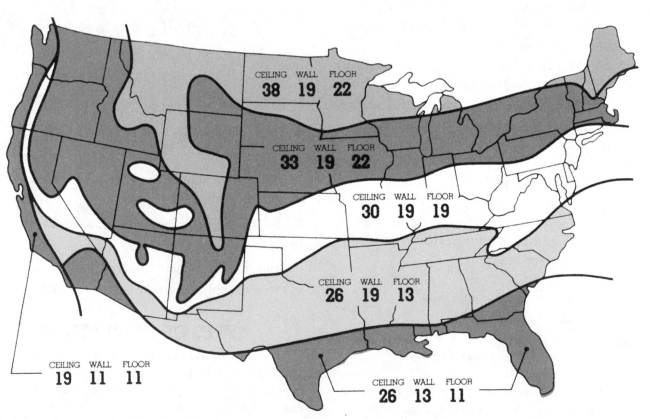

Fig. 13-6 *Owens-Corning Fiberglas minimum insulation recommendations for energy-efficient homes.*

Fig. 13-7 *Recommendations for insulating your home. Note the R values and where they are assigned.* (New York State Electric and Gas.)

Provide proper ventilation. Natural venting is important. It controls moisture and relieves heat buildup in the summer (Fig. 13-8). In some cases, there isn't enough natural ventilation. A fan usually is needed to exhaust an attic during summer. This is when heat builds up and causes the air conditioner to work harder (Fig. 13-9). The airspace of the attic is often 60°F hotter than outside. Removing the heated air reduces the heat transfer to the home. Keep in mind that power venting is not an effective solution in conjunction with ridge venting. Power venting isn't always effective with vents located close to the unit.

Crawl space Crawl spaces need ventilation (Fig. 13-10). There should be vents (two with moisture seals) in crawl spaces. Sealed and unsealed crawl spaces are treated differently. Vents in the basement wall help to circulate the outside air and draw out moisture. This moisture is generated through the exposed earth in the crawl space. Four vents are needed for no moisture seal.

Figure 13-11 shows some of the points that may need special attention. Note how the vapor barrier is installed at these points. Figure 13-12 shows the best insulation for desired results. Note that some houses are now built with 2- × 6-inch instead of 2- × 4-inch studs in the walls. This allows for more insulation. Note how 1-inch foam sheathing board is used on the

outside. This gives the added insulation needed for 2 × 4 studs.

INSTALLING INSULATION

Installing Insulation in Ceilings

Push batts up through joist spaces. Pull them back down even with bottom of the joists (Fig. 13-13). This ensures full R-value performance because it prevents compression of the insulation. When high-R batts are used, the batts can expand over the tops of the joists. Compressing the batt reduces its insulating value. Overlap the top plate. Do not block eave vents. Make sure that the ends butt tightly against each other. Obstructions (such as electrical boxes for light fixtures) should have insulation installed over them.

> **Caution:** Insulation must be kept at least 3 inches away from recessed light fixtures.

Installation Safety

You should keep in mind some of the health hazards involved with installing Fiberglas. It has tiny fibers of resin (glass) that can be breathed in and cause trouble in your lungs. Just be careful, and take the proper precautions. If you work with Fiberglas all day, make sure that you wear a mask over your nose and mouth.

EAVE VENT

GABLE VENT

ROOF VENT

CUPOLA VENT

RIDGE VENT

FOUNDATION VENT

WIND TURBINE

160° | 120°

145° | 115°

105°

85° | 75°

82° | 72°

80° | 70°

NOT VENTILATED | VENTILATED

SUMMER

Fig. 13-8 *Location of vents for adequate ventilation.* (New York State Electric and Gas.)

SOFFIT

Fig. 13-9 *Attic ventilation—forced air removal.* (New York State Electric and Gas.)

WITH MOISTURE SEAL

NO MOISTURE SEAL

VAPOR BARRIER

AT LEAST TWO VENTS REQUIRED

AT LEAST FOUR VENTS REQUIRED

1 FT² OF VENT FOR EACH 1500 FT² OF CRAWL SPACE

1 FT² OF VENT FOR EACH 150 FT² OF CRAWL SPACE

Fig. 13-10 *Crawl space ventilation.* (New York State Electric and Gas.)

Notice that the person in the photos in the following pages is wearing long sleeves and gloves. This will prevent the itching that this type of material can cause. You should wear goggles—not just glasses. These will keep the fibers out of your eyes and prevent damage to your vision.

A hat and proper long pant legs are also called for in this job. Proper shoes are needed because you are working around a construction area. The nails are still exposed in some types of work, so make sure that your shoes can protect your feet from the nails.

The usual safety procedures should be followed in using a ladder and climbing around an attic or any-

VAPOR BARRIER

END OF INSULATION
BLANKET PUSHED UP,
OR SHORT PIECE AT HEADER

SILL SEALER

VAPOR BARRIER

Fig. 13-11 *Note placement of the vapor barrier in the installation of insulation.* (New York State Electric and Gas.)

place that may be elevated from floor level. A hard hat helps to keep you from getting roofing nails in your scalp when you are installing insulation near the underside of a roof.

Installing Insulation in Unfloored Attics

Install faced building insulation in attics that have no existing insulation (Fig. 13-14). The vapor barrier should face down toward the warm-in-winter side of the structure.

Use unfaced insulation when you are adding to existing insulation (Fig. 13-15). Batts or blankets can be laid either at right angles or parallel to existing insulation.

Start from the outside, and work toward the center of the attic (Fig. 13-16). Lay insulation in long runs first, and use leftovers for shorter spaces. Be sure to butt insulation tightly at joints for a complete barrier.

Insulation should extend far enough to cover the top plate, but it should not block the flow of air from the eave vents (Fig. 13-17).

Insulation should be pushed under wiring unless the wiring will compress the insulation. The ends of batts or blankets should be cut to lie snugly around cross-bracing and wiring. Fill spaces between the chimney and the wood framing with unfaced Fiberglas insulation.

Scuttle holes, pulldown stairways, and other attic accesses also should be insulated. Insulation can be glued directly to scuttle holes. Pulldown stairways may need a built-up framework to lay batts on and around.

Installing Insulation in Floored Attics

If the attic is not used for storage, unfaced batts or blankets may be laid directly on top of the floor (Fig. 13-18).

Floorboards must be removed before insulation is installed if the attic is to be used for storage. Since the flooring will be replaced, the R value will be limited. It is limited to that of the amount of uncompressed insulation that can be fitted between the joists. Remove the flooring in the attic. It then should be insulated as shown in Fig. 13-19. Replace the floorboards and stairsteps.

Installing Insulation in Floors

Floors over unheated areas should be insulated. The recommended insulation R values should be used (Fig. 13-20). With faced insulation, the vapor barrier should be installed face up. This is toward the warm-in-winter side of the house. Insulation should overlap the bottom plate. It also should overlap the band joist (Fig. 13-21).

(A)

R-44
WITH 2 × 6
OR 2 × 8
CEILING JOISTS

6½", R-22

6½", R-22

Use
6½" thick R-22 Fiberglas
insulation between
joists and 6½" thick R-22
Fiberglas insulation
across the joists.

R-30
WITH 2 × 4
TRUSS
CONSTRUCTION

6", R-19

3½", R-11

Use
3½" thick R-11 Fiberglas
insulation between ceiling
joists and 6" thick R-19 Fiberglas
insulation across the joists.

R-38
WITH 2 × 6
CEILING JOISTS

6", R-19

Use
6" thick R-19 Fiberglas
insulation between
ceiling joists and 6" thick R-19
Fiberglas insulation across the joists.

R-19
WHERE 2 × 6
FRAMING EXISTS

New construction techniques
may use 2 × 6
studs on 24" centers. This
permits installation of
6" thick R-19
Fiberglas insulation
between studs.

6", R-19

(B)

R-30
WITH 2 × 6
CEILING JOISTS

6", R-19

3½", R-11

Use
6" thick R-19 Fiberglas
insulation between
ceiling joists and 3½" thick R-11
Fiberglas insulation across the joists.

R-13
WHERE 2 × 4
FRAMING EXISTS

3½" FULL WALL, R-13

R-13 is the maximum standard
insulation product for use in
conventional sidewall
construction. Use
R-13 Full Wall Fiberglas
insulation between studs.

R-30
WITH 2 × 6
OR 2 × 8
CEILING JOISTS

10", R-30

Use
10" thick R-30 Fiberglas
insulation. It compresses
between ceiling joists and expands
over them to form a continuous
insulation blanket.

R-19
WHERE 2 × 4
FRAMING EXISTS

3½" FULL WALL,
R-13

When even greater R-values
are desired in walls which
have 2 × 4 framing, use
3½" thick R-13
Full-Wall Fiberglas
insulation between studs
and 1" foam sheathing
board on exterior.

1" FOAM
SHEATHING
BOARD

Fig. 13-12 *A. Optimal insulation for ceilings and attics. B. Optimal insulation for sidewalls.* (Certain-Teed.)

Fig. 13-13 *Installing insulation in ceilings. Note the long-sleeved shirt, gloves, hat, respirator, and goggles. (Owens-Corning.)*

Fig. 13-16 *Work toward the center of the attic, laying long runs of insulation first. Fill in with short pieces. Be sure to butt insulation tightly at all joists. Insulation should cover top plate but must not block eave vents. Insulation must be kept at least 3 inches away from recessed light fixtures. (Owens-Corning.)*

Fig. 13-14 *Install insulation with a vapor barrier facing into attics with no existing insulation. The vapor barrier should face the warm-in-winter side of the structure. (Owens-Corning.)*

Fig. 13-17 *If the attic is used for storage, the floorboards must be removed. The flooring will limit the R value to the amount of uncompressed insulation that can fit between the joists. (Owens-Corning.)*

Fig. 13-15 *Use insulation without a vapor barrier facing when adding to existing insulation. Batts or blankets can be laid at right angles or parallel to existing insulation. (Owens-Corning.)*

Fig. 13-18 *If the attic is not used for storage, unfaced batts or blankets may be laid directly on top of the floor. (Owens-Corning.)*

Fig. 13-19 *Once the flooring has been removed, the attic should be insulated in this manner.* (Owens-Corning.)

Fig. 13-20 *Floors over heated areas should be insulated to the R values recommended.* (Owens-Corning.)

Fig. 13-21 *Note how the insulation vapor barrier is installed in the crawl space.*

VAPOR BARRIER

Bow supports Fiberglas insulation can be supported by heavy-gauge wires. The wire may be bowed or wedged into place under the insulation.

Crisscross wire support Insulation also may be supported by lacing wires around nails located at intervals along the joists.

Chicken-wire support Chicken wire nailed to the bottom of floor joists also will support the insulation.

Insulating Basement Walls

Masonry walls in conditioned areas may be insulated. Install a furring strip or 2×4 framework. Nail the top and bottom strips in place. Then install vertical strips 16 or 24 inches on center (O.C.). Next, cut small pieces of insulation. Push the pieces in place around the band joist. Be sure to get insulation between each floor joist and against the band joist.

Install either faced or unfaced insulation. Unfaced insulation will require a separate vapor barrier. Do not leave faced insulation exposed because the facing is flammable. It should be covered with a wall covering. Cover the insulation as soon as it has been installed.

Insulating Crawl Spaces

Measure and cut small pieces of insulation to fit snugly against the band joist. This is done so that there is no loss of heat through this area. Push pieces into place at the end of each joist space.

> **Caution:** Do not use faced insulation; the facing is flammable.

Use long furring strips. Nail longer lengths of insulation to the sill. Extend the insulation 2 feet along the ground into the crawl space. Trim the insulation to fit snugly around the joists.

On walls that run parallel to the joists, it is not necessary to cut separate header strips. Simply use longer lengths of insulation, and nail (with furring strips) directly to the band joist. Install the insulation. Lay the polyethylene film under the insulation and the entire floor area. This prevents ground moisture from migrating to the crawl space. Use 2- $\times$ 4-inch studs or rocks to help hold the insulation in place.

Installing Insulation in Walls

Insulating walls can pay dividends. Figure 13-22 shows that the sidewalls lose 15.2 percent of the heat lost in a ranch house. In a two-story house, where there is more wall area, the loss is 30.8 percent. This means that it is very important to insulate walls properly.

ONE-STORY HOME
(30 x 40 feet)

CEILING
48.1% of heat loss

AIR CHANGE
2/hour
21.8% of heat loss

WINDOWS AND DOOR
14.9% of heat loss

SIDEWALLS
15.2% of heat loss

TWO-STORY HOME
(30 x 30 feet)

CEILING
17.2% of heat loss

AIR CHANGE
2/hour
16.2% of heat loss

DOORS AND WINDOWS
35.8% of heat loss

SIDEWALLS
30.8% of heat loss

Fig. 13-22 *Heat loss in a one-story and a two-story house.* (New York State Electric and Gas.)

NO DIAGONAL WALL BRACING REQUIRED WITH PLYWOOD PANEL SIDING

INSULATION AS REQUIRED

PLYWOOD PANEL SIDING

Fig. 13-23 *Placement of insulation inside a wall between studs.* (American Plywood Association.)

Figure 13-23 shows how insulation fits into the wall area. It is placed between the studs. The outside wood is also an insulation material.

Insulation for walls can be R-19. Use a combination of R-11 insulation with 1-inch (R-8) high-R sheathing or R-13 insulation with ¾-inch (R-8) high-R sheathing on the interior or exterior (Fig. 13-24). R-13 insulation also may be used with ⅝-inch high-R sheathing only on the exterior.

In placing R-30, use a single-layer blanket. Or you can use R-19 and R-11 blankets. This combination is shown in Fig. 13-25.

High-R batts (R-26, R-30, and R-38) are made in full 16- and 24-inch widths. They expand over the tops of the joists. This presents a continuous barrier to heat

Fig. 13-24 *R-19 insulation can be obtained by combining R-11 insulation with 1-inch (R-8) high-R sheathing or combining R-13 insulation with ¾-inch (R-8) sheathing on the exterior. R-13 insulation also may be used with ⅝-inch high-R sheathing only on the exterior.* (Owens-Corning.)

Fig. 13-25 *R-30 can be obtained by using a single R-30 blanket or R-19 and R-11 Fiberglas blankets in combination.* (Owens-Corning.)

Fig. 13-26 *Single-layer Fiberglas high-R batts (R-26, R-30, and R-38) are made in full 16- and 24-inch widths so that they expand over the tops of the joists for a continuous barrier to heat flow. (Owens-Corning.)*

Fig. 13-27 *R values for floors can be achieved with a single layer of Fiberglas blanket insulation.*

flow (Fig. 13-26). Floors can have R values. Choose the right single layer of blanket insulation (Fig. 13-27).

Cutting fiberglas insulation   Measure the length of Fiberglas insulation needed. Place the insulation on a piece of scrap plywood or wallboard. Compress the material with one hand. Cut the material with a sharp knife (Fig. 13-28).

Fig. 13-28 *Cutting Fiberglas insulation. (Owens-Corning.)*

Installing faced insulation   Install faced insulation using either the insert or faced stapling method. Cut lengths 1 inch longer than needed. This is done so that facing may be peeled back at the top and bottom. This makes it possible to staple to plates. Make sure that the insulation fills the entire cavity (Fig. 13-29).

Fig. 13-29 *Installing faced insulation.*

Insert stapling   Insulation is pushed into the stud or joist cavity. Vapor barrier flanges are stapled to the sides of the studs. Make sure that flanges do not gap. This will allow vapor to penetrate (Fig. 13-30).

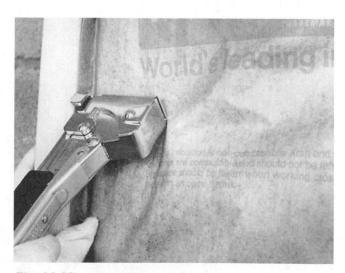

Fig. 13-30 *Insert stapling. (Owens-Corning.)*

Faced stapling Vapor barrier flanges are overlapped. They are then stapled to the edge of the stud. Flanges must be kept smooth. This is necessary for proper application of the wall finish (Fig. 13-31).

Installing unfaced insulation Wedge unfaced batts into the cavity. Make sure that they fit snugly against the studs. Also fit the top and bottom plates snugly (Fig. 13-32).

Fig. 13-31 Faced stapling. (Owens-Corning.)

Fig. 13-32 Installing unfaced insulation. (Owens-Corning.)

Fig. 13-33 Filling small gaps. (Owens-Corning.)

Filling small gaps Hand stuff small gaps around window and door framing. This can be done with scrap pieces. Cover the stuffed areas with a vapor-resistant material (Fig. 13-33). Insulation must fit snugly against both the top and bottom plates. It also should fit against the sides of the stud opening (Fig. 13-34). Improperly installed, it will permit heat to escape.

Insulating behind wiring and pipes Insulation should be fitted behind or around heat ducts. It also should fit closely around pipes and electrical boxes. This will help to keep pipes from freezing. It also prevents unnecessary heat loss. Repair any tears in the vapor barrier of faced insulation. This prevents condensation problems (Fig. 13-35).

Filling narrow stud spaces You can cut unfaced insulation to stud width. Wedge the insulation into place. This area must be covered with a vapor barrier. With faced insulation, make sure that the vapor barrier is stapled properly to the studs (Fig. 13-36).

Fig. 13-34 Close doesn't count. (Owens-Corning.)

Fig. 13-35 Insulating behind wiring and pipes. (Owens-Corning.)

Fig. 13-36 *Filling narrow stud spaces.* (Owens-Corning.)

Fig. 13-37 *Insulation around wiring.* (Owens-Corning.)

Insulating around wiring   Split unfaced insulation, and install it on both sides of the wiring. Install faced insulation behind the wiring. Pull it through to permit stapling (Fig. 13-37).

VAPOR BARRIERS AND MOISTURE CONTROL

Excess moisture may be a concern in a well-insulated house because excess air changes have been eliminated. Therefore, some consideration must be given to the tight house. The insulation has been located properly. It keeps the inside air inside. This means that it doesn't require too much energy to reheat it. It doesn't have to be recooled too much during the summer.

Condensation

A house must *breathe*. Condensation that is not caused by insulation can be controlled. Condensation is sim-

ply water vapor. Water in the form of a fog comes from a variety of sources in a home. It is produced in baths, in the kitchen, and in the laundry. It will distribute itself throughout the house. Unless it is controlled, it can condense. Just as frost forms on the outside of an iced glass on a warm day, vapor condenses on house surfaces. Slowly but surely, it can cause damage.

Moisture control   There are two ways to control moisture. They are elimination and ventilation. *Elimination* means to get rid of the source of moisture. This can be leaks, wet ground in crawl spaces, or small leaks in the structure.

Ventilation means simply getting rid of moisture. It carries excess humidity outside. This is done with vent fans or hoods in the kitchen. The bathroom may be vented to the outside. Automatic clothes dryers must be vented to the outside. They are a major source of moisture.

Vent fans at high-moisture-producing locations are helpful, for example, the fan over a kitchen range. The bathroom may have a fan to remove moisture. Any of these fans can be equipped with a humidistat that will turn it on when the humidity gets to a preset level.

Attic ventilation   Attic ventilation is very important for many reasons. It helps to eliminate moisture in the attic. That moisture can leak into the living quarters and ruin the ceiling. Or it can rot studs or other parts of the house. One way to provide ventilation is to place vents in the ends of the roof structure (Fig. 13-38). They can be various shapes. Some buildings use cupolas to allow air to move upward and out. Some use a vented ridge section to allow hot air to escape in the summer and a constant flow of air all during the year.

These openings must be covered to prevent insects, birds, snow, and rain from entering. There are a number of materials used to cover the openings. Table 13-1 lists some of them.

Keep in mind that it takes 1 square foot of inlet area and 1 square foot of outlet area for each 600 square feet of ceiling area. At least half the vent should

TABLE 13-1 *Attic Ventilation Coverings*

Type of Covering	Size of Covering
1/4-inch hardware cloth and rain louvers	2 × net vent area
8-mesh screen	1 1/4 × net vent area
1/4-inch hardware cloth	1 × net vent area
8-mesh screen and rain louvers	2 1/4 × net vent area
16-mesh screen	2 × net vent area
16-mesh screen and rain louvers	3 × net vent area

NOTE: There is 1 ft² inlet and 1 ft² outlet for each 600 ft² of ceiling area, with at least half of the vent area at the top of the gables and the balance at the eave.

(A)

NOT VENTILATED   VENTILATED

WINTER

(B)

Fig. 13-38 *A. Spacing of attic vents. B. Ventilation can control condensation.*

THERMAL CEILINGS

Thermal ceiling panels have insulating values up to R-12. They are the only types of insulation that can be used for some types of homes. Take, for instance, the A frame. The only way the ceiling can be insulated is

Fig. 13-39 *Thermal ceiling panels are easy to install.* (Owens-Corning.)

with thermal ceiling panels (Figs. 13-39 and 13-40). These types of ceiling panels have a finish that is pleasing. They also are sound-absorbing. They can be used in attics being turned into living space or attics opened up to the rest of the home. If the attic is already finished, this may be the only way to insulate.

A number of sizes are available, with different types of finishes on the face. There are sizes and faces to suit almost every installation requirement. They come in sizes from 2 × 4 feet up to 4 × 16 feet. Figure 13-41 shows some of the applications for these panels.

Fig. 13-40 *Typical installations of thermal ceiling panels.* (Owens-Corning.)

Fig. 13-41 *Thermal ceiling panels for cathedral ceilings and regular ceilings.* (Owens-Corning.)

Installing Thermal Ceiling Panels

There are a number of ways to install thermal ceiling panels.

Grid system   This uses the suspended interlocking metal grid system. It can be installed on the 2- × 4-foot or 4- × 4-foot system. A larger-faced grid system is needed for 3-inch R-12 panels (Fig. 13-42A).

Solid-wood beams   Solid beams, as in Fig. 13-42B, are used in many homes. The beam becomes a support system for the insulation panels. The panels can be removed and replaced if they are damaged.

False-bottom beams   A system of 1- × 6-inch false-bottom, or box, beams provides the beauty of beams without the expense. When wood-beam support is used, wood block spacers must be installed between the existing ceiling or joists and the beam. This is done so that the panels are not compressed when the beams are nailed up.

Suspended beams   False beams also may be used to lower a ceiling like a grid system. Beams are suspended. They use screw eyes attached to the existing ceiling joists.

Decorative Beams

The suspended-beams method of putting up ceiling insulation is easy. In Fig. 13-43, note how the plastic beams are cut to length with a knife.

Make a sketch on paper showing the pattern you want for your beams. Now it is simple to cut the beams to size. Use a knife or sharp tool with a razor blade.

Measure the exact length you need. Cut the beams to size (Fig. 13-44). Use a ruler or chalk line to mark

(A)

(B)

(C)

(D)

Fig. 13-42 *Methods of supporting thermal ceiling panels: A. grid system; B. solid-wood beams; C. false-bottom beams; D. suspended beams.* (Owens-Corning.)

Fig. 13-43 *Cutting a plastic beam.*

Fig. 13-45 *Note the groove in the beam. You may want to glue a furring strip in it.*

Fig. 13-46 *Method of supporting beam while the adhesive sets.*

where you want to install the beam. Beams can be glued to other material. They can be glued to the ceiling with an adhesive. You might want to glue a piece of furring strip down the groove of the beam (Fig. 13-45). This will allow you to attach screw hooks. They will be wired to the old ceiling later (see Fig. 13-42D). Angles can be cut in the beam with a miter box.

If you want to glue the beam up, use a brace such as that shown in Fig. 13-46. This can be done if you want to make a ceiling resembling the one in Fig. 13-42B. Nicks and scratches in this type of plastic foam beam can be touched up with shoe polish. Since the beam is made of foam, it is also very good insulation.

Fig. 13-44 *A ceiling using plastic foam beams.*

STORM WINDOWS

A single pane of glass is a very efficient heat transmitter. Window and door areas approach 20 percent of the sidewall area of the average home. It is important that heat loss from this source be minimized.

Properly fitted storm windows are a must for a well-insulated home. They come in styles suitable to the design of any home.

Older windows have storms fitted on the outside. This can cause some reduction in heat loss. In fact, an old window with a triple-deck outside storm window reduces heat loss by 67 percent. Adding another layer of insulating glass or plastic, as in Fig. 13-47, makes it possible to reduce heat loss by up to 93 percent. An inside storm window is easily added to any window. Figures 13-48 through 13-54 show how an inside storm window is made and placed into position. Also check Table 13-2.

Weather-stripping a window There are three basic types of weather stripping for double-hung windows. The foam-rubber type is probably the best; it is the

TABLE 13-2 *Storm Windows*

Window (Glass Area Only)	Approximate R of Unit*
Single glazing	0.88
Double glazing with 1/4-inch air space	1.64
Double glazing with 1/2-inch air space	1.73
Single glazing plus storm window	1.89
Double glazing plus storm window	2.67

Courtesy of New York State Electric and Gas Corp.

*R value for vertical air space is for air space from 3/4 to 4 inches thick. R values for glass are actual for type of window listed. R (resistance) indicates the amount of heat a material will prevent from passing through it in a given time. The higher the R value, the more heat the material will hold back, and hence, the better the insulation of that material.

Fig. 13-47 *Standard window with storm window on the exterior and an "insider" on the inside.* (Plaskolite.)

TRIPLE-TRACK OUTSIDE STORM

PRIMARY WINDOW

THE INSIDER

Fig. 13-48 *Measuring for the inside storm window.* (Plaskolite.)

Fig. 13-49 *Cutting the inside window.* (Plaskolite.)

Fig. 13-50 *Cutting the trim for the window.* (Plaskolite.)

easiest to work with (Fig. 13-55). Each of the weatherstripping materials comes in coils. It often comes in complete cut-to-size kits. Metal-frame windows (such as casement, awning, or jalousie styles) use invisible vinyl tape, joint-sealing self-adhesive foam rubber, or casement aluminum strip.

Installing weather stripping on a window Metal strip insulation should be nailed at three positions. The sash channels inside the frame (making sure that you leave the pulleys near the top uncovered) are one location. Underneath the upper sash on the inside is another location. Figure 13-56 shows how the spring metal type of window is insulated with weather stripping.

Vinyl and adhesive types of weather stripping can be installed as shown in Fig. 13-57. Vinyl strips should be nailed in place at three locations (Fig. 13-58):

Fig. 13-51 *Placing the trim around the plastic.* (Plaskolite.)

CLEAR RIGID IN-SIDER SHEET

SNAPPED OPEN

CLOSED

VINYL MOUNTING TRIM

WINDOWSILL TRIM

JOINER STRIP

Fig. 13-52 *Pieces that go into the making of a storm window for the inside.* (Plaskolite.)

Fig. 13-53 *Installing the window.*

Fig. 13-54 *Placing the plastic inside the window.*

1. On the outer part of the bottom rail of the upper sash

2. On the same location on the lower sash

3. On the outer surface of the parting strips

There are other methods that can help. Invisible polyethylene tape replaces the old type of caulk. This fits around the windows to seal them inside (see Fig. 13-57).

Fig. 13-55 *Basic types of strip or roll weather stripping.*

THIN SPRING-METAL INSULATION

(A)

VINYL INSULATION

(B)

FOAM-RUBBER INSULATION
WITH ADHESIVE BACKING

(C)

UPPER-SASH
BOTTOM RAIL

UPPER-SASH
BOTTOM RAIL

(A)

(B)

LOWER-SASH
BOTTOM RAIL

(C)

Fig. 13-56 *Vinyl tape and self-adhesive foam-rubber strip installation.*

STORM DOORS

In most of the country, it is necessary to install storm doors. These fit on the outside and serve as screen doors during the summer. The glass and screen can be taken out and exchanged as needed (Fig. 13-59). Of course, the doors aren't too effective unless they have door closers to make sure that they close (Fig. 13-60). Storm doors usually come in a prehung package. They are easily installed simply by placing screws through the holes and checking for plumb. They are designed for the do-it-yourselfer. In most cases, homes do not come with storm doors. The first owner has to install them. This is also true of the mailbox.

Take a look at Table 13-3. It shows you just how important the storm door really is. It can make quite a difference in heat loss.

TABLE 13-3 *Storm Doors*

Door Type	Approximate R of Unit*
Solid wood, 1-inch	1.56
Solid wood, 2-inch	2.33
Solid wood, 1-inch plus metal/glass storm door	2.56
Solid wood, 2-inch plus metal glass/storm door	3.44
Solid wood, in-inch plus wood/glass (50%) storm door	3.33
Solid wood, 2-inch plus wood/glass (50%) storm door	4.17
Doors with rigid insulation core	up to 7

Courtesy of New York State Electric and Gas Corp.
*R (resistance) indicates the amount of heat a material will prevent from passing through it in a given time. The higher the R value, the more heat the material will hold back, hence the better the insulation. To find the R value of a building material, multiply above R value by actual thickness of the material.

SEALANTS

There are a number of materials on the market for sealing air leaks in a house. Leaks appear around doors, windows, and chimneys. There are a number of caulking compounds made to fill these gaps. Some caulks will last longer than others. However, one of the best appears to be a paintable silicone sealant made by Dow-Corning (Fig. 13-61). These sealants are usually packaged in 13-ounce tubes suitable for use in a caulk-

ing gun. Some, of course, are in tubes for use inside the house.

Acrylic latex sealant This sealant cures to a rubbery seal. Used inside and outside, it can seal any joint up to $\frac{1}{2} \times \frac{1}{2}$ inch in size. It is also used between common construction materials where movement is expected. It can be cleaned from tools or hands with water. Do not use under standing water or when rain is expected. It has an expected life of 8 to 10 years and is paintable in 30 to 60 minutes after application.

Fig. 13-57 *Applying clear tape around windows.* (Manco Tape.)

Fig. 13-58 *Metal strip insulation nailed at three positions: A. sash channels inside the frame; B. underneath the upper sash; C. on the lower sash bottom rail.*

Vinyl latex caulk   This is a middle-priced sealant. It has the same advantages and restrictions as the acrylic latex sealant. Life expectancy is 3 to 5 years.

Butyloid rubber caulk   Especially useful in narrow openings up to ⅜ inch, this is used where movement is expected. It can be used in glass-to-metal joints. It is also used between overlapping metal. It can be used to seal gutter leaks or under metal or wooden thresholds. This type of caulk is useful under shower door frames. Life expectancy of the caulk is 7 to 10 years.

Roof cement   Roof cement may be used on wet or dry surfaces. Cold weather application is possible. It stops leaks around vent pipes, spouts, valleys, skylights, gutters, and chimneys. This cement can be used to tack down shingles. It is also useful for sealing cracks in chimneys or foundations. Roof cement comes in black. It is not paintable.

Concrete patch   This is a ready-made concrete and mortar patching compound. It contains portland cement. This patching compound is quick and easy. It is used to repair cracks in concrete sidewalks, steps, sills, or culverts. It is also recommended for cracked foundations, tuck pointing, and stucco repair.

WINTERIZING A HOME

There are a number of ways to winterize a home. They all deal with insulation and stopping air leaks. Check the 33 ways listed in Table 13-4. Note how effective they are at saving energy.

Seal all outside doors, including basement doors, with weather-stripping material. In some cases, you can use old carpet strips.

Put masking tape around moving parts of windows. Caulk around window and door frames, including those in the basement. You can stop drafts under doors by placing rugs at the bottom.

Fig. 13-59 *Metal storm door with full glass panel.*

Fig. 13-60 *Storm door closer (top) and spring for keeping the door from being blown off its hinges.* (Stanley Tools.)

INSULATING FOAM SEALANT

Energy conservation is not the only use for sealants. They can be used for a number of purposes in any construction job. Low-expansion rigid foam can be obtained in a pressurized can and applied to foundations, electrical outlets, and switches, and in the attic, as well as everywhere an electrical cable goes through a stud or rafter (Fig. 13-62). The foam can be used to prevent rodents, insects, and dust from entering a building. Once applied, it expands and fills in a space about five to six times its original bead. When applied around the electrical receptacle and switch boxes, it expands to fill in the area between the box and the drywall to seal out dust, ants, and other insects. When using foam, be

Fig. 13-61 *Sealant samples ¼ inch wide by ¼ inch deep after 600 hours in an accelerated weathering machine. This is the equivalent of more than 1 year weathering outdoors in Florida. Only the paintable silicone sealant seems to have held up.* (Dow-Corning.)

sure to wear vinyl gloves; the foam is hard to remove from your skin. The application releases a flammable gas. Do not smoke while applying the foam. Use in a well-ventilated area with proper respiratory protection. The product is extremely sticky and difficult to remove from the skin. Use acetone to remove it. Acetone can be bought in liquid form, or you can use nail polish remover that contains acetone for a satisfactory job of removal.

The foam, which takes about an hour to cure, is waterproof and airtight, bonds to most materials, cures rigid, and trims easily. It is sandable and paintable, and can be used for interior or exterior areas. It fills, seals, and insulates. Figure 13-62 shows how the expanded foam is used to seal the hole that was drilled for Romex cable.

CHAPTER 13 STUDY QUESTIONS

1. Why weren't older homes insulated?
2. What does insulation do?
3. What is a vapor barrier?
4. What is meant by moisture control?
5. What does *R value* mean?
6. What is an insulation batt?
7. Why do crawl spaces have to be insulated?
8. What is the purpose of ventilation?

TABLE 13-4 *Some Energy-Saving Ideas*

33 Ways To Winterize Your Home

KEEPING THE COLD AIR OUT

1. Seal all outside doors, including basement doors, with weather-stripping material. In some cases, you can use old carpet strips.
2. Put masking tape around moving parts of windows. Caulk around window and door frames, including those in the basement. You can stop drafts under doors by placing rugs at the bottom.
3. Check pipes entering your home. You can keep the cold air out by packing rags around them.
4. Check light bulb fixtures for air leaks.
5. Put tape over unused keyholes.
6. Make sure unused flue or chimney covers fit tightly.
7. Keep fireplace dampers closed tightly when not in use.
8. Seal your foundation and sill plate with caulking material, insulation or rags.

KEEPING THE WARM AIR IN

9. Install storm windows and storm doors. If you don't have storm windows, you can substitute plastic sheeting but make sure it's tacked tightly all around the edges.
10. Install insulation between warm and cold areas. Begin by insulating the attic floor.
11. Wherever possible, carpet floors. If your attic floor can't be insulated, lay down a carpet.
12. Close off rooms you don't use, particularly those with the biggest windows and the largest outside walls.

MAINTAINING YOUR HEATING SYSTEM

13. Keep your heating system clean.
14. Make sure there is nothing blocking your registers, radiators or baseboard heaters.

15. Keep cold air returns clear.
16. Change or clean furnace filters monthly.
17. Have a qualified person adjust your heating system.

ADJUSTING YOUR LIVING HABITS

18. Keep the thermostat set at 68 degrees during the day. If this seems too cold for you, try wearing a sweater.
19. Set the temperature back at least 3 degrees at night.
20. If you're going away for a few days, set the thermostat at 60 degrees before you leave.
21. Try lowering the temperature in those rooms you don't spend much time in by adjusting registers, radiators or thermostats.
22. Keep humidifiers at the 30 percent mark or place pans of water on warm air registers or radiators. You'll feel more comfortable at relatively lower temperatures simply by maintaining the right humidity in your home.
23. Cover windows with drapes or curtains.
24. Open your drapes during the day to let the sunshine in and close them at night to keep the cold air out.
25. Try locating your furniture away from cold outside walls and windows.

SOME DO'S

26. Do fix leaky faucets, especially hot water taps.
27. Do use cold water for clothes washing.
28. Do turn off the lights, TV, radio or record player when not needed.

SOME DON'TS

29. Don't permanently fasten windows and doors shut—they may be needed for an emergency.
30. Don't use kitchen appliances to heat your home.
31. Don't use portable heaters as the main source of heat—be particularly cautious with oil or gas space heaters not vented or vented to your chimney.
32. Don't seal off attic ventilation.
33. Don't put insulation over recessed light fixtures.

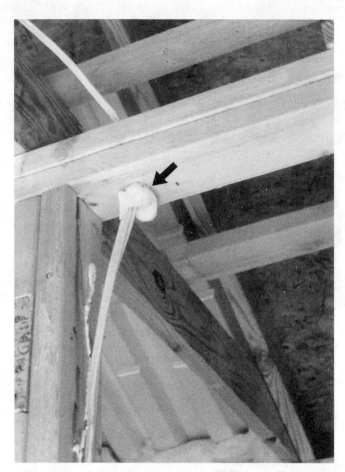

Fig. 13-62 *Expandable foam used to fill the hole in an electrical wiring installation. It can serve to insulate, seal out insects, and prevent fire movement as well.*

9. Which way should the vapor barrier face in an insulated home?

10. What are floorboards?

11. What does faced insulation mean?

12. What is the meaning of unfaced insulation?

13. Why can't you place faced insulation close to a chimney or source of heat?

14. Why should floors over unheated basements be insulated?

15. How do you cut Fiberglas insulation?

16. What is insert stapling?

17. How do you fill narrow stud spaces with insulation?

18. What is condensation?

19. How do you prevent condensation?

20. If you have a house with 1,200 square feet of ceiling space and need to ventilate it, how much attic ventilation space would be needed if you used No. 8 mesh hardware cloth to cover the vent areas?

14
CHAPTER

Interior Walls and Ceilings

ONCE THE OUTSIDE WALLS ARE FINISHED, AND after the doors and windows have been installed, it is time to concentrate on the interior walls and ceilings. This makes the building weathertight and protects the interior from weather damage. This is important because most interior materials are not weatherproof. Then the interior can be finished on bad days, when rain or snow makes work on the exterior impractical.

Usually, ceilings are finished first; then the walls. Floors are completed last. Keep in mind that workers will drop tools and materials as they work; it usually does no damage to drop tools on unfinished floors.

Ceilings and walls may be finished the same way. This is so because the same materials can be used for both walls and ceilings. Several steps are involved. It is common for carpenters to apply vapor barriers. Other items to be installed include insulation, plumbing, and electrical wiring. The carpenter usually does not do the plumbing and wiring. However, the carpenter does install the insulation.

The most common interior material is gypsum board or drywall. Gypsum board (drywall) is made from a chalklike paste of gypsum. It is spread evenly between two layers of heavy paper. When the gypsum hardens, it forms a rigid board. The paper cover adds to its strength and durability. Finish (e.g., paint, wallpaper, or other types of treatment) may be applied directly to the paper. Wood panels or other materials may be applied over the gypsum board. Gypsum board is also called by other names, such as *Sheetrock* and *drywall*. Other interior wall panels are made of wood, plaster, and hardboard. Wooden boards are also used for interior walls.

Carpenters use special skills in covering walls and ceilings. These skills are related to working the various materials. Plaster is held up by strips called *lath*. Lath can be made of several things. Wood boards, special drywall, or metal mesh may be used. Carpenters often nail the wood edge guides and lath. However, the plasterer, not the carpenter, puts up the plaster.

Skills carpenters need to prepare internal walls and ceilings include

- Measuring and cutting wall materials
- Installing insulation
- Applying gypsum board
- Installing paneling
- Prepping walls for plaster
- Putting up board panels
- Applying ceiling materials

SEQUENCE

Ceilings are put up before either walls or floors in most instances. The carpenter usually does not select the interior materials. However, the carpenter must plan the way in which the work is done. To do this, the carpenter must know the various processes. The carpenter also must use the right process to do a particular procedure. The sequence for the carpenter to use is as follows:

1. Make sure the material and the fasteners are available.
2. Plan the correct work sequence.
3. Put insulation in the walls.
4. Put in the wall material.
5. Put in trim and molding.

PUTTING INSULATION IN WALLS

All exterior walls are insulated. Insulation is commonly used in interior walls. Interior wall insulation reduces sound transmission between rooms or apartments. However, the insulation must be put in before the walls are covered.

Loose-woven fibers are usually used for insulation. The material comes in standard-width rolls or strips. These are made to fit between the wall studs. There are two commonly used standard widths. One is for walls with studs that are 16 inches on center (O.C.). The other is for walls with studs that are 24 inches O.C. The insulation is usually made from glass or mineral fibers that are woven into a thick mat. The mat then is glued to a heavy-paper backing that is a little wider than the mat. In this way, the edges can be used to nail the insulation to the studs. The insulation may come in long rolls, or it may come in precut lengths. The precut lengths are called *batts*.

The paper side of the insulation batt is located so that it is toward the living space (Fig. 14-1). The paper backing may also be coated with metal foil. The metal foil helps to reduce the energy loss (Fig. 14-2).

Before putting up the insulation, the material must first be cut to fit. The batts are cut to the proper length for the walls. The batts are then placed between the studs. The tops of the batts should firmly touch the top plate in the walls. The batts should reach fully from top to the bottom. They should reach without being stretched. Either nails or staples may be used as fasteners. The fasteners may be driven on the inside of the

Fig. 14-1 *The paper back on the insulation should face the living space. It also acts as a vapor barrier.*

Fig. 14-2 *Installing foil-backed insulation between the studs. Note that the foil is toward the living space. It is also the vapor barrier.*

studs. The edges of the paper are sometimes lapped on the edges of the studs. Fasteners are started at the top of the material. The nails or staples should be placed about 12 inches apart. Fasteners are alternated from side to side (Fig. 14-3).

Fig. 14-3 *Insulation is stapled between the studs.*

INSTALLING A MOISTURE BARRIER

Once the insulation has been installed, the installer should check the plans carefully. Vapor barriers are often put up on the inside. Plastic film from 2 to 4 mils thick (0.002 to 0.004 inches thick) may be used. It is fastened across the interior studs as in Fig. 14-4. Moisture barriers are essential in cold climates. They keep cold air and moisture from entering the building. Some sheathing can act as a vapor barrier. Also, the paper or foil back on the insulation acts as a vapor barrier. However, vapor barriers are needed on all outside walls and may also be used on interior walls. Vapor barriers may be made of plastic film, builder's felt, insulation paper, or sheathing.

PUTTING UP GYPSUM BOARD

Earlier it was mentioned that gypsum board also is called *Sheetrock* or *drywall*. It remains, probably, the most common interior wall material. There are several advantages to using drywall: It is quick and easy to apply, reducing labor costs; it is not an expensive material; and there is no need to wait for it to dry.

Drywall is available in several sizes and thicknesses. Widths of 16, 24, and 48 inches are standard. Lengths of 48 and 96 inches are used. One of the most often used sizes is the standard 4- × 8-foot sheet. This large size produces fewer seams that have to be finished.

Fig. 14-4 *Plastic film may be applied over the inside of the studs and insulation as a vapor barrier.* (Forest Products Laboratory.)

Labels in figure: EXTERIOR WALL, VAPOR BARRIER, CEILING MOLDING, BLOCKING BETWEEN STUDS, SHEATHING, WOOD PANELING, BASE, BLIND AND FACE NAIL, INSULATION

Edges of the sheets are finished or shaped in various ways. For most houses, the tapered edge is used (Fig. 14-5). The edge is tapered to make a sunken bed for the seam. In this way, the edges may be covered and hidden. The edges are first coated with plaster and tape. Later coats are added to make a flat, smooth surface (Fig. 14-6). This flat surface then can be covered, papered, or painted without seams showing through.

ROUNDED
TAPERED
BEVELED
SQUARE EDGE
TONGUE AND GROOVE

Fig. 14-5 *Standard edges for drywall.* (Gypsum Association.)

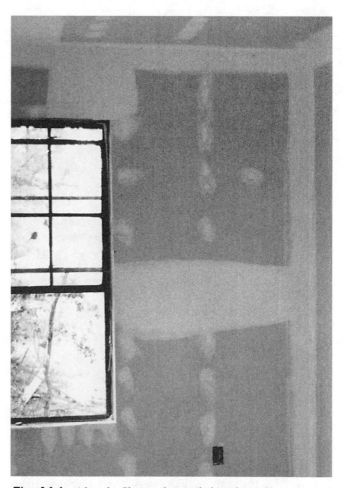

Fig. 14-6 *After the Sheetrock is nailed in place, the seams and nail dimples are covered.*

Drywall is supported by nails, screws, or adhesives. A special hole called a *dimple* is made for nails and screws. This hole is made by the hammer on the last stroke. The hole, or dimple, is filled with plaster or drywall compound. The plaster helps by covering the nail and screw to give a smooth surface. Figure 14-7 shows the dimple. On ceilings, nails or screws are also used with adhesives; however, not as many fasteners are needed. Adhesives are not used alone on ceilings.

Fig. 14-7 *A dimple, or depression, is made around the nailhead. It is later covered with plaster or drywall compound.* (Forest Products Laboratory.)

Using adhesives rather than nails speeds up the ceiling job. However, some building codes prohibit the use of adhesives without nails. In such cases, the work is still fast because fewer fasteners are used. Panels can be attached with the adhesives. Then the nails are added. No special holding is needed for nailing.

Thickness of drywall sheets can vary. Drywall is very fire-resistant. Layers can be used to increase the resistance to fire. Several layers can increase the strength. They will also reduce the transmission of noise between rooms. However, most construction is done with one thickness (single ply). There are three common thicknesses of drywall sheets. These are ⅜, ½, and ⅝ inch. In some locations, ceilings are made of ⅜-inch thickness. This reduces the weight being held. However, local building codes often detail what thickness can be used. The thicker the drywall sheet, the heavier and more difficult it is to support while being nailed or glued.

Sheets of drywall can be used as a base for other types of walls. Drywall sheets are often used for plaster lath. The sheets may be solid, or they may have holes (perforations) to aid in holding the plaster (Fig. 14-8). The thinner drywall sheets are used for backing thin finished panels. Several types of surface finishes are available for drywall.

Fig. 14-8 *Drywall sheets used for plaster lath.*

There are two ways to apply drywall sheets. These are called *horizontal* (or *parallel*) and *perpendicular* applications. This refers to the direction of the long side with respect to the studs or joists. It does not refer to the direction with respect to the floor. For example, in perpendicular applications the long side is perpendicular, or at right angles, to the stud or joist. Horizontal application means that the long side runs in the same direction as the stud or joist (Fig. 14-9).

Putting Up the Ceiling

Drywall application usually starts with attaching the thinner sheets to the ceiling. Then the walls are covered with a thicker sheet of drywall. Drywall should be put on a ceiling in a horizontal application. Most manufacturers recommend that the first piece be placed in the middle of the ceiling. Then additional sheets are nailed up. A circular pattern from the center toward the walls is used. This procedure makes sure that the joists are properly spaced. The drywall acts as a brace to hold the joists in place. Joist spacing is important. Edges and ends of the drywall sheets can be joined only on well-spaced joists. Nonsupported joints will move. Movement can ruin the taped seams. This will give a bad appearance.

Measure and locate first piece Before the first sheet is applied, two base walls are selected. One wall should be parallel to the joists (Fig. 14-10). The distance to the approximate center of the room is measured. Intervals of 4 feet are marked from this base wall. Next, the distance from the second wall to the center is found. Intervals of 4 feet are also marked off

SINGLE NAILING

NOT LESS THAN 3/8"
FROM EDGES OR ENDS

PARALLEL CEILING APPLICATION

CEILING JOISTS

FLOATING ANGLES

NAILS 7" O.C.

OMIT NAILS HERE

STUDS

NAILS 8" O.C.

OMIT NAILS HERE

GYPSUM BOARD
(PERPENDICULAR APPLICATION)

GYPSUM BOARD
(PARALLEL APPLICATION)

Fig. 14-9 *Drywall sheets can be applied either parallel or perpendicular.*
(Gypsum Association.)

FIRST
BASE
WALL

SECOND
BASE
WALL

JOISTS
16" O.C.

FOURTH
WHOLE
SHEET

WALL

4'

THIRD
WHOLE
SHEET

SECOND
WHOLE
SHEET

FIRST
SHEET

ONLY
CUTTING
ON TWO
WALLS

WALL

Fig. 14-10 *For ceilings, the first sheet is near the center. Measure so that whole sheets can be used toward the two base walls.*

because the standard sheet size is 4 × 8 feet. The first sheet is located so that whole sheets can be applied from the center to that one wall. Cutting is done only on one side of the room. This saves time in trimming and piecing.

Before the first panel is applied, measure the distances between the joists at the center. The distance should be the proper 16 or 24 inches O.C. If the joists are spaced correctly, the first panel may be applied. The center nails should be driven first. Then joists should be spaced properly by slight pressure. Edges of the panels should rest on the centers of the joists. This provides a nail base for both panels.

Strongbacks Braces across the ceiling joints are called *strongbacks*. They are made for two reasons:

- They help to space the joists properly. To space the joists, a flat board is nailed across the tops of them.
- They help to even up the bottom edges of the ceiling joists. One of the disadvantages of drywall is that studs and joists must be very even. Drywall cracks easily when it is nailed over uneven joists or studs.

Two boards are used to make a strongback. The bottom board is laid flat across the tops of the joists. Use a 2 × 4. The end of the board is nailed to the joist nearest the wall. Two 16d common nails should be used. The proper distance (16- or 24-inch marks) is measured. Pressure is applied to move the joist to the proper spacing. Then the joist is nailed at the proper spacing. This is done to each joist across the distance involved. This braces the joists into the proper spacing on centers.

The strongback is completed with the second brace (Fig. 14-11). The second piece is made from a 2- × 4- or 2- × 6-inch board turned on edge. The edge provides a brace to even the joists to the same height. One end of the second brace is nailed to the end of the

Fig. 14-11 *Strongbacks help to even up and space joists. The flat piece is nailed first.* (Forest Products Laboratory.)

first brace. The carpenter should then stand on the flat piece over each joist. The carpenter's weight is used to force the joists to an even line. When the joists are even, the piece is nailed in place.

Applying Ceiling Sheets

Drywall sheets being nailed to the ceiling must be supported. Special braces are used in this nailing process. These temporary supports or braces are called *cradles*. Figure 14-12 shows a cradle being used. Cradle braces are shaped like a large T. The leg of the cradle is about ⅛ to ¼ inch longer than the ceiling height. In this way, the cradle can be wedged in place to hold the panel. Once the sheet has been nailed, the brace can be removed and used again.

T-BRACES — 1X4S NAILED TO 2X4 UPRIGHTS THAT ARE 1/2" LONGER THAN FLOOR-TO-CEILING HEIGHT— EASE INSTALLATION.

Fig. 14-12 *A T-shaped cradle can be used to hold up pieces for nailing.* (Gold Bond.)

Usually, several workers are involved; some workers will hold materials in place, while others will nail. Scaffolds also can be used to let the carpenters stand at a good working height. Stilts, as in Fig. 14-13, are also used. However, workers on stilts should not try to lift materials from the floor. A worker on stilts cannot bend or lean over. That person can only brace, nail, or plaster seams and joints. A two-person team is usually called for in ceiling work.

Fasteners First fasten the sheet at the center of the panel. Then fasteners are placed toward the edges. Fasteners are driven at intervals on ceilings. Spacing depends on the type of Sheetrock used. Nails, screws, and staples are all used.

Special nails are used for Sheetrock. Some are smooth, and some have ridges on the shank of the nail. Special screws are also used, driven by an attachment that releases the screw when it is in place. Nails and screws for Sheetrock are shown in Fig. 14-14A. As a rule, the nail or screw should penetrate into solid wood

(B)

Screw Guide

(C)

Fig. 14-14 *B. A magnetic sleeve to guide screws. Match the drive bit and screw. C. Pull the sleeve over the screw. The screw now can be driven easily.*

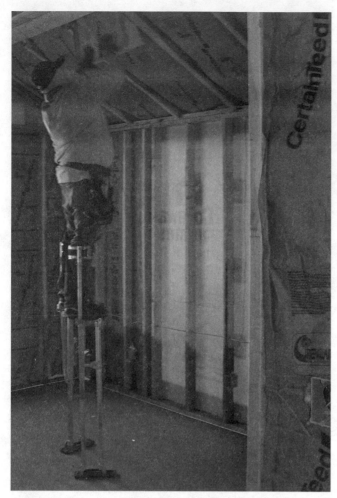

Fig. 14-13 *Stilts are often used for nailing and finishing ceilings.* (Goldblatt Tools.)

about 1 inch. Thus, for ⅜- inch drywall, a nail 1⅜ inch long would be recommended. When staples are used, the crown of the staple should be perpendicular to the joist or stud.

Edge spacing Seams of the sheets should be staggered. The edges of one junction should not align with the edges of another junction. This is part of a process

called *floating*. Figure 14-14B shows the use of a magnetic sleeve screw guide. Figure 14-15 shows how edges are floated or staggered. This allows each sheet to reinforce the next sheet. It makes a stronger, more rigid wall or ceiling. Floating reduces expansion and contraction of the walls and ceilings. With this controlled, the finished seams are less likely to crack. By eliminating the nails, the sheets are allowed to move or "float."

TYPE S

TYPE S

TYPE S-12

TYPE S-12

TYPE G

TYPE W

TYPE S

TYPE S

Fig. 14-14 *A. Typical nails and screws for Sheetrock. Each is used for a certain application.* (Gypsum Association.)

PERPENDICULAR CEILING APPLICATION

CEILING JOISTS

NAILS 7" O.C.

GYPSUM WALLBOARD

FLOATING ANGLES
(OMIT NAILS)

SINGLE NAILING

NOT LESS THAN 3/8"
FROM EDGES OR ENDS

PARALLEL CEILING APPLICATION

FLOATING ANGLES

CEILING JOISTS

NAILS 7" O.C.

OMIT NAILS HERE

STUDS

NAILS 8" O.C.

OMIT NAILS HERE

GYPSUM BOARD
(PERPENDICULAR)

GYPSUM BOARD
(PARALLEL APPLICATION)

Fig. 14-15 *Floating-angle construction helps to eliminate nail popping and corner cracking. Fasteners at the intersection of walls or ceiling are omitted. Note that seams are not matched.* (Gypsum Association.)

Cutting Gypsum Board

Gypsum board is easily cut. Accurate measurements are made before cutting. It is a good idea to measure at least twice to be sure. Most often the measurements are checked by measuring from two different locations.

Cutting straight lines Straight lines are cut after being carefully marked. A pencil line is made on the good, finished side of the sheet (Fig. 14-16). This line then is scored with a sharp knife. The cut should go through the paper and slightly into the gypsum core.

The board or sheet then is placed over a board or a rigid back. The carpenter sometimes can use a knee for

a back (Fig. 14-17). The board then is snapped or broken on the scored line. The paper on the backside then is cut with a knife. This finishes the cut (Fig. 14-18).

Cutting openings Openings should be measured and marked carefully. Then the lines are heavily scored. Next, the piece may be strongly tapped with a hammer. If a Sheetrock hammer is used, the hatchet end may be used. The cutting edge is placed into the score. The blade then is pushed through with even pressure. The backside is cut with a knife, and the section falls free.

Another method is to use a special device to cut the holes. This device resembles a box with teeth on the

Fig. 14-16 *Sheetrock is marked and scored with a knife for cutting.* (Gold Bond.)

Fig. 14-17 *After scoring, Sheetrock can be "cut" by breaking it over the knee or a piece of lumber.* (Gold Bond.)

edges. The teeth are placed in the outlet box in the wall. The approximate center is marked on the panel. The panel is put in place. The handle of the special tool is

Fig. 14-18 *After scoring and breaking, the paper back may be cut. This separates the pieces.*

forced into the marked center. Prongs are engaged into a slot in the teeth. The teeth quickly cut through the gypsum core. The tooth plate and core are simply pulled out. Thin saws, knives, and punches are also commonly used. Care should be taken with any method. Edges should be as smooth and even as possible.

Applying Wall Sheets

Wall panels usually are applied horizontally to the studs. This means that the long edge is parallel to the studs. It also means that the long edge will be vertical or perpendicular to the floor.

Note that one corner of the first sheet is not nailed. It is held in place when the second piece at that corner is applied. This technique is also used on ceilings (Fig. 14-19). This is part of the process of floating. Floating allows for expansion and contraction of the walls. It helps keep seams from cracking.

Most ceilings are slightly more than 8 feet from the subfloor. This slight distance gives some working space and clearance. With this extra distance, standard 8-foot lengths will not catch. The drywall should be butted against the ceiling. It should be raised off the floor when applied. To start, one edge of the panel is laid on the floor. The top is laid against the wall studs. The sheet of drywall then is raised off the floor. Then it is nailed in place. A kick lever, as in Fig. 14-20, is used. Kick levers may be purchased or may be made from lumber. A carpenter can step on the pedal to raise the panel. By using a kick lever, the carpenter keeps both hands free for nailing and holding (Fig. 14-21).

OMIT FASTENERS
IN UNDERLYING
BOARD ONLY

WALL FRAMING

CEILING FRAMING MEMBER

UNDERLYING BOARD

7" FOR SINGLE NAILING
11" TO 12" FOR DOUBLE NAILING
OR SCREWS

8" FOR SINGLE NAILING
11" TO 12" FOR DOUBLE NAILING
OR SCREWS

WALL FRAMING MEMBER

Fig. 14-19 *Techniques for floating corners.* (Gypsum Association.)

Fig. 14-20 *A kick lever.* (Goldblatt Tools)

CORNER STUDS

CORNER BEAD

PLASTER BASE

Fig. 14-22 *Corners are butted together. A metal bead is then applied to protect the corner.* (Forest Products Laboratory.)

Fig. 14-21 *A kick lever being used to raise a panel in place.* (Goldblatt Tools.)

The first piece is usually placed at a corner of the room. The work is done from corner to corner. The carpenter does not start in the middle of a wall. Nails, staples, or screws are driven at 8-inch intervals on the studs.

Corner treatments Outside edges and corners are reinforced with metal strips. Outside edges and corners are easily damaged and broken. The metal strips help to prevent damage to these exposed edges. The metal strips are called *beads*. The most commonly used bead is the corner bead as shown in Figs. 14-22 and 14-23. Corners are butted together. The bead is then used to cover the corner. The reinforcement bead protects the corner or edge. The bead helps to prevent ugly damage to the edge or corner.

In most instances, wood is not covered with plaster. Plaster and wood expand and contract at different rates. The difference in movement of these two materials causes cracking. The wood is covered with drywall. This reduces cracking.

Inside corners normally are not reinforced with beads. They are taped just as are other seams. The taping and beading process will be explained later.

Double-Ply Construction

Two layers of drywall are sometimes used. The second layer increases the ability of the wall or ceiling to resist fire and reduces noise transfer from above, in the attic or second floor of multistory buildings. It also helps to reduce the amount of noise transmitted from one room to another when used on adjoining rooms.

Double-ply ceilings are glued and then nailed. However, the nails are driven from 16 to 24 inches apart. The length of the nail is also increased. A longer nail is needed so that it will penetrate through the second thickness of the drywall.

CORNERBEAD (Numbers indicate width of flanges,
i.e.—118 is 1 1/8-inches wide flange)
CB—100 × 100
CB—118 × 118
CB—114 × 114
CB—100 × 114
CB—PF (Paper flange,
steel corner,
combination
bead)

"L" BEAD (Numbers indicate thickness of
board to be used)
L-38
L-12
L-58
L-34

"LK" BEAD (For use with Kerfed jamb)
LK

"U" BEAD (Numbers indicate thickness of
board to be used, i.e.—38 is 3/8 inches)
U-38
U-12
U-58
U-34

"LC" BEAD (Numbers indicate thickness of
board to be used)
LC-38
LC-12
LC-58
LC-34

Fig. 14-23 *Metal trim and casing.* (Gypsum Association.)

Joints of each layer should overlap. The joint of the top layer should occur over a solid sheet. Figure 14-24 shows this. Adhesives are applied as in Fig. 14-25. The sheets of drywall should be firmly pressed in place. It is best to nail the glued sheets immediately. If they must be left, a temporary brace should be used. A brace will hold the panels firmly until the glue dries. Nails are driven when the brace is removed.

Finishing Joints and Seams

Drywall as a base   Drywall is often used as a base. Either panels and/or plaster may be applied over it. However, in most cases, it is finished directly. It may be painted or papered. Before gypsum board is finished, the seams should be smoothly covered. Some building codes require seams to be covered even when

FINISH LAYER 3/8" or 1/2"
TAPERED EDGE GYPSUM BOARD

CEILING JOISTS

BASE LAYER 3/8" OR 1/2"
BACKER BOARD OR GYPSUM BOARD

LAMINATING ADHESIVE
(APPLY WITH NOTCHED
TROWEL OR MECHANICAL
SPREADER)

WOOD FRAMING MEMBERS

FASTENERS (SINGLE NAILS SHOWN)

BASEBOARD

JOINT TREATMENT
(ALL JOINTS AND CORNERS)

Fig. 14-24 *Double-ply application. Note that joints are not in the same place for each layer.* (Gypsum Association.)

Fig. 14-25 *Notched spreaders are used to apply adhesive for ceiling sheet lamination.*

the drywall is merely a base, such as in garages or utility buildings. The covered seams increase the fire-resistive properties.

Successive coats of plaster and special tape are used for seams. The tape may be either paper or Fiberglas mesh (Fig. 14-26). Figure 14-27 shows how the layers are applied. Covering seams are referred to as *finishing joints*. The process is often called *taping and bedding* (Fig. 14-28). The seams on all corners and edges are taped and bedded with plaster. However, when paneling or plaster is applied, it is not always necessary to tape the joints.

The carpenter usually does not finish drywall seams. This is done by workers in the trowel trades. Plasterers or Sheetrock workers do most of this unless you are building a house or storage space for yourself.

Fig. 14-26 *Seams are finished in layers.* (Gypsum Association.)

(A)

(B)

(C)

Fig. 14-27 *A. Spot nailheads with a first coat of compound either as a separate operation before joint treatment or after applying tape. B. The first coat of joint compound fills the channel formed by the tapered edges of the wallboard. C. Tape is embedded directly over the joint for the full length of the wall. Smooth joint compound around and over the tape to level the surface.*

Putting Up Gypsum Board   443

(D)

(E)

(F)

Fig. 14-27 *D. The first finishing coat is applied after the first coat has dried. Apply it thinly, and feather out 3 to 4 inches on each side of the joint. Apply a second coat to the nailheads if needed. E. The second finishing coat is applied when the last coat has dried. Spread it thinly, feathering out 6 to 7 inches on each side of the joint. Finish nail spotting may be done at this time. F. After 24 hours, smooth the finished joints with a damp sponge. If light sanding is required, use a respirator to avoid inhaling the dust.* (Gold Bond.)

Fig. 14-28 *Taping and bedding Sheetrock seams.*

Ceiling Panels

Callbacks are one of the biggest and most expensive concerns of drywallers. They are concerned about their reputation. But with the new Sheetrock-brand interior ceiling panel, there is less worry about those callbacks brought on by sagging ceilings and panel and joint cracking. The typical ceiling panel of ⅝-inch-thick Sheetrock weighs 70 pounds. The ½-inch panel weighs 50 pounds. This is almost a 30 percent reduction in weight. This means less to ship and handle. All this can be translated into savings if the ½-inch panel doesn't sag noticeably. Fiberglass-reinforced panels allow the use of ½-inch panels because they have less sag associated with them than the ⅝-inch size.

Sag-resistant panels are designed for parallel or perpendicular application to framing components spaced up to 24 inches O.C. with a maximum of 2.2 pounds per square foot insulation loading and wet texturing for ceiling application. For single-layer wood-framed ceilings, nails are spaced 7 inches O.C., and 1¼-inch type W screws are spaced 12 inches O.C. Adhesive and/or nail-on fastening improves bond strength and reduces face nailing.

Steel furring channels can be used if spaced a maximum of 24 inches O.C. Pay careful attention to framing construction and alignment. Problems will "telegraph" through the board if the framing is not true. Excessively long drying times might also result in problems with the ceiling finish, such as joint banding and staining. You should ensure proper ventilation to remove excess moisture during and after finishing. Supplemental heat or dehumidification might be required. In some instances, you may open the windows to allow the airflow to aid in drying and removing the moisture if it is not too humid outside.

When panels are used as a base for water-based spray-applied finish, the weight of overlaid insulation should not exceed 2.2 psf. Thorough ventilation should be ensured to dry a texture finish.

TUB AND SHOWER WALL PREPARATION

The walls around tubs and showers require careful preparation. This is to prevent harmful effects from water and water vapor. Walls are finished with tile, panels, and other coverings. Water-resistant gypsum board can be used as a wall base. Special water-resistive adhesives may also be used. Edges and openings around pipes and fixtures should be caulked. A waterproof, non-hardening caulking compound should be used. The caulking should be flush with the gypsum. The wall finish should be applied at least 6 inches above tubs. It should be at least 6 feet above shower bases (Fig. 14-29). Figure 14-30 shows a tub support board. It is made from 2- × 4-inch lumber and is nailed over the gypsum wallboard.

Figure 14-31 is another technique for preparing walls. In this situation, gypsum wallboard is not used.

Fig. 14-30 *Tubs are supported on the walls. Note the two layers of gypsum board and the gap used. (Gypsum Association.)*

Instead, a vapor barrier, water-resistant sheathing paper or plastic film is applied directly over the studs. A metal spacer is used around the edge of the tub, as shown in the figure. Next, metal lath is applied over the vapor barrier. The wall then is finished by applying plaster. A water-resistive plaster should be used for the tub or shower area.

Special tub and shower enclosures are also used. These may be made of metal or fiberglass (Fig. 14-32). These are framed and braced to manufacturer's recommendations. No base wall or vapor barrier is needed.

Fig. 14-29 *Water-resistive wall finishes should be applied around bathtubs and showers. (Gypsum Association.)*

Fig. 14-31 *Vapor barrier and metal lath directly over studs. (Forest Products Laboratory.)*

PANELING WALLS

Prefinished walls are often finished with standard-size panels. The panels are made of wood, fiberboard, hardboard, or gypsum board. The surfaces of these panels can be made to resemble wood or tile. Paint, stain, or varnish is not needed. This gives a fast finish with little labor. These panels can also be covered with wallpaper or plastic laminates. Paint and other materials lend an extremely wide range of appearances. Wood panels are also used widely and provide a wide range of wood grains and finishes. Most panels are prefinished. However, unfinished panels are also used. Stains may then be applied to customize the appearance of the interior.

All panels, finished or prefinished, come in standard 4- × 8-foot sizes. They can be put up with nails, screws, or glue. Special nails and screws are available, colored to match the surface finish. Plain nails are also used. Casing or finish-head nails are recommended. These should be set below the surface and filled. This filling should match the color and texture of the surface. Filling material is usually available in stick form and is easily applied over the nail heads.

Apply adhesives in patterns, as shown in Fig. 14-25. On studs where panels join, two lines of adhesives are used.

Panels are available in various materials and range in thickness from ⅛ to 1 inch. As a rule, wood paneling ¼ inch thick needs no base wall. However, the wall is more substantial if a base wall is used, especially with thinner panels. The wall is stronger and more fire-resistant. Many building codes require gypsum base

(A)

(B)

Fig. 14-32 *A. Special tub and shower units are framed and braced by carpenters. B. This shower stall is made of Fiberglas. No base wall is needed. (Cori Industries.)*

walls. Gypsum drywall is the most common base wall for panels. Some codes require it to have the joints covered. The ⅜-inch thickness is commonly used.

Edges and corners Panel edges and corners are finished in several ways. Figure 14-33 shows several methods. Divider clips are used between panels. To install the panel, a special trick is often used. The edge trim is nailed down first. Then one edge of the panel is inserted into one clip. The second edge is moved away from the wall. Pressure is applied against the edge.

DIVIDER   OUTSIDE CORNER

INSIDE CORNER   END CAP

Fig. 14-33 *Joining edges and corners of paneling.*

This causes the panel to bow slightly. The second edge of the panel then can be slipped into the second clip. When the pressure is released, the edge will spring into the clip. The panel will be held flat against the wall. The panel then can be pressed against the adhesives or nailed in place.

Spacing and nailing When panels are fastened directly to the base walls, the panels should be spaced carefully. A line may be snapped at 4-foot intervals to show where panels should join. These lines then are used for guides because they show where the adhesives are applied. They also show where the panels are applied. Chalk lines are snapped at 4-foot intervals on vertical or horizontal references.

Nail wooden paneling in place. Nails are placed approximately 4 inches O.C. around the edges. Nails should be placed at 8-inch intervals on the inside studs, but keep in mind that panels made of laminates, hardboards, and other materials may require special methods. Always read the manufacturer's suggestions; they usually describe special methods and materials.

Panels over studs First, crooked or uneven studs must be straightened. They can be planed or sawed to make a flat base. Sometimes, when panels cannot be joined over studs, special boards are added between the studs. This is often true for vertical applications of gypsum board. Fire stops and special studs may be added.

Panels over masonry Basements are good examples of panels being used over concrete walls. Basements are paneled in both new and remodeling jobs.

Special nail bases must be used. These nail bases are made from wooden boards called *furring strips*. Furring strips are fastened to the walls first. Top and bottom plates are also needed. Masonry nails or screw anchors are used. They may be added horizontally or vertically.

Vapor barriers are used to keep out the damp feeling of a basement. They should be used in most geographic locations. Apply the barriers over the masonry or over the furring strips. In either case, insulation also should be added. Place the insulation between the furring strips (Figs. 14-34 through 14-37). For concrete blocks, masonry screws are better than nails. Nails driven into blocks can crack or break the thin walls of the blocks. To use masonry screws, match the drill-bit size with the screw size. Then select the correct length. For example, if you are attaching ¾-inch lath, use flat-head screws that are at least 1 inch longer than the lath. The screw should be at least 1¾ inches long. Hold the lath in place, and drill the hole through the lath and into the concrete, as in Fig. 14-35B. It is easier if you use two electric drills for the job. The carbide-tipped drill is used for the holes in the first one. The second one holds the drive bit that matches the slot on the screw (see Fig. 14-35C). Then drive the screw through the lath and into the concrete (see Fig. 14-35D). After insulation and vapor barriers have been installed, the wall cover can be applied. Use either adhesives or nails.

Fig. 14-34 *A vapor barrier is applied over a concrete wall. This is a must for best paneling.* (Masonite.)

Board Walls

Boards can be used to finish walls. Figure 14-38 shows a boarded interior. The boards may be wood, plywood, or composition. The boards may be prefinished or unfinished. Most are shaped on the edges for joining. Tongue-and-groove joints and rabbeted joints are common.

(A)

(B)

(C)

(D)

Fig. 14-35 *A. Furring strips are attached to the concrete wall. They are aligned with the level. They form a nail base for the panels. B. A matching drive bit, masonry screw, and pilot-hole bit. Used correctly, this ³⁄₁₆-inch screw grips with 360 pounds of tension. C. Hold the lath in place. Drill the pilot hole through the wood and into the concrete. D. Drive the screw into the concrete.* (Masonite.)

Boards laid horizontally (parallel to the floor) are braced in several places by the studs. For vertical-board walls, the boards are nailed to the top and bottom plates. Sometimes more nail base is needed. Then special headers or fire stops can be used (Fig. 14-39).

Tongue-and-groove joints are strong. Also, nails can be hidden in the tongues of the boards. In this way, the nails are covered by the next board. Boards are tapped into place with a block and hammer (Fig. 14-40). Boards may be applied horizontally, vertically, or diagonally (Figs. 14-41 through 14-43). However, as in Fig. 14-39, special bracing may be needed for vertical or diagonal boards.

PLASTERED WALL PREPARATION

Carpenters do not generally apply plaster to walls. However, carpenters sometimes prepare the wall for the plaster. This includes nailing edge strips around the walls, doors, and windows. These strips are called *grounds*. Grounds may be made of wood or metal. They are used to help judge the thickness of the plaster and also guide its application.

Carpenters often apply lath that holds up the plaster. The most common lath is made from drywall. Both solid and perforated drywall is used. Figure 14-44 shows perforated drywall used for lath. Metal mesh, as in Fig. 14-45, also is used. Wooden lath is seldom used because of cost and because it takes longer to install.

Fig. 14-36 *Insulation is added between the furring strips.*

Fig. 14-38 *Boards may be used for attractive interior walls.*
(Weyerhauser.)

Fig. 14-37 *Finally, the panels are nailed in place. A level may be used to align the panels.* *(Masonite.)*

Nailing Plaster Grounds

Plaster grounds are strips of wood or metal that are nailed around the edges of a wall. Grounds are also nailed around openings, corners, and floors. The frame of the door or window is usually put up first. Grounds are nailed next to the frames. Sometimes the edges of the window or door casing serve as the grounds.

Grounds on interior doors and openings Plaster usually is applied before doors are installed. A temporary ground is nailed in place. Figure 14-46 shows grounds around an interior opening. Two methods are

TOP AND BOTTOM PLATES
ARE NAIL BASE

FIRE STOP
MAY BE
ADDED AS
EXTRA NAIL
BASE

NOTE: Nails are driven in edge of board. They will not show.

Fig. 14-39 *Nailing vertical board walls.*

used. The standard width of the plastered wall is 5¼ inches. A piece of lumber 5¼ inches wide may be used. Note that it is centered on the stud and nailed in place. Thus the outside edges of the board form the grounds. These help to guide application of the plaster. The other method is shown in Fig. 14-46B. This method uses less material. Two strips are nailed in place as shown. Here, the carpenter must be careful with the measurements.

Fig. 14-40 *Boards may be forced into place using a scrap block and hammer. The scrap blocks protect the edges.*

BLOCK

Fig. 14-41 *Vertical board walls accent height.* (California Redwood Association.)

Fig. 14-42 *Horizontal wood boards provide a natural look that goes well with modem designs.* (Potlatch.)

Fig. 14-43 *Diagonal boards can give dramatic wall effects.* (California Redwood Association.)

Fig. 14-44 *Most plaster lath for houses is made of perforated gypsum board.*

Plaster normally is applied to a certain thickness for both the lath and the plaster together. Two thicknesses are often used, either ¾ or ⅞ inch. Local building codes and practices determine which is used. If ⅜-inch drywall lath is used for a ⅞-inch wall, then ½ inch of plaster must be applied.

Plaster is applied in two or three coats:

•   The first coat is called the *scratch coat*. The scratch coat is the thickest coat. It is put on and allowed to harden slightly. It is then scratched to make the next coat hold better.

Fig. 14-45 *Wire lath for plaster.*

- The second coat may be either the *brown coat* or the *finish coat*. The finish coat is a thin coat. In most residential buildings, it is ⅛ inch thick or less. For this application, the guides become like screeds in concrete work.

- The third coat is put on as a flat, even surface.

Plaster ground strips are left in place after the plaster is applied. They become the nail base for the molding and trim. Trim is applied around ceilings, floors, and door and window openings. The grounds on interior door openings are usually removed. Then the jambs for the interior doors are built or installed.

Plastering normally is done by people in the trowel trades. Two types of finish coats may be put on by these workers. The *sand float finish* is a textured finish. This can then be painted. The *putty finish is* a smoother finish. It is commonly used in kitchens and bathrooms. In these locations, the plaster is often painted with a gloss enamel. This hard finish makes the wall more water-resistive. In addition, an insulating plaster can be used. It is made of vermiculite, perlite, or some other insulating aggregate. It may also be used for wall and ceiling finishes. Figure 14-47 shows plaster being applied with

(A)

(B)

(C)

Fig. 14-46 *Plaster grounds around openings: A. recommended for doors; B. recommended for windows; C. temporary.* (Forest Products Laboratory.)

a trowel. Figure 14-48 shows a worker applying plaster with a machine.

FINISHING MASONRY WALLS

Masonry walls may consist of brick, stone, or concrete block. They are not comfortable walls. Although common in basements, they are cold, and they sometimes

Fig. 14-47 *Plaster being applied with a trowel.* (Gold Bond.)

Fig. 14-48 *Plaster being applied by machine.* (Gypsum Association.)

CEILING TILE INSTALLATION

Ceiling tiles are installed over gypsum board, plaster, joists, or furring strips. Ceiling tile is available in a wide variety of surface textures and appearance. It is also available in a variety of sizes. A standard tile is 12 × 12 inches. However, sizes such as 24 × 24, 24 × 48, and 16 × 32 inches are also available. Appearance, texture, light reflection, and insulation should be considered. Tiles also can aid in absorbing and deadening sound. This characteristic is called the *acoustical quality*. Other factors in choosing tile include fire resistance, cost, and ease of installation.

Ceiling tile can be used for new or old ceilings. Ceiling tile may add a new decorative element, or it can improve an old appearance. Special decorative trim, as in Fig. 14-49, can be added. These artificial beams and corner supports are made of rigid foam. They greatly enhance a room's appearance. Figure 14-50 shows the overall appearance.

Fig. 14-49 *Special decorative trim can be used with ceiling tile.* (Armstrong Cork.)

Tiles over Flat Ceilings

This procedure is good for either new or old ceilings. Tiles usually are applied with a special cement. Interlocking tiles may also be used in this situation. To install the ceiling tiles:

sweat. This can give the room a cold, clammy feeling. To make masonry walls comfortable, they are finished with care. A vapor barrier and insulation are installed and covered with paneling or drywall, as explained earlier.

Fig. 14-50 *A basement area using ceiling tile and artificial beams and posts.* (Armstrong Cork.)

- First, brush the loose dirt and grime from the ceiling.

- Then locate the center of the ceiling. Snap chalk lines on the surface of the ceiling to provide guides.

- Next, check the squareness of the lines with the walls. Make sure that the tiles on opposite walls are equally spaced. If the space is not even, tiles on each wall should be equally trimmed. Tiles should not be trimmed on only one side of the ceiling. The tile on both sides must be trimmed an equal amount. This makes the center symmetric and pleasing in appearance.

- Apply cement to the back of a tile. The cement may be spread evenly or spotted. Use the chalk lines as guides. Then press each tile firmly in place. The work should progress from the center out in a spiral.

Furring Strips Used to Install Ceiling Tile

Small tiles are attached to ceiling joists. However, furring strips are also needed to provide a nail base. Figure 14-51 shows how the furring strips provide the nail base. Small strips of wood 1 × 2 or 1 × 3 inches are most often used.

These are nailed with one nail at each joist. They are perpendicular to the joists, as shown. Insulation may be added to conserve energy or deaden sound. The furring strips are spaced to the length of the tile. Each edge of the tile should rest in the center of a strip.

Staples, adhesives, or nails are used to fasten tile. The nails or staples are applied to the inside of the tongue. In this way, the next tile covers the nail so that it is not seen.

Suspended Ceilings Installation

Suspended ceilings most often use larger panels. Panels that are 24 × 24 or 24 × 48 inches are common. The panels are suspended into a metal grid. This grid is composed of special T-shaped braces. These are called *cross T's* and *runners*. The grid system is suspended on wires from the ceiling or the joists.

Suspended ceilings are often used for remodeling. They also give a pleasing appearance to new builds. These ceilings are economical, quick, and easy to install. They provide good sound-deadening qualities. They also provide access to pipes and utilities. Lighting fixtures usually are inserted into the grid system. In this way, no special brackets are used.

JOIST

1 X 2 OR 1 X 3
NAILING STRIP

WALL

ACOUSTIC OR
OTHER TYPE OF
CEILING TILE

ADD INSULATION WHEN
ADDITIONAL SOUND
RESISTANCE IS REQUIRED

Fig. 14-51 *Furring, or nailing, strips are used as a nail base for ceiling tile. Nails are driven in the edges so that they will not show.* (Forest Products Laboratory.)

A suspended ceiling is installed in five steps:

1. Nail the molding to the wall at the proper ceiling height. This supports the panels around the edge of the room.

2. Attach hanger wires to the joists. These are usually nailed in place at 4-foot intervals.

3. Fasten the main runners of the metal grid frame to the hanger wires. Check to make sure that the main runners are suspended at the proper distance. Or the main runners may be nailed directly to the joists (Figs. 14-52 and 14-53).

4. Snap the cross-T into place between the runners. The cross-T and runners form a grid (Fig. 14-54).

5. Lay the ceiling panels into the grid (Fig. 14-55).

Concealed Suspended Ceilings

In concealed suspended ceilings, a special technique is used to hide the metal grid. Panels with a grooved edge are used (Figs. 14-56 and 14-57). The metal then holds the panels from the groove. In this system, panels are installed in much the same manner as regular suspended ceilings, but the metal grids fit into the grooves. Once the tiles are in place, no grid system is visible.

Environmental Concerns

• Lead-based paints

• Formaldehyde-based glue in cabinets and furniture

Fig. 14-52 *Main runners can be suspended from the joists.* (Armstrong Cork.)

Fig. 14-53 *Main runners also can be nailed directly to joists.* (Armstrong Cork.)

Fig. 14-54 *Cross-T pieces then are attached to the runner to form a grid.* (Armstrong Cork.)

Fig. 14-55 *Last, the panels are laid in place. The pattern of the tile helps to disguise the grid.* (Armstrong Cork.)

- Condensation
- Plumbing leaks—mold
- Radon

CHAPTER 14 STUDY QUESTIONS

1. What materials are used for interior walls?

Fig. 14-56 *Panels with grooved edges may be used for a system where no grid shows.* (Armstrong Cork.)

Fig. 14-57 *No metal grid shows when grooved-edge tile is used.* (Armstrong Cork.)

2. What are the advantages of gypsum wallboard?
3. Why is gypsum wallboard used under other panels?
4. How are vapor barriers installed?
5. What is the sequence for covering a wall?
6. What are the three methods of applying wall panels?
7. Why should a carpenter know how plaster is applied?
8. What are some other names for gypsum board?
9. Why are strongbacks used?
10. Why are ceiling panels started near the center of the room?
11. What is floating?
12. How is gypsum board cut?
13. How are ceiling panels held for nailing?
14. How are wall panels held for nailing?
15. What are the nailing intervals for walls and ceilings?
16. What is done for bath or kitchen walls?

17. What is done to edges and corners?

18. What ways are used to hang ceiling tile?

19. How are boards for walls positioned in place?

20. Why should ceiling tiles be started in the exact center?

15
CHAPTER

Interior Finishing

THE LAST PART OF A HOUSE BUILDING PROJECT IS THE interior. The frame has been built and covered. Exterior walls and roof are complete. Exterior doors and windows are done. The interior walls and ceilings also have been built, but they have not been finished. Also, windows and doors on interior sections have not been installed.

Finishing the building interior consists of several steps:

• Interior doors are installed

• Moldings and trim are applied around the insides of the windows and around the interior doors

• Cabinets are built or installed

• Cabinets are needed in bathrooms and kitchens

• Other special cabinets or shelves (called built-ins) are constructed

• Plumbing installation is finished

• Woodwork is painted or stained

• Walls and ceilings are finished

• Wiring is connected to the electrical switches, fixtures, and outlets and to built-ins

• Appliances such as dishwashers, ovens, lights, and other things are installed and/or connected

• A final floor layer is finished to complete the interior

Skills needed to finish the interior include being able to

• Install cabinets in kitchens and baths

• Build shelves and cabinets

• Apply interior trim and molding

• Apply finishes

• Paint or paper walls

• Install floor materials

SEQUENCE

Generally speaking, the sequence can be varied. However, a few factors must be considered in planning the sequence. The first factor is the type of floor involved. Some buildings are constructed on a slab. Here, the interior may be completed before the finish flooring is applied. However, with a wooden frame, two layers of flooring are laid. Then the second layer of flooring may be applied before the interior is finished. The second layer, however, is not finished until later. For a frame building on a wooden (frame) floor, the general sequence should be

1. Install cabinets in kitchen and baths.
2. Install interior doors.
3. Trim out interior doors and windows.
4. Paint or stain wood trim and cabinets.
5. Finish walls with paint, texture paint, or paper.
6. Install electrical appliances.
7. Lay the finish floor.
8. Finish floors by sanding, staining, varnishing, or laying linoleum or carpet.

INTERIOR DOORS AND WINDOW FRAMES

Interior door units are installed to complete the separation of rooms and areas. Then the insides of the windows must be cased or trimmed. Trim is applied to all interior door units. The trim is needed to cover framing members. It also seals these areas from drafts and airflow.

Standard Sizes of Doors

Doors can be obtained from the mills in stock sizes. They are less expensive than when they are made by hand. Stock sizes of doors cover a wide range, but those used most commonly are

• 2 feet, 4 inches × 6 feet, 8 inches

• 2 feet, 8 inches × 6 feet, 8 inches

• 3 feet, 0 inches × 6 feet, 8 inches

• 3 feet, 0 inches × 6 feet, 8 inches

• 3 feet, 0 inches × 7 feet, 0 inches

These sizes are either 1⅜ (interior) or 1¼ inches (exterior) thick.

Interior Doors

Openings for interior doors are larger than the door. They are higher and wider than the actual door width. Frames are usually 3 inches higher than the door height. They are usually 2½ inches wider than the door width. This provides space for the members of the door. Interior door frames are made up of two side pieces and a headpiece. These cover the rough opening frame. The pieces that cover these areas are called *jambs*. They consist of side jambs and head jambs. Other strips are installed on the jamb. These are called *stops*. The stops, also called *door stops*, form a seat for the door. It can latch securely against the stops. Most jambs are made in one piece, as in Fig. 15-1A. However, two- and three-piece adjustable jambs are also

Fig. 15-2 *Door jambs are shimmed in place with wedges, usually made from shingles.* *(Forest Products Laboratory.)*

Fig. 15-1 *Types of interior door jambs.* *(Forest Products Laboratory.)*

made. Adjustable jambs are used because they can be adapted to a variety of wall thicknesses (see Fig. 15-1B). Interior door frames can be purchased with the door prehung and ready for installing. Sometimes, however, the carpenter must cut off the door at the bottom. This is done to give the proper clearance over the finished floor material.

Door frames The door unit is assembled first. Then it is adjusted vertically and made square. This is called being *plumb*. The adjustment is made with small wedges, usually made from shingles, as shown in Fig. 15-2. The jamb is plumbed vertically and made square. Then nails are driven through the wedges into the studs as shown. Any necessary hinging and adjusting of the door height are done at this point. For instructions on installing door hinges, see Chapter 9.

Trim is attached after the door frame is in place. The door is first hung properly. Then trim, called *casing*, may be applied. Casing is the trim around the edge of the door opening. Casing is also applied over interior door and window frames. Casing is nailed to the jamb on one side. On the other side it is nailed to the plaster ground or the framing stud. It should be installed to run from the bottom of the finished floor. Note in Fig. 15-3 that two procedures may be used for the top casing. The top casing piece can be elaborately shaped. Other pieces of molding are used to enhance

the line. This is often done to fit historical styles. A slight edge of the jamb is exposed, as shown in Fig. 15-3C.

Next, the door stop may be nailed in place. The door is closed. The stop should be butted close to the hinged door. However, a clearance of perhaps ¹⁄₁₆ inch should be allowed. The door stop is usually 1½ or 3 inches wide. It is wide enough for two nails to be used. The finishing nails may be driven as shown. When a stop is spliced, a beveled cut in either direction is made (Fig. 15-4). Casing may be shaped in several ways, as shown in Fig. 15-5.

Door details Two interior door styles are the flush and the panel door. Other types of doors are also commonly used. They include folding doors and louvered doors. These are known as *novelty doors*.

Most standard interior doors are 1⅜ inches thick. They are used in common widths. Doors for bedrooms and other living areas are 2 feet, 6 inches wide. Bathrooms usually have doors with a minimum width of 2 feet, 4 inches. Small closets and linen closets have doors with a minimum width of 2 feet. Novelty doors come in varied widths for special closets and wardrobes. They may be 6 feet or more in width. In most cases, regardless of the door style, the jamb, stop, and casing are finished in the same manner.

The standard height for interior and exterior doors is 6 feet, 8 inches. However, for upper stories, 6 feet, 6 inch doors sometimes are used.

Fig. 15-3 *Two types of joints for door casing. Note the decoration in part C. This does not change the joint used.*

Fig. 15-4 *Miter joints should be used to splice stop and casing pieces where necessary.*

COLONIAL   RANCH

Fig. 15-5 *Two common casing shapes: A. colonial; B. ranch.* (Forest Products Laboratory.)

The flush interior door is usually a hollow-core door. This means that a framework of some type is covered with a thin outside layer. The inside is hollow but may be braced and stiffened with cores. These cores are usually made of cardboard or some similar material. They are laid in a zigzag or circular pattern. This gives strength and stiffness but little weight. Cover layers for hollow-core doors are most often hardwood veneers. The most commonly used wood veneers are birch, ma-

hogany, gum, and oak. Other woods are used occasionally as well. Doors mayalso be faced with hardboard. The hardboard may be natural, or it may be finished in a variety of patterns. These patterns include a variety of printed wood-grain patterns.

Door hinges Doors should be hinged to provide the easiest and most natural use. Doors should open or swing in the direction of natural entry. If possible, the door should open and rest against a blank wall. It should not obstruct or cover furniture or cabinets. It should not bar access to or from other doorways. Doors never should be hinged to swing into hallways. Doors should not strike light fixtures or other fixtures as they are opened.

Door hardware Hardware for doors includes hinges, handles or locksets, and strike plates. Hinges are sold separately. Door sets include locks, handles, and strike plates. They are available in a variety of shapes and classes. Special locks are used for exterior doors. Bathroom door sets have inside locks with safety slots. The safety slot allows the door to be opened from the outside in an emergency. Bedroom locks and passage locks are other classes. Bedroom locks are sometimes keyed. Often, they have a system similar to a bathroom lock. They are also available without keyed or emergency opening access. Passage "locks" cannot be locked.

To install door sets, two or more holes must be drilled. Most locksets today feature a bored set. This is perhaps the simplest to install. A large hole is drilled first. It allows passage of the handle assembly. The hole is bored in the face of the door at the proper spacing. The second hole is bored into the edge of the door. The rectangular area for the face plate is routed or chiseled to size. Variations occur from manufacturer to manufacturer. The carpenter should refer to specifications with the locks before drilling holes. The door handle should be 36 to 38 inches from the floor. Other dimensions and clearances are shown in Fig. 15-6. A machine used to cut mortises and bores for locks is shown in Fig. 15-7.

Fig. 15-6 *Interior door dimensions. (Forest Products Laboratory.)*

Fig. 15-7 *A machine used for cutting mortises and bores for locks. (Rockwell International Power Tool Division.)*

A second type of lockset is called a *mortise lock*. Its installation is shown in Fig. 15-8. For a mortise lock, two holes must be drilled in the face of the door. One is for the spindle. The other is for the key. Also, the area inlet into the edge of the door is larger. This requires more work. As shown in the figure, setting the face plate also requires more work than the other style.

Hinge installation Interior doors need two hinges, but exterior doors need three. The third hinge offsets

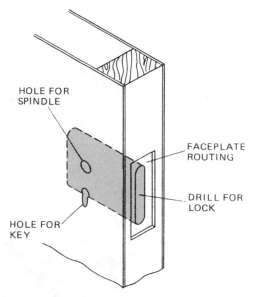

Fig. 15-8 *Mortise lock preparation. (Forest Products Laboratory.)*

the tendency of doors to warp. They tend to do this because the weather on the outside is very different from the "weather" on the inside.

Special door hinges should be used in all cases. Loose-pin *butt* hinges should be used. For 1⅜-inch-thick exterior doors, 4- × 4-inch butt hinges should be used. For 1¾-inch-thick interior doors, 3½- × 3½-inch butt hinges may be used.

First, the door is fitted to the framed opening. The proper clearances shown in Fig. 15-6 are used. Hinge halves are then laid on the edge of the door. A pencil is used to mark the location. The portion of the edge of the door is inlet or removed. This is called *cutting a gain*. A chisel is often used. For many doors, though, gains are cut using a router with a special attachment.

The hinge should be placed square on the door. Both top and bottom hinges should be placed on the door. The door then is blocked in place. The hinge location on the jamb is marked. The door is removed, and the hinges are taken apart. The hinge halves are screwed in place. Next, the gain for the remaining hinge half is cut on the door jamb. The door then is positioned in the opening. The hinge pins are inserted into the hinge guides. The door can then be opened and swung. If the hinges are installed properly, the door will swing freely. Also, the door will close and touch the door stops gently. If the door tends to bind, it should be removed and planed or trimmed for proper fit. Also note that a slight bevel is recommended on the lock side of the door. This is shown in Fig. 15-9.

Strike plate installation The strike plate holds the door in place. It makes contact with the latch and holds

(A)

(B)

Fig. 15-9 *A. Door details.* (Forest Products Laboratory.) *B. Using templates and fixtures to cut hinge gains.* (Rockwell International Power Tools Division.)

it securely. This holds the door in a closed position (Fig. 15-10). To install the strike plate, close the door with the lock and latch in place. Mark the location of the latch on the door jamb. This locates the proper position of the strike plate. Outline the portion to be removed with a pencil. Use a router or a chisel to remove the portion required. Note that two depths are required for the strike plate area. The deeper portion receives the lock and latch. It may be drilled the same diameter as that used for the lockset.

Door stops Stops are first nailed in place with nail heads out. In this way, their positions can be changed after the door is hung. Once the door has been hung, the stops are positioned. Then they are securely nailed in place. Use 6d finish or casing nails. The stops at the lock side should be nailed first. The stop is pressed gently against the face of the door. The pressure is ap-

Fig. 15-10 *Strike plate details.*

plied to the entire height. Space the nails 16 to 18 inches apart in pairs, as in Fig. 15-10.

Next, nail the stop behind the hinge side. However, a very slight gap should be left. This gap should be approximately the thickness of a piece of cardboard (½₂ inch). A matchbook cover makes a good gauge. After the lock and the hinge door stops have been nailed in place properly, the door is checked. When the clearances are good, the head jamb stop is nailed.

Doors

Paneled doors   Paneled (or sash) doors are made in a variety of panel arrangements—horizontal, vertical, or a combination of both. A sash door has for its component parts a top rail, a bottom rail, and two stiles, which form the sides of the door. Doors of the horizontal type have intermediate rails forming panels. Panels of the vertical type have horizontal rails and vertical stiles forming the panels.

The rails and stiles of a door are generally mortised and tenon, the mortise being cut in the side stiles as shown in Fig. 15-11. Top and bottom rails on paneled doors differ in width, the bottom rail being wider. Intermediate rails are usually the same width as the top rail. Paneling material usually is plywood that is set in grooves or dadoes in the stiles and rails, with the molding attached on most doors as a finish.

Fig. 15-11   *Door construction showing mortise joints.*

Flush doors   Flush doors are usually perfectly flat on both sides. Solid planks are rarely used for flush doors. Flush doors are made up with solid or hollow cores, with two or more plies of veneer glued to the cores.

Solid-core doors   Solid-core doors are made of short pieces of wood glued together with the ends staggered very much as in brick laying. One or two plies of veneer are glued to the core. The first section (about ⅛ inch) is applied at right angles to the direction of the core, and the other section (⅛ inch or less) is glued

Fig. 15-12   *Construction of a laminated or veneered door.*

with the grain vertical. A ¾-inch strip (the thickness of the door) is glued to the edges of the door on all four sides. Figure 15-12 shows how this type of door is constructed.

Hollow-core doors   Hollow-core doors have wooden grids or other honeycomb material for the base, with solid wood edging strips on all four sides. The face of this type door is usually three-ply veneer instead of two single plies. This hollow-core door has a solid block on both sides for installing doorknobs and to permit the mortising of locks. The honeycomb-core door is for interior use only.

Louver doors   This type of door can have fixed or adjustable louvers. They may be used as an interior door, room divider, or closet door. The louver door comes in many styles, as shown in Fig. 15-13. An exterior louver door may be used (which is called a *jalousie door*). This door has adjustable louvers that are usually made of wood or glass. Since there is little protection against winter winds, a solid storm window is made to fit over the louvers to give added protection.

WINDOW TRIM

There are two common ways to finish window interiors. The window interior should be finished to seal out drafts and air currents. Trim is applied to block off the openings. The trim gives the window a better appearance because it is the trim that completely covers the rough opening.

Fig. 15-13 *Various styles of louvered doors.*

Finishing Wooden-Frame Windows

The traditional frame building has wooden double-hung windows. The frame of the window consists of a jamb around the sides and top. The bottom piece is a sill. The bottom sill is sloped. Water on the outside will drain to the outside. The frame covers the inside of the rough opening. However, there is still a gap between the window frame and the interior wall. This must be covered to seal off the window from air currents. Two ways are commonly used to trim, or case, around a window.

Making a window stool   A separate ledge can be made at the bottom on the inside (Fig. 15-14). The bottom piece extends like a ledge into the room. This bottom piece is called a *stool*. However, a finishing piece is applied directly beneath it. This is called the *apron* (see Fig. 15-14). The sides and top of the window are finished with casing.

Casing a window   The second method of finishing windows also uses trim. However, trim or casing is applied completely around the opening. Shown in Fig. 15-15, this method requires fewer pieces and takes less time to install.

Sometimes the carpenter also nails the stops in place around the window jamb. The stop in the window provides a guide for moving the window. It also forms a seal against air currents. To install the window stop, the window is fully closed. Next, the bottom stop is placed against the window sash. If the window is unpainted, $\frac{1}{32}$ inch (about the thickness of a stiff piece of cardboard) is left between the stop and the sash. The

Fig. 15-14   *Stool and apron window trim.* (Forest Products Laboratory.)

stop is nailed in place with paired nails. This was also done for doors. Next, the side stops are placed on the side jamb. The same amount of clearance is allowed. The stops on each side are nailed in place. The nail heads are left protruding. Then the window is opened and shut. If the window slides evenly and smoothly, the stops then are nailed down in place. If the window binds, it means that not enough clearance was provided. If the window wobbles, too much clearance was provided. In such cases, pull the nails and reposition the stops.

Fig. 15-15 *Finishing a window with casing. (Forest Products Laboratory.)*

Finishing Metal Window Frames

More metal-framed windows are being used to aid in reducing the expense associated with replacement of the rotten wood trim and repainting (Fig. 15-16). The metal frame is nailed to the frame members of the rough opening. This is done after the wall has been erected. Later, the window is sealed with insulation, caulk, or plastic foam. The window is finished in two

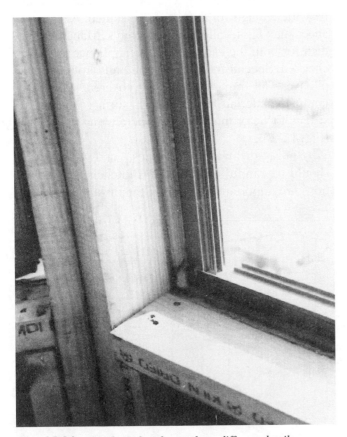

Fig. 15-16 *Metal window frames have different details.*

Fig. 15-17 *For metal windows, the jamb is butted next to the window unit. The jamb then is nailed to the rough opening frame, and the window is cased.*

steps. The inside of the rough opening is framed with plain boards. These boards are trimmed flush with the finish wall. Then casing is applied to cover the gap between the wall and the window framing (Fig. 15-17).

For a different appearance, the window may be cased with a stool and apron (Fig. 15-18). A flush-width board is used on the sides and top. However, for the bottom, a wider board is used. The ends are notched to allow the side projections, as in the figure. For a more finished appearance, the edges can be trimmed with a router. The apron then is added to finish out the window.

CABINETS AND MILLWORK

Millwork is a term used for materials made at a special factory or mill. It includes both single pieces of trim and big assemblies. Interior trim, doors, kitchen cabinets, fireplace mantels, china cabinets, and other units are all millwork. Most of these items are sent to the building ready to install. So the carpenter must know how to install the units.

However, not all cabinets and trim items are made at a mill. The carpenter must know how to both construct and install special units. This is called *custom* work. Custom units usually are made with a combination of standard-dimension lumber and molding or trim pieces. Also, many items that are considered millwork do not require a highly finished appearance. For example, shelves in closets are considered millwork, but they generally do not require a high degree of fin-

UNIT
SIZE

ROUGH OPENING =
UNIT SIZE + 1/2"

STOOL

APRON

Fig. 15-18 *Stool and apron detail for an aluminum window.*
(ALCOA.)

ish. On the other hand, cabinets in kitchens or bathrooms do require better work.

Various types of wood are used for making trim and millwork. If the millwork is to be painted, pine or other soft woods are used most often. However, if a natural wood finish is applied, hardwood species generally are preferred. The most common woods include birch, mahogany, and ash. Other woods such as poplar and boxwood are also used. These woods need little filling and can be stained for a variety of finishes.

Installing Ready-Built Cabinets

Ready-built cabinets are used most in kitchens and bathrooms. They may be made of metal, wood, or wood products. The carpenter should remember that overhead cabinets may be used to hold heavy dishes and appliances. Therefore, they must be solidly attached. Counters and lower cabinets must be strong enough to support heavy weights. However, they need not be fastened to the wall as rigidly as upper units. Ready-builts are obtained in widths from 12 to 48 inches. The increments are 3 inches. Thus cabinets of 12, 15, 18, and 21 inches and so forth are standard. They may be obtained easily. Figure 15-19 shows some typical kitchen cabinet dimensions.

Ordering cabinets takes careful planning. There are many factors for the builder or carpenter to consider. The finish, the size and shape of the kitchen, and the dimensions of built-in appliances must be considered. For example, special dishwashing units and ovens are often built in. The cabinets ordered should be wide enough for these to be installed. Also, appliances may be installed in many ways. Manufacturer's data for both the cabinet and the appliance should be checked. Special framing may be built around the appliance. The frame then can be covered with plastic laminate materials. This provides surfaces that are resistant to heat, moisture, and scratching or marring (Fig. 15-20).

Sinks, dishwashers, ranges, and refrigerators should be carefully located in a kitchen. These locations are important in planning the installation of cab-

Fig. 15-19 *Typical cabinet dimensions.*

(A)

(B)

Fig. 15-20 *A. Appliance is built in to extend. B. Appliance is built in flush.*

inets. Plumbing and electrical connections also must be considered. In addition, natural and artificial lighting can be combined in these areas.

Cabinets and wall units should have the same standard height and depth. It would be poor planning to have wall cabinets with different widths in the same area.

Five basic layouts are used commonly in the design of a kitchen. These include the sidewall type, as shown in Fig. 15-21A. This type is recommended for small kitchens. All elements are located along one wall.

The next type is the parallel or Pullman kitchen (see Fig. 15-21B). This is used for narrow kitchens and can be quite efficient. Arrangement of sink, refrigerator, and range is critical for efficiency. This type of kitchen is not recommended for large homes or families. Movement in the kitchen is restricted, but it is efficient.

The third type is the L shape (see Fig. 15-21C). Usually, the sink and the range are on the short leg. The refrigerator is located on the other. This type of arrangement allows for an eating space on the open end.

The U arrangement usually has a sink at the bottom of the U. The range and refrigerator should be located on opposite sides for best efficiency (see Fig. 15-21D).

The final type is the island kitchen. This type is becoming more and more popular. It promotes better utilization and has a better appearance. This arrangement makes a wide kitchen more efficient. The island is the central work area. From it, the appliances and other work areas are within easy reach (see Fig. 15-21E).

Screws should be used to hang cabinet units. The screws should be at least 3-inch No. 10 screws. The screws reach through the hanging strips of the cabinet. They should penetrate into the studs of the wall frame. Toggle bolts could be used when studs are inaccessible. However, the walls must be made of rigid materials rather than plasterboard.

To install wall units, one corner of the cabinet is fastened. The mounting screw is driven firm; the other end then is plumbed level. While someone holds the cabinet, a screw is driven through at the second end. Next, screws are driven at each stud interval.

Fig. 15-21 *Basic kitchen layouts: A. sidewall; B. parallel* (Pullman); *C. L shape; D. U shape; E. island.* (Forest Products Laboratory.)

When installing counter units, care must be taken in leveling. Floors and walls are not often exactly square or plumb. Therefore, care must be taken to install the unit plumb and level. What happens when a unit is not installed plumb and level? The doors will not open properly, and the drawers will stick and bind.

In order to have leveled cabinets, shims and blocking should be used where necessary. Shingles or planed blocks may be used as shims and are inserted beneath the cabinet bases.

First, install the base unit and the wall unit. Then place the countertop in its location. Countertops may be supplied as prefabricated units. These include dashboard and laminated tops. After the countertops are applied, they should be protected. Cardboard or packing should be put over the top. It can be taped in place with masking tape and removed after the building is completed. Sometimes the cabinets are hung but not finished. The countertop is installed after the cabinets are properly finished.

Ready-built units may be purchased in three ways:

- They may be purchased assembled and prefinished. Counters are usually not attached. The carpenter must install the counters and protect the finished surfaces.

- Cabinets may be purchased assembled but unfinished. These are sometimes called *in-the-white* and are very common. They allow all the wood-

work and interior to be finished in the same style. Such cabinets may be made of a variety of woods. Birch and ash are among the more common hardwoods used. The top of the counter is usually provided but not attached. Also, the plastic laminate for the countertop is not provided. A contractor or carpenter must purchase and apply this separately.

- The cabinets are purchased unassembled. The parts are precut and sanded. However, the unit is shipped in pieces. These are put together and finished on the job.

Cabinets are sometimes a combination of special and ready-built units. Many builders use special crews that do only this type of finish work. The combination of counter types gives a specially built look at the least cost.

Kitchen Planning

In most homes, the kitchen is the focal point for family life (Fig. 15-22). Some sources estimate that family members spend up to 50 percent of their time in the kitchen. For this reason, this is generally the most used and most remodeled room of a home. Remodeling calls for a carpenter to be able to rearrange the kitchen according to the desires of the owners.

The kitchen is a physically complex area in that it sustains heavy traffic flow from people using it and

Fig. 15-22 *Efficient arrangement of a kitchen cooking area.* (Jenn-Air.)

passing through it. It contains hot and cold water sources, drains, plumbing, and electrical outlets and fixtures and is exposed to high moisture, splashing water and other liquids, items that are intensely hot, and items that are very cold. In addition, activities involve sharp instruments, and considerable forces may be exerted in blending, rolling, pressing, cutting, and shaping. Other activities include mixing, washing, transferring, and storing.

Four major functions must be considered in kitchen planning: cooking, storing, eating, and entertaining. While most people consider cleaning up as part of cooking, it should be considered separately. The kitchen also may be the base of operations for several other functions done by one or more family members—studying, using a computer, home-office work, sewing, and laundry. A larger kitchen sometimes may include home entertainment equipment or be used as an area for children.

The work area in a modern kitchen must be considered in laying out the cabinets, refrigerator, range, and sink, as well as the dishwasher. The work triangle is very important because this is where most of the food preparation is done (Fig. 15-23).

Cabinets may be finished in a variety of ways. They should be coordinated with the kitchen fixtures. Wooden cabinets with white plastic laminate covers can be made or purchased. These lend a European style that contrasts with the white and warm wood

Fig. 15-23 *Location of the work triangle for three different kitchen arrangements.*

Fig. 15-24 *Plywood base installed for the application of countertop tile.* (American Olean Tile.)

tones. Figure 15-24 shows the rough layout before the tile is placed on the countertop. Figure 15-25 shows how the finished countertop looks. As you can see from these illustrations, the cabinets can be made on the job or purchased and the finishing touches made by the carpenter on the job.

Countertops Countertops may be built as part of the cabinet and then covered with almost any counter material, or they may be purchased as slabs cut to the right length. Openings in either type may be cut out. These slabs may have a splashboard molded in as part of the slab, or the splashboard may be added later. Some building codes require all sinks and basins to have splashboards; others do not (Fig. 15-26).

Kitchen light Kitchens need lots of light to be most useful and enjoyable. It is a good idea to have a good

Fig. 15-25 *Finished countertop.* (American Olean Tile.)

Fig. 15-26 *Note the splashboards on these countertops.*

Fig. 15-27 Ceramic tile is used on the floors and the walls as well as the countertop to provide a durable kitchen with plenty of light. (American Olean Tile.)

general light source combined with several additional lights for specific areas. Areas in which lights are necessary include the sink, cooktop, under cabinets, and where recipes or references are kept. Figure 15-27 is a good example of a kitchen with a skylight and plenty of cooking and work areas.

Kitchen safety Unfortunately, kitchens are the scene of many accidents. Sharp objects, wet floors, boiling pots, and open flames are all potential hazards.

Cooktops never should be installed under windows because most people hang curtains, shades, or blinds in them. These can catch fire from the flames or heating elements. In addition, people also lean over to look out a window or to open or shut it. This puts them directly over the cooktop, where they could be burned.

Cooktops should have 12 to 15 inches of counter around all sides. This keeps the handles of the pots from extending beyond the edges, where they can be bumped by passing adults or grabbed by curious children. Either situation can cause bad spills and nasty burns.

Using revolving shelves or tiered and layered swingout shelves can reduce falls caused by trying to reach the back of top shelves. Heavy objects should be stored at or near the level at which they are used. Above all, they should not be stored high overhead or above a work area.

Making Custom Cabinets

Custom cabinets are special cabinets that are made on the job. Many carpenters refer to these and any type of millwork as *built-ins* (Fig. 15-28). These jobs include building cabinets, shelves, bookcases, china closets, special counters, and other items. A general sequence can be used for building cabinets. The base is con-

Fig. 15-28 Typical built-ins.

structed first. Then a frame is made. Drawers are built and fitted next. Finally, the top is built and laminated.

Pattern layout Before beginning a cabinet, a layout of the cabinet should be made. This may be done on plywood or cardboard. However, the layout should be done to full size, if possible. The layout should show sizes and construction methods involved. Figure 15-29 shows a typical layout.

There are at least two ways to build custom cabinets:

- The first involves cutting the parts and assembling them *in place*. Each piece is cut and attached to the next piece. When the last piece is done, the cabinet is in place. The cabinet is not moved or positioned.

- The second procedure calls for all parts to be cut first. The cabinet unit is assembled in a convenient

TOP VIEW

LAP JOINT

END PANEL (1/4" PLY)

CORNER MOLD

ALTERNATIVE TOP VIEW

1/4" PLY END PANEL

LAP JOINT

ALTERNATIVE TOP VIEW

3/4" PLY END PANEL

LAP JOINT

RAIL   STILE

ALTERNATIVE JOINTS

RAIL   STILE

DOUBLE DOWEL

GAIN IS COVERED BY DRAWER LIP

RAIL

STILE

ADD FACING STRIP BETWEEN STILES

DRAWER

RAIL

DRAWER GUIDE SUPPORT

DOOR

SHELF

STILE

BOTTOM

TOP

SPLASH BOARD

WALL

Fig. 15-29 *A cabinet layout. Dimensions and joints also would be added.*

place. Then it is moved into position. The cabinet is leveled and plumbed and then attached.

Several steps are common to cabinet making. A bottom frame is covered with end and bottom panels. Then partitions are built, and the back top strip is added. Facing strips are added to brace the front. Drawer guides and drawers are next. Shelves and doors complete the base.

Making the base The base for the cabinets is made first. Either 2- × 4- or 1- × 4-inch boards may be used for this. No special joints are used (Fig. 15-30). Then the end panel is cut and nailed to the base. The toeboard, or front, of the base covers the ends. The end panel covers the end of the toeboard. A temporary brace is nailed across the tops of the end panels. This braces the end panels at the correct spacing and angles.

Cut the bottom panels next. The bottom panels serve as a floor over the base. The partition panels are placed next. They should be notched on the back top to allow them to be positioned with the back top strip.

Nail the back top strip between the end panels as shown. Then the temporary brace may be removed.

The partition panels are placed into position and nailed to the top strip. For a cabinet built in position, the partitions are toenailed to the back strip. The process is different if the cabinet is not built in place. The nails are driven from the back into the edges of the partitions. The temporary braces may be removed.

Cutting facing strips Facing strips provide the unit with a finished appearance. They cover the edges of the panels, which are usually made of plywood. This gives a better and more pleasing appearance. The facing strips also brace and support the cabinet. And the facing strips support the drawers and doors. Because they are supports, special notches and grooves are used to join them.

The vertical facing strip is called a *stile*. The horizontal piece is called the *rail*. They are notched and joined as shown in Fig. 15-31. Note that two types of rail joints are used. The flat or horizontal type uses a notched joint. The vertical rail type uses a notched lap joint in both the rail and the stile.

As a rule, stiles are nailed to the end and partition panels first. The rails then are inserted from the rear.

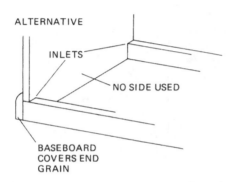

Fig. 15-30 *The base is built first. Then end panels are attached to it.*

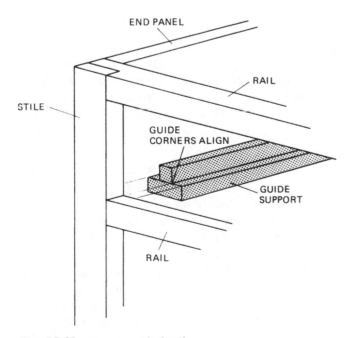

Fig. 15-31 *Drawer guide detail.*

Glue may be used, but nails are preferred. Nail holes are drilled for nails near board ends. The hole is drilled slightly smaller than the nail diameter. Finishing nails should be driven in place. Then the head is set below the surface. The hole is later filled.

Drawer guides   Once the stiles and rails are made and installed, the drawer assemblies are made. The first element of the drawer assembly is the drawer guide. This is the portion into which the drawer is inserted. The drawer guides act as support for the drawers. They also guide the drawers as they move in and out. The drawer guide is usually made from two pieces of wood. It provides a groove for the side of the drawer. The side of the guide is made from the top piece of wood. It prevents the drawer from slipping sideways.

A strip of wood is nailed to the wall at the back of the cabinet. It becomes the back support for the drawer guide. The drawer guide then is made by cutting a bottom strip. This fits between the rail and the wall. The top strip of the guide is added next (see Fig. 15-31). It becomes the support for the drawer guide at the front of the cabinet. Glue and nails are used to assemble the unit. The glue is applied in a weaving strip. Finish nails then are driven on alternate sides about 6 inches apart to complete the assembly. Each drawer should have two guides. One is on each side of the drawer.

There are three common types of drawer guides that are made by carpenters:

- Side guides
- Corner guides
- Center guides

Special guide rails, usually made of metal with ball bearings, also may be purchased. These include special wheels, end stops, and other types of hardware. A typical set of purchased drawer guides is shown in Fig. 15-32.

The corner guide has two boards that form a corner. The bottom edge of the drawer rests in this corner (Fig. 15-33A). The side guide is a single piece of wood nailed to the cabinet frame. As in Fig. 15-33B, the single piece of wood fits into a groove cut on the side of the drawer. The guides also serve as a support.

For the center guide, the weight of the drawer is supported by the end rails. However, the drawer is kept in alignment by a runner and guide (see Fig. 15-33C). The carpenter may make the guides by nailing two strips of wood on the bottom of the drawer. The runner is a single piece of wood nailed to the rail.

In addition to the guides, a piece should be installed near the top. This keeps the drawer from falling

Fig. 15-32 *A built-in unit with factory-made drawer guides.* (Formica.)

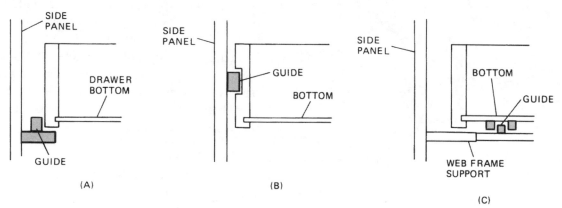

Fig. 15-33 *Carpenter-made drawer guides: A. corner guide; B. side guide; C. center guide.*

as it is opened. This piece is called a *kicker* (Fig. 15-34). Kickers may be installed over the guides, or a single kicker may be installed.

Drawer guides and drawer rails should be sanded smooth. Then they should be coated with sanding sealer. The sealer should be sanded lightly, and a coat of wax should be applied. In some cases, the wood may be sanded and wax applied directly to the wood. However, in either case, the wax will make the drawer slide better.

Making a drawer   Most drawers made by carpenters are called *lip drawers*. This type of drawer has a lip around the front (Fig. 15-35). The lip fits over the drawer opening and hides it. This gives a better appearance. It also lets the cabinet and opening be less accurate.

The other type of drawer is called a *flush drawer*. Flush drawers fit into the opening. For this reason, they must be made very carefully. If the drawer front is not accurate, it will not fit into the opening. Binding or large cracks will be the result. It is far easier to make a lip drawer.

Drawers may be made in several ways. As a rule, the procedure is to cut the right and left sides as in Fig. 15-36. The back ends of these have dado joints cut into them. Note that the back of the drawer rests on the bottom piece. The bottom is grooved into the sides and front piece as shown.

The front of the drawer should be made carefully. It takes the greatest strain from opening and closing. The front should be fitted to the sides with a special joint. Several types of joints may be used. These are shown in Fig. 15-37.

KICKER—
SIDE OR
MIDDLE

DRAWER
BACK   SIDE

DOUBLE OR
FALSE FRONT
(OPTIONAL)

LIP OF DRAWER
COVER OPENING

Fig. 15-34 *Kickers keep drawers from tilting out when opened.*

MOLDED EDGE
CUT INTO SOLID
PIECE

Fig. 15-35 *Lip drawers have a lip that covers the opening in the frame. Drawer fronts can be molded for appearance.*

The highest-quality work can feature a special joint called a *dovetail joint*. However, this joint is time-consuming and expensive to make and cut. As a rule, other types of dovetail joints are used. The dovetailed dado joint is often used.

Drawers are made after the cabinet frame has been assembled. Fronts of drawers may be of several shapes. The front may be paneled, as in Fig. 15-38A. Or it can be molded, as in Fig. 15-38B. Both styles can be made thicker by adding extra boards.

The wood used for the drawer fronts is selected and cut to size. Joints are marked and cut for assembly. Next, the groove for the bottom is cut.

Then the stock is selected and cut for the sides. As a rule, the sides and backs are made from different wood than the front. This is to reduce cost. The drawer front may be made from expensive woods finished as required. However, the interiors are made from less expensive materials. They are selected for straightness and sturdiness. Plywood is not satisfactory for drawer sides and backs.

The joints for the back are cut into the sides. The joints should be cut on the correct sides. Next, the grooves for the drawers are cut in each side. Be sure that they align properly with the front groove.

Next, the back piece is cut to the correct size. Again, note that the back rests on the bottom. No groove is cut. All pieces then are sanded smooth.

A piece of hardboard or plywood is chosen for the bottom. The front and sides of the drawer are assembled. The final measurements for the bottom are made. Then the bottom piece is cut to the correct size. It is lightly sanded around the edges. Then the bottom is inserted into the grooves on the bottom and front. This is a trial assembly to check the parts for fit. Next, the drawer is taken apart again. Grooves are cut in the sides for the drawer guides. Also, any final adjustments for assembly are made.

The final assembly is made when everything is ready. Several processes may be used. However, it is best to use glue on the front and back pieces. Sides should not be glued. This allows for expansion and contraction of the materials.

The drawer is checked for squareness. Then one or two small finish nails are driven through the bottom

Fig. 15-37 *Joints for drawer fronts.*

into the back. One nail should be driven into each side as well. This will hold the drawer square. The bottom of the drawer is numbered to show its location. A like number is marked in the cabinet. This matches the drawer with its opening.

Cabinet door construction   Cabinet doors are made in three basic patterns. These are shown in Fig. 15-39.

A common style is the rabbeted style. A rabbet about half the thickness of the wood is cut into the edge of the door. This allows the door to fit neatly into the opening with a small clearance. It also keeps the cabinet door from appearing bulky and thick. The outside edges then are rounded slightly. The door is attached to the frame with a special offset hinge (Figs. 15-40 and 15-41).

The molded door is much like the lip door (see Fig. 15-39B). As can be seen in the figure, the back side of the door fits over the opening. However, no rabbet is cut into the edge. In this way, no part of the door fits inside the opening. The edge is molded to reduce the apparent thickness. This can be done with a router. Or it may be done at a factory where the parts are made. This method is becoming more widely used by carpenters. Its advantages are that it does not fit in the opening, that means no special fitting is needed, and that special hinges are not required. The appearance gives extra depth and molding effects. These are not available for other types of cabinets.

Both styles of door are commonly paneled. Paneled doors give the appearance of depth and contour. Doors may be made from solid wood. However, when cabinets are purchased, the panel door is very common. The panel door has a frame much like that of a screen door. Grooves are cut in the edges of the frame pieces. The panel and door edges are then assembled and glued solidly together. Cross sections of solid and built-up panels are shown in Fig. 15-42.

Flush doors fit inside the cabinet. These appear to be the easiest to make. However, they must be cut very

Fig. 15-38 A. These drawer types cover the opening. The accuracy needed for fitting is less. B. Molding can be used to give a paneled effect.

carefully. They must be cut in the same shape as the opening. If the cut is not carefully made, the door will not fit properly. Wide or uneven spaces around the door will detract from its appearance.

Sliding doors Sliding doors are also used widely. These doors are made of wood, hardboard, or glass. They fit into grooves or guides in the cabinet (Fig. 15-43). The top groove is cut deeper than the bottom groove. The door is installed by pushing it all the way to the top. Then the bottom of the door is moved into the bottom groove. When the door is allowed to rest at the bottom of the lower groove, the lip at the top still will provide a guide. Also, special devices may be purchased for sliding doors.

Making the countertop Today, most cabinet tops are made from laminated plastics. There are many trademarks for these. These materials are usually ¹⁄₁₆ to ⅛ inch thick. The material is not hurt by hot objects and does not stain or peel. The laminate material is very hard and durable. However, in a thin sheet it is not strong. Most cabinets today have a base top made from plywood or chipboard. Usually, a ½-inch thickness or more is used for countertops made on the site.

Also, specially formed counters made from wood products may be purchased. These tops have the plastic laminates and the mold board permanently formed into a one-piece top (Fig. 15-44). This material may be purchased in any desired length. It is then cut to shape and installed on the job.

Top pieces are cut to the desired length. They may be nailed to the partitions or to the drawer kickers on the counter. The top should extend over the counter approximately ⅜ inch. Next, sides or rails are put in place around the top. These pieces can be butted or rabbeted as shown in Fig. 15-45. These are nailed to the plywood top. They also may be nailed to the frame of the counter. Next, they should be sanded smooth. Uneven or low spots are filled.

Particle board is commonly used as a base for countertops. It is inexpensive, and it does not have any grain. Grain patterns can show through on the finished surface. Also, the grain structure of plywood may form pockets. Glue in pockets does not bond to the lami-

Fig. 15-39 Cabinet door styles: A. rabbeted; B. molded; C. flush.

Fig. 15-40 *A colorful kitchen with rabbeted lip drawers and cabinets. Note the lack of pulls.* (Armstrong Cork.)

Fig. 15-41 *How drawers and doors are undercut so that pulls need not be used.*

Fig. 15-42 *A. Cross section of a solid door panel. B. Cross section of a built-up door panel.*

nate. Particle board provides a smooth, even surface that bonds easily.

Once the counter has been built, the top is checked. Also, any openings should be cut. Openings can be cut for sinks or appliances. Next, the plastic laminate is cut to rough size. Rough size should be to ¼ inch larger in each dimension. A saw is used to cut the laminate, as in Fig. 15-46.

Next, contact cement is applied to both the laminate and the top. Contact cement should be applied with a brush or a notched spreader. Allow both surfaces to dry completely. If a brush is used, solvent should be kept handy. Some types of contact cement

are water-soluble. This means that soap and water can be used to wash the brush and to clean up.

After the glue has dried, the surface is shiny. Dull spots mean that the glue was too thin. Apply more glue over these areas, and allow it to dry.

NOTE: Gap is included so that door can be lifted up.

LIFT UP

SWING OUT

METAL OR PLASTIC TRACK

(A) (B) (C)

Fig. 15-43 *Sliding doors. A. Doors may slide in grooves. B. Doors may slide on a metal or plastic track. C. To remove, lift up, and swing out.*

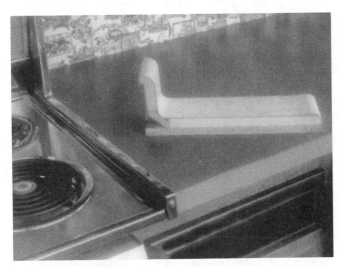

Fig. 15-44 *Countertops may be flat or may include splashboards.*

EDGE INLET

TOP

SOLID BLOCK ON BOTTOM

SAW CUTS

COUNTERTOP

NAIL, SINK HEAD

EDGE

TOP

RABBET

NAIL, SINK HEAD

EDGE

Fig. 15-45 *Blocking up counter edges.*

It usually takes about 15 or 20 minutes for contact cement to dry. As a rule, the pieces should be joined within a few minutes. If they are not, a thin coat of contact cement is put on each of the surfaces again.

To glue the laminate in place, two procedures may be used. First, if the piece is small, a guide edge is put in place. The straight piece is held over the area, and the guide edge is lowered until it makes contact. The entire piece is lowered into place. Pressure is applied from the center of the piece to the outside edges. Hands may be used, but a roller is better.

For larger pieces, a sheet of paper is used. Wax paper may be used, but almost any type of paper is acceptable. Allow the glue to dry first. Then the paper is placed on the top. The laminate is placed over the paper. Then the laminate is positioned carefully. The paper is gently pulled about 1 inch from beneath the laminate. The position of the laminate is checked. If it is in place, pressure may be applied to the exposed edge. If it is not in place, the laminate is moved until it is in place. Then the exposed edge is pressed until a bond is made. Then the paper is removed from the entire surface. Pressure is applied from the middle toward the edges (Fig. 15-47).

Trim for laminated surfaces Trim the edges. The pieces were cut slightly oversize to allow for trimming. The tops should extend over the sides slightly. The tops and corners should be trimmed so that a slight bevel is exposed. This may be done with a special router bit, as in Fig. 15-48A. It may also be done with a sharp and smooth file, as shown in Fig. 15-48B.

Fig. 15-46 *Cut into the laminate to avoid chipping.*

The back of most countertops has a raised portion called a *splashboard*. Splashboards and countertops may be molded as one piece. However, splashboards are also made as two pieces, in which case metal cove and cap strips are applied at the corners. Building codes may set a minimum height for these splashboards. The Federal Housing Authority (FHA) requires a minimum height of 4 inches for kitchen counters.

Installing hardware Hardware for doors and drawers means the knobs and handles. These are frequently called *pulls*. A number of styles are available. As a rule, drawers and cabinets are put into place for finishing. However, pulls and handles are not installed. Hinges are applied in many cases. In others, the doors are finished separately. However, pulls are left off until the finish is completed.

Drawer pulls are placed slightly above center. Wall cabinet pulls are placed in the bottom third of the doors. Door pulls are best put near the opening edge. For cabinet doors in bottom units, the position is different. The pull is located in the top third of the door. It is best to put it near the swinging edge. Some types of hardware, however, may be installed in other places for special effects (Fig. 15-49). These pulls are installed in the center of the door panel.

To install pulls, the location is first determined. Sometimes a template can be made and used. Whatever method is used, the locations of the holes are found. They are marked with the point of a sharp pencil. Next, the holes are drilled from front to back. It is a good idea to hold a block of wood behind the area. This reduces splintering.

There are other types of hardware. These include door catches, locks, and hinges. The carpenter always should check the manufacturer's instructions on each.

As a rule, it is easier to attach hinges to the cabinet first. Types of hinges are shown in Fig. 15-50. Types of door catches are shown in Fig. 15-51.

Shelves

Most kitchen and bathroom cabinets have shelves. Also, shelves are used widely in room dividers, bookshelves, and closets. There are several methods of shelf construction that may be used. Figure 15-52 shows some types of shelf construction. Note that each of these allows the shelf location to be changed.

For work that is not seen, shelves are held up by *ledgers*. Figure 15-53 shows the ledger method of shelf construction.

Special joints may also be cut in the sides of solid pieces. These are types of dados and rabbet joints. Figure 15-54 shows this type of construction.

As a rule, a ledger-type shelf is used for shelves in closets, lower cabinets, and so forth. However, for exposed shelves, a different type of shelf arrangement is used. Adjustable or jointed shelves look better on bookcases, for instance.

Facing pieces are used to hide dado joints in shelves (Fig. 15-55). The facing may also be used to make the cabinet flush with the wall. The facing can be inlet into the shelf surface.

Applying Finish Trim

Finish trim pieces are used on the bases of walls. They cover floor seams and edges where carpet has been laid. Also, trim is used around ceilings, windows, and other areas. As a rule, a certain procedure is followed for cutting and fitting trim pieces. Outside corners, as

Fig. 15-47 *Applying plastic laminate to countertops. A. Apply the edge strip. B. Trim the edge strip flush with the top. C. Apply glue. Lay a slip sheet or sticks in place when the glue is dry. D. Position the top. Remove the slip sheet or sticks. E. Apply pressure from center to edge. F. Trim the edge at a slight bevel. G. To cut holes for sinks, first center a punch for drilling. H. Next, drill holes at corners.* (Formica.)

(I)

(J)

(K)

(L)

(M)

Fig. 15-47 *Applying plastic laminate to countertops. I. Then cut out the opening.* (Rockwell International Power Tool Division.) *J. Mitered corners can be held with special clamps placed in cutouts on the bottom.* (Formica.) *K. Ends also may be covered by splashboards.* (Formica.) *L. Lay sink in opening.* (Formica.) *M. Ends may be covered by strips.* (Formica.)

(A)

(A)

(B)

Fig. 15-48 *A. Edges may be trimmed with edge trimmers or routers.* (Rockwell International Power Tool Division.) *B. Edges also may be filed. Note direction of force.*

Fig. 15-49 *A. Pulls may be located in the center for dramatic effect. B. Regular pull location.*

Fig. 15-50 *Hinge types.*

Fig. 15-51 *Types of cabinet door catches.*

in Fig. 15-56, are cut and fit with miter joints. These are cut with a miter box (Fig. 15-57).

However, trim for inside corners is cut with a different joint. This is done because most corners are not square. Miter joints do not fit well into corners that are not square. Unsightly gaps and cracks will be the result of a poor fit. Instead, a coped joint is used (see Fig. 15-56).

For a coped joint, the first piece is butted against the corner. Then the outline is traced on the second piece. A scrap piece is used for a guide. The outline is then cut using a coping saw (Fig. 15-58). The coped joint may be used effectively on any size or shape of molding.

Slotted bookshelf standards and clips are ideal if you want to adjust bookcase shelves, but they add to total cost.

If you use wood or metal pegs set into holes, you must be sure to drill holes at the same level and 3/8 inch deep.

Fig. 15-52 *Methods of making adjustable shelves.*

MAKE STRIP
1. BORE
2. SAW

Fig. 15-53 *Ledgers or cleats are used for shelves and steps where appearance is not important.*

APPLYING FINISH MATERIALS

Frequently, the millwork is finished before the wall surfaces. Many wood surfaces are stained rather than painted. This enhances natural wood effects. Woods such as birch are commonly stained to resemble darker woods such as walnut, dark oak, and pecan. These stains and varnishes are easily absorbed by the wall. They are put on first so that they do not ruin the wall finish.

Paint for wood trim is usually a gloss or semigloss paint. These paints are more washable, durable, and

INLET SUPPORT

EXTERNAL SUPPORT

Fig. 15-55 *Facing supports for shelves and cabinets.*

costly. Flat or nonglare paints are used widely on walls.

Paints, stains, and varnishes are often applied by spraying. This is much faster than rolling. When spray

DADO JOINT— WITH FASCIA STRIP

RABBET JOINT

Fig. 15-54 *Types of dado joints.*

Fig. 15-56 *Outside corners of trim are mitered. Inside corners are coped.*

Fig. 15-57 *A miter box used for cutting miters.*

Fig. 15-58 *Making a coped joint: A. tracing the outline; B. cutting the outline; C. nailing the molding in place.*

equipment is used, there is always an overspray. This overspray would badly mar a wall finish. However, the wall finish can be put on easily over the overspray (Fig. 15-59).

To prepare wood for stain or paint, first sand it smooth. Mill and other marks should be sanded until they cannot be seen. Hand or power sanding equipment may be used.

Applying Stain

Stain is much like a dye. It is clear and lets the wood grain show through. It simply colors the surface of the wood to the desired shade. It is a good idea to make a test piece. The stain can be tested on it first. A scrap piece of the same wood to be finished is used. It is sanded smooth, and the stain is applied. This lets a worker check to see if the stain will give the desired appearance.

Stain is applied with a brush, a spray unit, or a soft cloth. It should be applied evenly with long, firm strokes. The stain is allowed to sit and penetrate for a few minutes. Then it is wiped with a soft cloth. It is

also a good idea to check the manufacturer's application instructions. There are many different types of stains. Thus there are several ways of applying stains.

The stain should dry for a recommended period of time. Then it should be smoothed lightly with steel wool. Very little pressure is applied. Otherwise, the color will be rubbed off. If this happens, the place should be retouched with stain.

The stain is evenly smoothed. Then varnish or lacquer may be applied with a brush or a spray gun. Today, most interior finishes are sprayed (see Fig. 15-59).

Applying the Wall Finish

Two types of wall finishes are used commonly. The first is wall paint. The other is wallpaper.

Painting a wall The wood trim for doors and so forth is finished. Then the walls may be painted. Most wall paint is called *flat paint*. This means that it does not shine. It makes a soft, nonglare surface. Most wall

Fig. 15-59 *Cabinets are stained before walls are painted or papered. Note the overspray around the edges.*

paints are light-colored to reflect light. Before painting, all nail holes or other marks are filled. Patching paste should be used for this.

There are several methods of applying paint. It may be brushed, sprayed, or rolled. Also, special types of paint may be applied. Plain paint is a liquid or gel that gives a smooth, flat finish. However, special "texture" paints are also used. Sand-textured paints leave a slightly roughened surface. This is caused by particles of sand in the paint. Also, thick mixtures of paints are used. These may be rolled on to give a heavy-textured surface. Thicker paints can be used for a shadowed effect.

Applying wallpaper   Wallpaper is not used widely in buildings today. It is used mostly as accenting or in small areas. The "paper" may not be paper at all. It may be various types of vinyl or plastic films. Also, a mixture of plastic and paper is common. In either case, the wall surface should be prepared. It should be smooth and free of holes or dents. As a rule, a standard single roll of wallpaper covers about 30 square feet of wall area. A special wall sealer coat called *size* or *sizing* is used. This may be purchased premixed and ready to use. Powder types may be purchased and

mixed by the worker. In either case, the sizing is painted on the walls and allowed to dry. As it dries, it seals the pores in the wall surface. This allows the paste or glue to hold the paper to the wall.

Choose a corner for a starting point. It should be close to a window or door. The width of the wallpaper is measured out from the corner. One inch is taken from the width of the wallpaper. Make a small mark at this distance. The mark should be made near the top of the wall. For a 27-inch roll, 26 inches is measured from the corner. A nail is driven near the ceiling for a chalk line. The chalk line is tied to this nail at the mark. The end of the chalk line is weighted near the floor. The line is allowed to hang free until it comes to a stop. When still, the line is held against the wall. The line is snapped against the wall. This will leave a vertical mark on the wall (Fig. 15-60A).

Next, several strips of paper are cut for use. The distance from the floor to the ceiling is measured. The wallpaper is unrolled on the floor or a table. The pattern side is left showing. The distance from the floor to the ceiling is laid off and 4 inches is added. This strip is cut, and then several more strips are cut.

The first precut strip is laid on a flat surface. The pattern should be face up. The strip is checked for appearance and cuts or damage. Next, the strip is turned over so that the pattern is face down. The paste is applied with a brush (see Fig. 15-60B). The entire surface of the paper is covered. Next, the strip is folded in the middle. The pattern surface is on the inside of the fold. This allows the worker to hold both the top and the bottom edges (see Fig. 15-60C).

The plumb line is used as a starting point. The first strip is applied at the ceiling. About 1 inch overlaps the ceiling. This will be cut off later. A stiff bristle brush is moved down the strip (see Fig. 15-60D). Take care that the edges are aligned with the chalk line. About 1 inch will extend around the corner. This will be trimmed away later. This is an allowance for an uneven corner.

The entire strip then is brushed to remove the air bubbles. The brush is moved from the center toward the edge. The second strip is prepared in the same manner. However, care is taken to be sure that the pattern is matched. The paper is moved up or down to match the pattern. The edge of the second strip should exactly touch the edge of the first piece.

Next, the second piece is folded down. The edge is exactly matched with the edge of the first piece. The edge also should exactly touch the edge of the previous piece. Then any bubbles are smoothed out. The process is repeated for each strip until the wall is finished. Then the wall is trimmed.

Fig. 15-60 *Putting wallpaper on walls. A. A chalk line and plumb are used to mark the starting point. B. After the paper has been measured and cut, paste is brushed on the back. C. The bottom and top edges are used to carry the pasted wall covering. D. Strips are lapped at the ceiling and brushed down. E. Use a putty knife or straightedge as a guide to trim overlaps. F. Cut corners diagonally at windows and doors.*

The extra overlaps at the ceiling and at the base are cut. A razor blade or knife is used with a straightedge as a guide (see Fig. 15-60E).

A new line is plumbed at each new corner. The first strip on each corner is started even with the plumb line. In this way, each strip is aligned properly. The corners will not have unsightly gaps or spaces.

A diagonal cut is made at each corner of a window. About ½ inch is allowed around each opening. The di-

agonal cut forms a flap over the molding. This allows the opening to be cut for an exact fit (see Fig. 15-60F).

FLOOR PREPARATION AND FINISH

Finish the floor last. This really makes sense if you think about it. After all, people will be working in the building. They will be using paint, varnish, stain, plaster, and many other things. All these could damage a finished floor if they spilled.

Keep in mind that workers will be entering and leaving the building. Mud, dirt, dust, and trash will be tracked into the building. It is very difficult to paint or plaster without spilling. A freshly varnished floor or a newly carpeted one could be easily ruined. So the floor is finished last to avoid damage to the finished floor.

Flooring over a concrete slab   Special methods are sometimes needed to lay flooring over concrete. Figure 15-61 shows some of the steps in this process. Also this was discussed in Chapter 3.

Laying Wooden Flooring

For all types of floors, the first step is to clean them (Fig. 15-62). The surface is scraped to remove all plaster, mud, and other lumps. Then the floor is swept with a broom.

Generally, there are three shapes of wooden flooring. The first is called *strip flooring* (Fig. 15-63). The second is called *plank flooring* (Fig. 15-64). The third is *block* or *parquet flooring* (Fig. 15-65). The blocks may be solid, as in Fig. 15-66A. Here, the grain runs in one direction, and the blocks are cut with tongues and grooves. The type shown in Fig. 15-66B may be straight-sided. The blocks are made of strips with the grain running in the same direction. Such blocks may have a spline at the bottom or a bottom layer.

The type of block shown in Fig. 15-66C is made of several smaller pieces glued to other layers. The bottom layer is usually waterproof.

Most flooring has cut tongue-and-groove joints. Hidden nailing methods are used so that the nails do not show when the floor is done. Strip and plank flooring also will have an undercut area on the bottom. This undercut is shown in Fig. 15-63. The undercut, or hollow, helps to provide a stable surface for the flooring. Small

(A)

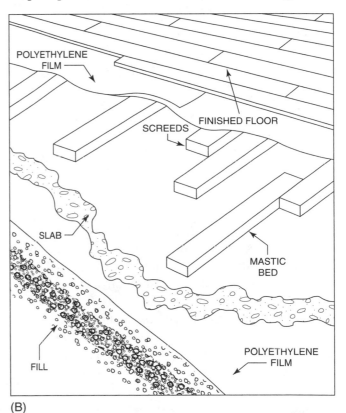

(B)

Fig. 15-61   Laying wood floors on concrete. A. Plywood-on-slab method of installing strip flooring. B. Screeds method of installing strip flooring on slab. C. Use of the power nailer for installing strip flooring.

(C)

Fig. 15-62 *Subflooring must be scraped and cleaned.* (National Oak Flooring Manufacturers Association.)

Fig. 15-64 *Plank flooring.*

Fig. 15-63 *Strip flooring.* (Forest Products Laboratory.)

bumps will not make the piece shift. Strip flooring has narrow, even widths with tongue-and-groove joints on the ends. The strips may be random end-matched. Plank flooring comes in both random widths and lengths. It may be drilled and pegged at the ends. However, today, most plank flooring has fake pegs that are applied at the factory. The planks are then nailed in much the same manner as strip flooring. Block flooring may be either nailed or glued, but glue is used most often.

Most flooring is made from oak. Several grades and sizes are available. However, the width or size is largely determined by the type of flooring.

Carpeting can cost much more than a finished wood floor. Also, carpeting may last only a few years. As a rule, wood floors last for the life of the building. It is a good idea to have hardwood floors underneath carpet. Then the carpet can be removed without greatly reducing the resale value of the house. Besides, the trends for floor coverings change with time. It takes about 15 years for a new home to look worn and in need of remodeling. A change from carpet to wooden floors is one of those changing trends that alternate periodically. Kitchen cabinets and counter tops are often "updated" depending on the occupants of the home. Bathroom fixtures and cabinets change with time and as the owners' taste changes with the times.

Preparation for Laying Flooring

Manufacturers recommend that flooring be placed inside for several days before it is laid. The bundles should be opened and the pieces scattered around the room. This lets the wood reach a moisture content similar to that of the room. This will help to stabilize the flooring. If the flooring is stabilized, the expansion and contraction will be even.

Check the subfloor for appearance and evenness. The subfloor should be cleaned and scraped of all de-

Fig. 15-65 *A parquet floor of treated wood blocks.* (Perma Grain Products.)

Fig. 15-66 *Wood block (parquet) flooring: A. solid; B. splined; C. substrated. (Forest Products Laboratory.)*

posits and swept clean. Nails or nail heads should be removed. All uneven features should be planed or sanded smooth (see Fig. 15-62).

A vapor barrier should be laid over the subfloor. It can be made of either builder's felt or plastic film. Seams should be overlapped 2 to 4 inches. Then chalk lines should be snapped to show the centers of the floor joists (Fig. 15-67).

Installing wooden strip flooring Wooden strip flooring should be applied perpendicular to the floor joists (Fig. 15-68). The first strip is laid with the grooved edge next to the wall. At least ½ inch is left between the wall and the flooring. This space controls expansion. Wooden flooring will expand and contract. The space next to the wall keeps the floor from buckling or warping. Warps and buckles can cause air gaps beneath the floor. They also ruin the look of the floor. The space next to the wall will be covered later by molding and trim.

The first row of strips is nailed using one method. The following rows are nailed differently. In the first row, nails are driven into the face (Figs. 15-69 and 15-70). Later, the nail is set into the wood and covered.

Hardwood flooring will split easily. Most splits occur when a nail is driven close to the end of a board. To prevent this, drill the nail hole first. It should be slightly smaller than the nail.

The next strip is laid in place as shown in Fig. 15-71. The nail is driven blind at a 45- or 50-degree angle as shown. Note that the nail is driven into the top corner of the tongue. The second strip should fit firmly against the first layer. Sometimes, force must be used to make it fit firmly. A scrap piece of wood is placed over the tongue, as shown in Fig. 15-72. The ends of the second layer should not match the ends of the first layer. The ends should be staggered for better strength and appearance. The end joints of one layer should be at least 6 inches from the ends of the previous layer. A nail set should be used to set the nail in place. Either the vertical position or the position shown in Fig. 15-72 may be used. Be careful not to damage the edges of the boards with the hammer.

The same amount of space (½ inch) is left at the ends of the rows. End pieces may be driven into place with a wedge. Pieces cut from the ends can become the first piece on the next row. This saves material and helps to stagger the joints. Figures 15-73 and 15-74 show other steps.

Wooden plank flooring Wooden plank flooring is installed in much the same manner. Today, both types generally are made with tongue-and-groove joints. The joints are on both the edges and ends. The same allowance of ½ inch is made between plank flooring and the walls. The same general nailing procedures are used.

Wood-block floors There are two types of wood-block floors. The first type is like a wide piece of

(A)

Fig. 15-68 *Strip flooring is laid perpendicular to the joists.* (Forest Products Laboratory.)

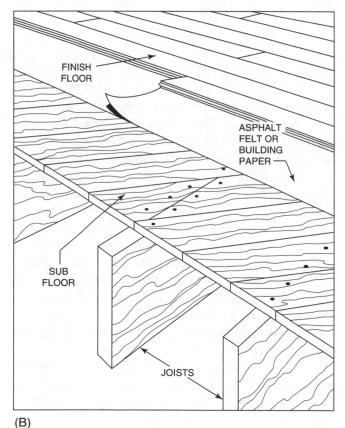

(B)

Fig. 15-67 *A. Moisture retarder using two layers of asphalt felt or building paper; B. Wood joist construction using square-edge board subfloor.*

board. The second type is called *parquet*. Parquet flooring is made from small strips arranged in patterns. Parquet must be laminated to a base piece. Today, both types of block floors are often laminated. When laminated, they are plywood squares with a thick veneer of flooring on the top. Usually, three layers of material make up each block.

Fig. 15-69 *Establishing starter line for nailing first strip.*

Fig. 15-70 *In the first strip, nails are driven into the face.* (Forest Products Laboratory.)

Fig. 15-71 *Nailing strips after the first. (Forest Products Laboratory.)*

Fig. 15-72 *A scrap block is used to make flooring fit firmly. This prevents damage to either tongues or grooves.*

Most blocks have tongue-and-groove edges. Common sizes are 3-, 5-, and 7-inch squares. With tongue-and-groove joints, the blocks can be nailed. However, today, most blocks are glued. When they are glued, the preparation is different than for plank or strip flooring. No layer of builder's felt or tar paper is used. The blocks are glued directly to the subfloor or to a base floor. As a rule, plywood or chipboard is used for a base under block flooring.

Blocks may be laid from a wall or from a center point. Some patterns are centered in a room. Then blocks should be laid from the center toward the edges. To lay blocks from the center, first the center point is found. Then a block is centered over this point. The outline of the block is drawn on the floor. Then a chalk line is snapped for each course of blocks. Care is taken that the lines are parallel with the walls.

Sometimes blocks are laid on the floor proceeding from a wall. The chalk lines should be snapped for each course from the base wall.

Blocks are often glued directly to concrete floors. These floors must be carefully and properly made. Moisture barriers and proper drainage are essential. When there is any doubt, a layer process is used. First, a layer of mastic cement is applied. Then a vapor barrier of plastic or felt is applied over the mastic. Then a second layer of mastic is applied over the moisture barrier. The blocks then are laid over the mastic surface.

Wooden blocks also may be glued over diagonal wood subflooring. The same layering process as above is used.

Fig. 15-73 *Use of the power nailer for installing strip flooring.*

FACE NAILS

Fig. 15-74 *Countersink screws in plank flooring, cover with plugs.*

FINISHING FLOORS

Floors are finished after walls and trim. When resilient flooring is applied, no finish step is needed. However, floors of this type should be cleaned carefully.

Finishing Wood Floors

Sanding is the first step in finishing wood floors. A special sanding machine is used, such as the one in Fig. 15-75. When floors are rough, they are sanded twice. For oak flooring, it is a good idea to use a sealer coat next. The sealer coat has small particles in it. These help to fill the open pores in the oak. One of two types of sealers or fillers should be used. The first type of sealer is a mixture of small particles and oil. This is

Fig. 15-75 *After the floor is laid, it is sanded smooth using a special sanding machine.* (National Oak Flooring Manufacturers Association.)

rubbed into the floor and allowed to sit a few minutes. Then a rotary sander or a polisher is used to wipe off the filler. The filler on the surface is wiped off. The filler material in the wood pores is left.

Floors may be stained a darker color. The stain should be applied before any type of varnish or lacquer is used. It is a good idea to stain the molding at the same time. The stain is applied directly after the sanding. If a particle-oil-type filler is used, the stain should be applied after the filler.

The second filling method uses a special filler varnish or lacquer. This also contains small particles that help to fill the pores. It is brushed or sprayed onto the floor and allowed to dry. The floor then is buffed lightly with an abrasive pad.

After filling, the floor should be varnished. A hard, durable varnish is best. Other coatings generally are not satisfactory. Floors need durable finishes to avoid showing early signs of wear.

Molding can be applied after the floors are completely finished. The base shoe and baseboard are installed as in Fig. 15-76. Many workers also finish the molding when the floor is finished.

Fig. 15-76 *Base molding: A. square-edged base; B. narrow ranch base; C. wide ranch base; D. installation; E. a cope joint.* (Forest Products Laboratory.)

Base Flooring for Carpet

Often two floor layers are used, but neither is the finished floor surface. The first layer is the subfloor, and the next layer is a base floor. Both base and subflooring may be made from underlayment. Underlayment may be a special grade of plywood or chipboard. In many cases, nailing patterns are printed on the top side of the underlayment.

It is a good idea to bring underlayment into the room to be floored and allow it to sit for several days exposed to the air. This allows it to reach the same moisture content as the rest of the building components.

Base flooring is used when added strength and thickness are required. It is also used to separate resilient flooring or other flooring materials from concrete or other types of floors. It provides a smooth, even base for carpet. It is much cheaper to use a base floor than a hardwood floor under carpets. The costs of nailing small strips and of sanding are saved.

INSTALLING CARPET

Carpet has become popular for many reasons. It makes the floor a more resilient and softer place for people to stand. Also, carpeted floors are warmer in winter. Carpeting also helps to reduce noise, particularly in multistory buildings, and it has an insulation value in houses that are built over an unheated basement.

To install carpet, several factors must be considered. First, most carpet is installed with a pad beneath it. When carpet is installed over a concrete floor, a plastic film also should be laid. This acts as an additional moisture seal. The carpet padding may then be laid over the film.

The first step in laying carpet is to attach special carpet strips. These are nailed to the floor. They are laid around the walls of a room to be carpeted. These carpet strips are narrow, thin pieces of wood with long tacks driven through them. They are nailed to the floor approximately ¼ inch from the wall (Fig. 15-77). The carpet padding then is unrolled and cut. The padding extends only to the strips.

Next, the carpet is unrolled. If the carpet is large enough, it may be cut to fit exactly. However, carpet should be cut about 1 inch smaller than the room size to allow for stretching.

The carpet is wedged between the carpet strip and one wall. A carpet wedge is used, as in Fig. 15-78. The carpet then is smoothed toward the opposite wall. All wrinkles and gaps are smoothed and removed. Next, the carpet is stretched to the opposite wall. The person

Fig. 15-77 *Carpet strips hold the carpet in place.*

Fig. 15-78 *Carpet edges are wedged into place. Base shoe molding then may be added.*

installing the carpet generally will walk around the edges pressing the carpet into the tacks of the carpet strips. This is done after each side is wedged. After the ends are wedged, the first side is attached. The same process is repeated. After the first side is wedged, the opposite side is attached.

To seam a carpet, special tape is used. This tape is a wide strip of durable cloth. It has an adhesive on its upper surface. It is rolled out over the area or edge to be joined. Half the tape is placed underneath the carpet already in place. The carpet then is pressed firmly onto the adhesive. Some adhesive tapes use special heating tools for best adhesion. The second piece of carpet is carefully butted next to the first. Be sure that no great pressure is used to force the two edges together. No gaps should be wider than 1/16 inch. The edges should not be jammed forcefully together either. If edges are jammed together, lumps will occur. If wide gaps are left, holes will occur. However, the nap or shag of the carpet will cover most small irregularities.

To cut the carpet for a joint, first unroll the carpet. Carefully size and trim the carpet edges as straight as possible. The joint will not be even if the edges are ragged. Various tools may be used. A heavy knife or a pair of snips may be used effectively.

Metal end strips are used where carpet ends over linoleum or tile. The open end strips are nailed to the floor. The carpet is stretched into place over the points on the strip. Then the metal strip is closed. A board is laid over the strip (Fig. 15-79). The board is struck sharply with a hammer to close the strip. Do not strike the metal strip directly with the hammer. Doing so will leave unsightly hammer marks on the metal.

Fig. 15-79 *A metal binder bar protects and hides carpet edges where carpet ends over linoleum or tile.*

RESILIENT FLOORING

Resilient flooring is made from chemicals rather than from wood products. Resilient flooring includes compositions such as linoleum and asphalt tile. It may be laid directly over concrete or base flooring. Resilient flooring comes in both sheets and square tiles. Frequently, resilient flooring sheets will be called *linoleum carpets.*

Installing Resilient Flooring Sheets

The first step is to determine the size of the floor to be covered. It is a good idea to sketch the shape on a piece of paper. Careful measurements are made on the floor to be covered. Corners, cabinet bases, and other features of the floor are included. It is a good idea to take a series of measurements. They are made on each wall every 2 or 3 feet. This is so because most rooms are not square. Thus measurements will vary slightly from place to place. The measurements should be marked on the paper.

Next, the floor is cleaned. Loose debris is removed. A scraper is used to remove plaster, paint, or other materials (see Fig. 15-62). Then the area is swept. If necessary, a damp mop is used to clean the area. Neither flooring nor cement will stick to areas that are dirty. Next, the surface is checked for holes, pits, nail heads, or obstructions. Holes larger than the diameter of a nail are patched or filled. Nails or obstructions are removed.

Most rooms will be wider than the roll of linoleum carpet. If not, the outline of the floor may be transferred directly to the flooring sheet. The sheet then may be cut to shape. The shaped flooring sheet then is brought into the room. It is positioned and unrolled. A check is made for the proper fit and shape. Any adjustments or corrections are made. Then the sheet is rolled up approximately halfway. The mastic cement is spread evenly (about 3/32 inch thick) with a toothed trowel. The unrolled portion is rolled back into place over the cemented area. Then the other end is rolled up to expose the bare floors. Next, the mastic is spread over the remaining part of the floor. The flooring then is rolled back into place over the cement. The sheet is smoothed from the center toward the edge.

However, most rooms are wider than the sheets of flooring. This means that two or more pieces must be joined. It is best to use a factory edge for the joint line. Select a line along the longest dimension of the room, as shown in Fig. 15-80. On the base floor, measure equal distances from a reference wall as shown. These are the same width as the sheet. Then snap a chalk line for this line. Often more than one joint or seam must be used.

The same measuring process is used. Measurements are taken from the edge of the first sheet. However, for smaller pieces, a different process is used. The center of the last line is found. A line is snapped at right angles. This shows the pattern for the pieces of flooring that must be cut. The second line should run

Fig. 15-80 *Floor layout for resilient flooring.*

The labels in the figure read:

LINE AND FACTORY EDGE FOR MAIN JOINT

WALL

WIDTH OF FLOORING PIECE

MEASURE AT SEVERAL POINTS

NOTE: This width may be snapped on second piece.

the entire width of the room. A carpenter's square may be used to check the squareness of the lines.

The two chalk lines are now the reference lines. These are used for measuring the flooring and the room area. Measurements to the cabinets or other features are made from these lines. Take several measurements from the paper layout. Walls are seldom square or straight. Frequent measurements will help to catch these irregularities. The fit will be more accurate and better.

Next, in a different area, unroll the first sheet of flooring to be used. Find the corresponding wall and reference line. The factory edge is aligned on the first line. The dimensions are marked on the resilient flooring. The necessary marks show the floor outline on the flooring sheet. A straight blade or linoleum knife is used to make these cuts. For best results, a guide is used. A heavy metal straightedge is used for a guide when cutting. A check is made to be sure that there is nothing underneath the flooring. Anything beneath it could be damaged by the knife used to cut the flooring.

Next, the cut flooring piece is carried into the area. It is unrolled over the area, and the fit is checked. Any adjustments necessary are made at this point. Next, the material is rolled toward the center of the room. The area is spread with mastic. The flooring is rolled back into place over the cement. The sheet is smoothed as

before. Do not force the flooring material under offsets or cabinets. Make sure that the proper cuts are made.

The adhesive should not be allowed to dry for more than a few minutes. No more than 10 or 15 minutes should pass before the material is placed. A heavy roller is recommended to smooth the flooring. It should be smoothed from the center toward the edges. This removes air pockets and bubbles.

Where seams are made, a special procedure is recommended. Unroll the two pieces of flooring in the same preparation area. The two edges to be joined are slightly overlapped (Fig. 15-81). Next, the heavy metal guide is laid over the doubled layer. Then both layers are cut with a single motion. The straightedge is used as a guide to make the straightest possible cut. Edges cut this way will match, even if they are not perfectly straight or square. Figure 15-82 shows how edges are trimmed.

Fig. 15-81 *Edges are overlapped during cutting. In this way, seams will match even if the cut is not perfectly straight.* (Armstrong Cork.)

Installing Resilient Block Flooring

Resilient block flooring is often called *tile*. This term includes floor tile, asphalt tile, linoleum tile, and others. As a rule, the procedure for all these is the same. The sizes of these tiles range from 6 to 12 inches square. To lay tile, the center of each of the end walls

Fig. 15-82 *Trimming the edges. A metal straightedge or carpenter's square is used to guide a utility knife.* (Armstrong Cork.)

Fig. 15-83 *Find the intersecting midpoints.* (Armstrong Cork.)

Fig. 15-84 *Lay tiles in place without cement to check the spacing.* (Armstrong Cork.)

is found (Fig. 15-83). A chalk line is tied to each of these points. Lines then are snapped down the middle of the floor. Next, the center of the first line is located. A square or another tile is used as a square guide. A second line is snapped square to the first line.

Next, a row of tiles is laid along the perpendicular chalk line as shown. They are not cemented. Then the distance between the wall and the last tile is measured. If the space is less than half the width of a tile, a new line is snapped. It is placed half the width of a tile away from the center line (Fig. 15-84). A second line is snapped half the width of a tile from the perpendicular line. The first tile then is aligned on the second snapped line. The tile now can be cemented. The first tile becomes the center tile of the room.

Another method is used if the distance at the sidewalls is greater than one-half the width of a tile. Then the tile is laid along the first line. In this way, no single tile is the center.

The first two courses are laid as a guide. Then the mastic is spread over one-quarter of the room area (Fig. 15-85). Lay the tiles at the center first. The first tiles should be laid to follow the snapped chalk lines. Tiles should not be slid into place. Instead, they are pressed firmly into position as they are installed. It is best to hold them in the air slightly. Then the edges are touched together and pressed down.

The first quarter of the room is covered. Only the area around the wall is open. Here, tiles must be cut to fit.

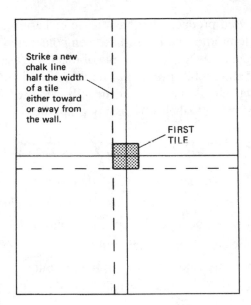

Fig. 15-85 *The next chalk line is used as a guide for laying tile. This method places one tile in the center.*

Cutting the tiles A loose tile is placed exactly on top of the last tile in the row (Fig. 15-86A). Then a third tile is laid on top of this stack. It is moved over until it touches the wall. The edge of the top of the tile becomes the guide (see Fig. 15-86B). Then the middle tile is cut along the pencil line as shown. This tile then will have the proper spacing.

A pattern is made to fit tile around pipes or other shapes. The pattern is made in the proper shape from paper. This shape is traced on the tile. The tile then is cut to shape.

LAYING CERAMIC TILE

Several types of tile are used commonly: ceramic tile, quarry tile, and brick. Tiles are desirable where water will be present, such as in bathrooms and wash areas. However, ceramic tiles also make pleasing entry areas and lobbies. Two methods are used for installing ceramic floor tile.

The first, and more difficult, method uses a cement-plaster combination. A special concrete is used as the bed for the tile. This bed should be mixed carefully, poured, leveled, and troweled smooth. It should be allowed to sit for a few minutes. Just before it hardens, it is still slightly plastic. Then the tiles are embedded in place. The tiles should be thoroughly soaked in water if this method is used. They should be taken out of the water one at a time and allowed to drain slightly. The tiles then are pressed into place in the cement base. All the tiles are installed. Then special grout is pressed into the cracks between the tiles. This completely fills the joints between the tiles. The grouted joints should be cleaned and tooled within a very few hours of installation.

Adhesives are most widely used today. Adhesives are much easier to use. The adhesives are spread evenly on the tiles. This is much like laying resilient

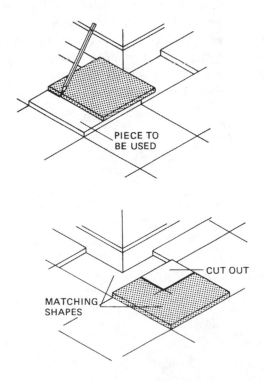

Fig. 15-86 *Cutting and trimming tile to fit: A. marking tile for edges; B. marking tile for corners.*

tile. The adhesive should be one that is recommended for use with ceramic tiles. After the adhesive has been spread, the tiles are placed. Then a common grout is forced between the tiles. It is wiped and allowed to dry.

Tiles are available as large individual pieces. They are also available as assortments. Small tiles may be pre-attached to a cloth mesh. This mesh keeps the tiles arranged in the pattern. This also makes the tile easy to space evenly.

Environmental Concerns

Adhesives Some of these adhesives are not acceptable for use in enclosed places.

- There are legitimate concerns for the chemical composition of the adhesives used in the building of a house. There are adhesives used on the floors (installing tile) and in placing the wallboards on the walls and ceiling.

- Many people have allergies that are aggravated by the adhesives used in applying wallboard and paneling, wallpaper, linoleum, and even hardwood flooring.

- Formaldehyde vapors usually are given off (called *off-gassing*) by kitchen cabinets and the glue or mastic used in the attachment process for countertops.

Interior doors are also suspect in the types of glues used to construct them.

There are a number of national and local directories of "green" building products. Keep in mind that using the green product in the proper way can make a difference in the quality of life of the residents of a house. There are wrong ways to use the green products that can do more harm than good. The main emphasis is to avoid products that have toxic or harmful emissions long after the house has been considered ready for occupancy. A good example of this is the use of recycled automobile tires to make rubber flooring. These products should not be used in fully enclosed indoor spaces owing to their off-gassing properties.

CHAPTER 15 STUDY QUESTIONS

1. How is expansion controlled on strip flooring?
2. When is a vapor barrier used under finish flooring?
3. How are board ends kept from splitting?
4. How are floor tiles laid out?
5. Why are floor tiles started in the center?
6. When is strip flooring laid? When is it finished?
7. How are seams in linoleum cut?
8. When is door trim applied?
9. What is the easiest way to lay ceramic tile?
10. How is interior trim used on metal window frames?
11. What are built-ins?
12. What is the sequence for hanging doors?
13. What are three ways of making shelves?
14. List the steps in building a cabinet counter.
15. How are ready-built cabinets hung?
16. How is plastic laminate applied?
17. Why are coped joints used?
18. Why are coatings smoothed from the center toward the edges?
19. When should floors be stained?
20. What are the advantages of lip drawers?

16

Special Construction Methods

HERE ARE MANY JOBS FOR CARPENTERS TO DO IN any type of building. They build the frames and cover them; however, the carpenter often must do other jobs. The regular methods are not always used for the frames. Special jobs include fireplaces, chimneys, and stairs. The post-and-beam method of building is also included.

Carpenters need special skills to do special jobs. Almost every building involves one or more of these special jobs.

Specific skills that can be learned in this chapter include

- Laying out, cutting, and installing stair parts
- Designing, cutting, and making fireplace frames according to the manufacturer's or builder's specifications
- Planning and building basic post-and-beam frames

STAIRS

Carpenters should know how to make several types of stairs. Main stairs should be pleasing and attractive. This is so because they are visible in the living or working areas. They also should be safe, sturdy, and easy to climb. It is a good idea to make them as wide as possible. Service stairs are not as visible. They are used mainly to give occasional access or entry. They are used in basements, attics, and other such areas. Service stairs are not as wide or attractive as main stairs. Service stairs can be steeper and harder to climb. However, all stairs should be sturdy and safe to use.

Stair Parts

The carpenter most often will build stairs around a notched frame (Fig. 16-1). This frame is called a *carriage*. The carriage is also called a *stringer* by some carpenters. The carriage should be made of 2- × 10-inch or 2- × 12-inch boards. In this way, one solid board extends from the top to the bottom. The other stair parts then are attached to the carriage.

The part on which people step is called the *tread*. The vertical part at the edge is called a *riser*. A stair unit is made of one tread and one riser. The *unit run* is the width of the tread. The *unit rise* is the height of the riser. The tread usually is rounded on the front edge. The rounded edge also extends over the riser. The part that hangs over is called the *nose* or *nosing* (Fig. 16-2A).

Stairs also must have a handrail. Where a stair is open, a special railing is used. The fencelike supports are called *balusters*. The handrail or banister rests on the balusters. The end posts are called *newels* (see Fig. 16-2B).

Fig. 16-1 *A notched frame called a* carriage *is the first step in building stairs.*

Some stairs rise continuously from one level to the next. Other stairs rise only partway to a platform. The platform is called a *landing*. Often landings are used so that the stairs can change direction (Fig. 16-3).

Stair Shapes

Stairs are made in several shapes (Fig. 16-4). Stairs may be in open areas, or they may have walls on one or both sides. Stairs with no wall or one wall on the side are called *open stairs*. Stairs that have walls on both sides are called *closed stairs*.

Straight-run stairs The straight run is the simplest stair shape. This type of stair rises in a straight line from one level to the next. The stairs may be open or closed (see Fig. 16-4A and B).

L-shaped stairs Figure 16-4C shows L-shaped stairs. These stairs rise in two sections around a corner. To make these, the carpenter builds two sets of straight runs. The first is to the platform or landing. The second is from the landing to the next level.

U-shaped stairs The stairs in Fig. 16-4D are U-shaped stairs. They are much like L-shaped stairs. However, two corners are turned instead of one. As a rule, a platform or landing is used at each corner. Again, the carpenter makes straight-run stairs from one platform to the next.

(A)

(B)

Fig. 16-2 *Stair parts: A. risers and treads; B. railing. The newels, balusters, and rails together are called a* balustrade. *(Forest Products Laboratory.)*

Winders Some stairs have steps that fan or turn as in Fig. 16-4E. No landing is used to turn a corner. These turning steps are called *winders* or *winding stairs*. These are the hardest to make, and they take special

Fig. 16-3 *A landing is used to change direction.*

framing. As a rule, the stairs are pie-shaped only on corners. However, some special jobs may feature the true winding staircase.

Winders are considered attractive by many people. However, carpenters make winders only when there is not enough room for a landing. The winder is not as safe as a regular landing.

Stair Design

Most stairs are made with 2- × 12-inch carriages rising through a framed opening. As a rule, the framed opening is made when the floors are framed. The carpenter may check for proper support and framing before making the carriage attachments.

Of course, carpenters build stairs as the designer or architect specifies. However, carpenters should know the fundamentals of stair design. Sometimes architects or designers will not specify all the details. In these cases, carpenters must use good practices to complete the job.

Stairs should rise at an angle of 30 to 35 degrees. The minimum width of a tread is about 9 inches. There is no real maximum limit for tread width. However, a long step results when the tread is very wide. As a rule, treads 9 inches wide are used only for basements and service stairs. For main stairs, tread width normally is 10 to 12 inches.

The riser for most main stairs should not be very high. If the riser is too high, the stairs become steep and hard to climb. As a rule, 8¼ inches is considered the maximum riser height. Less than that is desirable. For main stairs, 7 to 7½ inches is both more comfortable and safer.

Find the number of steps The amount of rise per step determines the number of steps in stairs. If a 7-inch rise is desirable, then the total rise is divided by 7 inches. The total rise is the total distance from the floor level of one landing to the next. A two-story

Fig. 16-4 *A. Straight-run stairs do not change direction. This stair is part open. B. This stair will be closed. It will have walls on both sides. C. A long L-shaped stair.* (Forest Products Laboratory.) *D. U-shaped stairs.* (Forest Products Laboratory.) *E. A winder saves space. It does not use a landing to change direction.* (Forest Products Laboratory.)

building with an 8-foot ceiling is an example. The width of the floor joists and second-story finished flooring is added. The second-floor joists are 2 × 10 inches. The total rise would be about 8 feet, 10 inches. For 7-inch risers, this distance (8 feet, 10 inches) is divided by 7:

$$8'10" = 106"$$

Height/rise = number of steps

$$\frac{106}{7} = 15.14$$

15 would be the number of steps

In practice, most stairs in houses have 13, 14, or 15 steps. Fourteen steps (a riser of about 7½ inches) is probably the most common.

Using ratios To find the step width, a riser-to-tread ratio is used. There are three rules for this ratio. Remember, one tread and one riser are considered one unit (Fig. 16-5).

1. Unit rise + unit run = 17 to 18.
2. Two unit rises + one unit run = 24 to 25.
3. Unit rise × unit run = 72 to 75.

Fig. 16-5 *A unit is one run and one rise.*

Now, the total length of run is divided by the number of steps. If this length is 13 feet, 0 inches, then the step width would be 156 inches divided by 15. This would be approximately 10⅜ inches.

These two numbers (7 inches for the riser and 10⅜ inches for the tread) are used to check the stair shape. To apply the rules:

1. Allowable unit rise + unit run = 17 to 18. Actual sum is

$$7 \text{ inches} + 10\text{⅜ inches} = 17\text{⅜}$$

This figure is OK.

2. Allowable length of two unit rises + one unit run = 24 to 25. Actual length is

$$2 \times 7 = 14 \text{ inches}$$
$$1 \times 10\text{⅜} = +10\text{⅜ inches}$$
$$24\text{⅜}$$

This is OK.

3. Allowable value of unit rise × unit run = 72 to 75. Actual value is

$$7 \times 10\text{⅜} = 72\text{⅝}.$$

This is also OK.

Sometimes, two or three combinations have to be tried to get a good design. Experience has shown that these ratios give a safe, comfortable, and well-designed stair.

Headroom Headroom is the distance between the stairs and the lowest point over the opening. The minimum distance should be above 6 feet, 4 inches. It is better to have 6 feet, 8 inches. The headroom can be quickly checked. The carpenter can lay a board at the desired angle. Or a line can be strung. The distance between the board or line and the lowest point can be measured.

Stair width The width of the stair is the distance between the rail and the wall or baluster. This is the *walking area*. This area between the rail and the wall is called the *clear distance*. Narrow stairs are not as safe as wide stairs. Also, it is hard to move furniture and appliances in and out through narrow stairs. The width of the stair is a major factor.

The minimum clear distance allowed by the Federal Housing Administration (FHA) is 2 feet, 8 inches on main stairs. Sometimes 2 feet, 6 inches is used on service stairs. However, added width is always desirable.

Tread width on winders should be equal to the regular width. Winder treads are checked at the midpoint. Or the tread width can be checked at the regular walking distance from the handrail. For example, tread width on a stair is 10⅜ inches. Then the tread width on the middle of the winder also should be 10⅜ inches (Fig. 16-6).

Fig. 16-6 *In a winder, tread widths should be equal on the main path.* (Forest Products Laboratory.)

Sequence in Stair Construction

Carpenters should follow a given sequence in building stairs. The stair is laid out on the carriage piece. The step (riser and tread) notches are cut. Two carriage pieces are used for stairs up to 3 feet in width. Three or more carriage boards are used for stairs that are wider than 3 feet, 0 inches. One is on each side, and one is in the center.

Carriage boards are often positioned as walls and floors are framed. However, the stairs are not finished until the walls have been finished. Often 2-inch lumber is used for temporary steps during construction (Fig. 16-7). Stairs that rest on concrete floors are not built until the concrete floor is finished.

Three methods of making stairs are used. The first two methods require the carpenter to make the carriage. The treads are finished later. In the first method, regular lumber is used for a tread. These treads are later covered with finished hardwood flooring.

Fig. 16-7 *Two-inch lumber is used for temporary steps during construction. It can be left as a base for finish treads.*

In the second method, hardwood treads are nailed directly to the carriage. These boards are usually oak that is 1⅛ or 1¹⁄₁₆ inches thick.

The third method is called a *housed carriage*. The carriage board has grooves to receive the treads and risers. As a rule, these pieces are precut at a factory or mill. The treads are inserted from the top to the bottom. Glue and wedges are used to wedge the stairs tightly in place (Fig. 16-8). This type of stair is very sturdy and normally does not squeak.

Fig. 16-8 *Housed carriage stairs: A. housing; B. tread detail.* (Forest Products Laboratory.)

Carriage Layout

Riser height and tread width are found from the ratios. The example given used a 7-inch riser with a 10⅜-inch tread. A 2- × 12-inch board of the proper length is selected. The board is laid across the stair area to see whether the board is long enough. If so, the board is placed on a sawhorse or work table.

A framing square is used as in Fig. 16-9. The square is placed at the top end of the carriage as shown. The short blade of the square is held at the corner. The outside edge is used for these measurements. The height of the riser is measured down from the cor-

Fig. 16-9 Lay out the carriage with a framing square.

ner on the outside tongue. The blade is swung until the tread width is shown on the scale. This is shown when the tread width (10⅜ inches) meets the edge of the board. Lines are drawn along the edges of the squares.

This process is repeated for all the steps. On the last step, two things are done. First, a riser is marked for the first step height (Fig. 16-10). Then the height is adjusted to allow for the tread thickness. This must be done. When the tread is nailed to the carriage, it will raise the riser distance. For example, suppose a carriage is cut so that the full first step rises 7 inches. Then a board 1⅛ inches thick is added for the tread. This would make the total riser height 8⅛ inches. So the thickness of the tread is subtracted as in Fig. 16-10.

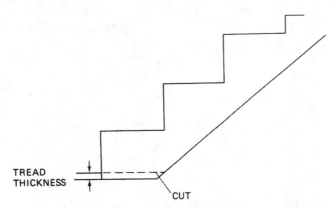

Fig. 16-10 The carriage bottom must be trimmed to make the first rise correct.

Notches are cut. A bayonet saw or a handsaw should be used. It is not a good idea to use a circular saw. The blade does not cut vertically through the wood (Fig. 16-11).

For some outside stairs, the carriage is cut differently. The layout is the same. However, the steps are dadoed into place, as in Fig. 16-12. To do this, the dado is cut on each side of the line. Cut the dado one-third to one-half the thickness of the carriage. Standard 2-inch lumber is only 1½ inches thick. The dado should be cut ½ inch (one-third the thickness) to ¾ inch (one-half the thickness) deep.

Fig. 16-11 A circular saw should not be used because it will cut past the line.

Fig. 16-12 Exterior stair construction: A. dadoed carriage; B. cleated carriage.

Frame the Stairs

First, the ledger for the top is cut. Then it is nailed in place at the top of the landing or second story. The ledger is made from 2- × 4-inch lumber. It is nailed with 16d common nails (Fig. 16-13). Next, a kicker is cut from 2- × 4-inch lumber. It is nailed (16d common nails) at the bottom, as in Fig. 16-13. Then notches are cut in the carriage for the kicker and ledger. Next, a backing stringer is nailed on the walls. This should be a clear piece of 1-inch lumber. It gives a finished appearance to the stairs. Use 8d common nails (Fig. 16-14). Next, the carriage is nailed to the ledgers, kickers,

Fig. 16-13 *A. A kicker anchors the bottom of the carriage. B. The ledger anchors the top.* (Forest Products Laboratory.)

wall, and backing strips. Heavy spikes are recommended. Next, the outside carriage is nailed to the ledger and kicker.

The outside finish board is nailed to the outside of the carriage. If the finish board is 2 inches thick, 16d finish nails should be used. If the outside board is 1-inch lumber, 8d nails should be used. After the wall is complete, the newel is nailed in place. A finish stringer can be used as in Fig. 16-15. The finish stringer is cut from a 1- × 12-inch board. A finish stringer is blocked and nailed in place as shown. Finish nails should be used.

Temporary steps are now removed from the carriage. As a rule, cutting is avoided on stair pieces. Tread overhang less than 1½ inches is ignored. The overhang becomes the nosing. However, excess width is cut from the back side.

In hardwoods, it's a good idea to drill holes for nails. Doing this keeps the nails from splitting the board. The drilled hole should be slightly smaller than the nail. The nails should be set and filled after they are driven. Of course, only finish nails are used.

First, the bottom riser is nailed in place. It must be flush with the top of the carriage riser. It is normal to have a small space at the bottom of the riser. The space at the bottom can be covered by molding. The riser is always laid first. A convenient number of steps is worked at one time. This includes all the risers that can be reached easily. Probably two or three steps can be worked at the same time. Next, the bottom tread is placed in position. It is moved firmly against the riser. The tread is drilled and

Fig. 16-14 *Backing or finish stringers on a wall.* (Forest Products Laboratory.)

nailed as needed. Treads are laid until one is on each placed riser. Then the worker moves up the steps. Again, a convenient number of risers and treads are laid. This process is repeated until the stair is finished.

The baluster is started when all the treads are installed. The ends of balusters are round like dowels. Remember, holes are located by the center of the hole. The centers of the holes for the balusters are located. They are spaced and located carefully. The holes are drilled to receive the baluster ends. The top newel post is attached as specified by the manufacturer. Next, the bottom newel post is fastened in place. Large spikes should be used. It is best to attach the newel post to the

Fig. 16-15 *A finish stringer on open stairs.* (Forest Products Laboratory.)

Fig. 16-16 *Templates are used to rout the housed stringer on the site.* (Rockwell International Power Tool Division.)

subflooring. Sometimes, carpenters must first chisel an area in the finished floor. This receives the bottom of the newel post. Next, glue is put on the round bottoms of the balusters. These are inserted in the holes drilled in the treads. Glue is never put in the hole. All the bottom pieces are glued and placed. Glue then is put on the top stems of the balusters. Then the top rail, or banister, is laid over the tops of the balusters. It is tapped firmly into position and allowed to dry. Then finish caps, molding, handrails, and trim are installed.

Install Housed Carriages

To install housed carriages, it is best to work from the back. Here, the carpenter should start at the top and work toward the bottom. The first riser is applied and lightly wedged. Then the tread is attached. The wedge is coated with glue and tapped firmly into place. Then the riser wedge is coated with glue. It is then wedged firmly into place (see Fig. 16-8A).

Wedged and glued stairs squeak less than other types. Stair squeaks are usually caused by loose nails. Since the closed carriage has fewer nails, it squeaks less.

Although this type of stair is commonly precut, carpenters sometimes cut these on the site. Special template guides and router bits should be used (Fig. 16-16).

STAIR SYSTEMS

There is a faster way of making a stair system. You might be able to locate a lumber dealer with all the supplies you need to produce a top-quality stairway. Such manufacturers as L. J. Smith Company in Ohio make it possible for the do-it-yourselfer or the carpenter who

wants to improve his or her skills to quickly produce a stairway that meets all standard requirements.

Two types of systems are available: the post-to-post system and the over-the-post system. Figure 16-17 shows the post-to-post system with the necessary parts identified. Check Fig. 16-18 for the guidelines on ordering a system. As you can see in this type of system, there are no curved railings to bother with; just use the correct upeasing or rail drop to take care of the changes in direction for the railing. Notice in this type of system that the posts have the knob or ball on top exposed as part of the decorative features of the stairway.

If railings must be bent, bending rails and bending molds make for a curved balustrade system that is very practical and inexpensive. The bending rails can be glued successfully to a 30-inch radius on the rake and a 36-inch radius on the level, provided that installation instructions are accurately followed. Installation sheets come with the wooden pieces. The rail is made up on the job by gluing either seven or nine pieces and using a polyvinyl bending mold because the glue does not stick to the Enduromold surface of the mold (Fig. 16-19). The molds are available in 8-foot lengths and fit the contours of the rail. By applying the glue, aligning the pieces of wood, and then placing the bending mold over the outside while clamping the rail to the stairs near the steps, the rail will be bent according to the curve in the stairway. Figure 16-20 shows a curved staircase and the starting step made of solid oak. The system can be made of beech, oak, poplar, or cherry.

The L. J. Smith system uses a Conect-A-Kit method of assembling and installing handrail fittings. No bolts are required. All the Conect-A-Kit fittings have a base with machined pockets and a removable top lid (or bottom filler) for easy installation (Fig. 16-21). All assembly hardware is concealed within the base of each fitting. Most of these fittings are manu-

Fig. 16-17 *Post-to-post system. Note the labeled parts. This is a combination stairway to show the possible number of parts needed and where they fit to make a complete system.* (L. J. Smith.)

factured with a 45-degree precision miter cut on the end or ends so that they can be used in combination with other fitting components to build a landing fitting assembly. The mitered ends are easily square cut for attachment to a straight handrail (Fig. 16-22).

There are a number of advantages to the kit assembly method. You can make left or right turns with the same fitting. Fittings can be used to make up a variety of fitting combinations (Fig. 16-23). Most joint connections are made on top of the rail system for better access. No rail bolts are required. The hardware included with each fitting provides better strength and tighter connections.

	ITEM	GUIDELINES
		Support System
1	TREADS	Select one **tread** for each step. For a stair open on one side order miter-returned (MRT) and add 1 1/4" to the stringer to stringer measurement. For a stair open both sides order MRT both and use the finished stringer to stringer measurement (measured outside to outside).
2	RISERS	Select one **riser** for each step. Select one more **riser** than treads per each flight because of landing tread (see #3). (Landing tread replaces the nosing over the last riser).
3	LANDING TREAD	Select sufficient lineal footage for the entire balcony and width of stairs at each landing (8090-5 is used with 3 1/2" or wider newels - check the systems catalog for newel dimensions).
4	COVE MOULD	Select sufficient lineal footage to go under all tread nosing (including miter-returns) and under all landing tread.
		Balustrade
5	UTILITY NEWEL	For a 30" - 34" rake rail height, use the shortest available **utility newel**. For a 34" - 38" rake rail height, use the longest available **utility newel**. Select one for each side with handrail.
6	INTERMEDIATE LANDING NEWEL (Rake-to-Rake)	Select an **intermediate landing newel** with 14 1/2" face when not using a landing fitting assembly. If a landing fitting assembly is used, select the 5" face **intermediate landing newel** (see pages 9 - 19 for newel length requirements at level landings, 2-winder landings, and 3-winder landings).
7	SECOND FLOOR LANDING NEWEL (Rake-to-Level)	Select the **2nd floor landing newel** with 11" face when not using a landing fitting assembly (36" balcony height). 42" balcony height requires the use of a landing fitting assembly. Use the shortest available **utility newel** for surface mount when using a landing fitting assembly. Use the longest available **utility newel** if the newel extends below the floor surface (with landing fitting assembly).
8	BALCONY NEWEL (On the Level Run)	Match the **balcony newel(s)** to the **2nd floor landing newel(s)**. For a run of 10 feet or more, use a newel at the midpoint or every 5 to 6 feet. Place a newel at every corner.
9	HALF NEWEL OR	Select the **half-newel** of the same style as the other *full* newels on the balcony. (i.e. 11" face or 5")
10	ROSETTES	Select the **round rosette** for all level run rail connections into a wall. Select the **oval rosette** for all angled rail connections into a wall (when the rail meets the wall on a rake).
11	NEWEL MOUNTING HARDWARE	Select one of the **newel mounting kits** for each newel post (if desired).
12	RAKE BALUSTERS (Open Stairs)	For 30" - 34" handrail height compliance, use the **34" baluster** for the 1st baluster on the tread and use **38"** for the 2nd and 3rd balusters on the tread. If using 3 balusters per tread, and a fitting, substitute a **42"** for the **3rd baluster** under each landing fitting assembly. For 34" - 38" handrail height compliance, use the **38" baluster** for the 1st baluster on the tread and use the **42" baluster** for the 2nd baluster on the tread. If using 3 balusters per tread, use the **38" baluster** for the 1st and 2nd balusters on the tread, and use the **42" baluster** for the 3rd baluster on the tread.*
13	RAKE BALUSTERS (Kneewall Stairs)	Select the shortest **baluster** available at a rate of 2 per tread. Place on 4" to 6" centers.
14	LEVEL RUN BALUSTERS	Use the **38" baluster** for all **36"** level runs/balconies. Use the **42" baluster** for all **42"** level runs/balconies. Place on 4" to 6" centers. Subtract one **baluster** from the calculated total to account for the end of the run. Subtract one **baluster** for each newel post on the level run. Do not however, subtract one for the newel post beneath the landing fitting at the 2nd floor landing.
15	HANDRAIL	Select **handrail** at a rate of 13" per each tread and include enough for all level runs.
16	SHOERAIL (Kneewall-Rake)	Select **shoerail** at a rate of 13" per each tread.
17	SHOERAIL (Level Runs)	Select **shoerail** to cover all balcony landing tread (if desired).
18	FILLET	Select enough **fillet** to fill all plowed handrail and all shoerail.
19	PLUGS	Select two **wood plugs** for every newel (depending on installation requirements).
20	DOUBLE-END SCREW	Select one **Dowel-Fast**™ double-end wood screw for each baluster. This is optional but highly recommended. Double-end wood screws are not needed for balusters installed within shoerail.
21	BRACKETS (Open Stairs)	Select one **bracket** for each tread (if desired).
22	HANDRAIL MOUNTING HDWE	Select one **Flush Mount Kit** or **Rail & Post Fastener** to fasten the handrail to newels.

* Note: when using 3 balusters per tread for 34" - 38" rail height, the **42" baluster** may not be long enough for use under a landing fitting assembly.

Fig. 16-18 *Post-to-post system guidelines for ordering a system.* (L. J. Smith.)

Upeasings, Caps, and Quarterturns

Upeasings and overeasings are used when rise changes. The level quarterturn is used without a newel.

The quarterturn with cap is used with a newel. The tandem cap allows for additional newels to be inserted into a long, level stretch of rail for strength. Figure 16-24 shows over-the-post upeasings, caps, and quarterturns. Figure 16-25 illustrates the applications of an

LJ-6910BM is a polyvinyl bending mold in a reusable form. Glue does not stick to Enduromold surface. Available in 8' lengths.

Fig. 16-19 *Profiled bending rails, shoerails and fillets and over-the-post volute, turnout, and starting easing with cap.* (L. J. Smith.)

(A)

(B)

Fig. 16-20 *A. A curved stairway with an over-the-post system and closed-in stairs. B. Note the oak starting step.*

Climbing Volutes

The Climbing Volutes are an addition to our line of code compliance products to provide a more attractive alternative to the extra long newel post on the starting step. The Climbing Volutes are also available in the following <u>non-plowed</u>, handrail profiles. Each volute requires 4 or 5 balusters. Climbing Volutes are a one-piece fitting, and are <u>not</u> part of the Conect-A-Kit fitting line.

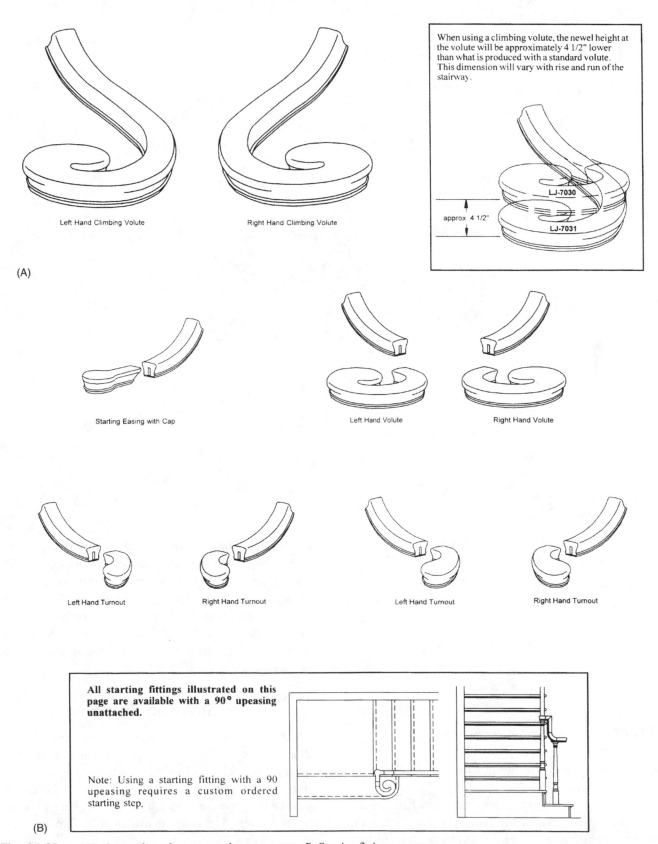

When using a climbing volute, the newel height at the volute will be approximately 4 1/2" lower than what is produced with a standard volute. This dimension will vary with rise and run of the stairway.

LJ-7030

approx 4 1/2"

LJ-7031

Left Hand Climbing Volute

Right Hand Climbing Volute

(A)

Starting Easing with Cap

Left Hand Volute

Right Hand Volute

Left Hand Turnout

Right Hand Turnout

Left Hand Turnout

Right Hand Turnout

All starting fittings illustrated on this page are available with a 90° upeasing unattached.

Note: Using a starting fitting with a 90 upeasing requires a custom ordered starting step.

(B)

Fig. 16-21 *A. Climbing volutes for an over-the-post system. B. Starting fittings.* (L. J. Smith.)

Quarterturn With Cap

Top Lid

Hardware Included:
for joining
end(s) to
handrail
or a fitting

for attaching
fitting to
newel post

(C)

Fitting Base
With
Machined
Pockets

Pre-drilled
Hole on Each
End

Upeasing

Upeasings
Have Square
Cut Ends

Bottom
Filler

Hardware Included:
for joining
end(s) to
handrail
or a fitting

Machined
Slot on Each
End

Fitting Base With
Machined Pockets

Fig. 16-21 *C. Exploded view of two Conect-A-Kit fittings.* (L. J. Smith.)

**1 Individual
Conect-A-Kit Fittings**

Quarterturn
With Cap

Rail Drop

Upeasing

**2 Attach Conect-A-Kit
Fittings Together**

**3 Landing Fitting
Final Assembly**

Fig. 16-22 *Building landing fitting assemblies.*
(L. J. Smith.)

1 From The Box

Quarterturn With Cap

2 Square Cut End(s)

3 Assemble to Rail End(s)

Fig. 16-23 *Traditional stairway applications.* (L. J. Smith.)

over-the-post system. Guidelines for ordering the system are given in Fig. 16-26.

A number of post styles are shown in Fig. 16-27. Notice the differences in the two systems. The over-the-post system has a dowel rod sticking up to fit into

the rail, whereas the post-to-post system has a knob or ball on top, with the exception of the square top. The rail fits against the post or newel in this style and terminates against the wall in a rosette (see Fig. 16-31). Wall rails and wall rail brackets are shown in Fig. 16-31, as is the quarterturn and starting easing. A handrail is also required, in some housing codes, on the wall side of the staircase (Fig. 16-28).

Starting Steps

All starting steps are shipped with necessary cove and shoe molding. All starting steps are available in longer lengths, 48 inches being the standard. Single bullnose starting steps are reversible and can be job-cut for shorter lengths. Tread is $1\frac{1}{32} \times 11\frac{1}{2}$ inches. Total rise is $8\frac{1}{32}$ inches. The riser measures $\frac{3}{4} \times 7$ inches (Fig. 16-29).

Handrail Profile											
	LJ-6010	LJ-6010P	LJ-6210	LJ-6210P	LJ-6400	LJ-6400P	LJ-6109	LJ-6109P0	LJ-6109P1	LJ-6519	LJ-6519P

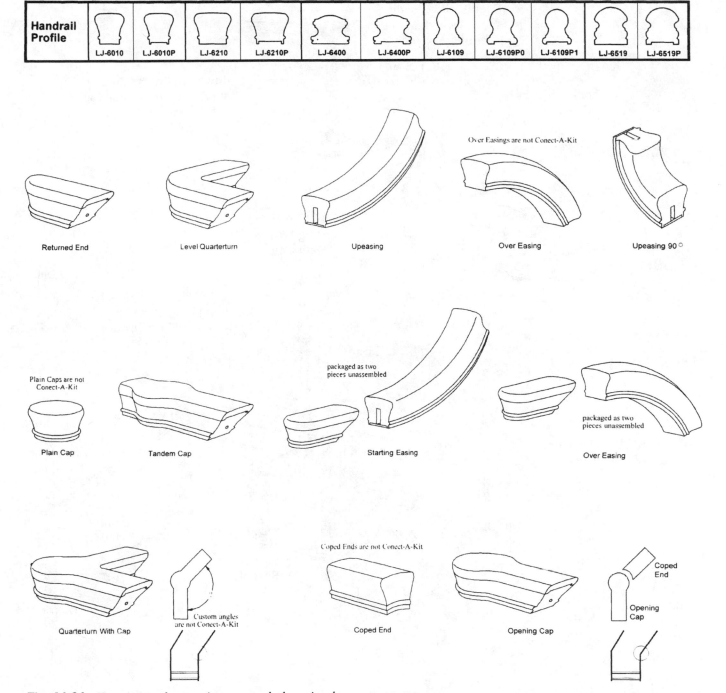

Returned End   Level Quarterturn   Upeasing   Over Easing   Upeasing 90°

Over Easings are not Conect-A-Kit

Plain Caps are not Conect-A-Kit

Plain Cap   Tandem Cap   Starting Easing   Over Easing

packaged as two pieces unassembled

packaged as two pieces unassembled

Quarterturn With Cap   Coped End   Opening Cap

Custom angles are not Conect-A-Kit

Coped Ends are not Conect-A-Kit

Coped End   Opening Cap

Fig. 16-24 *Upeasings and overeasings are used where rise changes.* (L. J. Smith.)

Treads and Risers

The false-tread kit has three possible applications: open treads, left hand; open treads, right hand; and closed treads. Various treads and risers are shown in Fig. 16-30.

Rosettes and Brackets

Rosettes are the round or oval pieces used against the wall that terminate the rail in a fitting arrangement (Fig. 16-31).

In some cases, wall rails are needed. Going down basement steps is a common location. Some commu-

nities have housing codes for new houses that include a handrail against the wall opposite the handrail on the open side of the staircase. The rails are available in 8-foot lengths with various quarterturns and starting easings as well as overeasings. Three shapes are shown here that represent the typical wall rail.

Figure 16-32A is an example of an over-the-post style with the handrail terminating in a rosette. However, Fig. 16-32B illustrates the post-to-post style with a turn. A well-chosen combination of materials prop-

Intermediate Landing
Newel with Landing
Fitting Assembly
(2 Upeasings, Quarterturn
With Cap, Rail Drop)

Landing Fitting
Assembly

Level
Quarterturn

Rail Drop

Upeasing

Utility Newel with Landing
Fitting Assembly (Tandem Cap,
Rail Drop, Upeasing)

Utility Newel and
Starting Easing
With Cap

Utility Newel with
Landing Fitting Assembly
(Tandem Cap, Rail Drop, Upeasing)

Intermediate Landing
Newel with Landing
Fitting Assembly
(2 Upeasings, Quarterturn
With Cap, Rail Drop)

34" and 38" Balusters, or
38" and 42" Balusters

Starting
Easing
With Cap

Utility
Newel

38" or 42"
Balusters

Utility Newel
with Left
Hand Turnout

Double End Bullnose
Starting Step

Utility Newel with Right
Hand Volute

Fig. 16-25 *Applications of an over-the-post system.* (L. J. Smith.)

	ITEM	GUIDELINES
		Support System
1	STARTING STEP	For use with volutes and turnouts. Select for bullnose on one or both ends (match the floor plan on p. 38 of the Systems Catalog). Measure finished stringers from outside to outside.
2	TREADS	Select one **tread** for each step (except the starting step). For a stair open one side order miter-returned (MRT) and add 1 1/4" to stringer to stringer measurement, then refer to the next longer standard length available. For a stair open both sides order MRT both and refer to stringer to stringer measurement (measured outside to outside).
3	RISERS	Select one **riser** for each step (except the starting step). Select one more **riser** than treads per each flight because of landing tread (see #4). Landing tread replaces nosing over last riser.
4	LANDING TREAD	Select sufficient lineal footage for the entire balcony and width of stairs at each landing. (8090-5 is used with 3 1/2" or wider newels - check the systems catalog for newel dimensions).
5	COVE MOULD	Select sufficient lineal footage to go under all tread nosing (including miter-returns) and under all landing tread.
		Balustrade
6	STARTING FITTING	Select either a **volute, turnout, or starting easing w/cap**. Choose a **climbing volute** to eliminate the need for an unusually long starting newel.
7	UTILITY NEWEL	Use everywhere except at the intermediate landing corner of an L-shaped stair.
8	UTILITY NEWEL (50")	Use for **balcony newel(s)** that will extend below the floor surface. Also use under a starting easing w/cap when a starting step is not used and the rake handrail height is 34" or higher.
9	INTERMEDIATE LANDING NEWEL	Use the **57" intermediate landing newel** at the intermediate landing corner of an L-shaped stair. Use the **65" intermediate landing newel** in 2-winder landing situations. Use the **73" intermediate landing newel** in 3-winder situations.
10	LEVEL RUN NEWEL (Utility Newel)	If the balcony is 10 feet or longer, use the **utility newel** every 5 or 6 ft. under a tandem cap. Place a newel at every corner under a quarterturn with cap. Use the **50" utility newel** if the newel is to extend below the 2nd floor surface.
11	HALF NEWEL OR	Select the **half-newel** of the same style as the other *full* newels on the balcony. (i.e. utility newels)
12	ROSETTES	Select the **round rosette** for all level run rail connections into a wall. Select the **oval rosette** for all angled rail connections into a wall (when the rail meets the wall on a rake).
13	NEWEL MOUNTING HARDWARE	Select one of the **newel mounting kits** for each newel post (if desired). The Shortest and Longest Utility Newels are packaged with newel mounting hardware.
14	BALUSTERS FOR VOLUTES AND TURNOUTS	Volutes require **4 or 6 - 1 1/4" or 4 - 1 3/4" balusters**. Turnouts require **2 - 1 1/4" or 1 - 1 3/4" baluster(s)**. For 30" - 34" rake rail height, use **38" balusters** under all volutes and use **42" baluster(s)** under all turnouts. For 34" - 38" rake rail height, use **42" balusters** under all volutes and turnouts. Climbing Volute Requirements: For 30" rake rail height, use **3 - 34" and 1 or 2 - 38" baluster(s)**; For 34" rake rail height, use **3 - 38" and 1 or 2 - 42" baluster(s)**; For 36" rake rail height, use **4 - 42" balusters**.
15	BALUSTERS FOR STARTING EASING WITH CAP (Open Stairs)	Use **1 - 38" baluster** for 30" - 34" handrail height compliance. Use **1 - 42" baluster** for 34" - 38" handrail height compliance.
16	RAKE BALUSTERS (Open Stairs)	For 30" - 34" handrail height compliance, use the **34" baluster** for the 1st baluster on the tread and use **38"** for the 2nd and 3rd balusters on the tread. If using 3 balusters per tread, substitute a **42"** for the **3rd baluster** under each landing fitting assembly. For 34" - 38" handrail height compliance, use the **38" baluster** for the 1st baluster on the tread and use the **42" baluster** for the 2nd baluster on the tread. If using 3 balusters per tread, use the **38" baluster** for the 1st and 2nd balusters on the tread, and use the **42" baluster** for the 3rd baluster on the tread.*
17	RAKE BALUSTERS (Kneewall Stairs)	Use the **34" baluster** at a rate of 2 per tread. Place on 4" to 6" centers. Subtract one baluster from the calculated total as the starting newel replaces the first baluster.
18	LEVEL RUN BALUSTERS	Use the **38" baluster** for all 36" level runs/balconies. Use the **42"** for all **42"** level runs/balconies.** Place on 4" to 6" centers. Subtract one **baluster** from the calculated total to account for the end of the run. Subtract one **baluster** for each newel post on the level run. Do not however, subtract one for the newel post beneath the landing fitting at the 2nd floor landing.
19	HANDRAIL	Select **handrail** at a rate of 13" per each tread and include enough for all level runs.
20	HANDRAIL FITTINGS (Landing Fitting Components)	Match each corner of the floor plan to a corresponding plan in the systems catalog. Specify each **landing fitting component** needed to construct the landing fitting assembly. Assembled goosenecks have been replaced with components.
21	HANDRAIL FITTINGS (Over half-newel)	Select the **opening cap** which corresponds to the rail to cover each half-newel (i.e. LJ-6010 requires LJ-7019). This will be cut on the job.
22	HANDRAIL FITTINGS (Miscellaneous)	Each newel or half-newel must be covered with a **fitting**. Select the fittings from the drawings in the systems catalog.
23	SHOERAIL (Kneewall-Rake)	Select **shoerail** at a rate of 13" per each tread.
24	SHOERAIL (Level Runs)	Select **shoerail** to cover all balcony landing tread (if desired).
25	FILLET	Select enough **fillet** to fill all plowed handrail and all shoerail.
26	PLUGS	Select two **wood plugs** for every newel (depending on installation requirements).
27	DOUBLE END SCREW	Select one **Dowel-Fast** ™ double-end wood screw for each baluster. This is optional but highly recommended. (Not for use with shoerail)
28	BRACKETS (Open Stairs)	Select one **bracket** for each tread (if desired).

* Note: when using 3 balusters per tread for 34" - 38" rail height, the **42" baluster** may not be long enough for use under a landing fitting assembly. ** An Over The Post rake rail height of 34" - 38" requires **42" balusters** for 36" and 42" level balconies

Fig. 16-26 *Guidelines for ordering an over-the-post system. (L. J. Smith.)*

Stair Systems 517

Post to Post Styles

Shortest Utility Newel (5" top square) - Use this newel for the starting newel on all stairs with a 30" - 34" rake rail height. May use this newel as a balcony newel that is surface mounted.

Longest Utility Newel (5" top square) - Use this newel as a starting newel for stairs with a 34" - 38" rake rail height and for balcony newels that will extend below the floor surface.

2nd Floor Landing Newel (11" top square) - Use this newel for the 2nd floor landing newel when not using a landing fitting assembly. Will achieve a 36" balcony railing height.

Intermediate Landing Newel (14 1/2" top square) - Use this newel for level intermediate landings with no landing fitting assembly.

Intermediate Landing Newel (29" bottom square) - Use this newel for level landings with a landing fitting assembly.

Intermediate Landing Newel (37" bottom square) - Use this newel for intermediate landings with 2-winder treads and a landing fitting assembly.

Intermediate Landing Newel (45" bottom square) - Use this newel for intermediate landings with 3-winder treads and a landing fitting assembly.

Over the Post Styles

Shortest Utility Newel (43") - Use this newel under all starting fittings and as a balcony newel that is surface mounted. (See "Longest Utility Newel" for exception on a starting easing with cap).

Longest Utility Newel (50") - Use this newel under a starting easing with cap when a starting step is not used and the rake handrail height is 34" or higher. May also be used as a balcony newel that will extend below the floor surface.

Intermediate Landing Newel (57") - Use this newel for level intermediate landings.

Intermediate Landing Newel (65") - Use this newel for intermediate landings with 2-winder treads.

Intermediate Landing Newel (73") - Use this newel for intermediate landings with 3-winder treads.

Fig. 16-27 *Posts for post-to-post style and over-the-post style stairways.* (L. J. Smith.)

LJ-5015 LJ-5200 LJ-5035 LJ-5141 LJ-5067 LJP-2015 LJP-20158 LJT-2015 LJF-2015 LJ-5300 LJ-53008 LJP-2005 LJP-20058 LJT-2005 LJF-2005 LJ-5005 LJ-50058 LJ-5060 LJ-5360

All balusters are available in 3 lengths: 34", 38", 42"

The following guidelines will achieve a 32" minimum/36" maximum rake rail height

Rake Balusters:

For 30" - 34" rake rail height, use 34" for the 1st baluster on the tread and 38" for the 2nd and 3rd balusters on the tread. If using 3 balusters per tread, use a 42" baluster under each landing fitting assembly (formerly called gooseneck fittings).

For 34" - 38" rake rail height, use 38" for the 1st baluster on the tread and 42" for 2nd baluster on the tread. If using 3 balusters per tread, use 38" for the 1st and 2nd balusters on the tread and use 42" for the 3rd baluster on the tread. Note: when using 3 balusters per tread, the 42" baluster may not be long enough for use under a landing fitting assembly (formerly called gooseneck fittings).

Use the 34" baluster for all kneewall stairs.

Balcony Balusters:

Use 38" balusters for all 36" level balconies and use 42" balusters for all 42" level balcony rails (exception: an Over The Post rake rail height of 34" - 38" requires 42" balusters for all 36" and 42" level balconies).

Starting Fitting Balusters:

For 30" - 34" rake rail height, use 38" balusters under each volute and starting easing with cap, and use 42" balusters under each turnout.

For 34" - 38" rake rail height, use 42" balusters under all starting fittings.

Climbing Volute requirements: for 30" rake rail height, use 34" and 38" balusters; for 34" rake rail height, use 38" and 42" balusters; for 36" rake rail height, use 42" balusters.

Select a handrail that corresponds to the baluster choice.

Fig. 16-28 *Balusters and handrails.* (L. J. Smith.)

For Use with Volutes and Turnouts

LJ-8010 and LJ-8015 are for use with LJ-6010(P), LJ-6210(P) turnouts or standard volutes, and LJ-6109 (P0,P1) turnouts.

LJ-8310 and LJ-8315 are for use with LJ-6519(P), LJ-6400(P) turnouts or standard volutes, LJ-6109(P0,P1) standard volutes, and all climbing volutes.

For Use with Square Top & Box Newels, and 90° Starting Fittings

LJ-8040 and LJ-8060 are for use when placing any square top starting newel at the second riser. Also used with Box Newels and standard stair parts.
LJ-8440, LJ-8640, LJ-8460, and LJ-8660 are for use with situations where the starting fitting utilizes a 90° upeasing, as illustrated to the right. Use LJ-8440 and LJ-8460 with LJ-6010(P), LJ-6210(P) turnouts or volutes, and LJ-6109(P0,P1) turnouts. Use LJ-8640 and LJ-8660 with LJ-6519(P), LJ-6400(P) turnouts or volutes, and LJ-6109(P0,P1) volutes.

For Use with Box Newels

LJ-8030 and LJ-8050 are for use with Box Newels LJ-4075 and LJ-4091.

Fig. 16-29 *Starting steps.* (L. J. Smith.)

False Ends

LJ-8079
False Tread Kit

6"   11"   1/2"

14"   9"

The False Tread Kit includes all that is needed to handle an open tread right hand situation, or an open tread left hand situation, or a closed tread situation.

8"

LJ-8076
False Riser Only

6"   1/2"

1 1/4"   1 13\16"

LJ-8078
False Tread Mould
(Lineal)

The False Tread Kit may be used for one of three applications

Open Treads LH   Open Treads RH   Closed Treads

6"

LJ-8010-2
False End Starting Step of LJ-8010
LJ-8310-2
False End Starting Step of LJ-8310

LJ-8010-2 is used when carpeting center of stairs when a volute or turnout is used. LJ-8010-2 is reversible for L.H. or R.H. applications. Shipped as a set with cove and shoe moulding.

Treads, Risers and Mouldings

Plain Treads, Risers, and Mitered Return (one end) Treads are available in these lengths: 36", 42", 48", 54, 60", 72".

LJ-8070
Plain Tread
1 1/32" x 10 1/2" or 11 1/2"

LJ-8070MRT *
Mitered Return Tread
Specify right or left

LJ-8070MRT **
Mitered Return Tread
Both Ends

LJ-8075
Riser
3/4" x 7 1/2" or 8"

LJ-8090-5 is used with newels that are 3 1/2" or wider.

LJ-8090
Landing Tread
3 1/2"

LJ-8090-5
Landing Tread
5 1/2"

LJ-8080
Lineal Nosing
1 1/16" x 1 1/4" x RL

LJ-8080-13
Tread Return Nosing
1 1/16" x 1 1/4" x 13"

LJ-8422
Shoe Mould
1/2" x 3/4"

LJ-8095
Cove Mould
5/8" x 13/16"

* Add 1 1/4" to measurement across finished stringers (outside to outside) for one end miter returned treads.

48"   46 3/4"

Example: LJ-8070
11 1/2" x 48"
MR Right Hand

** Measure across finished stringers (outside to outside) for both end miter returned treads.

50 1/2"   48"

Example: LJ-8070
11 1/2" x 48"
MR Both Ends

Fig. 16-30 *Treads and risers.* (L. J. Smith.)

Rosettes and Brackets

Use LJ-7026 and LJ-7027 with LJ-6005, LJ-6010(P), LJ-6210(P), LJ-6109(P0,P1), LJ-6519(P), LJ-6400(P).

LJ-7026

LJ-7027

LJ-7028
11 1/2" x 8 1/8" x 5/16"

LJ-7029
11 1/2" x 8 1/8" x 5/16"

Wall Rails

LJ-6039
1 3/4" Round

LJ-6040
1 3/4" x 1 5/8"

LJ-6041
2" Round

LJ-6042
2 1/4" x 1 9/32"

Wall Rail Brackets

LJ-3002
Brass Plated
Wall Rail Bracket

LJ-3027
Polished Solid Brass
Wall Rail Bracket

Wall Rail Fittings

For use with LJ-6040 wall rail

LJ-8040

LJ-7023
Quarterturn

LJ-7024
Starting Easing

LJ-7025
Over Easing

For use with LJ-6041 wall rail

LJ-6041

LJ-7001
Quarterturn

LJ-7002
Starting Easing

LJ-7003
Over Easing

For use with LJ-6042 wall rail

LJ-6042

LJ-7005
Quarterturn

LJ-7006
Starting Easing

LJ-7007
Over Easing

Fig. 16-31 *Rosettes, brackets, wall rail brackets, wall rails, and wall rail fittings.* (L. J. Smith.)

erly assembled and installed can make a difference in the way a house is valued. The stairway can become the focal point of the interior and as much a part of the decorations as furniture, drapes, and carpets.

FOLDING STAIRS

Folding stairs have a number of uses. They are used to reach the attic but fold up and out of the way when not in use. They are also handy whenever it is necessary to have access to an area for which there is not enough room for a permanent type of stairway (Fig. 16-33).

CLOSED

OPEN

Fig. 16-33 *The folding stairway fully extended. (Memphis Folding Stairs.)*

Locating the Stairway

Allow sufficient space for a safe landing area at the bottom of the stairway. Be sure that there is enough clearance for the swing of the stair as it is being unfolded to its full length (Fig. 16-34).

Making the Rough Opening

Cut and frame the rough opening through the plaster or ceiling material the same size as shown on the carton. Generally, the rough opening size of the stair as listed on the carton will be ½ inch wider and ⅝ inch longer

(A)

(B)

Fig. 16-32 *A. A complete over-the-post system showing the handrail, utility newel, balusters, starting easing with cap, shoerail, fillet, and rosette. B. A complete post-to-post system with handrail, intermediate landing newel, second-floor landing newels, balusters, shoerail, fillet, and rosette. (L. J. Smith.)*

ROUGH OPENING	CEILING HEIGHT	LANDING SPACE (1)
22 × 48"	8'5"	59"
22 × 54"	8'9"	67"
22 × 54"	10'	73"
*22 × 60"	8'9"	67"
*22 × 60"	10'	73"
25 1/2 × 48"	8'5"	59"
25 1/2 × 54"	8'9"	67"
25 1/2 × 54"		73"
*25 1/2 × 60"	8'9"	67"
*25 1/2 × 60"	10'	73"
**30 × 54"	8'9"	67"
**30 × 54"	10'	73"
**30 × 60"	8'9"	67"
**30 × 60"	10'	73"

Fig. 16-34 *Space needed for the stairway.* (Memphis Folding Stairs.)

Fig. 16-35 *Framing the support for the stairway.* (Memphis Folding Stairs.)

than the *actual net size* of the stairway. This allows for shimming and squaring the stair in the opening.

In most cases, stairways are installed parallel to ceiling joists (Fig. 16-35). However, in some cases, the stair must be installed perpendicular to the ceiling joist (Fig. 16-36).

Caution: If the house uses roof trusses, do not cut ceiling joists without engineering consultation and approval. If it is necessary to cut the ceiling joists or truss cord, watch out for electrical wiring, and be sure to tie the cut members to other joists or trusses with 2 × 6 or 2 × 8 headers forming a four-sided frame or stairwell

to install the stair. Keep the corners square to simplify the installation. Figures 16-35 and 16-36 show the frame that has to be built before installing the stair if the stair is put in after the house is constructed. The dark area shows the frame needed before installing the stair.

Figures 16-35 and 16-36 also show how to frame the rough opening for the stair. Installation parallel to existing joists requires only single joists and headers. Installation perpendicular to the joists requires double headers and joists. Make the ceiling joists and header sections from the same size lumber as the existing joists. When making double headers, fasten members

Fig. 16-36 *Rough opening for the stairway. (Memphis Folding Stairs.)*

together with 10d common nails. The double-joist sections shown in Fig. 16-36 must be long enough to be supported by a load-bearing wall at both ends.

Temporary Support for the Stairway

It is necessary to hold the stairway in the prepared rough opening by use of temporary boards that extend across the width of the rough opening and form a ledge of ½ inch at the main hinge end and a ledge of ⅞ inch at the pull cord end. These boards should be nailed securely enough to hold the weight in the rough opening.

> **Caution:** Do not place any weight on the stair at this time (Fig. 16-37).

Fig. 16-37 *Placement of temporary supports. (Memphis Folding Stairs.)*

Placing Stairway in the Rough Opening

Raise the stairway into the attic by turning it sideways. Pull it up through the rough opening, and then lower it carefully until the hinge end rests on the ½-inch ledge and the pull cord end rests on the ⅞-inch ledge. This can be done from above in the attic while a helper on the floor is needed to lower the stair ladder sections out

of the way for nailing the stair frame to the rough opening frame. As an additional safeguard against dropping the stair through the rough opening, it is best to secure it with several 8d common nails driven through the stair frame into the rough opening frame to help hold the stair in place. Do not drive these nails home so that they can be removed after permanent nailing is completed.

Be sure that the stair is square and level in the rough opening. Blocks of wood can be used for shims to straighten the stair frame in the event that it has become bowed in inventory. This is normal because these wood parts are subjected to strong spring tension sometimes several months before installation, but they can be straightened by the use of nails and shims.

The next step involves lowering the stairway and unfolding the stairway sections. Do not stand on the stair at this time. Use a step ladder or extension ladder.

Nail the sides (jambs) of the stairwell to the rough opening frame using 16d nails or 3-inch wood screws. Holes are provided in the pivot plate and the main header. Also nail through the main hinge into the rough opening header, and then complete the permanent nailing by nailing sufficient 16d nails to securely fasten the stair to all four sides of the opening. Remove the temporary support slats and the 8d finish nails used for temporary support.

Adjustments

Pull the stairway down. Open the stair sections, folding the bottom section under the middle section so that the top and middle sections form a straight line (Fig. 16-38). Apply pressure so that the hardware arms are fully extended. Maintain this pressure, and use a straightedge placed on top of the middle section (see Fig. 16-38). Slide the straightedge down until it contacts the floor. Measure from point A to the floor. Record on the top side of bottom section C. Using the same procedure, measure from bottom side B to the floor. Record on the bottom side D. Cut from C to D (see Fig. 16-38). It is possible for your landing area to be uneven owing to a floor drain, an unlevel floor, or other reasons. Be sure to measure both sides of the bottom section using these procedures. The bottom section should fit flush on the floor on both sides after cutting (Figs. 16-39 and 16-40). All joints should be tight at each section with weight on the stair (see Fig. 16-40).

When the stair is correctly installed, stand on the second step of the bottom section. Be sure that the stair is slanted from the ceiling to the floor and that all sections of the stair are in a completely straight line, as shown in Fig. 16-41. This should occur whenever the stair is used.

The feet of the stairway that rests on the floor always must be trimmed so that each part of the foot or bottom section of the stair always fits flush on the floor and rests firmly and snugly on the floor (Fig. 16-42). Failure to do so may produce undue stress on the components of the stairway and cause a break in the stairway. It could possibly cause bodily injury. Figure 16-43 shows the completed unit.

Fig. 16-39 *Bottom section measurements for cutting.* (Memphis Folding Stairs.)

Fig. 16-38 *Measuring the bottom section for fitting the floor.* (Memphis Folding Stairs.)

Fig. 16-40 *All joints should fit tightly, and a snug fit should be seen at the floor.* (Memphis Folding Stairs.)

526   Special Construction Methods

Fig. 16-41 *Proper and improper fits.* (Memphis Folding Stairs.)

Fig. 16-42 *Right and wrong way floor fit.* (Memphis Folding Stairs.)

FIREPLACE FRAMES

Although the carpenter is ordinarily not concerned with the building of the chimney, it is necessary to be acquainted with the methods of framing around it. Many contemporary homes have prefabricated fireplaces installed with their own set of instructions included for the carpenter and crew. As a rule, the carpenter does not make a fireplace. Fireplaces are usually made of brick, stone, or concrete block. These are constructed by people in the trowel trades. However, the carpenter must make framed openings for the fireplace and flues (or chimney). Instructions for prefab installation by carpenters include the following minimum requirements:

- No wooden beams, joists, or rafters should be placed within 2 inches of the outside face of a chimney. No woodwork should be placed within 4 inches of the back wall of any fireplace.

- No studs, furring, lathing, or plugging should be placed against any chimney or joints thereof. Wooden construction should be set away from the chimney, or the plastering should be directly on the masonry, on metal lathing, or an a noncombustible furring material.

- The walls of fireplaces should never be less than 8 inches thick if brick or 12 inches if stone.

Also, today many builders use metal fireplace units. The chimneys for these are made of metal too. These fireplaces and chimneys are installed in framed openings (Fig. 16-44).

Fireplaces give the appearance and feeling of comfort and security. Decorative fireplaces are low in heating efficiency. However, there are popular devices that help to offset their inefficiencies. Fireplaces can be made with various types of ventilators and reflector devices. These allow the heat to be directed and radiated into the living area for greater efficiency. Chimneys are also used to vent furnaces, water heaters, and other fuel-burning appliances. Often these vents are brought through the roof inside the chimney unit. This makes the roofline more attractive. Fewer unsightly vents pierce the roof structure.

CLOSED

OPEN

Fig. 16-43 *Fully extended stairway. (Memphis Folding Stairs.)*

Ceilings and Roofs Openings

Carpenters do the rough openings in ceilings and roofs. The carpenter makes these openings from the plans. These should be made as the roof and walls are constructed. They should be built before the fireplace is started. Special care must be taken to maintain proper clearances. A minimum distance of 2 inches is required between all frame members and the chimney. This allows for uneven settling of the chimney foundation. It also provides additional safety from fire.

Use the same sequence used for constructing stair openings for fireplaces. First, an inner joist is nailed in place. This becomes the side of the opening. Then a first header is nailed in place. This boxes in the sides of the rough opening. Then tail joists or rafters are nailed in place. After this step, second headers are nailed on the insides. Next, the outer joist or rafter is added to complete the rough opening. After the rough opening is framed, the deck for the roof is applied.

At this point, the fireplace, flue, and chimney should be built or put in place. Proper flashing techniques should be used where the chimney goes through the roof. After the flashing is installed, the roof is applied (Fig. 16-45).

Fireplace Types

Some fireplaces are installed after the house has been finished. They are usually the stand-alone type with a hole cut into the ceiling and roof to accommodate their particular instructions for a safe installation. They may be of the contemporary or traditional style and usually are chosen by the owner after the house has been declared safe for occupancy.

Several types and shapes of fireplaces are common. Others, such as the one in Fig. 16-46B, are used occasionally. These special units add beauty, warmth, and utility to a living area.

A shape that is used commonly is the standard rectangular shape. Another shape is the double opening, with open areas on the face and one side. See-through fireplaces have openings on both faces, as in Fig. 16-46. This fireplace serves two rooms, as shown.

Many free-standing fireplaces (in contemporary surroundings) are nothing more than a piece of furniture and are chosen for their ability to contribute to the decor of the room. Free-standing fireplaces are popular. They are usually metal units located over a hearth area on the floor (Fig. 16-47). Another version of the free-standing fireplace is shown in Fig. 16-48. This type of fireplace features an open fire without walls. A large hood is placed over the unit for the chimney.

Brick (Fig. 16-49) and natural stone are the most common materials used for hand-built fireplaces. However, some of these are metal units faced with brick and stone. The insides must be lined with firebrick. Most bricks will absorb moisture. Fire changes the moisture to steam. This expands and causes the cracks often seen in the brick. Firebrick is a special brick. It has been glazed so that it will not absorb moisture. Using firebrick, which does not have moisture, reduces cracking.

Many fireplaces are made from masonry materials. They are constructed piece by piece in a proper shape. However, getting the correct shape is very critical. A poorly designed fireplace will not draw the smoke. This will allow the smoke to enter the house or building and cause discomfort.

To make the design of the fireplace easier, builders often use metal units. There are two types of these units. One is a plain unit made of sheet metal. The sheet-metal unit is placed on a reinforced concrete hearth or bottom. A properly designed frame is built around it. The outside may be faced with brick or stone as desired.

Fig. 16-44 *Metal fireplace units must be enclosed by frames.* (Martin Industries.)

Another type of prefabricated metal fireplace has hollow chambers or tubes. These tubes pass through the heated portion. They allow air to be drawn in and heated. The heated air is circulated into the living area. These openings or tubes are vented to the front of the fireplace. Since heated air rises naturally, natural ventilation occurs (Fig. 16-50). These tubes can be connected to the regular furnace air system. Special blowers are also used to increase their heating effect. These systems make the fireplace more efficient.

General Design Factors

The design of a fireplace and chimney is extremely important. Certain practices must be followed. The heat from the burning fire will cause the smoke to rise. It must pass through the chimney to the outside. As a rule, the chimney and fireplace are built as a single unit. A special foundation is required for this combined unit. The chimney is a vertical shaft. The smoke from the burning fire passes through it to the outside. The chimney also may be used for vents or flues from furnaces and water heaters. Footings for the fireplace should extend beneath the frostline. Also, they should extend 6 inches or more beyond the sides of the fireplace. Chimney walls should be at least 4 inches thick. Manufacturers' specifications always should be followed closely.

The lower part of a fireplace is called the *hearth* (Fig. 16-51). The hearth has two sections. The first is the outer hearth. This is the floor area directly in front

(A)

(B)

Fig. 16-45 *A. Chimney construction above the roof. B. Various types and arrangements for prefab fireplaces.*

of the fireplace. The second part is the bottom of the chamber where the fire is built. The front hearth should be floored with fire-resistive material. Tile, brick, and stone are all used. These contrast with the regular floor. The contrast makes the hearth decorative as well as practical. Some fireplaces have a raised hearth (Fig. 16-52). The raised hearth is also an advantage in some cases. As in Fig. 16-53A, a special ash pit

may be built at floor level. This is an advantage when houses are built on concrete slabs. The raised hearth allows a cleanout unit to be installed. Notice how a roof peak higher than the top of the chimney can cause downdrafts (see Fig. 15-53B and C).

The back hearth, back wall, and sides of the fireplace are lined with firebrick. The back wall should be approximately 14 inches high. The depth from the

(A)

(B)

OPEN FRONT

OPEN FRONT AND SIDE

OPEN FRONT AND BACK

Fig. 16-46 *A. A see-through fireplace serves two rooms.* (Potlatch.) *B. Types of firplaces.*

Fig. 16-47 *A free-standing metal fireplace.* (Majestic.)

Fig. 16-48 *A free-standing fireplace that is open on all sides.* (Weyerhauser.)

Fig. 16-49 *This metal fireplace is faced with brick.*

front lintel to the back wall should be 16 to 18 inches (see Fig. 16-51). Sidewalls and the top portion of the back are built at an angle. This angle helps the brick to radiate and reflect heat into the room. The back wall should slope forward. The amount of slope is almost one-half the height of the opening. This slope reflects heat to the room. It also directs the smoke into the throat area of the fireplace.

Fireplace throat The throat is the narrow opening in the upper part of the fireplace (see Fig. 16-51). The throat makes a ledge called a *smoke shelf.* The smoke shelf does three things for the fireplace. First, it provides a barrier to downdrafts. The smoke and heated air rise rapidly in the chimney. This movement causes an opposite movement of cold air from the outside.

This is called a *downdraft* (see Fig. 16-53). The smoke shelf blocks the downdraft as shown. It also keeps the downdraft from disrupting the movement of the smoke. The throat narrows the area in the chimney.

A. Fan Kit. Increases the flow of heated room air.
B. Duct Kits. Divert heated air into adjacent rooms, even upstairs, with the flick of a lever.
C. Glass Enclosure Kits. To save warmed room air from going up the chimney. Tempered glass, framed in matte black or antique brass trim.
D. Outside Air Kits. Feed the fire with outside air, rather than warm room air.

☐ Heated air
☐ Room air
■ Outside air

Fig. 16-50 *A metal fireplace unit with an air circulation system.* (Majestic.)

This increases the speed of the smoke. The faster speed helps to create a good draft. Without proper throat design, the chimney will not draw well. The smoke will enter the living area.

Second, the smoke shelf provides a rain ledge. The ledge shields the burning fire from direct rainfall. Third, the throat and smoke shelf provide a fastening ledge for the damper.

The area of the throat should not be less than that of the flue. For best results, the throat area should be slightly bigger than the flue. For this reason, the throat opening should be long. Its length should be the full width of the fireplace. The throat width is approximately one-half the width of the flue. To have the same area, the throat must be longer than the flue.

Fireplace damper The narrowest part of the throat has a door. The door is called a *damper*. Special mech-

anisms keep the damper open in normal use. The damper should open as in Fig. 16-51. In this way, the door helps to form a wall at the bottom of the smoke shelf. When the fireplace is not being used, the damper may be closed. This keeps heated or cooled air from being lost through the flue.

Smoke chamber The *smoke chamber* is the area above the smoke shelf. The throat and the smoke shelf are the full width of the fireplace. The smoke chamber narrows from the width of the fireplace to that of the flue. This also helps to reduce the velocity of downdrafts. The effects of wind gusts are lessened here. A smoke chamber is necessary for proper draft. Smoke chambers should be plastered smooth with fire clay.

Chimney flue The chimney flue is the passage through the chimney. Its size is determined by the area of the fireplace opening. The flue area should be one-

Fig. 16-51 *Fireplace parts.*

tenth the opening area. Lower chimney heights are often used in modern buildings. Here, flues should be about 10 percent larger than this ratio.

A rule of thumb concerning flue areas is as follows: Fourteen square inches of flue are allowed per square foot of the opening of the main fireplace. For example, a fireplace opening has a height of 30 inches (2½ feet). The width is 36 inches (3 feet). The area of the fireplace opening would be 7½ square feet. The chimney flue area should be 14 square inches for every square foot. For an opening of 7½ square feet, the flue

size should be 7.5 × 14. This would be a total of 105 square inches. A good flue size would be 10½ × 10 inches.

Flues should be as smooth as possible on the inside to prevent buildup of soot, oily smoke, or tar. Fire clay or mortar is used most often to line flues. Industrial chimneys are exposed to chemical action. Special porcelain liners may be needed for these. For most buildings today, tile flue liners are used.

The drawing capability of a chimney is affected by the air differential. The air rising from the chimney is hot. The outside air is cold. The difference in the temperature of these two is the *air differential*. The greater the difference, the better the chimney will draw. Locating chimneys on outside walls does not help this. Locating them on inside sections conserves the heat of the air rising in the chimney. Thus chimneys located on the inside are usually more efficient.

Flues and chimneys Chimneys are made of several different materials. For standard masonry chimneys, brick, stone, and concrete block are all used. Special metal chimneys are also used. In many cases, the metal chimneys are encased with special coverings (Fig. 16-54). These covers must be built according to manufacturer's specifications. However, in most cases, metal standoffs must be used to hold the metal chimney away from the wood members. The chimney cover is framed from wood and siding materials. Some chimney covers have a fake brick or masonry appearance.

Many chimneys have special covers over them. These keep out rain or snow. Several methods are used to fasten these covers to the chimney. Some types slip inside the flue opening. Other types are fastened around the outside areas of the chimney by pressure bands. These are simply clamps that are tightened

Fig. 16-52 *A fireplace may have a raised hearth.*
(Martin Industries.)

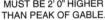

PRECAST
SLAB
(REINFORCED)

CLEAN-OUT
DOOR

ASH PIT

(A)

(B)

MUST BE 2' 0" HIGHER
THAN PEAK OF GABLE.

MINIMUM
HEIGHT 3' 0"

FLASHING

(C)

Fig. 16-53 *A. Proper construction blocks downdrafts that push smoke into a room. B. Note how a roof peak higher than the top of the chimney can cause downdrafts. C. Here, notice that the minimum height is 3 feet, and if the chimney is 10 feet or less, the chimney must be 2 feet higher than the peak of the gable.*

against the masonry of the chimney. Other types must be fastened by special expanding screw anchors. For these, holes are drilled into the brick or stone of the chimney. The special screw anchors are inserted and expanded firmly against the brick. This provides a firm anchor for regular bolts. These screw anchors always should be placed in the solid brick or stone. Placing them into the mortar will cause the joint to crack. This will loosen the masonry from its mortar bed. This weakens the chimney.

GAS VENTS

If you are building the house yourself, there are a number of things you must be aware of and capable of doing in order to complete the building and make it habitable. One of these is the installation of gas vents for hot water heaters, fireplaces, and other devices used in modern living. There are some dos and don'ts associated with the venting of exhaust gases in various appliances.

Round type B gas vents are for gravity draft venting of "listed" liquid propane (LP)– or natural gas–burning appliances. Appliances that might be so vented are those equipped with a draft hood, those designated as Category I, and those for which the installation instructions call for a permit venting with type B gas vents. These appliances include, but are not limited to, the following types: furnaces, boilers, water heaters, room heaters, unit heaters, duct furnaces, floor furnaces, and decorative appliances.

Fig. 16-54 *The sequence for installing special chimney covers.* (Martin Industries.)

Do not use type B gas vents for wall furnaces not "listed" for use only with type B gas vents. Do not use type B gas vents for Category II, III, or IV appliances or for any gas-burning appliance that requires either a pressure-tight or liquid-tight venting system. Do not use on any appliance for which the installation instructions call for taping or sealing the joints to prevent leakage.

All sizes of type B gas vents made by Eljer Manufacturing can be used in single- and multistory buildings. All type B gas vents can be used for both individual and multiple appliance venting. Eljer Manufacturing's type B gas vents are to be installed and used in accordance with *National Fuel Gas Code* standard NFPA No. 54 or standard for chimneys, fireplace and venting systems, NFPA No. 211. Type B gas vents also are suitable for use in existing, otherwise unused, masonry chimneys to protect the chimney from the damaging effects of moist combustion products from the appliances listed previously. In general, all type B gas-vent systems must include draft hoods at the appliance. However, certain advanced-technology gas appliances have flue outlet characteristics acceptable for venting with type B gas vents without draft hoods. Never use type B gas vents on any appliance that is not listed and approved for venting with type B gas vents.

A notice should be posted near the point where the gas vent is connected to the appliance with the following wording: "Connect this gas vent only to gas-burning appliances as indicated in the Installation Instructions. Do not connect incinerators or liquid or solid fuel–burning appliances."

Round Gas Vent

All joints in gas vents must be secured using the appropriate method. Draft hood connectors must be attached to the appliance outlet with screws. Single connectors, if used, must be secured to the appliance, to the gas vent, and to at all joints with three sheet-metal screws per joint. Table 16-1 lists methods of joining variously sized round vents. Figure 16-55 shows the fitting of gas vents. Figure 16-55A shows the 3- to 8-inch type of vent connected with six sheet-metal screws per joint, while Fig. 16-55B shows the

Fig. 16-55 *Round type B gas vent. (Eljer.)*

10- to 48-inch vents that must be joined with eight sheet-metal screws.

Air Supply

Gas appliances must have an adequate supply of air for combustion, vent operation, and ventilation (Fig. 16-56). Special provisions for bringing in outside air might be necessary in tight buildings or when appliances are in small rooms. Consideration must be given to climate in choice of the air-supply method. Check the local building code or *National Fuel Gas Code* for methods of air supply. Figure 16-56 shows typical air-supply requirements.

Fig. 16-56 *Typical air supply requirements. (Eljer.)*

Vent Connector Type and Size

Figure 16-57 shows the vent pipe connections that meet all building code and safety standards for use as gas appliance vent connectors and breechings.

TABLE 16-1 *Method of Joining and Clearance to Combustion*

Size	Shape	Method of Joining	Min. Clear. to Combust.
3–8" incl.	Round	Integral Coupler	1"
10–24" incl.	Round	6 sheet metal screws/joint	1"
26–48" incl.	Round	6 sheet metal screws/joint	2"

WARNINGS
Do not apply tape or sealant to joints of Eljer manufacturing type B gas vents. Do not use type B gas vent for: incinerators, or for appliances burning liquid or solid fuels, or for any appliance which needs a chimney such as "Factory-Built" or masonry.

Fig. 16-57 *Combined vent system using type B parts starting with draft hood connections. (Eljer.)*

Vent Location

Metal vents are suitable for indoors or outdoors (Figs. 16-58 and 16-59). Outdoor gas vents should be designed as close to maximum capacity as possible. Appliances served by an outside gas vent must have an air supply to the appliance room adequate to balance indoor and outdoor pressures. Otherwise, *stack action* of the heated building can cause reverse venting action when the appliance is off or operating only on its pilot.

In multifamily residential, high-rise, and many other types of buildings, vents are not permitted to penetrate the floor structure. They must be located in noncombustible shafts or chases with no openings except for inspection access (Fig. 16-60). Building code requirements in such cases must be followed carefully with respect to wall construction, access, clearance, support, initial penetration of breeching, and method of termination. See the NFPA Nos. 54 and 211 for such situations in which gas vents are installed.

Clearances and Enclosures

Type B gas vents that are from 3 to 24 inches in diameter can be installed with 1-inch minimum airspace clearance to combustibles. The 26- to 48-inch sizes require 2 inches of minimum airspace clearance to combustibles. These clearances are marked on all gas-carrying items. They apply to indoor and outdoor vents, whether they are open, enclosed, horizontal or vertical, or pass through floors, walls, roofs, or framed

Fig. 16-58 *Typical ceiling and roof penetrations. (Eljer.)*

*Maintain at least minimum clearance to combustibles, wires, and insulation.
1" – for 3" thru 24" sizes
2" – for 26" thru 48" sizes at walls, ceiling, roof, and penetrations.

Fig. 16-59 *Wall penetrations for type B vents. (Eljer.)*

spaces. The appropriate clearance should be observed to joists, studs, subfloors, plywood, drywall, plaster enclosures, insulating sheathing, rafters, roofing, and any

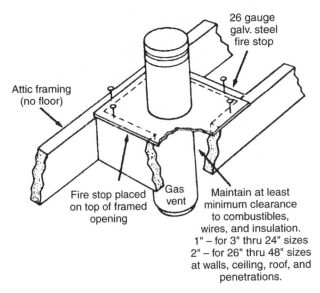

Fig. 16-60 *Fire stopping required for all ceiling and floor penetrations.* (Eljer.)

Fig. 16-61 *Fire stopping required for all ceiling and floor penetrations.* (Eljer.)

other materials classed as combustible. The required minimum airspace clearances apply also to electrical wires and any kind of building insulation. Keep electrical wires and building insulation away from the gas vents and out of their required airspace clearance.

Reduced clearances must be used with 4-inch oval gas vents in 2 × 4 stud walls. Gas vents can be fully enclosed or installed in the open. Single wall materials used as vent connectors are not permitted to be enclosed. There is no need to enclose gas vents when used as connectors under floors, in crawl spaces and basements, or in normally unoccupied or inaccessible attics. Enclosing the vertical portions of vents is recommended where they pass through rooms, closets, halls, or other occupied spaces.

Fire Stopping

All type B gas vents passing through floors or ceilings or in framed walls must be fire stopped at floors or ceilings using 26-gauge or heavier galvanized steel. Figure 16-61 illustrates how a fire stop is built. The fire stop must close the area between the outer wall of the pipe and the opening in the structure. In attics, the fire stop should be placed on top of the framed ceiling opening. Keep wires and insulation out of required airspace clearance around gas vents.

Fire stops can be used as vent-pipe supports provided that the aluminum inner wall of the gas vent is not penetrated or damaged by the method of attachment. See gas-vent piping later on to see how the vents must be secured. For gas vents within a shaft or chase, fire stopping is provided by the vertical walls of the shaft.

Use of Gas-Vent Fittings

Figure 16-59 shows adjustable lengths that can be telescoped over a fixed length to make up in between lengths of vent or connector. An adjustable length suspended below a support serves as an expansion joint between two fixed points of properly supported gas vent. Ordinarily, the adjustable length must be secured, but for expansion joints, it should just maintain good contact and a minimum 1.5-inch overlap. Do not use adjustable lengths to suspend any weight of pipe below. Elbows for sizes of 3 to 8 inches are fully adjustable. Elbows for sizes of 10 to 48 inches are 45-degree fixed only.

Most building codes require the use of full-size fittings for vent interconnections. Tees, elbows, increasers, and short lengths are especially designed to facilitate interconnections.

Tees used as fittings to start vertical vents must be tightly capped to prevent air leakage. Gas vents and vent connectors having more than 90-degree turns (an elbow and a tee) might require additional height or a size increase to compensate for added flow resistance. All unused openings in a gas vent must be sealed to prevent loss of effective vent action.

Minimum Gas-Vent Height

A minimum gas-vent height of 5 feet above the appliance draft hood is required (Fig. 16-62). Where the vent has a lateral or offset or serves multiple appliances, greater heights might be required for proper

Fig. 16-62 *Minimum height of vent above appliance.* (Eljer.)

Minimum vertical height 5 feet

venting. Special care must be taken that short gas vents on duct furnaces, unit heaters, and furnaces in attics have sufficient vent height to ensure complete venting. Check appliance the manufacturer's instructions and local codes for the required minimum heights. Along with these suggestions, minimum heights of gas vents also must comply with the rules mentioned in *Vent Termination.*

Support

Gas-vent piping must be securely supported. Lateral runs are to be supported at least every 5 feet; when offsets are necessary, adequate support above and below the offset is required. In addition, securing the offset elbow with three maximum 0.25-inch-long sheet-metal screws is recommended. Vertical runs fire stopped at 8- to 10-foot intervals need be supported only near the bottom. Tees used as vent inlets can be supported by sheet-metal plates or brackets. Plumber's tape can be used to space both horizontal and vertical piping. Short vents with less than 6 feet of vertical pipe below the flashing can be suspended with the flashing. The pipe can be supported by the storm collar resting on the top of the flashing. Use 0.29-inch-long sheet-metal screws to attach the storm collar to the pipe at the appropriate place. Gas vents supported only by the flashing must be guyed above or below the roof to withstand snow and wind loads. All gas vents extending above the roof more than 5 feet must be securely guyed or braced.

Indoors, gas vent 3 through 8 inches can be supported as shown in Fig. 16-63 and fire stopped as well

using the support plate as illustrated. Cut away all floors and ceilings to 1-inch minimum airspace clearance. The support plate will support 35 feet of vent.

Vent sizes of 10 to 24 inches can be supported within or adjacent to fireproof or noncombustible structures or shafts. See Fig. 16-60. Use a split plate support (10 to 24 inches) that fits around the groove near the end of the pipe and rests on masonry or a metal frame (Fig. 16-64A). Use bucket support for 26 inches and larger sizes, or suitable structural iron bands can be constructed using 16-gauge steel or heavier for the same purpose.

Fig. 16-64 *A. Split-plate suggested support detail. B. Typical expansion joint below each support.* (Eljer.)

Flashing

The roof opening should be located and sized such that the vent is vertical and has the required airspace clearance. The tall cone flashing is for flat roofs only (Fig. 16-65A). It is nailed in place through all four sides of the base flange. The adjustable roof flashing shown in

Fig. 16-63 *Storm collar on top of flashing.* (Eljer.)

Fig. 16-65 *A. Tall cone flashing for flat roofs only. B. Elements for proper roof flashing installation.* (Eljer.)

Fig. 16-65B is positioned with the lower portion of the base flange over roofing material. Nail through only the upper portion and sides of the base flange. Do not nail through the lower flange. Use nails with a neoprene washer, or cover the nail heads with siliconized caulk. Finish roofing around the flashing, covering the sides and upper areas of the flange with roofing material.

Gas Termination

Gas-vent piping must extend through the flashing to a height above the roof determined by rule 1 or rule 2 shown below. A storm collar is installed on the vent pipe over the opening between pipe and flashing. Siliconized caulk is used over the joint between the pipe and storm collar. The top is securely attached to the gas vent using the proper method for the model of pipe (see Figs. 16-67 and 16-68).

Termination height for gas-vent sizes of 12 inches and less can be in accordance with Table 16-2. Tops 14 inches and over in size and roof assemblies must be located in accordance with rule 2 below.

TABLE 16-2 *Minimum Height from Roof to Lowest Discharge*

Roof Pitch	Opening, Foot
Flat to 7/12	1.0
Over 7/12 to 8/12	1.5
Over 8/12 to 9/12	2.0
Over 9/12 to 10/12	2.5
Over 10/12 to 11/12	3.25
Over 11/12 to 12/12	4.0
Over 12/12 to 14/12	5.0
Over 14/12 to 16/12	6.0
Over 16/12 to 18/12	7.0
Over 18/12 to 20/12	7.5
Over 20/12 to 21/12	8.0

Rule 1: Tops for gas vents 12 inches and smaller
The cap is suitable for installation on listed gas vents terminating a sufficient distance from the roof so that no discharge opening is less than 2 feet horizontally from the roof surface, and the lowest discharge opening will be no closer than the minimum height specified in Fig. 16-66B. These minimum heights may be used provided that the vent is no less than 8 feet from any vertical wall.

Rule 2: Tops for gas vents 14 inches and larger
For installations other than covered by Table 16-2 in Fig. 16-66B or closer than 8 feet to any vertical wall, the cap should be not less than 2 feet above the highest point where the vent passes through the roof and at least 2 feet higher than any portion of a building within 10 feet. Vent caps 14 inches and larger and chimney-

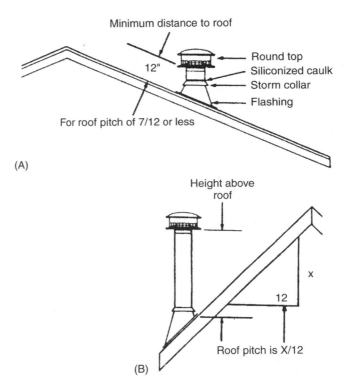

Fig. 16-66 *A. For roof pitches of 7/12 or less. B. Location rules for flat roofs to those with pitches of 21/12. (Eljer.)*

style roof housings must comply with rule 2 regardless of roof pitch.

These rules were established on the basis of tests conducted in accordance with American National Standard ANSI/UL 441.

Top Installation

Round tops for pipe sizes of 3 to 8 inches have a spring clip that locks into the upper end of the vent. The clip engages automatically when the top is pushed into the pipe (Fig. 16-67A). The top also will fit any sheet-metal pipe with full nominal inch dimensions. To attach securely, bend a 1-inch length of the pipe under end inward about 0.125 inch. The spring clip will lock under this bent edge (Fig. 16-67B). To remove the top from any pipe, pull up evenly on opposite sides of the skirt of the top.

Top Installation of 10- to 24-Inch Vents

The round tops for this size pipe feature an expanding collar that clamps to the inside of the gas vent on any other sheet-metal pipe of similar nominal size. To install, loosen the screw on top of the collar, and squeeze the bottom so that it will enter the pipe easily, as shown in Fig. 16-68A. Press down evenly on the skirt until it contacts the upper end of the pipe (see Fig. 16-68B).

Fig. 16-67 *A. Round top installation on Eljer Manufacturing's gas vent. B. Round top installation on single wall pipe.* (Eljer.)

Fig. 16-68 *A. Dome mounted on top of squeeze collar. B. Dome being mounted. C. Location of screws to hold dome in place.* (Eljer.)

Tighten the screw to expand the collar against the inner pipe (see Fig. 16-68C). Don't overtighten. Now attempt to lift the skirt to be sure the top is secure.

Checking Vent Operation

Complete all gas piping, electrical, and vent connections. After adjusting the appliance and lighting the main burner, allow a couple of minutes for warm-up. Hold a lighted match just under the rim of the draft

Fig. 16-69 *Match test for spillage.* (Eljer.)

hood relief opening. Proper venting will draw the match flame toward or into the draft hood. Improper venting is indicated by escape or spillage of the burned gas and will cause the match to flicker or go out. Smoke from a cigarette also should be pulled into the draft hood if the vent is drawing correctly (Fig. 16-69).

Painting

To prolong the life and appearance of the galvanized steel outer casing and other parts of the gas vents located outdoors, use proper painting procedures at the time of installation. Remove oil and dirt with a solvent. Paint first with a good-quality zinc primer or other primer recommended for use on galvanized steel. Next, apply an appropriate finish coat. Ordinary house paints might not adhere well and will not prevent underfilm corrosion, which leads to paint loosening and peeling.

POST-AND-BEAM CONSTRUCTION

The post-and-beam method is the oldest construction method used in the United States. It is still used, particularly where the builder wants large, open areas in a house. This type of framing is characterized by heavy timber posts often with intermediate posts between. It has been used since the early days of the 1600s by the first settlers. Warehouses, barns, and other large buildings are all examples of post-and-beam construction. Designers are now using post-and-beam construction for schools, homes, and office buildings as well. Post-and-beam construction can have wider, longer rooms and higher ceilings. Exposed beams, as in Fig. 16-70, can add beauty, line, and depth to a building. Beams can be laminated (engineered lumber) or boxed in several shapes. This makes unusually shaped buildings possible.

Fig. 16-70 *Post-and-beam construction allows wider, longer open areas.*

Wall supports Wall supports are widely spaced. Wide spacing makes possible many very large windows. Using many windows makes a building interior light and cheerful. Wooden beams give a natural appearance. These and natural wood finishes can add greatly to a feel of natural beauty. The actual construction methods are simple. Floors and roofs are constructed in a similar manner. Several different materials and techniques may be used with good results.

Roof decking and floors Roof decking and floors are done in much the same manner. Wooden planks, plywood sheets, and special panels all may be used on either floors or roofs. In many construction processes, the roof and ceiling are the same. This cuts down the complexity and cost. Only one layer is used instead of two or three. The walls may combine a variety of glass, stone, brick, wood, and metal materials. There are other names for the post-and-beam type of construction. It is sometimes called *plank and beam* when used in floors and roofs. For ceilings, the term *cathedral ceiling* is often used.

General Procedures

Regular footing and foundation methods are used. The floor beams or joists are laid as in Fig. 16-71. Decking is applied directly over the floor beams as shown. Most commonly used is grooved-plank decking and plywood sheets. As a general rule, any piece of decking should

Fig. 16-71 *Widely spaced floor or roof joists are decked with long planks.* (Weyerhauser.)

span at least two openings. Figure 16-72 shows how this works. End-joined boards may be used so that joints need not occur over beams. However, if end-joined pieces are not used, joints should be made over the floor beams.

Wall stud posts Wall pieces are nailed together while they are flat on the subfloor. Studs are spaced 12 to 24 inches apart.

Fig. 16-73 *Posts may be nailed to top plates and soleplate. However, note the greater space between posts than for conventional frames.*

Fig. 16-74 *Posts also may be anchored directly to floors. They must be anchored solidly.*

Nail the post-and-beam stud posts to the soleplate and top plates (Fig. 16-73). However, these posts should not be less than 4 inches square. They may be spaced 4 to 12 feet apart. This gives a much wider area between supports and allows larger glass areas without special headers to support the roof.

Stud posts also may be anchored directly to floors (Fig. 16-74). In either method, various metal straps and cleats are used. They anchor the stud posts to the floor or soleplate. The posts should be attached directly over the floor joists or beam. Where a soleplate is used, the stud posts may be attached at any point.

Stud posts should not be attached directly to concrete floors exposed to outside weather. A special metal bracket should be used. This bracket raises the post off the concrete, as shown in Fig. 16-75. These brackets must be used to prevent rot and deterioration of the post caused by moisture. Outside concrete retains moisture. Moisture leads to wood decay. Any wood piece that directly touches the concrete may decay. The special metal brackets make an airspace so

Fig. 16-75 *Foot plates help to reduce moisture damage.*

that the wood does not touch the concrete. The circulating air keeps the wood dry and reduces decay.

Beams Beams are supports for roof decking. Two systems are used:

- Beams may run the length of the building, as in Fig. 16-76A.
- Beams may run across the building, as in Fig. 16-76B.

Both straight gable and shed roof shapes are common. Other beams may be the frame of a roof and wall combined.

Beams may combine the wall studs and roof beams. The special beam shapes shown in Fig. 16-77 are typical. These beams combine interior beauty with structural integrity. They are common in churches, schools, and similar buildings.

There are four basic types of beams:

- Solid beams
- Laminated beams
- Bent laminated beams
- Plywood box beams

Solid beams are made of solid pieces of wood. Obviously, for larger beams, this becomes impractical. To make larger beams, pieces of wood are laminated and glued together.

Laminated beams may be used horizontally or vertically. To provide graceful curved shapes, beams are both laminated and curved to shape. These are built up of many layers of wood. Each layer is bent to the shape desired. These *bent laminated beams* are extremely strong and fire-resistant. They also provide graceful beauty and open space. *Laminated beams* have a more uniform moisture content. This means that they expand and contract less.

Plywood box beams provide a heavy appearance, great strength, and comparatively light weight. They are made as in Fig. 16-78. They have one or more wooden beams at the top and bottom. These beams are stiffened and spaced by lumber spacers as shown. However, the interiors are hollow. The entire framework then is covered with a plywood skin as shown. Although these beams appear to be solid beams, they are much lighter.

Beam systems also may be used. These use small beams between larger beams, and this allows large

(A)

(B)

Fig. 16-76 *A. Longitudinal beams run the length of a building. B. Cross-beams or transverse beams run across the width.*

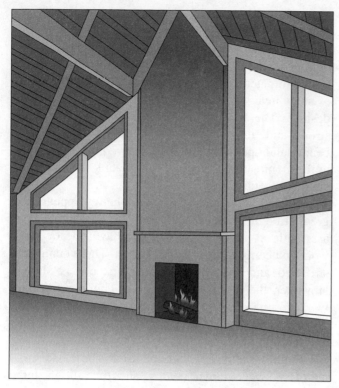

Fig. 16-77 *Special beam shapes may combine wall and roof framing.*

PRESSURE-LAMINATED LUMBER FLANGES, TOP FLANGE

STAGGER FLAGS

BUTT JOINT BETWEEN PLYWOOD WEBS

GLUE JOINT

PLYWOOD WEB OR "SKIN"

LUMBER STIFFENERS

LUMBER BEARING STIFFENER

Fig. 16-78 *Box beams look massive, but they are made of ply-wood over board frames.*

beams to be spaced wider apart. The secondary beams are called *purlins* (Fig. 16-79).

Anchor the beams to either posts or top plates. There are two ways to anchor beams:

• The first method is to use special wooden joints. These are cut into the beams, posts, and other

GIRDER

METAL HANGERS USED

BEAM

PURLIN

Fig. 16-79 *Post-and-beam roof frame.*

members. To prevent side slippage and other shear forces, special pegs, dowels, and splines are used. Many of the old buildings in Pennsylvania and other areas were built in this manner. However, today, this method is time-consuming and costly.

• The second method is used more often for contemporary housing. In this method, special metal or wooden brackets are used (Fig. 16-80). There are many types of bracket devices. They hold two or more pieces together and prevent slipping and twisting. They should be bolted rather than nailed. Nails will not hold the heavy pieces. Although lag bolts may be used, through bolts are both stronger and more permanent.

Ridge beams are used for gable construction using the post-and-beam method. Where beams are joined to the ridge beams, special straps are used (Fig. 16-81). These straps may resemble gussets and are applied to the sides. However, an open appearance without brackets can be created. The straps are inlet and applied across the top as shown.

Purlins are held by hangers and brackets like joist hangers. They are most often made of heavy metal. Most metal fasteners are made specifically for each job to the designer's recommendations.

Bearing plates Where several beams join, special allowances must be made for expansion and contraction. The surface of this joining area also should be enlarged. All beams then may have enough area to support them. Special steel plates are used for these types of situations. These are called *bearing plates*.

Fig. 16-80 *Special brackets are often used to hold beams and other parts together.*

Fig. 16-81 *Beams are tied to girders with gussets and straps.*

Bearing plates are used at tops of posts, bottoms of angular braces, and in other similar situations.

Decking Common decking materials include plywood boards and 2-inch-thick lumber. Decking is often exposed on the interior for a natural look. Both lumber and plywood should be grooved for best effect. The grooves should be on both the edges and ends. Great strength is needed because of the longer spans between supports.

Materials Materials used for roof coverings may include concrete and special insulated gypsum materials. Common post-and-beam construction has insulation applied on top of roof decking (Fig. 16-82). A moisture barrier should be used on top of the decking and beneath the insulation. A rigid insulation should be used on top of the decking. It should be strong enough to support the weight of workers without damage. As a rule, a built-up roof is applied over this type of construction. Roofs of this type generally are shed roofs with very little slope. When gables are used, a low pitch is deemed better. Built-up roofs may not be used on steep-pitched roofs.

However, on special shapes, several different roofing techniques may be used. Decking or plywood may be used to sheath the beams or bends. Metal or wooden roofing can be applied over the decking. Usually sheet copper, sheet steel, or wood shingles are used.

Stressed-skin panels Stressed-skin panels are large prebuilt panels. They are used for walls, floors, and roof decks. They can be made to span long distances and still give strong support. They are built in a factory and hauled to the building site. They cost more to make, but they save on construction time.

These panels are formed around a rigid frame with interior and exterior surfaces included. The panels also can include ducts, insulation, and wiring. Be sure to check local code requirements when placing wiring inside ductwork. One surface can be roof decking. The interior side can be insulating sound board to make the finished ceiling. The exterior could be heavy plywood for the roof decking. A standard composition roof could be applied over the panels. The same techniques can be used for wall and floor sections (Fig. 16-83). The panels usually are assembled with glue and such fasteners as nails and/or staples.

(A)

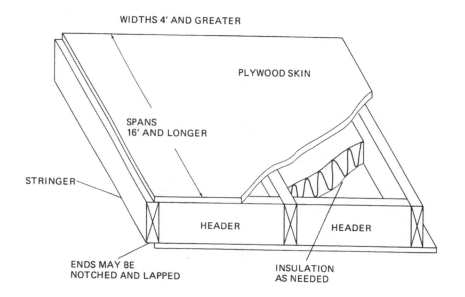

(B)

Fig. 16-82 *A. Insulation can be applied on top of either beams or roof decking. B. A composition roof then is laid.* (Gold Bond.)

DECKS

Patios can crack and become unsightly in time. They have to be removed and new slabs poured. This can be an expensive job. Most of the damage is done when the fill against the footings or basement walls begins to settle after a few years of snow and rain. One way to renew the usable space that was once the patio is to make a deck. And of course, if you are building, the best bet is to attach a deck onto the house while it is being constructed (Fig. 16-84).

Fig. 16-84 *A deck above grade level.* (Teco.)

There are a few things you should know before replacing a concrete patio slab with a wooden deck. Working drawings for new houses or for remodeling seldom show much in the way of details. The proposed

Fig. 16-83 *Stressed-skin panels are used on floors and roof decks.*

JOIST HANGER   POST CAP

Fig. 16-85 *Wood deck elevation.*
(Teco.)

POSTS TO BE EMBEDDED AT LEAST 1/3 OF LENGTH IN GROUND AND/OR 18"

location of any deck should be shown on the floor plan. Its exact size, shape, and construction, however, are usually determined on site after a study of the site conditions, the contours of the land, orientation of the deck, and the locations of nearby trees (Fig. 16-85). Decks must be built to the same code requirements as floor framing systems inside the house. Every deck has four parts: a platform, a floor frame, vertical supports, and guard rails.

Platform

The deck itself is built of water-resistant 2 × 4s or 2 × 6s laid flat. All lumber should be of such species as to ensure long duration. This means redwood, cedar, cypress, or pressure-treated wood certified as being suitable for exterior use.

If the decking has vertical graining, the grain runs up and down when the lumber is flat. This means that either side may be placed face up. If the decking has flat graining (the grain curves from side to side), each piece must be laid with the annular rings pointing down at the edges. Annular rings are produced as the tree grows. They cause wood to have hard and soft sections. Parts of these rings can be observed at the end of a piece of lumber.

Alternative platform layouts are shown in Figs. 16-86 through 16-90. These decks have the lumber and necessary connectors for making the deck that particular size. Decking may run parallel to the wall of the house, at right angles to it, or in a decorative pattern, such as a parquet or herringbone (Fig. 16-91).

Whatever the pattern, each piece of decking must be supported at both ends, and long lengths must be given intermediate support. Decking must run at right angles to the framing beneath it. To prevent water buildup from rain or snow, the boards are spaced about ¼ inch apart before they are nailed down. A quick way

to do the spacing without having to guess is to put a nail between the decking pieces so that they cannot be pushed together while being nailed.

Frame

The deck can be supported by its perimeter on posts or piers. This means that the frame will be a simple box-shaped arrangement. When the decking runs parallel to the wall of the house, supporting joists are attached with joist hangers to an edge joist bolted through sheathing into a header. Or joist hangers can be used to anchor the joist of the deck to the header of the house (Fig. 16-92). Refer to Figs. 16-87 through 16-90 for alternate ideas of making a deck frame. Single joists can be spaced up to 2 feet, 6 inches apart with the joist span no more than 6 feet.

Support

The other edges of a deck frame usually arc supported on 4 × 4 wood posts. Figure 16-93 shows a post joint assembly detail. Figure 16-94 shows exterior joist-to-beam connection, whereas beam-to-post connection is shown in Fig. 16-95. The posts rest on 2-inch wood caps anchored to the tops of concrete piers. Figure 16-96 shows one method used in anchoring. Caps may be omitted if posts are less than 12 inches long. The tops of the piers must be at least 8 inches above grade. If the side is not level, the lengths of posts must be varied so that their tops are at the same level to support the deck frame. The beams that span from post to post may be 4-inch timbers or a pair of 2-inch joists. The sizes of the beams and the spacing of posts have to be considered in the basic design at the beginning. Maximum spans of typical decks are shown in Table 16-3.

As you can see from the table, it is usually less expensive to use larger beams and fewer posts. The

BASIC WOOD DECK PLAN

LUMBER		CONNECTORS	
QTY.	TYPE	QTY.	TYPE
9	4x4 POSTS X LENGTH	9	POST CAPS
6	2x6 10/0 BEAMS	20	JOIST HANGERS
10	2x6 6/0 JOISTS		
25☆	2x6 10/0 DECKING	10	FRAMING ANGLES
*	2x6 STAIR TREAD CLEATS	*	3/8" × 4" BOLT
*	2x6 STAIR CARRIAGES	9	5- 1/2" x 6 1/2" PLYWOOD FILLER SHIMS
*	2x6 STAIR TREADS		
*	DETERMINED BY WIDTH & RISE OF STAIR		

☆ IF 2x4 DECKING IS USED QUANTITY REQUIRED IS 39

Fig. 16-86 *A. Basic wood deck plan.* (Teco.) *B. Adding to the basic plan can produce a porch. Notice the roof.*

posts need more work because they call for a hole to be dug and concrete to be mixed and poured. Plan for the spacing of posts before you begin the job. They have to be placed in concrete and allowed to set for at least 48 hours before you nail to them.

If you space the joists as shown in Fig. 16-86, make sure that they correspond to the local code, which should be checked before you start the job. The layouts in Figs. 16-86 through 16-90 are suggestions only and do not indicate compliance with any specific structural, code, service, or safety requirements.

TABLE 16-3 *Deck Beam Spacing*

Deck Width (feet)	Distance Between Posts		
	4 × 6 Beam	4 × 8 Beam	4 × 10 Beam
6	6'9"	9'0"	11'3"
8	6'0"	8'0"	10'0"
10	5'3"	7'0"	8'9"
12	4'6"	6'0"	7'6"

6'0" MODULE (ALL MEMBERS 2x6)

LUMBER		CONNECTORS	
QTY.	TYPE	QTY.	TYPE
12	4x4 POSTS X LENGTH	12	POST CAPS
8	2×6 10/0 BEAMS	30	JOIST HANGERS
15	2×6 6/0 JOISTS	12	5-1/2" x 6 1/2" PLYWOOD
37☆	2×6 10/0 DECKING		FILLER SHIMS

☆ IF 2x4 DECKING IS USED QUANTITY REQUIRED IS 58

Fig. 16-87 *A deck plan using 6-foot modules. (Teco.)*

ADDITION OF 5'0" MODULE
(ALL MEMBERS 2x6)

LUMBER		CONNECTORS	
QTY.	TYPE	QTY.	TYPE
12	4x4 POSTS X LENGTH	12	POST CAPS
6	2x6 15/0 BEAMS	28	JOIST HANGERS
14	2x6 6/0 JOISTS	12	5-1/2" x 6 1/2" PLYWOOD
41☆	2x6 10/0 DECKING		FILLER SHIMS

☆ IF 2x4 DECKING IS USED QUANTITY REQUIRED IS 38

Fig. 16-88 *A deck plan using 5-foot modules. (Teco.)*

9'0" x 5'0" MODULE (ALL MEMBERS 2x6)

LUMBER		CONNECTORS	
QTY.	TYPE	QTY.	TYPE
9	4x4 POSTS X LENGTH	9	POST CAPS
6	2x6 10/0 BEAMS	36	JOIST HANGERS
18	2x6 9/0 JOISTS	9	5-1/2" x 6 1/2" PLY-
37☆	2x6 10/0 DECKING		WOOD FILLER SHIMS

☆ IF 2x4 DECKING IS USED QUANTITY REQUIRED IS 58

Fig. 16-89 *A deck plan using 9- × 5-foot modules. (Teco.)*

8'0" x 4'0" MODULE (ALL MEMBERS 2 x 6)

LUMBER		CONNECTORS	
QTY.	TYPE	QTY.	TYPE
9	4x4 POSTS X LENGTH	9	POST CAPS
6	2x6 8/0 BEAMS	28	JOIST HANGERS
14	2x6 8/0 JOISTS	9	5-1/2" x 6 1/2" PLY-
34☆	2x6 8/0 DECKING		WOOD FILLER SHIMS

☆ IF 2x4 DECKING IS USED QUANTITY REQUIRED IS 52

Fig. 16-90 *A deck plan using 8- × 4-foot modules. (Teco.)*

Guard Rails

A deck more than 24 inches above grade must be surrounded with a railing. Supporting posts must be part of the deck structure. The railing cannot be toenailed to the deck. The posts may extend upward through the deck, or they may be bolted to joists below deck level (Fig. 16-97). Spacing of support every 4 feet provides a sturdy railing. Support on 6-foot centers is acceptable (Fig. 16-98).

Making a Hexagonal Deck   A hexagonal deck can add beauty to a house or make it the center of attraction for an area located slightly away from the house. Figure 16-99 shows the basics of the foundation and layout of stringers. Figure 16-100 shows the methods needed to support the deck. To make a hexagonal deck:

(A)

(B)

Fig. 16-91 *A. Parquet deck pattern.* (Western Wood.) *B. Nailing treated southern pine 2 × 4s to a deck attached to a house.* (Southern Forest Products Association.)

JOINT HANGER   JOIST-TO-BEAM CONNECTION

Fig. 16-92 *Hanger used to connect joists to a beam for a deck.* (Teco.)

1. Lay out the deck dimensions according to the plan, and locate the pier positions.

2. Excavate the pier holes to firm soil. Level the bottoms of the holes, and fill with gravel to raise the piers to the desired height. To check the pier height, lay a 12-foot stringer between piers, and check with a level.

3. When all piers are of equal height, place a 12-foot stringer in position. Cut and fit 6-foot stringers,

2–2×6

5¹⁄₂×6×¹⁄₂"
EXTERIOR GRADE PLYWOOD FILLER SHIM

POSTCAP

JOINT HANGER

2×6 JOIST

4×4 POST

Fig. 16-93 *Post joint detail assembly.* (Teco.)

Fig. 16-94 *Beam connection.* (Teco.)

Fig. 16-95 *Beam-to-post connection.* (Teco.)

BEAM

JOIST

HANGER

CAP

POST

PIER

Fig. 16-96 *Wood posts that support the deck rest on caps anchored to concrete piers. Piers may be bought precast, or you can pour your own.*

4' × 4' RAILING POST

2' × 8' FASCIA

3/8" LAG BOLT

RAILING POST

Fig. 16-97 *Connections of post for the railing and the fascia board. Note that lag bolts are used and not nails.* (Western Wood.)

and toenail them with 10d nails to the 12-foot stringer at the center pier. Use temporary bracing to position the stringers in the correct alignment. Cut and apply the fascia. Cut and apply the stringers labeled *B* in Fig. 16-99. Note the details in Fig. 16-100.

4. Drive the bracing stakes into the ground, and nail them to the stringers to anchor the deck firmly into position (Fig. 16-101).

Fig. 16-98 *A deck located well above grade.* (Western Wood.)

Fig. 16-99 *Layout of a hexagonal deck.* (Western Wood.)

5. Apply the decking. Begin at the center (see Fig. 16-101). The center edge of the first deck member is over the 12-foot stringer. Use 10d nails for spacing guides between deck members. Apply the remaining decking. Nail the decking to each stringer with two 10d nails. Countersink the nails. Check the alignment every five or six boards. Adjust the alignment by increasing or decreasing the width between deck members.

6. Tack the trim guide in place, and trim the edges, allowing a 2-inch overhang (Fig. 16-102). Smooth the edges with a wood rasp or file (Fig. 16-103).

Raised Deck

A raised deck (see Fig. 16-98) uses a different technique for anchoring the frame to the building. If the deck is attached to a building, it must be inspected by the local building inspector, and local codes must be

Fig. 16-100 *Supports for the deck.* (Western Wood.)

BRACING STAKES FOR ANCHORING TO GROUND

FASCIA

(A)

(B)

Fig. 16-101 *A. Anchoring the deck to the ground.* (Western Wood.) *B. Finished project.* (Southern Forest Products Association.)

checked for proper dimensions and proper spacing of dimensional lumber. The illustrations here are general and will work in most instances to support a light load and the wear caused by human habitation.

The deck shown in Fig. 16-106 should be laid out according to the sketch in Fig. 16-104. Figure 16-105 shows how the deck is attached to the existing house. Figure 16-106 shows how the decking is installed. Use 10d nails for spacing guides between the deck members. Nail the deck member to each stringer with two 10d nails. Countersink the nails. Check the alignment every five or six boards. Adjust the alignment by increasing or decreasing the width between the deck members.

The railing posts are predrilled, and then the fascia is marked and drilled for insertion of the ⅜- × 3-inch lag bolts per post. The railing cap can be installed with two 10d nails per post. This 12- × 12-foot deck takes about 5 gallons of Penta or some other type of wood preservative—that is, of course, if you didn't start with pressure-treated wood.

TRIM GUIDE

Fig. 16-102 *Trimming the overhang.* (Western Wood.)

Fig. 16-103 *The finished hexagonal deck. (Western Wood.)*

Fig. 16-104 *Supporting a raised 12- × 12-foot deck. (Western Wood.)*

Steps

If the deck is located at least 12 inches above grade, you will have to install some type of step arrangement to make it easy to get onto the deck from grade level. Steps also will make it easier to get off the deck.

By using the Teco framing angles, it is possible to attach the steps to the deck rather easily (see Fig. 16-106). These galvanized metal angles help to secure the steps to the deck without the sides of the steps being cut and thus weakened. Another method also can be used to attach steps to the deck. Figure 16-107 shows how step brackets (galvanized steel) are used to make a step without having to cut the wood stringers. They come in a number of sizes to fit almost any conceivable application.

CONCRETE PATIOS

Before you pour a patio, a number of things must be taken into consideration: size, location in relation to the house, shape, and last but not least, cost. Cost is a function of size and location. Size determines the amount of excavation, fill, and concrete. Location is important in terms of getting the construction materials to the site. If materials must be hauled for a distance, it naturally will cost more in time and energy, and therefore, the price will increase accordingly.

Sand and Gravel Base

If the soil is porous and has good drainage, it is possible to pour the concrete directly on the ground, if it is well tamped. If the soil has a lot of clay, and drainage is poor, it is best to put down a thin layer of sand or gravel before pouring the concrete (Fig. 16-108). Just before pouring the concrete, give the soil a light water sprinkling. Avoid developing puddles. When the earth is clear of excess water, you can begin to pour the concrete. In some instances, voids need to be filled before the concrete is poured. When these additional areas need attention, bring them up to the proper grade with granular material thoroughly compacted in a maximum of 4-inch layers.

If you have a subgrade that is water-soaked most of the time, you should use sand, gravel, or crushed stone for the top 6 inches of fill. This will provide the proper drainage and prevent the concrete slab from cracking when water gets under it and freezes. When you have well-drained and compacted subgrades, you do not need to take these extra precautions.

Expansion Joints

When you have a new concrete patio abutting an existing walk, driveway, or building, a premolded material, usually black and ½ inch thick, should be placed at the joints (Fig. 16-109). These expansion joints are placed on all sides of the square formed by the intersection of the basement wall or floor slab and the patio slab. Whenever a great expanse is covered by the patio, it is best to include expansion joints in the layout before pouring the concrete.

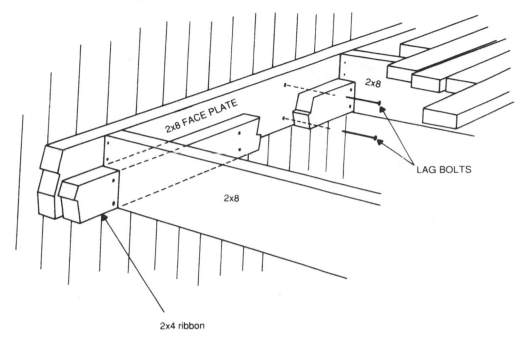

Fig. 16-105 *Notched stringers are used to attach to the nailing ribbon to the beam.* (Western Wood.)

Fig. 16-106 *Details of the raised deck.* (Western Wood.)

The Mix

If you are going to mix your own concrete, which is a time-consuming job, keep in mind that the concrete should contain enough water to produce a concrete that has relatively stiff consistency, works readily, and does not separate. Concrete should have a slump of about 3 inches when tested with a standard *slump cone*. Adding more mixing water to produce a higher slump than specified lessens the durability and reduces the strength of the concrete.

The *slump test* performed on concrete mix measures the consistency of the wet material and indicates if the mix is too wet or dry. The test is performed by

MODERN STYLE EASIER RISE

TECO FRAMING ANGLE (3")

3/8" x4 BOLT

2x6

2x6 SUBHEADER

2x6

2x4

1-1/4"
2 1/2"

TECO 3" FRAMING ANGLE

TECO 3"
FRAMING ANGLE

3"

STAIR & SUBHEADER ANCHORAGE

STAIR STEPS

DECK HEIGHT

A

B

C

STRINGER

TECO 3" FRAMING
ANGLES

TECO PC-2
POST CAP

TECO FRAMING ANGLE (1 1/2")
EACH SIDE OF SUBHEADER

Fig. 16-107 *Methods of mounting steps to a deck.* (United Steel Products.)

Fig. 16-108 *Using a wooden tamper to compact the sand or crushed stone before pouring the concrete.*

Fig. 16-109 *Expansion joints are used between large pieces.*

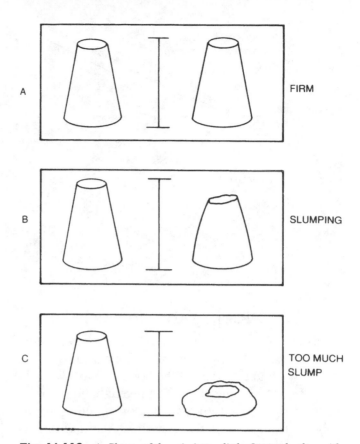

Fig. 16-110 *A. Slump of the mix is too little. It stands alongside the test cone. B. The slump is acceptable. C. The slump is too much.*

filling a bucket with concrete and letting it dry (Fig. 16-110). Then take another wet sample and dump it alongside. Test for the amount of slump the wet mix displays. It is obvious if the mixture is too wet: The cone will slump down or wind up in a mess, as shown in Fig. 16-110.

In northern climates where flat concrete surfaces are subjected to freezing and thawing, *air-entrained concrete* is necessary. It is made by using an air-entraining Portland cement or by adding an air-entrained agent during mixing. Before the concrete is poured, the subgrade should be thoroughly dampened. Make sure that it is moist throughout but without puddles of water.

Keep in mind that concrete should be placed between forms or screeds as near to its final position as practicable. Do not overwork the concrete while it is still plastic because an excess of water and fine materials will be brought to the surface. This may lead to scaling or dusting later when it is dry. Concrete should be spaced properly along the forms or screeds to eliminate voids or honeycombs at the edges.

Forms

Forms for the patio can be either metal or wood (Fig. 16-111). Dimensional lumber (2 × 4s or 2 × 6s) are usually enough to hold the material while it is being worked and before it sets up. Contractors use metal forms that are used over and over. If you are going to use the 2 × 4s for more than one job, you may want to oil them so that the concrete will not stick to them and make a mess of the edges of the next job. You can use old crankcase oil and brush it on the wooden forms.

Forms should be placed carefully because their tops are the guides for the screeds. Make sure that the distances apart are measured accurately. Use a spirit level to ensure that they are horizontal. If the forms are used on an inclined slab, they must follow the incline. Forms or curved patios or driveways are made from ½-inch redwood or plywood. If you want to bend the redwood, soak it in water for about 20 minutes before trying to bend it.

Place stakes at intervals along the outside of the forms, and drive them into the ground. Then nail the stakes to the forms to hold them securely in place. The tops of the stakes must be slightly below the edge of the forms so that they will not interfere with use of the strike-off board for purposes of screeding later.

Concrete levelled to top of form board (screed), form board, spreader, slope as required (usually 1/8" per foot), stakes 3'-0" OC, tamped earth or granular fill, screed rod, 2x4 side form boards.

Fig. 16-111 *Forms are made of 2 × 4s or 2 × 6s and staked to prevent movement when the concrete is added and worked.*

Placing the Joints

Joints are placed in concrete to allow for expansion, contraction, and shrinkage. It is best not to allow the slab to bond to the walls of the house but to allow it to move freely with the earth. To prevent bonding, use a strip of rigid waterproof insulation, building paper, polyethylene, or something similar. Also use the expansion-joint material where the patio butts against a walk or other flat surface (Fig. 16-112).

Wide areas such as a patio slab should be paved in 10- to 15-foot-wide alternate strips. A construction joint is made by placing a beveled piece of wood on

Construction joint. 2" lumber with beveled 1"x2" strip.

Fig. 16-113 *Construction joint.*

the side forms (Fig. 16-113). This creates a groove in the slab edges. As the intermediate strips are paved, concrete fills the groove and the two slabs are keyed together. This type of joint keeps the slab surfaces even and transfers the load from one slab to the other when heavy loads are placed on the slabs.

You have seen contraction joints, often called *dummy joints*, cut across a slab (Fig. 16-114). They are cut to a depth of one-fifth to one-fourth the thickness of the slab. This makes the slab weaker at this point. If the concrete cracks owing to shrinkage or thermal contraction, the crack usually occurs at this weakened section. In most instances, the dummy joint is placed in the concrete after it is finished off. A tool is drawn through the concrete before it sets up. It cuts a groove in the surface and drops down into the concrete about

Continuous rigid waterproof insulation board. Isolation.

Fig. 16-112 *Isolation joint prevents the slab from cracking the wall as it expands.*

DUMMY JOINT

1/5 TO 1/4 DEPTH OF SLAB

Fig. 16-114 *Dummy or contraction joint.*

½ inch. In case the concrete cracks later after drying, the crack usually will follow these grooves. These grooves are usually placed 10 to 15 feet apart on floor slabs or patios.

Pouring the Concrete

Most concrete is ordered from a ready-mix company and delivered with the proper consistency. If you mix your own, some precautions should be taken to ensure its proper placement. Concrete should be poured within 45 minutes of the time it is mixed. Some curing begins to take place after that, and the concrete may become too thick to handle easily. If you are pouring a large area, mix only as much as you can handle within 45 minutes. The first batch should be a small one that you can use for trial purposes. After the first batch, you can determine whether the succeeding batches require more or less sand. Remember, do not vary the water for a thicker or thinner concrete—only the amount of sand. Once you have established the workability you like, stay with that formula.

Pour the concrete into the forms so that it is level with the form edges. Then immediately after the first batch is poured into place, spade the concrete with an old garden rake or hoe. This will even out the wet concrete so that it is level with the form boards. Make sure that there are no air voids in the mix.

Use a striker board to level the concrete in the form. It usually requires two people to do this job—one at each end of the board. The idea is to get a level and even top surface on the poured concrete by using the board forms as the guide.

Draw the striker board across the concrete while using the form edges as the guides. Then you can seesaw the board as you move it across. Now it is obvious why the stakes for the forms had to be below the edges of the form boards: They should not interfere with movement of the striker board. The striker board takes off the high spots and levels the concrete. If the board skips over places where it is low and lacks concrete, fill in the low spots, and go over them again to level off the new material.

Once you have finished striking the concrete and leveling it off, you may notice a coat of water or a shiny surface. This may not be evenly distributed across the top, but wait until this sheen disappears before you do any other work on the concrete. This may take an hour or two; the temperature determines how quickly the concrete starts to set up. The humidity of the air and the wind are also factors. The concrete will begin to harden and cure (Fig. 16-115).

Finishing

When the water sheen and bleed water have left the surface of the concrete, you can start to finish the surface of the slab. This may be done in one or more ways depending on the type of surface you want. Keep in mind that you can overdo the finishing process.

Fig. 16-115 *Relative concrete strength versus curing method.*

You can bring water under the surface up to the top to cause a thin layer of cement to be deposited near or on the surface. Later, after curing, it becomes a scale that will powder off with use.

Finishing can be done by hand or by rotating power-driven trowels or floats. The size of the job will determine which to use. In most cases, you will have to rent the power-driven trowel or float. Therefore, economy of operation becomes a factor in finishing. Can you do the job by yourself, or will you need help to get it done before the concrete sets up too hard to work? You should make up your mind before you start so that you can have additional help handy or have the power-driven machine around to speed up the job.

The type of tool used determines the type of finish on the surface of the patio slab. A wood float puts a slightly rough surface on the concrete. A metal trowel or float produces a smooth finish. Extra-rough surfaces are produced by using a stiff-bristled broom across the top.

Floating

A float made of a piece of wood with a handle for use by hand is used to work the concrete surface. In some cases, it has a long handle so that the concrete can be worked by a person standing up and away from the forms (Fig. 16-116). A piece of plywood or other board can be used to kneel on while floating the surface. The concrete should be set up sufficiently to support the person doing the work. Floating has some advantages. It embeds the large aggregate (gravel) beneath the surface and removes slight imperfections such as bumps and voids. It also consolidates the cement near the surface in preparation for smoother finishes. Floating can be done before or after edging and grooving. If the lines left by the edger and groover is to be removed, floating should follow the edging and grooving operation. If the lines are to be left for decorative purposes or to provide a crack line for later movement of the slab, edging and grooving will have to follow the floating operation.

Fig. 16-116 *Construction details for a long-handled float.*

Troweling

Troweling produces a smooth, hard surface. It is done right after floating. For the first troweling, whether by hand or power, the trowel blade must be kept as flat against the surface as possible. If the trowel blade is tilted or pitched at too great an angle, an objectionable washboard or chatter surface will be produced. For first troweling, do not use a new trowel. An older one that has been broken in can be worked quite flat without the edges digging into the concrete. The smoothness of the surface can be improved by a number of trowelings. There should be a lapse of time between successive trowelings to allow the concrete to increase its set or become harder. As the surface stiffens, each successive troweling should be made by a smaller trowel. This gives you sufficient pressure for proper finishing.

Brooming

If you want a rough-textured surface, you can score the surface after it is troweled. This can be done by using a broom. Broom lines should be straight lines, or they can be swirled, curved, or scalloped for decorative purposes. If you want deep scoring of the surface use a wire-bristled broom. For a finer texture, you may want to use a finer-bristled broom. Draw the broom toward you one stroke at a time with a slight overlap between the edges of each stroke. The broom should be wet when it is first drawn across the surface. You can use a pail of water to wash off the excess concrete. Each time a completed stroke is accomplished, dip the broom into the water and shake off the excess water before drawing the broom across the surface again. Most patios, however, look better with a smooth, silky surface.

Grooving

To avoid random cracking owing to heaving of the concrete slab after it has cured, it is best to put in grooves after the troweling is done. Then cracks that form will follow the grooves instead of marring the surface of an otherwise smooth finish.

FENCES

Installing a fence calls for the surveyor's markers to be located in the corners of the property. Lot markers are made of ¾-inch pipe, or they could be a corner concrete monument. Many times they are buried as much as 2 feet under the surface. The plot plan may be of assistance in locating the surveyor's markers. Locating proper limits will make sure that the fence is on the land it is intended for and not on the property next door.

The local zoning board usually has limitations on the height of the fence and just where it can be placed on a property. If ordinances allow it to be placed exactly on the property line, be sure to check with the abutting neighbors and obtain their consent in order to prevent any future law suits.

In most states, it is generally understood that if the posts are on the inside facing your property and the fence on the other side, except for post-and-rail fencing, the fence belongs to you. Local officials will be more than happy to help you in regard to the details.

Installation

Most fences are variations of a simple post, rail, and board design. The post and rail support structure is often made of standard-dimension lumber, whereas the fenceboards come in different shapes and sizes, giving the fence its individual style.

On a corner of the lot, place a stake parallel to the surveyor's marker, and attach a chalk line to the stake. Pull the chalk line taut, and secure it at the opposite stake. Repeat until all boundaries are covered (Fig. 16-117). Repeat the procedure to measure the required setback from the original boundary, and restake accordingly. You are now ready to install the fence.

To get the fencing started correctly, start from one corner of the lot, and place the first post in that corner.

Fig. 16-117 *Layout of the lot and location of posts for the fence.*

Setting Posts

Setting posts requires patience and is the most critical aspect of fence building. Posts must be sturdy, straight, and evenly spaced for the fence to look and perform properly. Redwood, western cedar, or chemically treated pine can be used for the posts. Posts are commonly placed 8 feet on center. Mark the post locations with stakes.

Dig holes about 10 inches in diameter with a posthole digger. Holes dug with a shovel will be too wide at the mouth to provide proper support. Auger-type diggers work well in rock-free earth. If you are likely to encounter stones, use a clam-shell type of digger.

Set the corner posts first. String a line between corner posts to mark the fenceline, and align the inside posts. For a 5- or 6-foot fence, post holes should be at least 2 feet deep. A 3-foot hole is required for an 8-foot fence (Fig. 16-118).

POSTS AND POST HOLES

8'

CONCRETE

GRAVEL

6'

2 1/2"

5'

1 1/2"

CONCRETE

GRAVEL

6'

2 1/2"

Fig. 16-118 *Depths needed for mounting fence posts.* (California Redwood Association.)

Proper drainage in post holes eliminates moisture and extends the life of posts. Fill the bottom of holes with gravel. Large, flat base stones also aid drainage. Fill in with more gravel, 3 or 4 inches up the post.

For the strongest fence, set the posts in concrete. You may figure your concrete requirements at roughly an 80-pound bag of premixed concrete per post. If you extend the concrete by filling with rocks or masonry rubble, be sure to tap it down. Cleats or metal hardware can be used to strengthen fenceposts. Cleats are 2 × 4 wood scraps attached horizontally near the base of the post that provide lateral stability. Large lag screws or spikes partially driven into the post can be used to do the same job when posts are set in concrete.

Make sure that the concrete completely surrounds the post to make a collar. It doesn't have to be a full shell. This reduces the tendency to hold water and promote decay (Fig. 16-119). When you have placed the posts to the height desired, pour the concrete collar around the post. Be sure that the posts are perpendicular to the surface, and allow the concrete to cure at least 48 hours before attaching the fencing.

Attaching the Rails

Two or three horizontal rails run between the posts. The number of rails depends on the fence height. Rails 8 feet long are common because this length of 2 × 4 is readily available and provides enough support for most styles of fenceboards. Upper rails should rest on

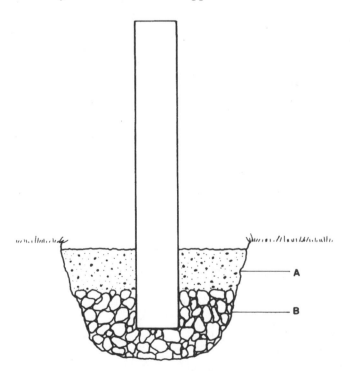

Fig. 16-119 *Location of concrete collar around the post at A. B is the gravel fill.*

the posts. Rails can be butt-joined and mitered at corners. Bottom rails can be toenailed into place, but the preferred method is to place a block underneath the joint for extra support. Metal hardware such as L brackets can be used to secure rails, but make sure that all metal fasteners including nails are noncorrosive (Fig. 16-120).

Attaching Fenceboards

Nailing the fenceboards in place is the easiest and most satisfying part of the project. Boards 1 inch thick and 4, 6, 8, or 10 inches wide are common. The dimensions of the fenceboards and the way they are applied will give the fence its character.

There are several creative ways to do this. For fenceboards 4 inches wide and less, use one nail per bearing. For wider fenceboards, use two nails per bearing. Do not overnail.

Picket fences leave plenty of room between narrow fenceboards. They are often used to mark boundaries and provide a minimal barrier. Just enough to keep small children or small dogs in the yard, they are typically 3 or 4 feet high (Fig. 16-121).

Board-on-board fences are taller and provide more of a barrier. Distinguished by the fenceboards' alternating pattern, board-on-board fences look the same from either side. This allows great flexibility in design and function. Depending on placement of fenceboards, this fence will block the wind and the view.

Lattice panels made from redwood lath or ½ × 2s can be used to create a lighter, more delicate fence. Lattice panels can be prefabricated within a 2 × 4 frame and then nailed to the posts.

Panel fences create a solid barrier. Panels are formed by nailing boards over rails and posts. By alternating the side that the fenceboards are nailed to at each post, you can build a fence that looks the same from both sides.

Stockade fencing is built level (Fig. 16-122). However, not all ground is flat. You can adjust for the slope of the ground by working downhill if possible. Fasten the section into a post that is plumb; then exert downward pressure on the end of the section before attaching it to the post (Fig. 16-123). This is called *racking*. When finished, all posts should follow the slope of the ground.

Nails and Fasteners

Use noncorrosive nails with redwood outdoors. Stainless steel, aluminum, and top-quality, hot-dipped gal-

TOP RAILS   BOTTOM RAILS

MITERED   BUTTED   BLOCK   TOE-NAIL   METAL

Fig. 16-120 *Attaching rails to the posts.* (California Redwood Association.)

PICKET   BOARD ON BOARD   LATTICE   PANEL

Fig. 16-121 *Various types of fence designs.* (California Redwood Association.)

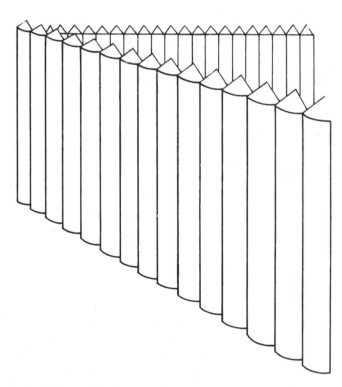

Fig. 16-122 *Stockade fence.*

vanized nails will perform without staining. Inferior hardware, including cheap or electroplated galvanized nails, will corrode and cause stains when in contact with moisture.

If you use redwood, do not finish off with varnishes and clear-film finishes, oil treatments, or shake and shingle-type paints.

Gates

Getting the gate to operate properly is another problem with fencing. By using the proper hardware and making sure that the gate posts are plumb and sturdy, it is possible to install a gate that will work for years (Fig. 16-124). To apply the hinges, position the straps approximately 4 inches from both the top and bottom of the gate on the side to be hinged (Fig. 16-125). Position the gate in the opening, and allow adequate clearance between the bottom of the gate and the ground. You can use a wooden block under the gate to ensure proper clearance and aid in installation. Mark and secure the vertical leaves to the fence posts. Drill pilot holes for the screws (Fig. 16-126). For flat gates to be mounted on a fence post, use an ornamental screw hook and strap hinge (Fig. 16-127). For gates framed with round posts, use the screw hook-and-eye hinge shown in Fig. 16-128.

FORCE DOWN HERE

Fig. 16-123 *Racking a fence to fit a sloping yard.*

Fig. 16-124 *Fence gate spring closer.*

Fig. 16-126 *Place a spacer block of wood under the gate while attaching the hinges.*

4"

4"

Fig. 16-125 *Allow 4 inches for the hinges.*

Fig. 16-127 *Attaching another type of hinge.*

To keep the gate closed, you can use a spring that is mounted such as the one shown in Fig. 16-129. Note that the spring is always inclined to the right. Place the spring as near vertical as conditions permit. Place the adjusting end at the top.

There are at least three ways to latch a gate. The cane bolt drops by gravity. It can be held in raised position, and the bolt cannot be moved while mounted. This drops down into a piece of pipe that is driven into the ground. Then it becomes very difficult to open the

Fig. 16-128 *The hook-and-eye hinge mounted to a round post.*

Fig. 16-129 *Spring closers.*

gate. This type usually is used when there are two gates and one is kept closed except when large loads are to pass through (Fig. 16-130). The middle latch is self-latching. It can be locked with a padlock. The ornamental thumb latch is a combination pull and self-latching latch with padlock capability (Fig. 16-131).

Energy Conservation

In order to make sure that the greatest efficiency is obtain from a fireplace, it is necessary to be able to turn it off and on or close the damper to make sure that cold

Fig. 16-130 *Gate hardware.*

Fig. 16-131 *Completed installation.*

air does not enter the room when the fireplace is not in operation.

Natural gas burners are usually placed in newer houses to improve on the amount of heat that is produced for heating the inside without causing a disturbance in the thermostat in other parts of the house. Conservation of energy is primary in the heating process, and the more efficient a fireplace is better. By using natural gas to provide the flame needed for heat and visual effects, it also makes for a more environmentally acceptable source of heat without the smoke and air pollution caused by burning wood or coal in the fireplace.

An electric heater inside the fireplace shell is another more acceptable heat source without the pollution.

CHAPTER 16 STUDY QUESTIONS

1. Why is firebrick used in fireplaces?
2. How many steps would be cut for stairs with a *rise* of 8 feet, 2 inches, a *run* of 14 feet, 0 inches, and a *riser* of 7 inches?

3. For the steps above, what would the tread width be?

4. What is the opening-to-flue ratio?

5. What would be a good flue size for a fireplace with an opening 30 inches high and 48 inches long?

6. What are the advantages of metal fireplaces?

7. What are the advantages of post-and-beam construction?

8. How are posts and beams connected?

9. Would you embed an exterior beam into concrete? Why or why not?

10. Why are laminated beams better than solid-wooden beams?

11. How can a stressed-skin panel be used for both roof decking and finished ceilings?

12. What kind of insulation is used over lumber roof decking?

13. What type of roof is not used on steep-pitched roofs?

14. What is a *housed carriage*?

15. What causes stairs to squeak?

16. Can stud posts be fastened at any point on a sub-floor?

17. What are two stairway post systems?

18. What is a *rosette*?

19. What is *reverse venting*?

20. How much clearance must you have in type B gas vents from 3 to 24 inches in diameter?

17

Maintenance and Remodeling

PLANNING IS AN IMPORTANT PART OF ANY JOB. THE job of remodeling is no exception. It takes a plan to get the job done correctly. The plan must have all the details worked out. This will save money, time, and effort. The work will go smoothly if the bugs have been worked out before the job is started.

In this chapter, you will learn how to

• Diagnose problems

• Identify needed maintenance jobs

• Make minor repairs

• Do some minor remodeling jobs

PLANNING THE JOB

Maintenance means keeping something operating properly. It means taking time to make sure that a piece of equipment will operate tomorrow. It means doing certain things to keep a house in good repair. Many types of jobs present themselves when it comes to maintenance. The carpenter is the person most commonly called on to do maintenance. This may range from the replacement of a lock to complete replacement of a window or door.

Remodeling means just what the word says. It means changing the looks and function of a house. It might mean that you have to put in new kitchen cabinets. The windows might need a different type of opening. Perhaps the floor is old and needs a new covering. The basement might require new paneling or tiled floors.

Working on a house when it is new calls for a carpenter who can saw, measure, and nail things in the proper place. Working on a house after it is built calls for many types of operations. The carpenter may be called on to do a number of different things related to the trade.

Diagnosing Problems

A person who works in maintenance or remodeling needs to know what the problem actually is. That person must be able to find out what causes a problem. The next step, of course, is to decide what to do to correct the problem.

If you know how a house is built, you should be able to repair it. This means that you know what has to be done to construct a wall properly or repair it if it is damaged. If the roof leaks, you need to know where and how to fix it.

In other words, you need to be able to diagnose problems in any building. Since we are concerned with residential types of buildings here, it is important to know how things go wrong in a well-built home. In addition, it is important to be able to repair them.

Before you can remodel a house or add on to it, you need to know how the original was built. You need to know what type of foundation was used. Can it support another story, or can the soil support what you have in mind? Are you prepared for the electrical loads? What about the sewage? How does all this fit into the addition plans? What type of consideration have you given the plumbing, drainage, and other problems?

Identifying needed operations If peeling paint needs to be removed and the wall repainted, can you identify what caused the problem? This will be important later when you choose another paint. Figure 17-1 shows what can happen with paint.

Paint problems by and large are caused by the presence of moisture. The moisture may be in the wood when it is painted, or it may have seeped in later. Take a look at Fig. 17-1 to see just what causes cracking and alligatoring. Also notice the causes of peeling, flaking, nailhead stains, and blistering.

Normal daily activity in the home of a family of four can put as much as 50 pints of water vapor into the air in one day. Since moisture vapor always seeks an area of lesser pressure, the moisture inside the house tries to become the same as the outside pressure. This equalization process results in moisture passing through walls and ceilings, sheathing, door and window casings, and the roof. The eventual result is paint damage. This occurs as the moisture passes through the exterior paint film.

Sequencing Work to Be Done

In making repairs or doing remodeling, it is necessary to schedule the work properly. It is necessary to make sure that things are done in order. For example, it is difficult to paint if there is no wall to paint. This might sound ridiculous, but it is no more so than some other problems associated with getting a job done.

It is hard to nail boards onto a house if there are no nails. Somewhere in the planning, you need to make sure that nails are available when you need them. If things are not planned properly—in sequence—you could be ready to place a roof on the house and not have the proper size nails to do the job. You wouldn't try to place a carpet on bare joists. There must be a floor or subflooring first. This means that there is some sequence that must be followed before you do a job or even get started with it.

FLAKING

CONDITION:

Siding alternatively swells and shrinks as moisture behind it is absorbed and then dries out. Paint film cracks from swelling and shrinking and flakes away from surface.

CORRECTION:

- All moisture problems must be corrected before surface is re-painted.
- Scrape and sand all peeling paint to bare wood including several inches around damaged areas. Feather edges.
- Apply primer according to label directions.
- Apply topcoat according to label directions.

BLISTERING

CONDITION:

Blistering is actually the first stage of the peeling process. It is caused by moisture attempting to escape through the existing paint film, lifting the paint away from the surface.

CORRECTION:

- All moisture problems must be corrected before repainting.
- Scrape and sand all blistering paint to bare wood several inches around blistered area.
- Feather or smooth the rough edges of the old paint by sanding.
- Apply primer according to label directions.
- Apply topcoat according to label directions.

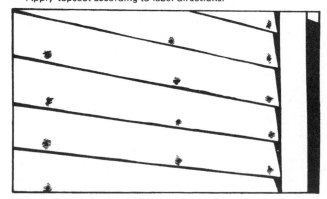

CRACKING AND ALLIGATORING

CONDITION:

Cracking and "alligatoring" are caused by (a) paint that is applied too thick, (b) too many coats of paint, (c) paint applied over a paint coat which is not completely dry, or (d) an improper primer.

CORRECTION:

- Removal of entire checked or alligatored surface may be necessary.
- Scrape and sand down the surface until smooth. Feather edges.
- Apply primer. Follow label directions.
- Apply topcoat. Follow label directions.

PEELING

CONDITION:

Peeling is caused by moisture being pulled through the paint (by the sun's heat), lifting paint away from the surface.

CORRECTION:

- All moisture problems must be corrected before surface is repainted.
- Remove all peeling and flaking paint.
- Scrape and sand all peeling paint to bare wood including several inches around damaged areas. Feather edges.
- Apply primer according to directions on label.
- Apply topcoat according to directions.

NAILHEAD STAINS

CONDITION:

Nailhead stains are caused by moisture rusting old or uncoated nails.

CORRECTION:

- All moisture problems must be corrected before repainting.
- Sand or wire brush stained paint and remove rust down to bright metal of nailhead.
- Countersink nail if necessary.
- Apply primer to nailheads. Allow to dry.
- Caulk nail holes. Allow to dry. Sand smooth.
- Apply primer to surface, following label directions.
- Apply topcoat according to label directions.

Fig. 17-1 *Causes of paint problems on houses.* (Grossman Lumber.)

Make a checklist to be sure that you have all the materials you need to do a job *before* you get started. If there is the possibility that something won't arrive when needed, try to schedule something else so that the operation can go on. Then you can pick up the missing part later when it becomes available. This means that sequencing has to take into consideration the problems of supply and delivery of materials. The person who coordinates this is very important. This person can make the difference between the job being a profitable one or a money loser.

MINOR REPAIRS AND REMODELING

When doors are installed, they should fit properly. This means that the closed door should fit tightly against the door stop. Figure 17-2 shows how the door should fit. If it doesn't fit properly, adjust the strike-jamb side of the frame in or out. Do this until the door meets the weather stripping evenly from top to bottom.

Fig. 17-2   Right and wrong ways of nailing the strike jamb. (General Products.)

Adjusting Doors

The strike jamb can be shimmed like the hinge jamb. Place one set of shims behind the strike-plate mounting location. Renail the jamb so that it fits properly. If it is an interior door, there will be no weather stripping. Figure 17-3 shows how to shim the hinges to make sure that the door fits snugly. In some cases, you might have to remove some of the wood on the door where the hinge is attached (Fig. 17-4). The figure shows how the wood is removed with a chisel. You must be

Fig. 17-3   Hinge adjustment for incorrectly fitting doors. Note the shim placement.

Fig. 17-4   Removing extra wood from a door. Be careful not to remove too much. (Grossman Lumber.)

careful not to remove the back part of the door along the outside of the door.

If the door binds or does not fit properly, you might have to remove some of the lock edge of the door. Bevel it as shown in Fig. 17-5 so that it fits. An old carpenter's trick is to make sure that the thickness of an 8d nail is allowed all around the door. This usually is enough space for the door to swell some in humid weather and not too much space when winter heat in the house dries out the wood in the door and causes it to shrink slightly.

Occasionally, it is necessary to remove the door from its hinges by removing the hinge pins. Take the door to a vise or workbench so that the edge of the door can be planed down to fit the opening. If the

Fig. 17-5   Beveling the lock stile of a door. (Grossman Lumber.)

Fig. 17-6 *Trimming the width of a door.* (Grossman Lumber.)

Fig. 17-7 *Adding a piece of plastic to a threshold to make sure that the bottom of the door fits snugly. This keeps rain from entering the room, as well as keeping out cold air in the winter.* (General Products.)

amount of wood to be removed from the door is more than ¼ inch, it will be necessary to trim both edges of the door equally to one-half the width of the wood to be removed. Trim off the wood with a smooth or jack plane, as shown in Fig. 17-6.

Sometimes it is necessary to remove a door from its hinges and cut off the bottom because the door was not cut to fit a room where there is carpeting. If this occurs, remove the door carefully from its hinges. Mark off the amount of wood that must be removed from the bottom of the door. Place a piece of masking tape over the area to be cut. Redraw your line in the middle of the tape. Tape the other side of the door at the same distance from the bottom. Set the saw to cut the thickness of the door. Cut the door with a power saw or handsaw. Cutting through the tape will hold the finish on the wood. You will not have a door with splinters all along the cut edge. If the door is cut without tape, it may have splinters. This can look very bad if the door has been prefinished.

If the top of the door binds, it should be beveled slightly toward the stop. This can let it open and close more easily.

In some cases, an outside door does not meet the threshold properly. It may be necessary to obtain a thicker threshold or a piece of plastic to fit on the bottom of the door, as shown in Fig. 17-7.

Adjusting Locks

Installing a lock can be as easy as following instructions. Each manufacturer furnishes instructions with each new lock. However, in some cases, you have to replace one that is around the house, and no instructions can be found. Figure 17-8 provides a step-by-

step method of placing a lock into a door. Note that the lockset is typical. It can be replaced with just a screwdriver. There are 18 other brands that can be replaced by this particular lockset.

Various locksets are available to fit the holes that already exist in a door. Strikes come in a variety of shapes too. You should choose one that fits the already-grooved door jamb.

A deadbolt type of lock is called for in some areas. This is where the crime rate is such that a more secure door is needed.

Figure 17-9 shows a lockset that is added to an existing lock in the door. This one is key-operated. This means that you must have two keys to enter the door. It is very difficult for a burglar to cause this type of lock bolt to retract.

In some cases, it is desired that once the door is closed, it is locked. This can be both an advantage and a disadvantage. If you go out without your key, you are in trouble. This is especially true if no one else is home. However, it is nice for those who are a bit absentminded and forget to lock the door once it is closed. The door locks automatically once the door is pushed closed.

Figure 17-10 shows how to insert a different type of lock, designed to fit into the existing drilled holes. All you need is a screwdriver to install it. Figure 17-11 shows how secure the lock can be with double cylinders to fit through the holes on the door jamb.

STEP 1: Prepare door; drill for bolt and lock mechanism.

STEP 2: Insert bolt and lock mechanism.

STEP 3: Engage bolt and lock mechanism; fasten face plate.

STEP 4: Put on clamp plate; Tighten screws.

STEP 5: "Snap-On" rose.

STEP 6: Apply knob on spindle by depressing spring retainer.

STEP 7: Mortise for latch bolt; fasten strike with screws.

Fig. 17-8 *Fitting a lockset into a door.* (National Lock.)

Fig. 17-9 *Key-operated auxiliary lock.* (Weiser Lock.)

Fig. 17-10 *Putting in a lockset.* (Weiser Lock.)

Fig. 17-11 *Strikes for single- and double-cylinder locks.*

About the only trouble with a lockset is loosening of the doorknob. It can be tightened with a screwdriver in most cases. If the lockset has two screws near the knob on the inside of the door, simply align the lockset and tighten the screws. If it is another type—with no screws visible—just release the small tab that sticks up inside the lockset; it can be seen through the brass plate around the knob close to the wooden part of the door. This will allow the brass plate or ring to be rotated and removed. Then it is a matter of tightening the screws found inside the lockset. Tighten the screws, and replace the cover plate.

The strike plate may work loose, or the door may settle slightly. This means that the striker will not align with the strike plate. Adjust the plate or strike screws if necessary. In some cases, it might be easier to remove the strike plate and file out a small portion to allow the bolt to fit into the hole in the door jamb.

If the doorknob becomes green or off-color, remove it. Polish the brass, and recoat the knob with a covering of lacquer and replace. In some cases, the brass is plated and will be removed with the buffing.

Here you may want to add a favorite shade of good metal enamel to the knob and replace it. This discoloring frequently occur in bathrooms, where the moisture attacks the doorknob lacquer coating.

Installing Drapery Hardware

One often-overlooked area of house building is the drapery hardware. People who buy a new house are faced with the question, "What do I do to make these windows attractive?" Installing window hardware can be a job in itself. Installing conventional adjustable traverse rods is a task that can be done easily by the carpenter, in some cases, or the homeowner. There are specialists who do these things. However, it is usually an extra service that the carpenter gets paid for after the house is finished and turned over to the homeowner.

Decorative traverse rods are preferred by those who like period furniture (Fig. 17-12). Installation of a traverse rod is shown step by step in Figs. 17-13 through 17-17.

Now that you have studied some of the different types of windows in Chapter 9, have you wondered how the draperies would be fitted? Figure 17-18 shows some of the regular-duty types of corner and bay window drapery rods. Note the various shapes of bay windows that are made. The rods are made to fit the bay windows.

Figure 17-19 shows more variations of bay windows. These call for some interesting rods. Of course, in some cases, the person looks at the cost of the rods and hardware and decides to put one straight rod across the bay and not follow the shape of the windows. This can be the least expensive, but it negates the effect of having a bay window in the first place.

Fig. 17-12 *An adjustable decorative traverse rod.*

Fig. 17-13 Install end brackets and center support.

Install end brackets above and 6" to 18" to the sides of casing. When installing on plaster or wallboard and screws do not anchor to studding, plastic anchors or other installation aids may be needed to hold brackets securely. Place center supports (provided with longer size rods only) equidistant between end brackets. Adjust bracket and support projection at screws. "A." *NOTE: Attach rosette screws to end brackets if not already assembled.*

Place end finials on rod. Finial with smaller diameter shaft fits in "inside" rod section, larger one fits in "outside" section. Place last ring between bracket and end finial.

Extend rod to fit brackets. Place rod in brackets so tongue on bottom of bracket socket fits into hole near end of rod. Tighten rosette screw.

Fig. 17-14 *Place rod in bracket.*

When center supports are used, insert top of rod in support clip flange. If outside rod section, turn cam with screwdriver counterclockwise to lock in rod. If inside rod section, turn cam clockwise.

Right-hand draw is standard. If left hand draw is desired, simply pull cord loop down from pulley wheels on left side of rod.

Fasten tension pulley to sill, wall or floor at point where cord loop falls from traverse rod. Lift pulley stem and slip nail through hole in stem. Pull cord up through bottom of pulley cover. Take up excess slack by pulling out knotted cord from back of overlap master slide. Re-knot and cut off extra cord. Remove nail. Pulley head may be rotated to eliminate twisted cords.

Fig. 17-15 *Install cord tension pulleys.*

Pull draw cord to move overlap master to end of rod. Holding cords taut, slide underlap master to opposite end of rod. Then fasten cord around locking fingers on underlap master to stop underlap master from slipping. Then center both masters.

Fig. 17-16 *Adjust the master slides.*

Maintenance and Remodeling

FINIAL SHAFT

TAB

PULLEY HOUSING

To train draperies, "break" fabric between pleats by folding the material toward the window. Fold pleats together and smooth out folds down the entire length of the draperies. Tie draperies with light cord or cloth. Leave tied two or three days before operating.

Slip off end finial. Remove extra ring slides by pulling back on tab and sliding rings off end of rod. Be sure to leave the last ring between pulley housing and end finial.

Fig. 17-17 *A. Remove unused ring slides. B. Training draperies.*

REGULAR DUTY CUSTOM CEILING SETS

ONE-WAY TRAVERSE SETS

| 1787R | RIGHT TO LEFT | 1787L | LEFT TO RIGHT | | | |

TWO CORD TRAVERSE SETS

WITH 1 PAIR MASTER CARRIERS
CENTER CLOSING

| 1788 | | | |

TWO CORD TRAVERSE SETS

WITH 2 PAIRS OF MASTER CARRIERS
CENTER CLOSING

| 1789 | | | |

CRATING CHARGE: 10% OF CURVED ROD ORDER   REGULAR DUTY CORNER, CURVED AND BAY WINDOW CUSTOM SETS

CURVED CORNER SETS

1½ TO 2½" (38 to 64 mm) CLEAR.

| 1790 | | | |

CORNER SETS

1½" TO 2½" (38 to 64 mm) CLEAR.
WITH 2 ONE-WAY RODS

| 1791 | | | |

CIRCULAR BAY SETS

1½ TO 2½" (38 to 64 mm) CLEAR.

| 1792 | | | |

CURVED COMBINATION SETS

INSIDE ROD: 1½ TO 2½" (38 to 64 mm) CLEAR.
OUTSIDE ROD: 3½ TO 4½" (89 to 114 mm) CLEAR.

| 1792P | | | |

CIRCULAR DOUBLE SETS

INSIDE ROD: 1½ TO 2½" (38 to 64 mm) CLEAR.
OUTSIDE ROD: 3½ TO 4½ (89 to 114 mm) CLEAR.

| 1792D | | | |

ANGULAR BAY SETS

1½ TO 2½" (38 to 64 mm) CLEAR.
WITH 2 BENDS

| 1793 | | | |

Fig. 17-18 *Regular-duty custom ceiling sets of traverse rods; regular-duty corner, curved, and bay window custom sets. (Kenny.)*

1793P

ANGULAR COMBINATION SETS

INSIDE ROD: 1½ TO 2½" (38 to 64 mm) CLEAR.
OUTSIDE ROD: 3½ TO 4½" (89 to 114 mm) CLEAR.
WITH 2 BENDS

ANGULAR DOUBLE BAY SETS

INSIDE ROD: 1½ TO 2½" (38 to 64 mm) CLEAR.
OUTSIDE ROD: 3½ TO 4½" (89 to 114 mm) CLEAR.
WITH 2 BENDS

1793D

SQUARE BAY SETS

1½" TO 2½" (38 to 64 mm) CLEAR.
WITH 2 BENDS

1794

SQUARE COMBINATION SETS

INSIDE ROD: 1½ TO 2½" (38 to 64 mm) CLEAR.
OUTSIDE ROD: 3½ TO 4½ (89 to 114 mm) CLEAR.

1794P

SQUARE DOUBLE BAY SETS

INSIDE ROD: 1½ TO 2½" (38 to 64 mm) CLEAR.
OUTSIDE ROD: 3½ TO 4½" (89 to 114 mm) CLEAR.

1794D

ANGULAR BAY SETS

1½ TO 2½" (38 to 64 mm) CLEAR.
WITH 4 BENDS

1795

ANGULAR COMBINATION SETS

INSIDE ROD: 1½ TO 2½" (38 to 64 mm) CLEAR.
OUTSIDE ROD: 3½ TO 4½" (89 to 114 mm) CLEAR.
WITH 4 BENDS

1795P

ANGULAR DOUBLE BAY SETS

INSIDE ROD: 1½ TO 2½" (38 to 64 mm) CLEAR.
OUTSIDE ROD: 3½ TO 4½" (89 to 114 mm) CLEAR.

1795D

Fig. 17-19 *More regular-duty corner, curved, and bay window custom rods.* (Kenny.)

578 **Maintenance and Remodeling**

Repairing Damaged Sheetrock Walls (Drywall)

In drywall construction, the first areas to show problems are over joints or fastener heads. Improper application of either the board or the joint treatment may be at fault. Other conditions existing on the job also can be responsible for reducing the quality of the finished gypsum board surface. A discussion of some of these conditions follows.

Panels improperly fitted

Cause   Forcibly wedging an oversize panel into place. This bows the panel and builds in stresses. The stress keeps it from contacting the framing (Fig. 17-20).

Fig. 17-20   *Forcibly fitted piece of gypsum board.* (U.S. Gypsum.)

Result   After nailing, a high percentage of the nails on the central studs probably will puncture the paper. This also may cause joint deformation.

Remedy   Remove the panel. Cut it to fit properly. Replace it. Fasten from the center of the panel toward the ends and edges. Apply pressure to hold the panel tightly against the framing while driving the fasteners.

Panels with damaged edges

Cause   Paper-bound edges have been damaged or abused. This may result in ply separation along the edge. Or it may loosen the paper from the gypsum core. Or it may fracture or powder the core itself. Damaged edges are more susceptible to ridging after joint treatment.

Remedy   Cut back any severely damaged edges to the sound board before application.

Prevention   Avoid using board with damaged edges that may easily compress. Damaged edges can take on moisture and swell. Handle Sheetrock with care.

Panels loosely fastened

Cause   Framing members are uneven because of misalignment or warping. If there is lack of hand pressure on the panel during fastening, loosely fitting panels can result (Fig. 17-21).

Remedy   When panels are fastened with nails, during final blows of the hammer, use your hand to apply ad-

Fig. 17-21   *Damaged edges and their effect on a joint.* (U.S. Gypsum.)

Fig. 17-22   *Apply hand pressure (arrow) while nailing the board to the stud.* (U.S. Gypsum.)

ditional pressure to the panel adjacent to the nail (Fig. 17-22).

Prevention   Correct framing imperfections before applying the panels. Use screws or adhesive method instead of nails.

Surface fractured after application

Cause   Heavy blows or other abuse has fractured finished wall surfaces. If the break is too large to repair with joint compound, do the following.

Remedy   In the shape of an equilateral triangle around the damaged area, remove a plug of gypsum. Use a keyhole saw. Slope the edges 45 degrees. Cut a corresponding plug from a sound piece of gypsum. Sand the edges to an exact fit. If necessary, cement an extra slat of gypsum panel to the back of the face layer to serve as a brace. Butter the edges and finish as a butt joint with joint compound.

Framing members out of alignment

Cause   Because of misaligned top plate and stud, hammering at points *X* in Fig. 17-23 as panels are applied on both sides of the partition probably will result in nail heads puncturing the paper or cracking the board. If framing members are more than ¼ inch out of alignment with adjacent members, it is difficult to bring panels into firm contact with all nailing surfaces.

Remedy   Remove or drive in problem fasteners and drive new fasteners only into members in solid contact with the board.

Prevention   Check the alignment of studs, joists, headers, blocking, and plates before applying panels. Correct before proceeding. Straighten badly bowed or crowned members. Shim out flush with adjoining surfaces. Use adhesive attachment.

Fig. 17-25 *Nailhead goes through but doesn't pull the board up tightly against the stud. The stud is prevented from meeting the nailing surface by a piece of bridging out of place.* (U.S. Gypsum.)

Remedy Allow the moisture content in the framing to stabilize. Remove the problem fasteners. Realign the bridging or whatever is out of alignment. Refasten with carefully driven screws.

Fig. 17-23 *If framing members are bowed or misaligned, shims are needed if the wall board is to fit properly.* (U.S. Gypsum.)

Members twisted

Cause Framing members have not been properly squared with the plates. This gives an angular nailing surface (Fig. 17-24). When panels are applied, there is a danger of fastener heads puncturing the paper or of reverse twisting of a member as it dries out. This loosens the board and can cause fastener pops.

Puncturing of face paper

Cause Poorly formed nailheads, careless nailing, excessively dry face paper, or a soft core can cause the face paper to puncture. Nailheads that puncture the paper and shatter the core of the panel are shown in Fig. 17-26. They have little grip on the board.

Fig. 17-26 *Puncturing of the face paper.* (U.S. Gypsum.)

Fig. 17-24 *Framing members improperly squared.* (U.S. Gypsum.)

Remedy Allow the moisture content in the framing to stabilize. Remove the problem fasteners. Refasten with carefully driven screws.

Prevention Align all the twisted framing members before you apply the board.

Framing protrusions

Cause Bridging, headers, fire stops, or mechanical lines have been installed improperly (Fig. 17-25). They may project out past the face of the framing member. This prevents the board or drywall surface from meeting the nailing surface. The result can be a loose board. Fasteners driven in this area of protrusion probably will puncture the face paper.

Remedy Remove the improperly driven fastener. Properly drive a new fastener.

Prevention Correcting faulty framing and driving nails properly produce a tight attachment. There should be a slight uniform dimple. Figure 17-27 illustrates the proper installation of fastener. A nailhead bears on the paper. It holds the panel securely against

Fig. 17-27 *Proper dimple made with the nailhead into the drywall board.* (U.S. Gypsum.)

the framing member. If the face paper becomes dry and brittle, its low moisture content may aggravate the nail cutting. Raise the moisture content of the board and the humidity in the work area.

Nail pops from lumber shrinkage

Cause Improper application, lumber shrinkage, or a combination of the two. With panels held reasonably tight against the framing member and with proper-length nails, normally only severe shrinkage of the lumber will cause nail pops. However, if panels are nailed loosely, any inward pressure on the panel will push the nailhead through its thin covering pad of compound. Pops resulting from nail creep occur when shrinkage of the wood framing exposes nail shanks and consequently loosens the panel (Fig. 17-28).

Fig. 17-28 *Popped nailhead. (U.S. Gypsum.)*

Remedy Repairs usually are necessary only for pops that protrude 0.005 inch or more from the face of the board (see Fig. 17-28). Smaller protrusions may need to be repaired if they occur in a smooth gloss surface or flat-painted surface under extreme lighting conditions. Those that appear before or during decoration should be repaired immediately. Pops that occur after 1 month's heating or more usually are caused wholly or partly by wood shrinkage. They should not be repaired until near the end of the heating season. Drive the proper nail or screw about 1½ inches from the popped nail while applying sufficient pressure adjacent to the nailhead to bring the panel in firm contact with the framing. Strike the popped nail lightly to seat it below the surface of the board. Remove the loose compound. Apply finish coats of compound and paint.

These are but a few of the possible problems with gypsum drywall or Sheetrock. Others are cracking, surface defects, water damage, and discoloration. All can be repaired with the proper tools and equipment. A little skill can be developed over a period of time. However, in most instances it is necessary to redecorate the wall or ceiling. This can become a problem of

greater proportions. It is best to make sure that the job is done correctly the first time. This can be done by looking at some of the suggestions given under "Prevention."

Installing New Countertops

A new countertop has to be constructed step by step, and many details must be checked before the completed installation can be considered ready for use. Today, most countertops are made from laminated plastic. There are many trademarks for these. These materials are usually 1/16 to 1/8 inch thick. The material is not hurt by hot objects and does not stain or peel. The laminate material is very hard and durable. However, in a thin sheet, it is not strong. Most cabinets have a base top made from plywood or chipwood. Usually, a ½-inch thickness or more is used for countertops made on the site.

Also, specially formed counters made from wood products may be purchased. These tops have the plastic laminates and the mold board permanently formed into a one-piece top (Fig. 17-29). This material may be purchased in any desired length. It then is cut to shape and installed on the job.

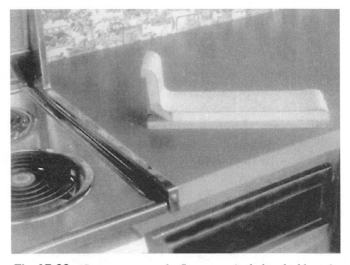

Fig. 17-29 *Countertops may be flat or may include splashboards.*

The top pieces are cut to the desired length. They may be nailed to the partitions or to the drawer kickers on the counter. The top should extend over the counter approximately ¾ inch. Next, sides or rails are put in place around the top. These pieces can be butted or rabbeted as shown in Fig. 17-30. They are nailed to the plywood top. They also may be nailed to the frame of the counter. Next, they should be sanded smooth. Uneven spots or low spots are filled.

Fig. 17-30 *Blocking up counter edges.*

Fig. 17-32 *Cutting plastic laminate with a saber saw. (Grossman Lumber.)*

Once the counter has been built, the top is checked. Also, any openings should be cut. Openings can be cut for sinks or appliances. Next, the plastic laminate is cut to rough size. Rough size should be ⅛ to ¼ inch larger in each dimension. A saw is used to cut the laminate, as shown in Figs. 17-31 through and 17-33.

Fig. 17-33 *Cutting laminate with a hand saw. (Grossman Lumber.)*

Particle board is commonly used for a base for a countertop. It is inexpensive, and it does not have any grain. Grain patterns can show through the finished surface. Also, the grain structure of plywood may form pockets. Glue in pockets does not bond to the laminate. Particle board provides a smooth, even surface that bonds easily.

Fig. 17-31 *Cut into the laminate to avoid chipping.*

Next, contact cement is applied to both the laminate and the top. Contact cement should be applied with a brush or a notched spreader. Allow both surfaces to dry completely. If a brush is used, solvent should be kept handy. Some types of cement are water-soluble. This means that soap and water could be used to wash the brush and to clean up.

When the glue has dried, the surface is shiny. Dull spots mean that the glue was too thin. Apply more glue over these areas. It usually takes about 15 to 20 minutes for contact cement to dry. As a rule, the pieces should be joined within a few minutes. If they are not, a thin coat of contact cement is put on each of the surfaces again.

To glue the laminate in place, two procedures may be used. First, if the piece is small, a guide edge is put in place. The straight piece is held over the area, and the guide edge is lowered until it contacts. The entire piece is lowered into place. Pressure is applied from the center of the piece to the outside edges. The hands may be used, but a roller is better (Fig. 17-34).

Fig. 17-34 *Placing the glue-coated plastic laminate on the bottom over a piece of paper.* (Grossman Lumber.)

For larger pieces, a sheet of paper is used. Wax paper may be used, but almost any type of paper is acceptable. The glue is allowed to dry first. Then the paper is placed on the top. The laminate is placed over the paper. The laminate is positioned carefully. The paper is gently pulled about 1 inch from beneath the laminate. The position of the laminate is checked. If it is in place, pressure may be applied to the exposed edge. If it is not in place, the laminate is moved until it is in place. Then the exposed edge is pressed until a bond is made. The paper is removed from the entire surface. Pressure is applied from the middle toward the edges (Fig. 17-35).

Trim for laminated surfaces The edges should be trimmed. The pieces were cut slightly oversize to al-

low for trimming. The tops should extend over the sides slightly. The tops and corners should be trimmed so that a slight bevel is exposed. This may be done with a special router bit, as shown in Fig. 17-36A. It also may be done with a sharp and smooth file, as shown in Fig 17-36B.

The back of most countertops has a raised portion

(A)

(B)

(C)

Fig. 17-36 *A. Edges may be trimmed with edge trimmers or routers.* (Rockwell International, Power Tool Division.) *B. Edges may also be files. Note direction of force. C. Marking around the sink rim for the cutout in the countertop.* (Grossman Lumber.)

Fig. 17-35 *Applying plastic laminate to countertops. A. Apply the edge strip. B. Trim the edge strip flush with the top. C. Apply glue. Lay a slip sheet or sticks in place when the glue is dry. D. Position the top. Remove the slip sheet or sticks. E. Apply pressure from center to edge. F. Trim the edge at slight bevel. G. To cut holes for sinks, first center-punch for drilling. H. Next, drill holes at corners.*

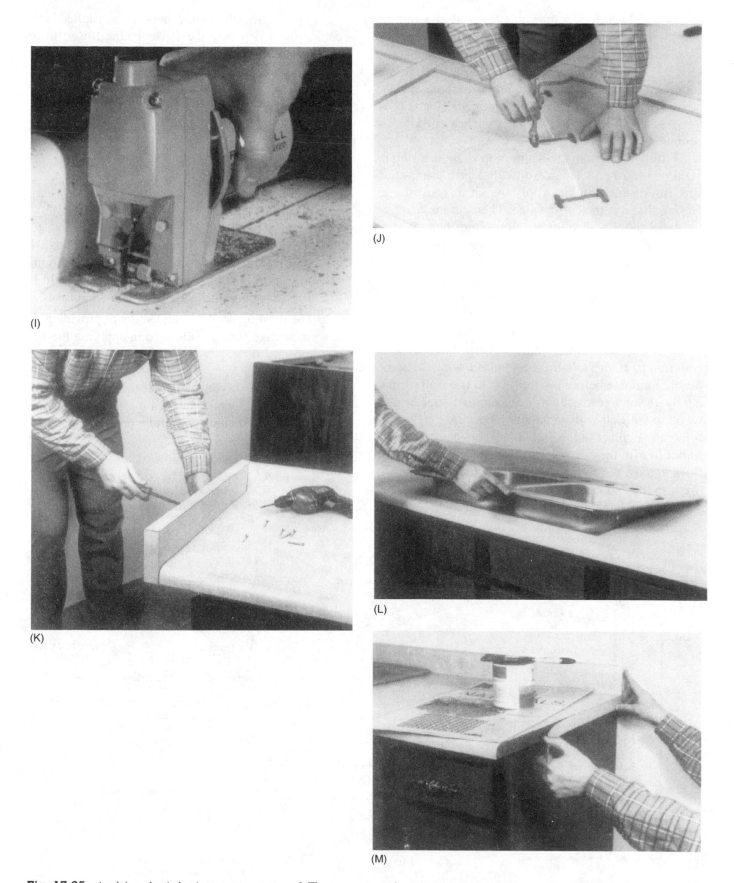

Fig. 17-35 *Applying plastic laminate to countertops. I. Then cut out opening.* (Rockwell International Power Tool Division.) *J. Mitered corners can be held with special clamps placed in cutouts on the bottom. K. Ends also may be covered by splashboards. L. Lay sink in opening. M. Ends may be covered by strips.* (Formica.)

called a *splashboard*. Splashboards and countertops may be molded as one piece. However, splashboards are also made as two pieces, in which case metal cover and cap strips are applied at the corners. Building codes may set a minimum height for these splashboards. The Federal Housing Administration (FHA) requires a minimum height of 4 inches for kitchen counters.

Fitting the sink in the countertop is another step in making a finished kitchen. You will need a clamp-type sink rim installation kit. Apply the kit to your sink to be sure that it is the correct size. Then cut the countertop hole about ⅛ inch larger all around than the rim. Make the cutout with a keyhole saw, a router, or a saber saw (see Fig. 17-36C).

Start the cutout by drilling a row of holes and leaving ⅟₁₆ inch to accommodate the leg of the rim. Install the sink and rim. Or you can go on to the backsplash.

The backsplash is a board made from ¾-inch plywood that is nailed to the wall. It is perpendicular to the countertop. Finish off the backsplash with the same plastic laminate before mounting it to the wall permanently. In some cases, it may be easier to mount the plywood to the wall with an adhesive because the drywall is already in place at this step in the construction. Use a router to trim the edges of the backsplash. It should fit flush against the countertop. You may want to finish it off with the end grain of the plywood covered with the same plastic laminate or with metal trim.

Figure 17-37 shows how the sink is installed in the countertop. Also note the repair method used to remove a bubble in the plastic laminate.

Repairing a Leaking Roof

In most areas when a reroofing job is under consideration, a choice must be made between removing the old roofing or permitting it to remain. It is generally not necessary to remove old wood shingles, old asphalt shingles, or old roll roofing before applying a new asphalt roof—that is, if a competent inspection indicates that

1. The existing deck framing is strong enough to support the weight of workers and the additional new roofing. This means that it also should be able to support the usual snow and wind loads.

2. The existing deck is sound and will provide good anchorage for the nails used in applying the new roofing.

Old roofing to stay in place If the inspection indicates that the old wood shingles may remain, the surface of the roof should be carefully prepared to receive the new roofing.

This may be done as follows:

1. Remove all loose or protruding nails, and renail the shingles in a new location.

Fig. 17-37 *Details of installing a sink in a laminated countertop. Note the method used to remove the bubble caught under the laminate.* (Formica.)

2. Nail down all loose shingles.

3. Split all badly curled or warped old shingles, and nail down the segments.

4. Replace missing shingles with new ones.

5. When shingles and trim at the eaves and rakes are badly weathered, and when the work is being done in a location subject to the impact of unusually high winds, the shingles at the eaves and rakes should be cut back far enough to allow for the application, at these points, of 4- to 6-inch wood strips, nominally 1 inch thick. Nail the strips firmly in place, with their outside edges projecting beyond the edges of the deck the same distance as did the wood shingles (Fig. 17-38).

6. To provide a smooth deck to receive the asphalt roofing, it is recommended that beveled-wood feathering strips be used along the butts of each course of old shingles.

Fig. 17-39 *Exposure of new shingles when reroofing.* (Bird and Son.)

Fig. 17-38 *Treatment of rakes and eaves when reroofing in a windy location.* (Bird and Son.)

Old roofing (asphalt) shingles to remain in place If the old asphalt shingles are to remain in place, nail down or cut away all loose, curled, or lifted shingles. Remove all loose and protruding nails. Remove all badly worn edging strips, and replace with new. Just before applying the new roofing, sweep the surface clean of all loose debris.

Square-butt strip shingles to be recovered with self-sealing square-butt strip shingles The following application procedure is suggested to minimize an uneven appearance of the new roof. All dimensions are given assuming that the existing roof has been installed with the customary 5-inch shingle exposure.

Starter course Cut off the tabs of the new shingle using the head portion equal in width to the exposure of the old shingles. This is normally 5 inches for the starter shingle (Fig. 17-39).

First course Cut 2 inches from the top edge of a full-width new shingle. Align this cut edge with the butt edge of the old shingle.

Second course Use a full-width shingle. Align the top edge with the butt edge of the old shingle in the next course. Although this will reduce the exposure of the first course, the appearance should not be objectionable because this area is usually concealed by the gutter.

Third course and all others Use full-width shingles. Align the top edges with the butts of the old shingles. Exposure will be automatic and will coincide with that of the old roof.

Old lock-down or staple-down shingles These shingles should be removed before reroofing. They have an uneven surface, and the new shingles will tend to conform to it. If a smoother-surfaced base is desired, the deck should be prepared as described under "Old Roofing to Be Removed" below.

New shingles over old roll roofing When new asphalt roofing is to be laid over old roll roofing without removing the latter, proceed as follows to prepare the deck:

1. Slit all buckles, and nail segments down smoothly.

2. Remove all loose and protruding nails.

3. If some of the old roofing has been torn away, leaving pitchy knots and excessively resinous areas exposed, cover these defects with sheet metal patches made from galvanized iron, painted tin, zinc, or copper having a thickness approximately equal to 26 gauge.

Old roofing to be removed   When the framing supporting the existing deck is not strong enough to support the additional weight of roofing and workers during application, or when the decking material is so far gone that it will not furnish adequate anchorage for the new roofing nails, the old roofing, regardless of type, must be removed before new roofing is applied. The deck thcn should be prepared for the new roofing as follows:

1. Repair the existing roof framing where required to level and true it up and to provide adequate strength.

2. Remove all rotted or warped old sheathing and replace it with new sheathing of the same kind.

3. Fill in all spaces between boards with securely nailed wood strips of the same thickness as the old deck. Or move existing sheathing together, and sheath the remainder of the deck.

4. Pull out all protruding nails, and renail sheathing firmly at new nail locations.

5. Cover all large cracks, slivers, knot holes, loose knots, pitchy knots, and excessively resinous areas with sheet metal nailed securely to the sheathing.

6. Just before applying the new roofing, sweep the deck thoroughly to clean off all loose debris.

Old built-up roofs

If the deck has adequate support for nails   When the pitch of the deck is below 4 inches per foot but not less than 2 inches per foot, and if the deck material is sound and can be expected to provide good nail-holding power, any old slag, gravel, or other coarse surfacing materials should be removed first. This should leave the surface of the underlying felts smooth and clean. Apply the new asphalt shingles directly over the felts according to the manufacturer's recommendations for low-slope application.

If the deck material is defective and cannot provide adequate security   All old material down to the upper surface of the deck should be removed. The existing deck material should be repaired. Make it secure to the underlying supporting members. Sweep it clean before applying the new roofing.

Patching a roof   In some cases, it is not necessary to replace the entire roof to plug a leak. In most instances, you can visually locate the place where the leak is occurring. There are a number of roof cements that can be used to plug the hole or cement the shingles down. In some cases, it is merely a case of backed-up water. To keep this from happening, heating cables may have to be placed on the roof to melt the ice. See Chapter 8 for more details.

Replacing Guttering

Guttering comes in both 4- and 5-inch widths. The newer aluminum gutter is usually available in 5-inch widths. The aluminum has a white baked-on finish. It does not require soldering, painting, or priming. It is easily handled by one person. The light weight can have some disadvantages. Its ability to hold ice or icicles in colder climates is limited. It can be damaged by the weight of ice buildup. However, its advantages usually outweigh its disadvantages. Not only is it lighter, prefinished on its exterior, and less expensive, but it is also very easy to put up.

The aluminum type of guttering is used primarily as a replacement gutter. The old galvanized type requires a primer before it is painted. In most cases, it does not hold paint even when primed. It becomes an unsightly mess easily with extremes in weather. It is necessary to solder the galvanized guttering. These soldered points do not hold if the weather is such that it heats up and cools down quickly. The solder joints are subject to breaking or developing hairline cracks that leak. These leaks form large icicles in northern climates and put an excessive load on the nails that support the gutter. Once the nails have been worked loose by expansion and contraction, it is only a matter of time before the guttering begins to sag. If water gets behind the gutter and against the fascia board, it can cause the board to rot. This further weakens the drainage system. The fascia board is used to support the whole system.

Figure 17-40A shows that the drainage system should be lowered at one end so that the water will run down the gutter to the downspout. About ¼ inch for every 10 feet is sufficient for proper drainage. Figure 17-40B shows a hidden bracket hanger. It is used to support the gutter. Figure 17-40C illustrates the method used to mount this concealed bracket.

The rest of Fig. 17-40 shows how the system is put together to drain water from the roof completely.

Fig. 17-40 *Replacing an existing gutter and downspout system.* (Sears, Roebuck and Co.)

Each of these pieces has a name. If you are planning a system, its cost is easily estimated by listing the various parts and the number of each part needed. Figure 17-41 shows the tools needed for installing the system. The caulking gun comes in handy to caulk places that were left unprotected by removal of the previous system. The holes for the support of the previous system should be caulked. Newer types of glues can be used to make sure that the connections of the inside and outside corners and the end caps are watertight. If you decide to get someone to replace old gutters, they may be made on site with a truck loaded with a roll of aluminum and a machine to shape the aluminum into guttering (see Fig. 17-41B). Figure 17-41C shows how the aluminum is formed into gutters. Installation can be done by one person on a ladder using an electric screwdriver (see Fig. 17-41D). Some special gutters are available with a top piece of aluminum that is formed so that it prevents leaves from clogging the drains. They are recommended where there are a lot of trees on the lot where the house is situated.

(A)

HAMMER   TAPE MEASURE

CAULKING GUN   PLIERS

TINSNIPS   HACKSAW

SCREWDRIVER   STRING   LADDER

TOOLS NEEDED

The only tools required in the installation of guttering are those that are commonly found in the home, such as the tools shown at the left. The instructions indicate what job each tool performs.

(B)

(C)

(D)

Fig. 17-41 A. *How to select the right fittings and the right quantity.* (Sears, Roebuck and Co.) *B. Extruding the one-piece gutter from a roll of aluminum located inside the truck. C. Shorter pieces ready for mounting on the facia board. D. Mounting the gutter using one person to do the job.*

Once the water leaves the downspout, it is spread onto the lawn or is conducted through plastic pipes to the storm sewer at the curb. Some locations in the country will not allow the water to be emptied onto the lawn. The drains to the storm sewer are placed in operation when the house is built. This helps to control the seepage of water back into the basement once it has been pumped out with a sump pump. The sump pump also empties into the drainage system and dumps water into the storm sewer in the street.

Extruding Gutters to Fit the House

Guttering, as previously mentioned, can be extruded in a seamless run from a roll of aluminum mounted in a truck (see Fig. 17-41B–D). The one-piece gutter eliminates problems with splicing and leaking that can develop in a conventional gutter installation, where pieces are connected by soldering or special glues. The seamless extruded on-site gutters have preformed corners that are sealed with the proper caulking for many years of trouble-free operation.

Specially designed guttering can be obtained that eliminates the problem with leaves clogging the system and making it useless. The leaf guard gutter is a patented process and is usually regarded as a replacement for existing gutters. The lead guard prevents leaves from entering the gutter and clogging it. These guards are especially useful in areas with a number of trees located near the house. Pine needles are highly acidic and can cause corrosion and leakage in some types of gutter systems if allowed to accumulate.

Replacing a Floor

Many types of floor coverings are available. You may want to check with a local dealer before deciding just which type of flooring you want. There are continuous rolls of linoleum, or there are 9- × 9-inch squares of tile. There are 12- × 12-inch squares of carpet that can be placed down with their own adhesive. The type of flooring chosen determines the type of installation method to use.

Staple-down floor This type of floor is rather new. It can be placed over an old floor. A staple gun is used to fasten down the edges.

Figures 17-42 through 17-44 show the procedure required to install this type of floor. This type of flooring is so flexible that it can be folded and placed in the trunk of even the smallest car. Unroll the flooring in the room, and move it into position (see Fig. 17-42). In the 12-foot width (it also comes in 6-foot width), it covers most rooms without a seam. In the figure, it is being laid over an existing vinyl floor. It also can be installed over plywood, particle board, concrete, and most other subfloor materials.

To cut away excess material (see Fig. 17-43), use a metal straightedge or carpenter's square to guide the utility knife. Install the flooring with a staple every 3 inches close to the trim (see Fig. 17-44). In this way, quarter-round trim can be installed over the staples, and they will be out of sight. Cement is used in places where a staple can't penetrate. For a concrete floor, use a special adhesive around the edges. Any dealer who sells this flooring stocks the adhesive.

The finished job looks professional, even when it is done by a do-it-yourselfer. This flooring has a built-in memory. When it was rolled face-side-out at the factory for shipment, the outer circumference of the roll was stretched. After it is installed in the home, the floor gently contracts, trying to return to the dimensions it had before it was rolled up. This causes any

Fig. 17-42 *Unroll the flooring and allow it to sit face up overnight before installing it.* (Armstrong Cork.)

Fig. 17-43 *Use a utility knife and carpenter's square to make sure that the cut is straight.* (Armstrong Cork.)

Fig. 17-44 *Place a staple every 3 inches along the kickboard. These staples will be covered by molding.* (Armstrong Cork.)

slack or wrinkles that might have been left in the flooring to gradually be taken up by the memory action.

No-wax floor   One of the first rooms that comes in for improvement is the kitchen. Remodeling may be a major project, but renewing a floor is fairly simple.

No-wax flooring comes in the standard widths—6 and 12 feet. In some instances, it doesn't need to be tacked or glued down. However, in most cases, it should be cemented down. It fits directly over most floors, provided that they are clean, smooth, and well bonded. Make sure that any holes in the existing flooring are filled and smoothed over. In the case of concrete basement floors, just vacuum or wash them thoroughly and allow them to dry.

The tools needed to install a new floor are a carpenter's square, a chalk line, adhesive, a trowel, and a knife or scissors.

The key to a perfect fit is taking accurate room measurements (Fig. 17-45). Diagram the floor plan on a chart, noting the positions of cabinets, closets, and doorways.

After transferring the measurements from the chart to the flooring material, cut along the chalk lines using a sharp knife and a straightedge. Transfer the measurements, and cut the material in a room where the material can lie flat. Cardboard under the cut lines will protect the knife blade.

Return the material to the room where it is to be installed. Put it in place. Roll back one-half of the ma-

Fig. 17-45 *Room measurements are essential to a good job.*

Fig. **17-46** *Spread the adhesive on half the floor, and place the flooring material down. Then do the other half.* (Armstrong Cork.)

terial. Spread the adhesive. Unroll the material onto the adhesive while it is still wet. Repeat the same steps with the rest of the material to finish the job (Fig. 17-46).

The finished job makes any kitchen look new. All that is needed in the way of maintenance is a sponge mop with detergent.

Floor tiles Three of the most popular kinds of self-adhering tiles were vinyl-asbestos, no-wax, and vinyl tiles that contain no asbestos filler. The benefit of no-wax tiles is obvious from the name. They have a tough, shiny, no-wax wear surface. You pay a premium price for no-wax tiles.

Vinyl tiles are not no-wax, but are easier to clean than the older vinyl-asbestos once were. Their mainte-nance benefits are the result of a nonporous vinyl wear surface that resists dirt, grease, and stains better than the older vinyl-asbestos type of tiles.

Any old tile or linoleum floor can be covered with self-adhering tiles, provided that the old material is smooth and well bonded to the subfloor. Just make sure that the surface is clean and that old wax is removed.

Putting down the tiles Square off the room with a chalk line. Open the carton, and peel the protective paper off the back of a tile (Fig. 17-47). Do one section of the room at a time. The tiles are simply maneuvered into position and pressed into place (Fig. 17-48). Border tiles for the edges of the room can be easily cut to size with a pair of ordinary household shears (Fig. 17-49). It doesn't take long for the room to take shape. There is no smell

Fig. 17-47 *Peel the paper off the back of a floor tile, and place the tile in a predetermined spot.* (Armstrong Cork.)

Fig. 17-48 *Do one section of the room at a time. The tiles can be maneuvered into position and pressed into place.* (Armstrong Cork.)

from the adhesive. The floor can be used as soon as it is finished. This type of floor replacement or repair is commonplace today.

Paneling a Room

The room to be paneled may have cracked walls. This means that you'll need furring strips to cover the old walls and provide good nailing and shimming for a smooth wall (Fig. 17-50). Apply furring strips (1 × 1s or 1 × 3s) vertically at 16-inch intervals for full-size 4-

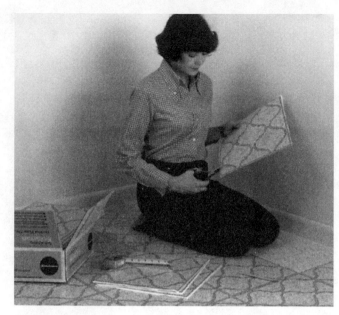

Fig. 17-49 *Border tiles for the edges of the room can be easily cut to size with a pair of household shears.* (Armstrong Cork.)

× 8-foot panels. Apply the furring strips horizontally at 16-inch intervals for random-width paneling. Start at the end of the wall farthest from the main entrance to the room. The first panel should be plumbed from the corner by striking a line 48 inches out. Trim the panel in the corner (a rasp works well) so that the panel aligns on the plumb line. Turn the corner, and butt the next panel to the first panel. Plumb in the same manner as the first panel.

To make holes in the paneling for switch boxes and outlets, trace the box's outline in the desired place on the panel. Then drill holes at the four corners of the area marked. Next, cut between the holes with a keyhole saw. Then rasp the edges smooth (Fig. 17-51).

Before you apply the adhesive, make sure that the panel is going to fit. If the panel is going to be stuck right onto the existing wall, apply the adhesive at 16-inch intervals, horizontally and vertically (Fig. 17-52A).

Spacing Avoid a tight fit. Above grade, leave a space approximately the thickness of a matchbook cover at the sides and a ⅛-inch space at top and bottom. Below grade, allow ¼ inch top and bottom. Allow not less than ¹⁄₁₆ inch space (the thickness of a dime) between panels in high-humidity areas (see Fig. 17-52B).

Vapor barrier A vapor barrier is needed if the panels are installed over masonry walls. It doesn't matter if the wall is above or below grade (see Fig. 17-52C). If the plaster is wet or the masonry is new, wait until it is thoroughly dry. Then condition the panels to the room.

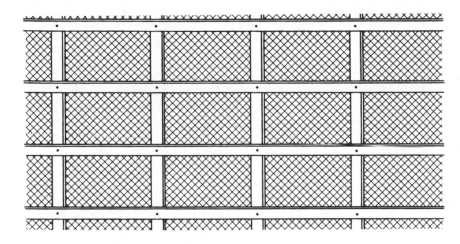

Fig. 17-50 *Applying vertical and horizontal furring strips on an existing wall. (Valu.)*

WALLBOARD SAW

Fig. 17-51 *Cutting holes in the paneling for electrical switches and outlets. (Valu.)*

(A)

1/8" OR 1/4"

MATCHBOOK COVER THICKNESS

1/8" OR 1/4"

(B)

(C)

Fig. 17-52 *A. Make sure that the panel fits before you apply adhesive. (Valu.) B. Allow space at top and bottom and at each side for the panel to expand. (Valu.) C. Apply a vapor barrier to the walls before you place the furring strips onto the wall. (Abitibi.)*

Adhesives Figure 17-53 shows a caulking gun being used to apply the adhesive to the furring strips. In some cases, it is best to apply the adhesive to the back of the panel. Follow the directions on the tube. Each manufacturer has different instructions.

Fig. 17-53 *Use a caulking gun to apply the adhesive.* (Valu.)

Hammer a small nail through the top of the panel. It should act as a hinge. Press the panel against the wall or frame to apply glue to both sides (Fig. 17-54A). Pull out the bottom of the panel and keep it about 8 to 10 inches away from the wall or framework. Use a wood block (see Fig. 17-54B). Let the adhesive become tacky. Leave it for about 8 to 10 minutes.

(A) (B) (C) (D)

Fig. 17-54 *A. Hammer a nail through the top of the panel. This will act as a hinge so that the panel can be pressed against the wall to apply the glue to both surfaces.* (Valu.) *B. Pull out the bottom of the panel, and keep it about 8 to 10 inches away from the wall with a wood block until the adhesive becomes tacky.* (Valu.) *C. Use a compass to mark panels so that they fit perfectly against irregular surfaces.* (Abitibi.) *D. Use a hammer and block of wood covered with a cloth to spread the adhesive evenly.* (Valu.)

Caution: Methods of application for some types of panel adhesives may differ. Always check the instructions on the tube.

Let the panel fall back into position. Make sure that it is correctly aligned. The first panel should butt one edge against the adjacent wall. The panel should be completely plumb. Trim the inner edge as needed. This is so that the outer edge falls on a stud.

Scribing Use a compass to mark panels so that they fit perfectly. It can be used to mark the variations in a butting surface (see Fig. 17-54C).

Nailing Use finishing nails (3d) or brads (1¼ inch) or annular hardboard nails (l inch). Begin at the edge. Work toward the opposite side. Never nail opposite ends first and then the middle. With a hammer and a cloth-covered block, hammer gently to spread the adhesive evenly (see Fig. 17-54D). If you are using adhesive alone to hold the panels in place, don't remove the nails until the adhesive is thoroughly dry. Then, after they're out, fill the nail holes with a matching putty stick. Moldings can be glued into place or nailed.

Installing a Ceiling

There are a number of methods used to install a new ceiling in a basement or a recreation room. In fact, some very interesting tiles are available for living rooms as well.

Replacing an old ceiling The first thing to do in replacing an old, cracked ceiling is to lay out the room (Fig. 17-55). It should be laid out accurately to scale. Use ½ inch = 1 foot, 0 inches as a scale. Do the layout on graph paper.

Then, on tracing paper, draw ½-inch squares representing the 12-inch ceiling tiles. Lay the tracing paper over the ceiling plan. Adjust the paper until the borders are even. Border pieces should never be less than half a tile wide. Use cove molding at the walls to cover the trimmed edges of the tile (Fig. 17-56). Tile can be stapled or applied with mastic. This is especially true of fiber tile. It has flanges that will hold staples (Fig. 17-57). You can staple into wallboard or into furring strips nailed to joists. When applying tile to furring strips, follow your original layout, but start in one corner and work toward the opposite corner.

If you apply tile with mastic, snap a chalk line as shown in Fig. 17-58. This will find the exact center of your ceiling. In applying the mastic to the ceiling tile, put a golf-ball-sized blob of mastic on each corner of the tile and one in the middle (Fig. 17-59).

Fig. 17-55 *Lay out your room accurately to scale on graph paper.* (Grossman Lumber.)

Fig. 17-56 *Draw ½-inch squares on tracing paper. Lay the tracing paper over the ceiling plan. Adjust the paper until the borders are even. Border pieces should never be less than half a tile wide.* (Grossman Lumber.)

Fig. 17-58 *If you are going to glue the tiles to the ceiling, snap a chalk line to find the exact center of the ceiling.* (Grossman Lumber.)

Fig. 17-57 *Fiber tile is stapled to furring strips applied to the joists.* (Grossman Lumber.)

Apply the first tile where the lines cross. Then work toward the edges. Set the tile just out of position, and then slide it into place, pressing firmly to ensure a good bond and a level ceiling (Fig. 17-60). Precut holes for lighting fixtures and pipes. Use a sharp utility knife. Always be sure to make the cutouts with the tile face up (Fig. 17-61).

The drop ceiling If an old ceiling is too far damaged to be repaired inexpensively, it might be best to install a drop ceiling. This means that the ceiling will be completely new and will not rely on the old ceiling for support. In this way, the old ceiling does not have to be repaired.

Fig. 17-59 *Place a blob of mastic at each corner and in the middle of the tile before sliding it into place on the ceiling surface.* (Grossman Lumber.)

Fig. 17-60 *Apply the first tile where the lines cross on the ceiling. Work toward the edges.* (Grossman Lumber.)

SHARP UTILITY KNIFE

MAKE CUTOUTS WITH TILE FACE UP.

Fig. 17-61 *Precut holes for lighting fixtures and pipes. Use a sharp utility knife. Always be sure to cut the tiles with the face up.* (Grossman Lumber.)

Figure 17-62 shows the first operation of a drop ceiling. Nail the molding to all four walls at the desired ceiling height. Either metal or wood molding may be used. Figure 17-63 shows how a basement dropped ceiling is begun the same way, by nailing molding at the desired height.

Fig. 17-62 *For a new suspended ceiling, nail molding to all four walls.* (Armstrong Cork.)

Fig. 17-63 *A suspended ceiling in a basement is begun the same way. Nail molding on all four walls.* (Armstrong Cork.)

The next step is to install the main runners on hanger wires, as shown in Fig. 17-64. The first runner is always located 26 inches out from the sidewall. The remaining units are placed 48 inches on center (O.C.), perpendicular to the direction of the joists. Unlike conventional suspended ceilings, this type requires no

Fig. 17-64 *Main runners are installed on hanger wires.* (Armstrong Cork.)

Fig. 17-66 *After all the main runners are in place, begin installing ceiling tile in a corner of the room. Note how the T slides into the tile and disappears.* (Armstrong Cork.)

Fig. 17-65 *In a basement installation, install the main runners by fastening them to hanger wires at 4-foot intervals.* (Armstrong Cork.)

Fig. 17-67 *When the main runners are in place, install cross T's between them. The T's have tabs that engage slots in the main runner to lock firmly in place. This type of suspended ceiling will have the metal showing when the job is finished.* (Armstrong Cork.)

complicated measuring or room layout. Figure 17-65 shows how the main runner is installed by fastening hanger wires at 4-foot intervals—the conventional method for a basement dropped ceiling.

After all main runners are in place, begin installing ceiling tile in a corner of the room. Simply lay the first 4 feet of tile on the molding, snap a 4-foot cross T onto the main runner, and slide the T into a special concealed slot on the leading edge of the tile (Fig. 17-66). This will provide a ceiling without the metal supports showing.

In the conventional method, as shown in the basement installation in Fig. 17-67, cross T's are installed between main runners. The T's have tabs that engage slots in the main runner to lock firmly in place. After

this is done, the tiles are slid into place from above the metal framework and allowed to drop into the squares provided for them.

In Fig. 17-68, tile setting is continuing. The tiles and cross T's are inserted as needed. Note how all metal suspension members are hidden from view in the finished portion of the ceiling.

The result of this type of ceiling is an uninterrupted surface. There is no beveled edge to produce a line across the ceiling. All supporting ceiling metal is concealed.

Fig. 17-68 *Continue across the room in this manner. Insert the tiles and cross T's.* (Armstrong Cork.)

One of the advantages of dropping a ceiling is the conservation of energy. Heat rises to the ceiling. If the ceiling is high, the heat is lost to the room. The lower the ceiling, the less is the volume to be heated, and the less energy is needed to heat the room.

Replacing an Outside Basement Door

Many houses have basements that can be made into playrooms or activity areas. The workshop that is needed but can't be put anywhere else probably will wind up in the basement. Basements can be very difficult to get out of if a fire starts around the furnace or hot water heater area. One of the safety measures that can be taken is placing a doorway directly to the outside so that you don't have to try to get up a flight of stairs and then through the house to escape a fire.

Figure 17-69 shows how an outside stairway can be helpful for any basement. This type of entrance or exit from the basement can be installed in any house. It takes some work, but it can be done.

Figure 17-70 shows how digging a hole and breaking through the foundation in stages will permit using tools to get the job done. A basement wall of concrete blocks is easier to break through than a poured wall.

Start by building the areaway. This is done by laying a 12- to 16-inch concrete footing. This footing forms a level base for the first course of concrete blocks (Fig. 17-71). Allow the footing to set for 2 to 3 days before laying blocks for the walls.

The areaway for a size C door is shown in Fig. 17-72. Figure 17-72A shows the starting course. Figure 17-72B shows how the next course of blocks is laid. The top course should come slightly above ground level (Fig. 17-73). It should be about 3 inches from the

Fig. 17-69 *Note the location of the outside basement door or entry on these drawings.* (Bilco.)

Fig. 17-70 *Excavate the area around the proposed entrance.* (Bilco.)

required areaway height as given in the construction guide provided with the door.

Build up the areaway to the right height in the manner shown. You may want to waterproof the outside of the new foundation. Use the material recommended by your local lumber dealer or mason yard.

Fig. 17-71 *Start the blocks on top of the footings.* (Bilco.)

BASEMENT WALL

12″
16″
16″
16″
16″

STARTING COURSE

(A)

16″
16″
16″
4″
16″

SECOND COURSE

(B)

Fig. 17-72 *A. The starting course, looking down into the hole. B. The second course, on top of the first, looks like this.* (Bilco.)

Now build a form for capping the wall (Fig. 17-74). Fill the cores of the top blocks halfway up with crushed balls of newspaper or insulation. The cap to be poured on this course will bring the areaway up to the required height.

When the cap is an inch or so below the desired height, set the door back in position with the header flange between the siding and the sheathing underneath. Make sure that the frame is square. Insert the mounting screws with the spring-steel nuts in the side pieces and the sill. Embed them in the wet concrete to hold the screws tightly (Fig. 17-75). Continue pouring the cap. Bring the concrete flush with the bottom of the sill and sidepiece flanges. Do not bring the capping be-

DOOR FRAME   HEADER   BEDPLATE

Fig. 17-73 *The finished block work ready for the cap of concrete.* (Bilco.)

Fig. 17-74 *Forms for the concrete cup are built around the concrete block.* (Bilco.)

low the bottom of the door. The door should rest on top of the foundation. With a little extra work, the cap outside the door can be chamfered downward as shown in Fig. 17-75. This ensures good drainage. Trowel the concrete smooth and level.

Fig. 17-75 *Anchor bolts are embedded into the concrete cap to allow attachment of the metal doors.* (Bilco.)

Fig. 17-76 *The stairs are installed inside the blocked-in hole.* (Bilco.)

Install a prehung door in the wall of the basement. These come in standard sizes. The door should be selected to fit the hole made in the wall. Use the widest standard unit that fits the entryway you have built.

Figure 17-76 shows how the steps are installed in the opening. The stringers for the steps are attached to the walls of the opening (Fig. 17-77).

Outside doors will resemble those shown in Fig. 17-78. There are a number of designs available for almost any use. Lumber for the steps is 2 × 10s cut to length and slipped into the steel stair stringers.

Seal around the door and foundation with caulk. Seal around the door in the basement and the wall. Allow the stairwell to *air out* during good weather by keeping the outside doors open. This will allow the moisture from the masonry to escape. After it has dried out, the whole unit will be dry.

CONVERTING EXISTING SPACES

There is never enough room in any house. All it takes is a few days after you unpack all your belongings to find out that there isn't enough room for everything. The next thing new homeowners do is look around at the existing building to see what space can be converted to other uses. It is usually too expensive to add on immediately, but it is possible to remodel the kitchen, porch, or garage.

Adding a Bathroom

As the family grows, the need for more bathrooms becomes very apparent. First, you look around for a place to put the bathroom. Then you locate the plumbing, and check to see what kind of a job it will be to hook up the new bathroom to the cold and hot water and to the drains. How much effort will be needed to hook up pipes and run drains? How much electrical work will be needed? These questions will have to be answered as you plan for the additional bathroom.

Start by getting the room measurements. Then make a plan of where you would place the various necessary items. Figure 17-79 shows some possibilities. Your plan can be as simple as a lavatory and water closet, or you can expand with a shower, a whirlpool bath, and a sauna. The amount of money available usually will determine the choice of fixtures.

Look around at books and magazines as well as literature from the manufacturers of bathroom fixtures. Get some ideas as to how you would rearrange your own bathroom or make a new one.

If you have a larger room to remodel and turn into a bathroom, you might consider what was done in Fig. 17-80. This is a Japanese bathroom dedicated to the art

Fig. 17-77 *Typical door installations.* (Bilco.)

SIZE SL
DOOR WEIGHT: 200 LB
STAIR STRINGERS: 22 LB

SIZE O
DOOR WEIGHT: 168 LB
STAIR STRINGERS: 33 LB

SIZE B
DOOR WEIGHT: 175 LB
STAIR STRINGERS: 39 LB

SIZE C
DOOR WEIGHT: 196 LB
STAIR STRINGERS: 44 LB

SIZE C WITH EXTENSION
EXTENSION WEIGHTS: 6" 20 LB,
12" 52 LB, 18" 70 LB, 24" 95 LB
SIZE E STRINGER EXTENSIONS: 18 LB

Fig. 17-78 *Various shapes and sizes of outside basement doors.* (Bilco.)

Fig. 17-79 *Floor plans for bathrooms.* (Kohler.)

of bathing. Note how the tub is fitted with a shower to allow you to soap and rinse on the bathing platform before soaking, as the Japanese do.

You may not want to become too elaborate with your new bathroom. All you have to do then is decide how and what you need. Then draw your ideas for arrangements. Check the plumbing to see if your idea will be feasible. Order the materials and fixtures, and then get started.

Providing Additional Storage

Cedar-lined closet Aromatic red cedar closet lining is packed in a convenient no-waste package that contains 20 square feet. It will cover 16⅓ square feet of wall space for lifetime protection from moths. In order to install the cedar, follow these steps (Fig. 17-81):

1. Measure the wall, ceiling, floor, and door area of the closet. Figure the square footage. Use the length times width to produce square feet.

2. Cedar closet lining may be applied to the wall either vertically or horizontally (Fig. 17-82). When applied to the rough studs, pieces of cedar must be applied horizontally.

3. When applying the lining to the wall, place the first piece flush against the floor with the grooved edge down. Use small finishing nails to apply the lining to the wall (Fig. 17-83).

Fig. 17-80 *Japanese bathroom design. The platform gives the illusion of a sunken bath. Many of these features can be incorporated into a remodeled room that can serve as a bathroom.* (Kohler.)

Fig. 17-81 *Cedar-lined closet. Aromatic red cedar serves as a moth deterrent.* (Grossman Lumber.)

Fig. 17-82 *Placing a piece of cut-to-fit red cedar lining vertically in a closet.* (Grossman Lumber.)

Fig. 17-83 *Work from the bottom up when applying the cedar boards.* (Grossman Lumber.)

4. When cutting a piece of cedar to finish out a course or row of boards, always saw off the tongued end so that the square sawed-off end will fit snugly into the opposite corner to start the new course (Fig. 17-84).

5. Finish one wall before starting another. Line the ceiling and floor in a similar manner. Each piece of cedar is tongued and grooved for easy fit and application (Fig. 17-85).

The cedar-lined closet will protect your woolens for years. However, in some instances, you might not have a closet to line. You might need extra storage space. In this instance, take a look at the next section.

Building extra storage space Any empty corner can provide the back and one side of a storage unit. This simple design requires a minimum of materials for a maximum of storage. Start with a 4-foot unit now and add 2 feet later (Fig. 17-86).

Fig. 17-84 *To start a new course, saw off the tongued end. The square sawed end will fit snugly into the opposite corner.* (Grossman Lumber.)

Fig. 17-85 *Finish one wall before starting another.* (Grossman Lumber.)

Fig. 17-86 *Extra storage space. The sliding-door closet and the single-door closet.* (Grossman Lumber.)

Fig. 17-87 *Details of door installation for the closets in Fig. 17-86.* (Grossman Lumber.)

Use a carpenter's level and square to check a room's corners for any misalignment. Note where the irregularities occur. The basic frame is built from 1- and 2-inch stock lumber. It can be adjusted to fit any irregularities in the walls or floor. Nail the top and shelf cleats into the studs. Cut the shelf from ½-inch plywood. Slip it into place, and nail it to the cleat. Attach the clothes rod with the conventional brackets. Fit, glue, and nail the floor, side, and top panels. Make them from ⅛-inch hardboard. You can buy metal or hardwood sliding doors and tracks. Or you can make the doors and purchase only the tracks. Doors are made of single sheets of ⅛-inch hardboard stiffened with full-length handles of 1 × 2 trim set on the edge. Metal glides on the ends of handles carry the weight of the door and prevent binding. Figure 17-87 shows some of the details for making the door operational.

A swinging-door unit is built almost like a double-door unit. The framework should be fastened to the wall studs where possible. Make the door frame of 1 × 3s laid flat with ⅛-inch hardboard (plain) on the face and ⅛-inch perforated hardboard on the inside. Again, a full-length 1 × 2 makes the handle.

Attach three hinges to the door. Make sure that the pins line up so that the door swings properly. Place the door in the opening, and raise it slightly. Mark the frame, and chisel out notches for the hinges.

Other types of storage space For adequate, well-arranged storage space, plan your closets first. Minimum depth of closets should be 24 inches. The width can vary to suit your needs. But provide closets with large doors for adequate access (Fig. 17-88).

Figure 17-89 shows some of the arrangements for closets. Use 2 × 4 framing for dividers and walls so that shelves may be attached later. Carefully measure and fit ⅜-inch gypsum board panels in place. Finish the interior or outside of the closet as you would any type of drywall installation. Install the doors you planned

Fig. 17-88 *A variation of cabinet storage designs.* (Grossman Lumber.)

Fig. 17-89 *Different plans and door openings for closets.* (Grossman Lumber.)

for. These can be folding, bifold, or a regular prehung type.

Remodeling a Kitchen In remodeling a kitchen, the major problem is the kitchen cabinets. There are any number of these available already made. They can be purchased and installed to make any type of kitchen arrangement desired.

The countertops have already been covered. The color of the countertop laminate must be chosen so that it will blend with the flooring and the walls. This is the job of the person who will spend a great deal of time in the kitchen.

Planning the kitchen The kitchen begins with a set of new cabinets for storage and work areas. The manufacturer of the cabinets will supply complete installation instructions.

Check the drawing, and mark on the walls where each unit is to be installed. Mark the center of the stud lines. Mark the top and bottom of the cabinet so that you can locate them easily at the time of installation.

As soon as the installation is completed, wipe the cabinets with a soft cloth dampened with water. Dry the cabinets immediately with another clean soft cloth. Follow this cleaning with a very light coat of high-quality liquid or paste wax. The wax helps to keep out moisture and causes the cabinets to wear longer.

Finishing up the kitchen After the cabinets have been installed, it is time to do the plumbing. Have the sink installed, and choose the proper faucet for the sink.

Kitchen floor Now it's time to put down the kitchen floor. In most cases, the flooring preference today is carpet, although linoleum and tile are also used. It is easier to clean carpet—just vacuuming is sufficient. It is quieter and can be wiped clean easily if something spills. If a total remodeling job has been done, it may be a good idea to paint or wallpaper the walls before installing the flooring. A new range and oven are usually in order, too. The exhaust hood should be properly installed electrically and physically for exhausting cooking odors and steam. This should complete the kitchen remodeling. Other accents and touches here and there are left up to the user of the kitchen.

Enclosing a Porch

One of the first things to do is to establish the actual size you want the finished porch to be. In the example shown here, a patio (16 × 20 feet) is being enclosed. A quick sketch will show some of the possibilities (Fig. 17-90). This becomes the foundation plan. It is drawn ¼ inch to equal 1 foot. The plan then can be used to obtain a building permit from the local authorities.

The floor plan is next (Fig. 17-91). It shows the location of the doors and windows and specifies their sizes.

A cross section of the addition or enclosure is next (Fig. 17-92). Note the details given here dealing with

Fig. 17-90 *Foundation plan for an addition to an already existing building: enclosing a patio.*

the actual construction. The scale here is ½ inch = 1 foot, 0 inches. Note that the roof pitch is to be determined with reference to the window location on the second floor of the existing building. Figure 17-95A shows what actually happened to the pitch as determined by the window on the second floor. As you can see from the figure, the pitch is not 5:12 as called for on the drawing. Because of the long run of 16 feet, the 2 × 6 ceiling joists, 16 inches O.C., had to be changed to 2 × 10s. This was required by the local code. It was a good requirement because with the low slope on the completed roof, the pileup of snow would have caused the roof to cave in (see Fig. 17-95A).

Elevations Once the floor plan and the foundation plan have been completed, you can begin to think about how the porch will look enclosed. This is where the elevation plans come in handy. They show you what the building will look like when it is finished. Figure 17-93A shows how part of the porch will look when it is extended past the existing house in the side elevation.

Figure 17-93B shows how the side elevation looks when finished. Note the storm door and the outside light for the steps.

The rear elevation is simple. It shows the five windows that allow a breeze through the porch on days when the windows can be opened. Figure 17-94 shows the enclosed porch viewed from the rear.

A side elevation is necessary to see how the other side of the enclosure will look. This shows the location of the five windows needed on this side to provide ventilation. Figure 15-95A illustrates the way the enclosure should look. Figure 17-95B shows how the finished product looks with landscaping and the actual

EXIST. WALLS

FAMILY ROOM

EXIST. DR.

EXIST. CONC. FLR. SLAB

8'-4½" X 4'-6¼" R.O.

8'-0"

8'-0"

5'-4"

2 8

CASEMENT WINDOWS

8'-4½" X 4'-6¼" R.O

10'-0"

10'-0"

20'-0"

Fig. 17-91 *Floor plan for an addition.*

FLOOR PLAN
¼"=1'-0"

3/8" PLYWOOD SHEATHING
15 # BLDG. PAPER
235 # ASPHALT SHINGLES

1X10

12

5

2X8-16"O.C

2X4-16"O.C.

2X8-16"O.C.

GUTTER

1X6

SOFFIT

2-2X4'S

MATCH EXIST. SIDING

2X4-16"O.C

7'-6" MIN.

LINE OF EXIST. HOUSE

EXIST. GRADE
BUILD UP GR.
FOR FOOTING

FLOAT TOP OF FTG.

1" RIGID INSULATION

FIN. FLR.

EXIST. 4"THK. CONC. SLAB

3'-6"

10" TRENCHED FOUND.

CROSS SECTION
½"=1'-0"

NOTE :
DETERMINE ROOF PITCH IN FIELD
WITH REFERENCE TO WINDOW LOCATION ON 2ND FLR.

Fig. 17-92 *Cross-sectional view of the addition.*

(A)

SIDE ELEVATION

(A)

(B)

Fig. 17-93 *A. Side elevation of the addition. B. What the side will look like when finished.*

REAR ELEVATION
¼" = 1'-0"

Fig. 17-94 *Rear elevation of the addition.*

(B)

Fig. 17-95 *A. Side elevation of the addition. B. What the side elevation will look like when finished.*

roofline created by the second-floor window location. With this low-slope roof, heating cables must be installed on the roof overhang. This keeps ice jams from forming and causing leaks inside the enclosed porch.

Once your plans are ready and you have all the details worked out, it is time to get a building permit. You have to apply and wait for the local board's decision. If you comply with all the building codes, you can go ahead. This can become involved in some communities. The building permit in Fig. 17-96 shows some of the details and some of the people involved in issuing a building permit. This building permit is for the porch enclosure shown in the previous series of figures.

Fig. 17-96 *Application for a building permit.*

Starting the project Once you have the building permit, you can get started. The first step is to dig the hole for the concrete footings or for the trench-poured foundation shown here. Once the foundation concrete has set up, you can add concrete blocks to bring the foundation up to the existing level. After the existing grade has been established with the addition, you can proceed as usual for any type of building. Put down the insulation strip and the soleplate. Attach the supports to the existing wall, and put in the flooring joists. Once the flooring is in, you can build the wall framing on it and push the framing up and into place. Put on the rafters and the roof. Get the whole structure enclosed with a nail base or plywood on the outside walls. Place the windows and doors in. Put on the roof and then the siding. Trim the outside around the door and windows. Check the overhang, and place the trim where it belongs. Put in the gutters, and connect them to the existing system. If there is to be electrical wiring, put that in next. In most porches, there is no plumbing, so you can go on to the drywall. Finish the ceiling and walls.

After the interior is properly trimmed, it should be painted. The carpeting or tile can be placed on the floor and the electrical outlets covered with the proper plates. The room is ready for use.

Finishing the project Throughout the building process, the local building inspector will be making calls to see that the code is followed. This is for the benefit of both the builder and the owner.

There is another *benefit* (to the local government) of the inspections and the building permit. Once the addition is inhabited, the structure can go on the tax roles, and the property tax can be adjusted accordingly.

ADDING SPACE TO EXISTING BUILDINGS

Adding space to an already-standing building requires some special considerations. You want the addition to look as if it "belongs." This means that you should have the siding the same as the original or as close as possible. In some cases, you may want it to look added on so that you can contrast the new with the old. However, in most instances, the intention is to match up the addition as closely as possible.

First, the additional space must have a function. You may need it for a den, an office, or a bedroom. In any of these instances, you don't need water or plumbing if you already have sufficient bathroom facilities on that floor. You will need electrical facilities. Plan the maximum possible use of the building, and then put in the number of electrical outlets, switches, and lights that you think you can use. Remember, it is much cheaper to do it now than after the wallboard has been put up.

Planning an Addition

In the example used here, we will add a 15- × 22-foot room onto an existing, recently built house. The addition is to be used as an office or den. It is located off the dining room, so it is out of the way of through-the-house foot traffic. The outside wall also will serve to deaden the sound. It was insulated when the house was built. This means that only three walls will have to be constructed. Make a rough sketch of what you think will be needed (Fig. 17-97) for an example. Note that a light switch has been added and that two lights over the bookcases will be added later. Note the outlets for electricity and heat. The windows are of two different

Fig. 17-97 *Floor plan for an addition.*

sizes. The local code calls for a certain amount of square footage in the windows for ventilation. One window has been taken from the existing upstairs bedroom. It is too large and will interfere with the roofline of this addition. A smaller window is put in up there, and the one used in the bedroom is moved down to the addition. Only two windows need to be purchased. Only one door is needed. These should be matched up with the existing doors and windows so that the house looks complete from the inside, too.

Elevations must be drawn up to show to the building inspector. You must get a building permit; therefore, elevation drawings definitely will be needed. Figures 17-98 through 17-100 show the front, rear, and side elevations. This information will help you to obtain a building permit. If you use Andersen window numbers, the town board will be able to see that these windows provide the right amount of ventilation.

To make the finished product fit your idea of what you wanted, it is best to write a list of specifications. Make sure that you list everything you want done and how you want it done. For example, the basement, siding, overhang, flooring, windows, door, walls, roof, and even the rafters should be specified.

Specifications

Addition for 125 Briarhill in the Town of Amherst (see drawings).

Basement

- Crawl space—skim coat of concrete over gravel. Drainage around footings and blocks to prevent moisture buildup.
- Blocks on concrete footings—42 inches deep.
- To be level with adjoining structure.
- To be waterproofed. Fill to be returned and leveled around the exterior.

Siding

- National Gypsum Woodrock Prefinished (as per existing).
- Size to match existing.
- Must match at corners and overlap in rear to look as if built with original structure.
- To be caulked around windows.

Fig. 17-98 *Front elevation.*

REAR ELEVATION
SCALE : APPROX. ¼"=1'-0"

Fig. 17-99 *Rear elevation.*

SIDE ELEVATION
SCALE : APPROX. ¼"=1'-0"

Fig. 17-100 *Side elevation.*

Overhang

- To match existing as per family room extension.
- Gutters front and rear to match existing. Downspouts (conduits) to match existing and to connect.
- Ventilation screen in Upson board overhang.
- Exterior fascia and molding to match existing.

Flooring

- Kiln-dried 2 × 10 (construction or better) 16 inches O.C.
- Plywood subfloor, ⅝-inch exterior, A–D.
- Plywood subfloor, ½-inch interior, A–D.
- Fiberglas insulation strip between soleplate and blocks.
- Must meet with existing room. Allowance made for carpeting.
- Must be level.

Windows

- Anderson No. 3456 for south wall.
- Replace existing window with Anderson No. 2832 and fir in. (See Fig. 17-100. Note size difference of windows.)
- Windows to be vinyl-coated and screened as per existing.
- Shutters on south window wall to match existing.

Door

- Interior with trim as per existing, mounted and operating.
- Size, 32 inches × 6 feet, 8 inches, solid, flush, walnut with brass lock and hardware.
- Dining room to be left in excellent condition.

Permits

- To be obtained by the contractor.

Walls

- Studs, 2 × 4 kiln-dried (construction or better), 16 inches O.C.
- Double or better headers at windows and door and double at corners where necessary.

Roof

- Composition shingles as per existing, black of same weight as existing.

- Felt paper under shingles and attached to ⅜-inch exterior plywood (A–D) sheathing.

Rafters

- Kiln-dried 2 × 4 truss type as per existing, 16 inches O.C.
- Roof type to be shown in attached drawings. Check with existing garage to determine type of construction, if necessary.

In order to add any information you may have left out, you can make drawings illustrating what you need. Figure 17-101 shows a cornice detail that ensures that there is no misunderstanding of what the overhang is. Other details are also present in this drawing. The scale of the cornice is 1½ inches = 1 foot, 0 inches.

The contractor is protected when the details of work to be done are spelled out in this way. There is little or no room for argument if things are written out. A properly drawn contract between contractor and owner also should be executed to make sure that both parties understand the financial arrangements. Figures 17-102 and 17-103 show the completed addition as it looks with landscaping added.

CREATING
NEW STRUCTURES

A storage building can take the shape of any number of structures. In most instances, you want to make it resemble some of the features of the home nearby.

Small storage sheds are available in precut packages from local lumber yards. Figure 17-104 shows three versions of the same package. It can be varied to meet the requirements of the buyer. Lawn mowers, bicycles, or almost anything can be stored in a structure of this type.

All you need is a slab to anchor the building permanently. In some instances, you may not want to anchor it, so you just drive stakes in the ground and nail the soleplate to the stakes. Most of the features are simple to alter if you want to change the design.

Before you choose any storage facility, you should know just how you plan to use it. This will determine the type of structure. It will determine the size and in some cases the shape. The greenhouse in Fig. 17-104 is nothing more than the rustic or contemporary shed with the siding left off. The frame is covered with glass, Plexiglas, or polyvinyl according to your taste or pocketbook.

COMP. SHINGLES

SHEATHING

INSULATION

GUTTER

CEILING

1"X6"

3/4" X 1 3/4" CROWN MOLDING

1" X 4"

UPSON BOARD WITH VENT. SCREEN

1'-0"

HEAD OF WD.

SHEATHING

INSULATION

SIDING

SCALE: APPROX. 1 1/2" = 1'-0"

CORNICE DETAIL

Fig. 17-101 *Cornice detail.*

Fig. 17-102 *Side elevation in finished form.*

Fig. 17-103 *Rear elevation in finished form.*

RUSTIC   GREENHOUSE   CONTEMPORARY

COVER WITH GLASS, PLEXIGLAS OR POLYVINYL

Fig. 17-104 *Various designs for storage facilities.* (TECO.)

Custom-Built Storage Shed

In some cases it is desirable to design your own storage shed. The example used here, shown in Fig. 17-105, was designed to hold lawn mowers and yard equipment. It was designed for easy access to the inside. Take a look at the overhead garage door in the rear of the shed. The 36-inch door was used along with the window to make it more appealing to the eye. It also has a hip roof to match the house to which it is a companion (Fig. 17-106).

The design is 10 × 15 feet (Fig. 17-107). This produces a 2:3 ratio that results in a pleasing appearance. A 3:4 ratio is also very common. The concrete slab was placed over a bed of crushed rocks and anchored by bolts embedded in the concrete slab. A 9- × 7-foot slab is tapered down from the floor to the yard. This allows rider lawn mowers to be driven into the shed. The outside pad also serves as a service center.

Wires serving the structure are buried underground and brought up through a piece of conduit. They enter the building near the small door. There is only one window, so the wall space can be used to hang yard tools. The downspouts empty into the beds surrounding the structure to water the evergreens. An automatic light switch turns on both lights at sundown and off again at dawn.

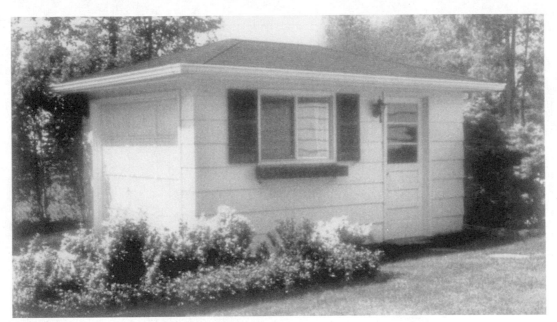

Fig. 17-105 *View of one end of a storage shed.*

SHINGLES (BLACK)
235 LB SEAL DOWN
OVER
½" C-D PLYWOOD AND 15 LB FELT UNDERLAY

HIP ROOF
2/12 PITCH

30'

6"

GUTTER

RAFTERS
2 X 6 16" O.C.
2 X 8 RIDGE BOARD

HEADERS
2 X 10 PBL
OVER DOORS
OVER WINDOW

STUDS
2 X 4 16" O.C.
TRIPLE CORNERS
WIND BRACES (KNOTCHED)

36 X 48
SLIDER
ALUMINUM

12 X 36
SHUTTERS

36" DOOR
3 PANES
LOCK & KEYS

DOWN SPOUT

12" PREFINISHED WOODROCK SIDING (WHITE)
OVER
⅜" C-D PLYWOOD NAILBASE & FELT PAPER

CONCRETE
PAD
EXTENDS 7'-0"

36"

GRADE

4" REINFORCED CONCRETE

2" GRAVEL

VIEW – FACING NORTH

Fig. 17-106 *View of the finished storage shed.*

The overhead door faces the rear of the property. This produced some interesting comments from the concrete installers, who thought the garage had been turned around by mistake. It does resemble a one-car garage, but it is specifically designed for the storage of yard equipment.

Don't forget to get a building permit. In some areas, even a tool shed requires a building permit. This doesn't necessarily mean that it goes on the tax rolls, but it does call for a number of inspections by the building inspector, which help to protect both the builder and the owner.

Buildings for storage take all shapes. They may be garages or barns. The design of a new structure should be chosen carefully to harmonize with the rest of the buildings on the property.

NORTH

36" ENTRY DOOR   36 X 48 SLIDER WINDOW

WEST

10'-0"

EAST

8' OVERHEAD DOOR

9'

7'

7'X9' CONCRETE PAD
4" REINFORCED ON
2" GRAVEL BASE

15'-0"

SOUTH

Fig. 17-107 *Outside dimension details.*

Environmental Concerns

Replacement parts should be given some consideration if they will allow a greener kitchen or living space. Such concerns may be

1. Glues used to attach countertops.

2. Paints used to color rooms.

3. Proper choice of appliances for their efficiency rating and energy consumption.

CHAPTER 17 STUDY QUESTIONS

1. What is the meaning of the word *maintenance*?

2. What do you need to know about a house before you start remodeling it?

3. How does moisture passing through the exterior paint cause problems with a house?

4. What is the importance of scheduling in getting a job done?

5. What is meant by shimming a strike jam of a door?

6. What is meant by side elevation?

7. What is a deadbolt lock? Where is it used?

8. What is a traverse rod? Where is it used?

9. What is another name for gypsum board?

10. How do you eliminate popped nails and their unsightliness?

11. Why do nails used to install drywall pop?

12. What do you use to cut plastic laminates for countertops?

13. How does contact cement work? Why do you have to wait for it to dry before you place the two pieces together?

14. What is meant by a course of shingles?

15. What are the two standard sizes of floor tiles?

16. What is the standard size of a piece of paneling?

17. What do you mean when you say the panel should be hung perfectly plumb?

18. What is a drop ceiling? Where would you use one?

19. What type of cedar is used to line a closet?

20. What is a bifold door? Where do you use them?

18
CHAPTER

The Carpenter and the Industry

THIS CHAPTER EXAMINES THE LATEST BUILDING methods and materials. Also discussed are

- Building codes
- Zoning provisions
- Trends in manufactured housing

You will learn

- How construction procedures are changing
- Why building codes are necessary
- How to fill out a building permit
- How factory-built commercial buildings are constructed
- How homes are constructed

BROADENING HORIZONS IN CARPENTRY

Carpentry, like other trades, is constantly changing. New materials are being introduced to replace old, time-honored ones that have become too expensive or scarce. New tools are available to work with new materials, and plastics have taken over where glass once reigned supreme. Reading trade journals and other printed material is absolutely necessary. Manufacturers usually include a set of instructions with each prefab unit or a newer type of product using a different kind of material.

New Building Materials

Figure 18-1 shows a house made of plywood. Only a few years ago all floors and walls were made of pieces of wood that measured 1 × 6 inches. Plywood has some special features. It is stronger, it has a good nail surface, and it is good insulation. It is easily placed in position, and it can be bought cheaply compared with single pieces of siding or flooring.

New materials are used as a nail base for siding. Carpenters have to adapt to these types of material and be able to handle them (Fig. 18-2).

Fig. 18-1 *Building materials used in construction are changing. Note the extensive use of plywood in a modern house.* (Western Wood Products.)

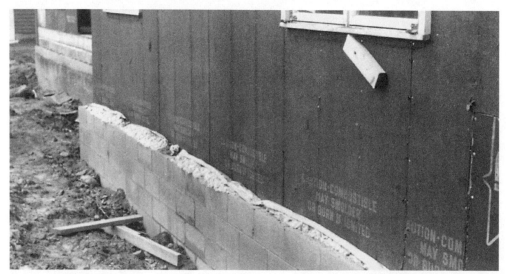

Fig. 18-2 *Carpenters must learn to adapt to new materials like this composition nail base used for siding.*

Changing Construction Procedures

Construction procedures are changing. Note that in Fig. 18-3 the wall construction is different from usual. This is a double-wall partition. It is used to separate rooms in apartments so that sound is not transmitted easily between rooms.

Figure 18-4 shows how adhesives can be used to apply panels. These panels are held in place with glue instead of nails. New clips are available for holding 2 × 4s and 2 × 10s. They are also designed to hold plywood sheets. Using these clips cuts installation time. They also make buildings stronger. The carpenter then becomes a valuable person on the job.

Fig. 18-3 Construction procedures are also changing. For example, double walls are used to help keep the noise down in apartments.

Innovations in Building Design

Architects are constantly coming up with new designs. These new designs call for different ways of doing things. The carpenter has to be able to work with new woods and new combinations of materials. Figure 18-5 shows one of the newer designs for a modern home. A mixture of brick and wood is used for the outside covering. The inside also calls for some new methods because the open ceiling is used here. Different window sizes call for a carpenter with the ability to innovate on the job. Doors are different from the standard types. This calls for an up-to-date carpenter or one who can adapt to the job to be done.

Fig. 18-4 Adhesives are used extensively in the building industry today. (U.S. Gypsum.)

Figure 18-6 shows a modern design for a condominium building. This uses many carpentry skills. Note the different angles and the wood siding. Even the fence calls for close following of drawings.

Different types of home units can draw on a carpenter's abilities. The skilled carpenter can adapt to the demands made by newer designs.

New materials and new ways of doing things are being developed. The carpenter today has to be able to adapt to the demands. The carpenter has to keep up-to-date on new materials and new techniques. New designs will demand an even more adaptable person in this trade in the future.

A person interested in doing something new and different surely can find it in carpentry.

BUILDING CODES AND ZONING PROVISIONS
Building Codes

Building codes are laws written to make sure that buildings are constructed properly. They are for the benefit of the buyer. They also benefit all the people in a community. If an expensive house is built next to a very inexpensive one, it lowers the property value of the expensive house. Codes are rules that direct people who build homes. They say what can and cannot be done with a particular piece of land. Some land is hard to build on. It may have special surface problems. There may be mines underneath. There are all kinds of things that should be looked into before building.

Fig. 18-5 *Modern construction using brick and wood. A number of various materials are used in house exteriors.*

Fig. 18-6 *Buildings of various shapes and sizes use wood today. The carpenter is needed to apply acquired skills in a building of this nature.*

If a single-family house is to be built, the building codes determine the location, materials, and type of construction. Codes are written for the protection of individuals and the community. In areas where there are no codes, there have been fires, collapsed roofs, and damage from storms.

Figure 18-7 shows inspections needed for a building permit. This is one way the community has to check on building. You have to obtain a permit to add onto a house. It is necessary to get a permit if you build a new house. Figure 18-8 shows the permit for a tool shed. One reason for requiring a building permit is to let the tax assessor know so that the property value can be changed.

Note in Fig. 18-8 how the application for a building permit is filled out. Note also the possibilities to be checked off. The town board is required to sign because it is responsible to the community for what is built and where.

In Fig. 18-7, look at the number of inspections required before the building is approved. Each inspection is made by the proper inspector. In this way, the person who buys the house or property is protected.

A certificate of occupancy is shown in Fig. 18-9. This is required before a person can move into a house. The certificate is given with the owner in mind. It means that the house has been checked for safety hazards before anyone is allowed to live in it.

Community Planning and Zoning

Zoning laws or codes are designed to regulate areas to be used for building. Different types of buildings can be placed in different areas for the benefit of everyone. Shopping areas are needed near where people live. Working areas or zones are needed to supply work for people. In most cases, we like to keep living and working areas separate. This is so because of the nature of the two types of building. Some people don't like to live near a factory with its smoke and noise.

Living near a sewage treatment plant also can be very unpleasant. Certain types of buildings should be near one another and away from living zones.

Some areas are designated as industrial. Others are designated as commercial. Still others are marked for use as residential areas. *Residential* means homes. Homes may be single houses or apartments. An industrial building cannot be built in a residential area. All over the United States, there are regional planning boards. They decide which areas can be used for what. Master plans are made for communities. Master plans designate where various types of buildings are located.

Community plans also include maps and areas outlined as to types of buildings. Parks are also designated in a community plan. Streets are given names, and maps are drawn for developments. A development

---

NOTICE
INSTRUCTIONS AND REQUIRED INSPECTIONS

A. Builder's name, phone number, property house number, and building permit number must be displayed on all construction or building projects.

B. A reasonable means of ingress must be provided to each structure and each floor. (Planks or other; and ladders or stairs from floor-to-floor.)

C. The following inspections are mandatory on all construction within the Town of Amherst:
1. FIRST WALL: When footing is ready to be poured, including trench pour.
2. PRE-LATH: Before insulating and after ALL electrical, telephone, plumbing, and heating rough work is complete, including metal gas vent and range-hood exhaust duct.
3. DRAIN TILE: Subfloor and/or footing drain tile, before pouring of concrete floor.
4. FINAL: Make application for Certificate of Occupancy, file copy of survey with application. Everything must be completed. All work must be finished, including sidewalks and drive aprons.) Each structure must have a house number displayed and visible from the street.

NOTE: An approved set of building plans and plot plans shall be made available to the Building Inspector at the above four inspections.

Call the Building Department 24 hours before you are ready for inspection.
Phone 555-6200, Ext. 42, 43, or 49

Fig. 18-7 *Notice of the inspections needed when a building permit is required for a house.*

Fig. 18-8 *Application for a building permit.*

means land to be developed for housing or other use. Figure 18-10 shows a typical plan for a residential development. Note how the streets are laid out. They are not straight rows. This has a tendency to slow down traffic. The safety of children is important in a residential area. It is best to have residential areas off main traveled roads or streets.

Overbuilding

Where there has been residential, commercial, and industrial overbuilding, the utilities have not been able to keep up. The additional activity puts a strain on the existing facilities. This is another reason for having a community plan.

TOWN OF AMHERST
CERTIFICATE OF OCCUPANCY

Date _____ June 25, 19XX _____   No. _____ 1800 _____

This Certifies that the building located at and known as
125 Briarhill

(NO.)   (STREET)

Sub Lot No. _____ 57 _____   Map Cover No. _____ 2259 _____

and used as _____ a single-family dwelling with private garage _____

(KIND OF OCCUPANCY)

has been inspected and the use thereof conforms to the Amherst Zoning Ordinance, and other

controlling laws and is hereby:

(   ) Approved

(x) Approved subject to _____ the planting of the required number of trees not _____

later than Oct. 15, 19XX.

Any alteration of the property or change in the use voids this certificate and a new certificate
will be required.

COMMISSIONER OF BUILDING

Fig. 18-9   *Certificate of occupancy.* *(Courtesy of the Town of Amherst, NY.)*

Fig. 18-10   *Map of a planned subdivision development.*

For example, there can be problems with sewage. Plants may not be big enough to handle the extra water and effluent. Storm sewers may not be available or may not be able to handle the extra water. Water can accumulate quickly if there is a large paved surface. This water has to be drained fast during a rainstorm. If it is not, flooding of streets and houses causes damage. Some areas are just too flat to drain properly. It takes a lot of money to build sewers.

Sewers are of two types. The sanitary sewer takes the fluid and solids from toilets and garbage disposals. This is processed through a plant before the water is returned to a nearby river or creek. A storm sewer is usually much larger in diameter than a sanitary sewer. It has to take large volumes of water. The water is dumped into a river or creek without processing.

Local communities can have a hand in controlling their growth and problems. They can form community planning boards. These boards enforce zoning requirements, and development of the community can progress smoothly.

TRENDS AND EFFECTS

Manufactured Housing

Some of the early attempts at manufacturing housing are shown in Fig. 18-11. Here are a number of houses that look alike. They have been made one after the other, as in a factory setup. They are made of poured concrete. The forms were moved from one place to the next, and concrete was poured to make a complete house. As you can see, the housing looks rather dull. It would be hard to find your house if you didn't know the house number.

Factory-produced buildings are relatively recent. They are made in a number of sizes and shapes. Some of them are used as office buildings, as in Fig. 18-12.

Large sections of the building are made in a factory. They are shipped to the construction site. There, they are bolted together. In this case, the parts make a very interesting building.

One of the advantages of this type of building is the minimum of waste and lost time in its manufacture. Construction workers do not have to move from one building to another but can do the same job day after day. They can work inside. After a while, the worker becomes very skilled at the job. Little material is wasted. Such things as plumbing, floors, walls, and electric facilities are included in the package.

Figure 18-13 shows another type of commercial building. This bank was built in sections inside a factory and then assembled on site. Everything is measured closely. This means that little time is wasted on the job. The building can be put together in a short time if everything fits. It is very important that the foundation and the water and plumbing lines are already in place in the floor.

Types of Factory-Produced Buildings

There are two types of factory-produced buildings. Modular buildings are constructed of modules. The module is completely made at the factory. A *module* is a part of a building, such as a wall or a room. The modular technique is very efficient. All it takes at the site is

Fig. 18-11 *Poured concrete houses. Example of an early type of manufactured house.* (Universal Form Clamp.)

Fig. 18-12 *Office building made in a factory and assembled at the site.* (Butler.)

Fig. 18-13 *Commercial building made in a factory and assembled on site.* (Butler.)

a crane to place the module where it belongs. Then the unit is bolted together. This type of construction can be used on hotels and motels. Housing of this type is very practical. It is used as a means of making dormitories for colleges.

In the other type of factory-produced building, only panels are constructed inside. The panels are assembled and erected at the site.

Some companies specialize in factory-produced buildings. Some are specialists in commercial and industrial buildings. Others are specialists in houses.

Some building codes will not allow factory-made housing. It is up to the local community whether or not this type of construction is allowed. In some cases, the low-cost-construction advantage is lost. The community can have code restrictions that make it expensive to put up such a house or building. In some cases, this is good for the community. The type of construction should blend in with the community. If not, the manufactured house can become part of a very big slum in a short time.

A builder who does not adhere to the zoning and building codes in a locality may have to tear down the building. With zoning and building codes, a building permit system is usually used. The permit system enables the local government to monitor construction. This will make sure that the type of building fits into the community plan.

Premanufactured Apartments

A combination of plant-built cores and precut woods is a key to volume production. This is especially true with apartment houses. Figure 18-14 shows a well-thought-out utility-core production line. Components are fed into the final assembly, where they are integrated into serviceable units. The units get to the job site in a hurry. Production scheduling then can be easily controlled.

Figure 18-15 shows that wood floor joists, precut in the plant, go together quickly. The frame is equipped with wheels. It is then turned over. Subflooring and finish flooring are applied. Then the floor is rolled to the main production line.

Figure 18-16 shows wall panels framed with 2 × 4s. The bathroom wall uses 2 × 6s to accommodate the stand pipe. For extra strength on the main floor units, special 3 × 4 studs are used. This size replaces the standard 2 × 6. This dimension will carry the added weight yet still fit precisely with the upper-floor cores.

After the drywall is applied to one side, the wall panels are stored. They are stored on wheeled carts. This means that they are ready to move to the production line when they are needed. The utility core is beginning to take shape. Carpenters nail the walls in place. Plumbing fixtures are installed before the end wall goes up.

Ceiling panels complete the framing. Next, the plumbing and wiring are done, and the furnace and water heater are installed. In the final step, the cabinets and appliances are installed. The interior walls are finished next (Fig. 18-17). Each unit is hooked up and tested before it leaves the plant.

The units are wrapped and shipped to the job site (Fig. 18-18). The cores are stacked on the foundation and hooked up to utilities. Local contractors add the interior and exterior walls. They use either panelized wood

Fig. 18-14 *Assembly line for premanufactured apartment house units.* (Western Wood Products.)

Fig. 18-15 *Lifting a floor section and turning it over for subflooring and finish flooring.* (Western Wood Products.)

Fig. 18-16 *Framing a wall panel.* (Western Wood Products.)

Fig. 18-17 *Ceiling panels are added to complete the framing.* (Western Wood Products.)

Fig. 18-18 *Units are wrapped before being shipped to the job site.* (Western Wood Products.)

framing or conventional construction. This conventional framing is precut to fit the house or apartment unit.

Designs can vary. They can be adapted to meet the floor plan and outside requirements. Figure 18-19 shows an example of manufactured apartments.

Manufactured Homes

One- and two-story homes can be constructed and completely finished in the plant. Interior finishing is completed on outdoor stations; then the homes are moved onto their concrete foundations. It requires unusually sturdy construction to move a home of this size. A web floor system of 2 × 4s and plywood is used for strength (Fig. 18-20). Two homes are built simultaneously, with floors built over pits so that workers can install heating ducts, plumbing, and electrical facilities from below.

While the floor system is under construction, carpenters are building wall panels from precut wood (Fig. 18-21). Wood-frame construction is used throughout the house.

At the building site, a giant machine places the foundation (Fig. 18-22). Then similar equipment moves the finished house from the factory onto the foundation (Fig. 18-23). This is done in just 1 hour and 10 minutes.

Fig. 18-19 Variations in design are possible in the premanufactured apartment house. (Western Wood Products.)

Fig. 18-20 Floor joists being laid for a manufactured house. (Western Wood Products.)

Although in-plant production is standardized to cut costs, the homes avoid the manufactured look. The wood-frame construction makes it easy to adapt the designs.

The advantages of this type of building are lower cost of materials, less waste, and lower-cost worker skills. It is a great advantage when it comes to maintaining production schedules. The weather does not play too much of a role in production schedules when the house is built indoors.

This type of construction is changing the way homes are being built. It requires a carpenter who can adapt to the changes.

The Green Home and the Carpenter

A contemporary carpenter must be knowledgeable in a number of subjects. With the advent of the "green" home, the carpenter must know which home incorporates smart design, technology, construction, and maintenance elements that lessen the negative impact of the home on the environment and improve the health of the people who live in it.

Exciting new technologies, products, and scientific breakthroughs are constantly emerging. The Inter-

Fig. 18-21 *Wall panels attached to the floor system.* (Western Wood Products.)

net is an ideal place to search for the latest improvements and technologies. Staying educated and updated on the latest green aspects of house building will make any carpenter a valuable member of the community. A truly green home will ensure the owner a healthier, high-performance environment that costs less to operate and has less environmental impact.

The U.S. Green Building Council is dedicated to making the green home accessible and enjoyable.

CHAPTER 18 STUDY QUESTIONS

1. What does the future look like for a carpenter?
2. What are the opportunities for a carpenter?
3. When are carpenters usually nonunion?
4. What is a premanufactured house?
5. What is a building code?
6. How do you obtain a building permit?
7. What does the term *community planning* mean?
8. What happens if a builder does not follow local building codes?
9. What is the advantage of making a house in a plant or factory?
10. What is meant by overbuilding?

Fig. 18-22 *A machine laying the foundation.* (Western Wood Products.)

Fig. 18-23 A machine moves the house over the foundation. *(Western Wood Products.)*

19

CHAPTER

Bathrooms

As with any house designing project, the bathroom must give way to the latest trends in design and appearance. A larger number of bathrooms are being built per home, and they include more open areas, more use of light, more outdoor views, and more features, such as

- Saunas
- Hot tubs
- Spas

Many are expanded to provide extra space within the bath area or even a private terrace, garden patio, or deck. Decks and enclosed decks may include a sauna, a Japanese *furo* tub, or a spa (Figs. 19-1 through 19-4).

Unusual materials and combinations of materials are also used on walls, for example, wood, stone, tile, plastic laminates, and wainscotting. Color plays a more important role as well, with traditional antiseptic whites and grays being replaced by bright chromatic

Fig. 19-1 *New trends in bathrooms include open space and comfort.* (American Olean Tile.)

Fig. 19-2 *This bath includes both a spa-tub combination and a glass-enclosed sauna.* (Kohler.)

Fig. 19-3 *A spacious bath with luxury and easy maintenance for shower, toilet, bidet, and twin vanities. The adjoining sunken bath overlooks a small garden.* (American Olean Tile.)

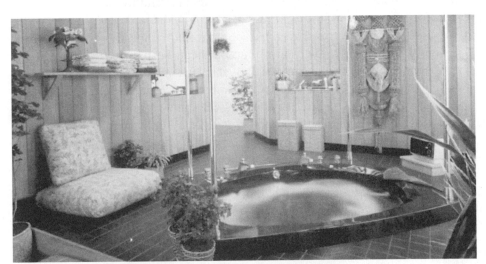

Fig. 19-4 *The gleaming smooth tile contrasts with the wooden walls.* (American Olean Tile.)

Fig. 19-5 *Contrasting tile colors.* (American Olean Tile.)

colors sharply contrasting with sparkling whites or the natural textures of wood and stone. Walls, floors, and ceilings now contrast and supplement each other (Fig. 19-5) and often match the decor of nearby rooms.

For any house to be inhabitable, all local codes require at least one bathroom. It is an essential part of any house. Bathrooms are designed with various objectives in mind. Some are large; others are small. Some have the minimum of fixtures, whereas others are very elaborate.

In this chapter you will learn

- How building codes influence bathroom design
- How plumbing, proper ventilation, and lighting are installed
- How to select the right toilet, tub, and shower
- How to space fittings
- How to select floors, fittings, countertops, and vanities

ROOM ARRANGEMENT

Current trends require larger bathrooms, but many new, smaller homes still use the standard size of 5 × 8 feet. These smaller sizes are appropriate for extra bathrooms used mainly by guests or for smaller homes. Size is relatively unimportant if the desired features can be arranged within the space. When adding hot tubs, saunas, and so forth, it may be necessary to make projecting bay windows or decks. The key to maximum appearance, utility, and satisfaction is not size but good arrangement.

While planning room arrangement, the typical pattern of movement in the room should be examined. Avoid major traffic around open doors or other projections. Also, consider the number of people who will be using the room at the same time. Many master bathrooms now have two lavatory basins because both members of the couple must rise, dress, and groom themselves at the same time. Having a slightly longer counter with two basins minimizes the frustrations of waiting or trying to use the same basin at the same time.

Access to the bathroom should be made from a less public area of the house such as a hall. Doors preferably should open into a bathroom rather than out from it. Also, it is a good idea to consider what is seen when the bathroom door is open. The most desirable arrangement is one in which the first view is of the vanity or basin area. Next, a view of the tub or bathing area is appropriate. If possible, the toilet should not be the first item visible from an open door.

Keep in mind that placing new fixtures close to existing lines and pipes minimizes carpentry and plumbing costs. Kitchen plumbing can be located near bathroom plumbing, or plumbing for two bathrooms can be located from the same wall (Fig. 19-6A). Locating the plumbing core in a central area is a good idea.

Water hammer is a pounding noise produced in a water line when the water is turned off quickly. The noise can be reduced by placing short pieces of pipe called *air traps* above the most likely causes of quick turnoffs: the clothes washing machine and the dishwasher. Air traps are rather simple to install and can help to quiet a noisy plumbing system.

Also, if future changes are anticipated, rough in the pipes needed for that installation, and cap them off. In this way, not all the walls will have to be opened later to make complicated connections. Upstairs and downstairs plumbing can be planned to run through the same wall. Again, this makes plumbing accessible and reduces the amount of carpentry and wall work required. Figure 19-6B shows a few bathroom layouts. These are provided here for reference. Consult the manufacturer for actual dimensions of the specific model to be installed in a house. These clearances are in compliances with most codes, but it is always best to check with the local authorities to make sure that the fixtures are installed within the code requirements.

(A)

(B)

WATER CLOSET
22"
36"
18"
18"
HEIGHT OF 14" TO 15"

LAVATORY
22"
30"
18"
6"

BATHTUB
34"
20"
8"

Fig. 19-6 *A. Locating a second bath near the first minimizes the plumbing needs and expense. B. Fixture clearances.*

FUNCTION
AND SIZE

Where the bathroom is large enough, and where two or more people are frequently expected to use it at the same time, compartments provide required privacy and yet accessibility to common areas at all times (Fig. 19-7). The bath and toilet are frequently set in separate compartments, which allows one to use either area in a degree of privacy while a lavatory or wash basin is used by someone else.

Fig. 19-8 Two smaller baths can be placed where one larger one once existed.

Fig. 19-7 By putting in compartments for showers, toilets, and bidets, more privacy results and permits the use by more than one person at a time.

Actually, there is nothing wrong with small bathrooms. There are certain advantages in having small bathrooms. They are more economical to build, they can accommodate several people, and they are easier to keep clean. Sometimes, instead of one large bathroom, the space can be converted into two smaller ones so that more family members can have access at the same time (Fig. 19-8). Small bathrooms can have all the main features that larger bathrooms have by careful use of space through built-in units, vanity counter cabinet space, and pocket doors for compartments. The comfort factors of ventilation, heating, and air conditioning and soundproofing, together with privacy and availability when needed, should be the main considerations.

A small bathroom easily can be made larger if it is located on an outside wall. Cantilever projections can be used to make windows into large bay windows (Fig. 19-9), or a deck can provide extra space. Also, a bathroom can be enlarged simply by taking other interior space to allow the construction of custom areas, the addition of privacy gardens or patios, or the addition of spas, saunas, and so forth.

Fig. 19-9 A cantilever can be used to extend a bathroom for a different result.

BUILDING CODES

Bathrooms can be complex; therefore, building codes may be involved. Codes may designate

- The types of floors
- The materials used
- The way things are constructed
- Where they are placed

There are usually good reasons for these regulations—even though they might not be obvious. Most cities require rigid inspections based on these codes.

Plumbing

Another complex area the builder must deal with is plumbing, perhaps the most obvious item affected by building codes. Rules apply to the size and type of pipe that can be used, placement of drains, and placement of shutoff valves.

Shutoff valves allow the water to be turned off to repair or replace fixtures. They are used in two places. The first controls an area. It usually shuts off the cold water supply to different parts of the house. A modern three-bedroom house built over a full basement typically would have three area valves—one for the master bath, one for the main bath, and one for the kitchen. Outdoor faucets may be part of each subsystem based on location, or they may be on a separate circuit. Second, each fixture, such as a hot water heater, toilet, lavatory, and so forth, will have a shutoff valve located beneath it for both hot and cold water lines. Most building codes require both kinds of valves.

Electrical

Many building codes specify three basic electrical requirements:

1. The main light switch must be located next to the door, but outside the bathroom itself.
2. The main light switch must turn on both the light and a ventilation unit.
3. At least one electrical outlet must be located near the basin, and it must be on a separate circuit from the lights. It also should have a ground-fault circuit interrupter (GFCI).

Ventilation

Ventilation is often required for bathrooms. It is a good idea and has many practical implications. In the past, doors and windows were the main sources of bathroom ventilation. They consumed no energy but allowed many fluctuations in room temperature. Forced ventilation is not required if the room has an outside window, but most codes require that all interior bathrooms (those without exterior walls or windows) have ventilation units connected with the lights.

Ventilation helps to keep bathrooms dry to prevent deterioration of structural members from moisture, rot, or bacterial action. It also reduces odors and the bacterial action that takes place in residual water and moisture.

Fans are vital in humid climates. They should discharge directly to the outdoors, either through a wall or through a roof, and not into an attic or wall space. Ventilation engineers suggest that the capacity of the fan should be enough to make 12 complete air changes each hour.

Spacing

Building codes also may affect the spacing of the fixtures, such as the toilet, tub, and wash basin. Figure 19-10 shows the typical spaces required between these units. It is acceptable to have more space but not less.

The purpose of these codes is to provide some minimum distance that allows comfortable use of the

Fig. 19-10 *Typical spacing requirements for bathroom fixtures.*

facilities and room to clean them. If there were no codes, some people might be tempted to locate facilities so close together that they could not be used safely or conveniently.

Other Requirements

Local codes might require the bathroom door to be at least two doors away from the kitchen. Some locations specify floors to be made of tile or marble, whereas others mandate tile or marble thresholds. Certain localities insist that a plastic film or vapor barrier be in place beneath all bathroom floors and that all basin counters have splash backs or splashboards.

Some specifications might be strict, requiring rigid enclosures on showers or prohibiting the use of glass in shower enclosures. Others might be open, stating the minimum simply to be a rod on which to hang a shower curtain.

FURNISHINGS

Furnishings make a bathroom either pleasant or drab. They include the fixtures, fittings, and vanity area. The vanity area consists of a lavatory or basin, lights, mirror, and perhaps a counter.

Fixtures

The term *bathroom fixtures* refers to just about everything in the room that requires water or drain connections, such as lavatories (or basins), toilets, bidets, tubs, and showers. Features to consider for each include color, material, quality, cost, and style.

Keep in mind that better-quality fixtures are more costly. Assuming three grades, the cheapest fixtures are not made to withstand long, heavy use. The difference between medium- and high-quality fixtures is the thickness of the plating and the quality of the exterior finish.

Toilets, bidets, and some lavatories are made of vitreous china, a ceramic material that has been molded, fired, and glazed. This material is hard, waterproof, easy to clean, and resists stains. It is very long-lasting; in fact, some china fixtures are still working well after 100 years. White is the traditional color, but most manufacturers provide up to 16 additional colors. Shopping around can give you many insights into colors and features available.

Toilet Selection

Toilets, or water closets, come in several different mechanisms and styles. The styles include those that fit in corners (Fig. 19-11A), rest on the floor, or have

Fig. 19-11 *A space-saving toilet designed to fit into a corner.*

their weight supported entirely by a wall. Corner toilets are designed to save space and are particularly functional in very small rooms. Even the triangular tank fits into a corner to save space. The wall-hung toilet is expensive and requires sturdy mounts in the wall. The most common is the floor-mounted toilet (Fig. 19-12A). Figure 19-12B gives a few more details about

For 4010 tanks: A = 254mm (10")
For 4014 tanks: A = 356 mm (14")
For all other tanks: A = 305mm (12")

WATER CLOSET	URINAL		BIDET
20" TO 24"	18"		14"
22" TO 29"	12" TO 24"		30"
20" TO 28"	24"	RIM HEIGHT	14"

Fig. 19-12 *A. Roughing in dimensions. B. Dimensions and sizes of the water closet, bidet, and urinal.* (American Standard.)

the water closet and urinal, if one is to be installed, and in some instances, the dimensions for the bidet are useful.

Floor-mounted toilets can be obtained in different heights. Older people generally find that an 18-inch-high toilet is easier to use than the 14-inch height of conventional units. Heights range from 12 to 20 inches, and some can be purchased with handles and other accessories for the ill or handicapped.

The siphon jet is perhaps the most common type and is most recommended. It is quiet and efficient. Most of the bowl area is covered by water, making it easier to clean. The siphon action mechanism is an improvement over the siphon jet. It leaves no dry surface, thus making it easier to clean. It is efficient, attractive, and almost silent. It is also the most expensive. Most builders recommend the siphon jet because it costs less.

Up-flush toilets are used in basements when the main sewer line is above the level of the basement floor. These require special plumbing and must be installed carefully.

Toilet Installation

Installing the two-piece toilet requires some special attention to detail to prevent leakage and ensure proper operation. The unit itself is fragile and should be handled with care to prevent cracking or breaking. Keep in mind that local codes have to be followed.

Roughing in Use Fig. 19-12 as a reference. Notice the distance from the wall to closet-flange centerline. The distance varies according to the unit selected. For instance, American Standard's 4010 tanks require distance A to be 254 millimeters, or 10 inches. Model 4014 needs 356 millimeters, or 14 inches, for distance A. All other tanks require 305 millimeters, or 12 inches. The tank should not rest against the wall. Also notice the location of the water supply.

Install the closet bolts as shown in Fig. 19-13. Install the closet bolts in the flange channel, turn 90 degrees, and slide into place 6 inches apart and parallel to the wall. Distance A shown here is the same as that in the preceding figure. Next, install the wax seal (Fig. 19-14). Invert the toilet on the floor (cushion to prevent damage). Install the wax ring evenly around the waste flange (horn) with the tapered end of the ring facing the toilet. Apply a thin bead of sealant around the base flange.

Position the toilet on the flange as shown in Fig. 19-15. Unplug the floor waste opening, and install the toilet on the closet flange so that the bolts project through the mounting holes. Loosely install the re-

Fig. 19-13 *Closet flange and bolts.* (American Standard.)

Fig. 19-14 *Installing the wax ring.* (American Standard.)

Fig. 19-15 *Positioning the toilet on the flange.* (American Standard.)

tainer washers and nuts. The sides of washers marked "This side up" *must* face up!

Install the toilet as in Fig. 19-16. Position the toilet squarely to the wall, and with a rocking motion, press the bowl down fully on the wax ring and flange. Alternately tighten the nuts until the toilet is firmly seated on the floor.

Bolt cap

Closet flange

Closet bolt

Washer   Nut

Fig. 19-16 *Installing the toilet. (American Standard.)*

Caution: Do not overtighten the nuts, or the base may be damaged. Install the caps on the washers, and if necessary, cut the bolt height to size before installing the caps. Smooth off the bead of sealant around the base. Remove any excess sealant.

Next, install the tank. In some cases, where the tanks and bowls use the speed connect system, the tank mounting bolts are preinstalled. Install the large rubber gasket over the threaded outlet on the bottom of the tank, and lower the tank onto the bowl so that the tapered end of the gasket fits evenly into the bowl water inlet opening (Fig. 19-17) and the tank mounting bolts go through the mounting holes. Secure with metal washers and nuts. With the tank parallel to the wall, alternately tighten the nuts until the tank is pulled down evenly against the bowl surface.

Tank mtg. bolts preinstalled

Rubber gasket

Washer

Nut

Fig. 19-17 *Installing the tank with preinstalled bolts. (American Standard.)*

Caution: Do not overtighten the nuts more than required for a snug fit.

In cases where the bolts are not preinstalled, start by installing the large rubber gasket over the threaded outlet on the bottom of the tank, and then lower the tank onto the bowl so that the tapered end of the gasket fits evenly into the bowl water inlet opening (Fig. 19-18). Insert the tank mounting bolts and rubber washers from the inside of the tank through the mounting holes; secure with metal washers and nuts. With the tank parallel to the wall, you then can alternately tighten the nuts until the tank is pulled down evenly against the bowl surface. Again, caution is needed to make sure that the nuts are not overtightened. Install the toilet seat according the manufacturer's directions.

Rubber washer

Tank mtg. bolts

Rubber gasket

Washer

Nut

Fig. 19-18 *Installing the tank without preinstalled bolts. (American Standard.)*

Connect the water supply line between the shutoff valve and tank water inlet fitting (Fig. 19-19). Tighten the coupling nuts securely. Check that the refill tube is inserted into the overflow tube. Turn on the supply valve and allow the tank to fill until the float rises to the shutoff position. Check for leakage at the fittings; tighten or correct as needed.

Adjustments   There are some adjustments that need to be made in most installations to ensure proper operation (Fig. 19-20):

1. Flush the toilet, and check to see that the tank fills and shuts off within 30 to 60 seconds. The tank water level should be set as specified by the mark on the inside of the tank's rear wall.

2. To adjust the water level, turn the water-level adjustment screw counterclockwise to raise the level and clockwise to lower the level.

Fig. 19-19 *Connecting to the water supply.* (American Standard.)

Fig. 19-20 *Making adjustments after installation.* (American Standard.)

3. To adjust the flow rate (tank fill time), turn the flow-rate adjustment screw clockwise to decrease the flow rate. This increases the fill time. Turn the adjustment screw counterclockwise to increase the flow rate or decrease the fill time.

4. Carefully position the tank cover on the tank.

5. The flush-valve float has been factory set and does not require adjustment. Repositioning the float will change the amount of water used, which might affect the toilet's performance.

Figure 19-21 identifies all the parts of the toilet. These illustrations and instructions are provided by American Standard.

Bidets

Bidets are common in Europe and are increasing in popularity in North America. They are used for sitz baths and are similar in shape and construction to a toilet (see Fig. 19-12B). The bidet is usually located outside the toilet, although combination units are available (Fig. 19-22). It is provided with both hot and cold water, and a spray or misting action is available as an extra component.

Fig. 19-21 *All parts of the water closet or 1.6-gallon-per-flush two-piece toilet.*

Fig. 19-22 *Toilet and bidet combinations are available.* (Kohler.)

Vanity Areas

Vanity areas include the lavatory, lights, mirror, and often a shelf or counter. Lavatories, or basins, can be obtained in a wide variety of shapes, sizes, and colors. Two basic styles are most common: counter basins and wall-mounted basins.

Basins and counters typically are located 31 to 34 inches above the floor, although they can be placed higher or lower. Most builders and designers suggest an 8-inch space between the top of the basin or counter and the bottom of any mirror or cabinet associated with it. Splash backs may or may not be required by local codes. They can be part of the basin or part of the counter.

Double basins should be widely separated. A minimum of 12 inches should separate the edges, but where space is available, this should be even greater. Basins should be located no closer than 6 to 8 inches from a wall (Figs. 19-23 and 19-24).

Countertop Basins

Countertop basin units are popular (Fig. 19-25). They have one or two basins and frequently incorporate

Fig. 19-23 *Double basins extend the vanity area so that it can be used by two people at one time—ideal for working couples.* (American Olean Tile.)

Fig. 19-24 *A double-basin counter combines crisp lines and easy maintenance.* (Formica.)

storage areas beneath them. Most basins are designed for counter use and are made from steel coated with a porcelain finish. They can be obtained in traditional white or in a variety of colors that will match the colors of the toilet, tub, and bidet.

Most basins used currently are self-rimming units that seal directly to the countertops for neater, quicker installations. Other available styles include basins that mount flush with the countertop and are sealed by a metal or plastic rim and recessed units that mount below the surface of the counter. Recessed units require more care to install and sometimes are difficult to keep clean. Note that the self-rimming unit in Fig. 19-26 also has a spray unit for hair care.

Countertop materials Countertops are made from a variety of materials. The most common are ceramic tile and plastic laminates. In recent years, the plastic laminates have changed from marble-like patterns to bright solid colors. While white remains a constant favorite, laminates are available in a wide variety of colors, including black, earth tones, and pastels. Countertops can be obtained in a variety of surface finishes and styles, including special countertops with molded splash back and front rims.

Ceramic tile also can be used for counters. It is has a hard, durable surface that is waterproof and easy to clean, and it has a beauty that does not fade or wear out. Figure 19-23 features tile counters. Figure 19-24 shows good use of plastic laminates for a countertop.

Other Materials for Countertops

Other materials commonly used for countertops include slate, marble, and sometimes wood. Butcher block construction is gaining in popularity and can be

Charleston, Dark
Oak with cane doors.

Nova, Honey Oak
with planked doors.

Fig. 19-25 *Cabinets topped by one or two basin counters are economical and practical. (NuTone.)*

Fig. 19-26 *Self-rimming basins may incorporate hair-grooming features. (Kohler.)*

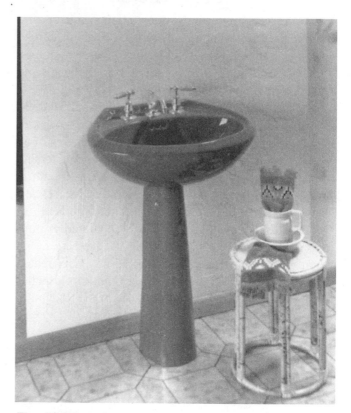

Fig. 19-27 *Pedestal basins are available in a wide variety of shapes and colors. (Kohler.)*

finished with either natural oils and waxes or be heavily coated with special waterproof plastic finishes. Counters need not be waterproof, but if they are not, water should be wiped up immediately.

Integral tops and basins Counters and basin can be made as one solid piece. The advantage of one-piece construction is that there are no seams to discolor or leak. Both fiberglass and synthetic marble are used for these units. Some provide a complete enclosure for the vanity unit. This protects walls and underlying structures from water damage while being striking to look at.

Wall-Mounted Basins

Wall-mounted basins may be entirely supported by the wall or may be placed to the wall and supported by a

RECTANGLE

SQUARE

MAY
HAVE SEAT OR STEP

RECEPTOR

(A)

LIGHTING TO
MATCH INTERIOR
FINISH

VENTILATION TO OUTSIDE

LIGHTING (FLUORESCENT
OR INCONDESCENT)

ELECTRICAL WITH
GFI PROTECTION

PLUMBING AS
PER CODE

FINISHED TO
MATCH FIXTURES

(B)

Fig. 19-28 *A. The three most common tub shapes are rectangular (made to fit into an alcove), square, and receptor. The square and receptor tubs have a longer tub area. B. Typical bathroom spacing.*

pedestal (Fig. 19-27). Other styles are supported by metal legs at the two front corners. Wall-hung basins placed in corners save space and allow easier movement in smaller bathrooms.

BATHING AREAS

A bathing area may be a tub, a shower, or a combination of both. There are many types and varieties of each, and custom units may be built for all type baths and combination baths.

Bathtubs

Many people like to soak and luxuriate in a tub. Tubs can be purchased in a variety of shapes and sizes (Fig. 19-28A) and can exactly match the color of the other fittings and fixtures in the bathroom. Tubs also can be purchased in shower-and-tub combinations that match the color of the other fixtures. Figure 19-28B shows a typical bathroom space.

Lighting Natural lighting by means of an exterior window or glazed openings is always desirable. A single overhead light fixture usually is not acceptable, and auxiliary lighting is required over the tub and/or shower. A light over the lavatory and the vanity counter and over any compartmentalized toilet spaces is also desirable. Keep in mind that the light fixture over the tub or shower should be resistant to water vapor.

Ventilation Bathrooms require either natural or mechanical ventilation in order to remove stale air and supply fresh air. A mechanical ventilation system may be used in place of natural ventilation. The ventilating fan should be located close to the shower and high on the exterior wall opposite the bathroom door. It should connect directly to the outside and be capable of providing five air changes per hour. The point of discharge should be 3 feet away from any opening that allows outside air to enter the building. Residential exhaust fans often are combined with light fixtures, a fan for a heater, or a radiant heat lamp.

Heating Heating may be supplied in the conventional manner through warm-air registers in the floor, hydronic or electrical baseboard units, or electrical resistance heaters in the wall.

Plumbing Plumbing walls should have sufficient depth to accommodate the required supply and waste lines and vents.

Finishes for the bathroom The finish used for the bathtub or shower enclosure should be moisture-resistant. All finishes should be durable, sanitary, and easy to clean, and the flooring should have a nonslip finish.

Whether or not to replace an old unit can be an important decision when remodeling. Old tubs can be refinished and built in to provide a newer and more modern appearance. Fittings can be changed and showers added, along with updated wall fixtures such as shelves and soap dishes. Sometimes the antique appearance is preferred, in which case reworking is more desirable than replacing.

Sunken tubs may be standard tubs with special framing to lower them below the surface of the floor.

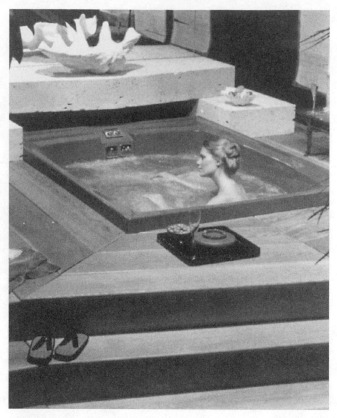

Fig. 19-29 *A pedestal can be built around a tub for many reasons. The pedestal can hide pumps, plumbing, and electrical support units.* (Jacuzzi.)

They may be custom-made or may incorporate specially manufactured tubs. Before installing a sunken tub, be sure that there is room beneath the bathroom floor. When space below is not available, the alternative is to raise the level of the floor in the remainder of the room. Of course, this presents considerable complications with existing doors and floors. One compromise is to construct a wide pedestal around the lip of the tub (Fig. 19-29). This pedestal then can become a sitting area or a shelf for various articles and even can have built-in storage.

Japanese tubs (*furos*) can be built-in and sunken (Fig. 19-30). The tub is simply a deep well that accommodates one or more people on a seat. The seat can be made of wood (the traditional Japanese style), tile, or other material. The soaker is immersed to the neck or shoulders for relaxation. The actual washing with soap or cleaners traditionally is done outside the tub area.

The standard rectangular tub is 60 inches long, 32 inches wide, and 16 inches high. It is enclosed or sided on one side but open at both ends and the remaining side. This shape was designed to fit into an alcove as shown.

Receptor tubs (Fig. 19-31) are squarish, low tubs ranging in height from 12 to 16 inches. Rectangular

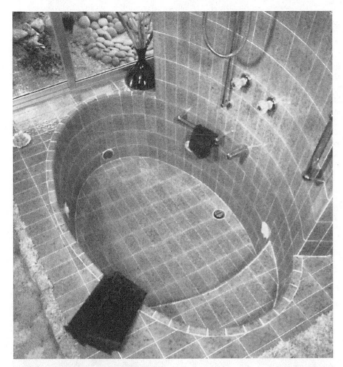

Fig. 19-30 *Furos, or Japanese-style tubs, are becoming increasingly popular.* (American Olean Tile.)

shapes makes them ideal for corner replacement. They are approximately 36 inches long and 45 inches wide. Square tubs are similar to receptor tubs in that they can be recessed easily into corners and alcoves. Some have special shelves set into corners, and some incorporate controls in these areas. Square tubs are approximately 4 to 5 feet square, increasing in 3-inch increments. The receptor tub has a diagonal opening, whereas the square tub may have a truly square basin. The disadvantages of a square tub are that it requires a larger

Fig. 19-31 *This receptor tub provides a longer than normal bathing area but requires less space and water than a full square tub. It also incorporates a hydromassage unit.* (Kohler.)

Fig. 19-32 *Tubs can be custom-built to any size and shape. This tile tub features the same color and style of tile for the tub, walls, counter, and floor.* (American Olean Tile.)

volume of water and getting in and out is sometimes difficult, particularly for elderly people.

Custom tubs (Fig. 19-32) often feature striking use of ceramic tiles. Tiles can be used in combination with metal, stone, and wood. Specially designed custom showers can have stone and glass walls with low tile sides and are used for sitting and storage.

A spa, also called a *hydromassage*, *whirlpool*, *water jet*, or *hot tub*, requires extra space to house the jet pump mechanism and the special piping and plumbing. Also, wiring is needed to power the motors. These hydromassage units are available in a variety of sizes and can be incorporated into standard tubs. If they are to be installed in addition to the tub, extra space is required. People with smaller bathrooms find that the combination tub and hydromassage unit (Fig. 19-33) is satisfactory.

Fig. 19-33 *The hydromassage action can be combined with more conservative settings to look like a conventional tub.* (Jacuzzi.)

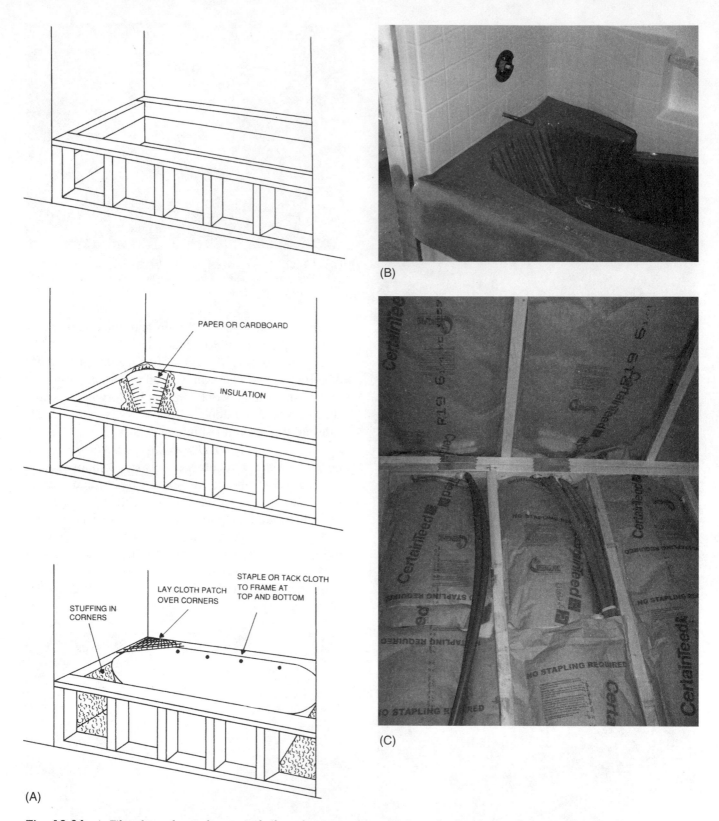

(B)

(A)

(C)

Fig. 19-34 *A. Fiberglass tubs can be custom-built to almost any shape: (1) A wooden frame gives dimension and support; (2) insulation or cardboard defines the approximate shape and contour; and (3) a layer of fiberglass is applied. It can be tinted any color to match decor. B. The tub is covered with plastic to protect it from various falling objects. C. New plastic tubing is substituted for copper. Note the metal plates to protect the tubing from dry wall nails. Red tubing is used for hot water and blue tubing is used for cold water supply to the tub.*

(D)

(E)

Fig. 19-34 *D. View of water supply for the tub/shower. The flexible plastic tubing eliminates the need for a number of copper fittings. E. A closer view of the mixer (tub/shower) for the water. Note the copper tubing used from the mixer to the shower.*

Specially shaped tubs can be built by making a frame (Fig. 19-34) that is lined or surfaced with a material such as plywood, which generally conforms to the size and shape desired. Make sure that the frame and lining will hold the anticipated weight and move-

ment. Next, tack or staple a layer of fiberglass cloth to the form. Pull the cloth to form the shapes around the corners that are desired. If additional support is desired during the shaping process, corner spaces can be filled in and rounded with materials such as fiber insulation. The tub will not need support in the corners, and the fiberglass material itself will be strong enough once completed.

After the cloth has been smoothed to the shape desired (a few seams are all right and will be sanded smooth later on), coat the cloth with a mixture of resin. Tint the resin the color desired for the tub, and use the same color for all coats. Allow the first coat to harden and dry completely. This will stiffen the cloth and give the basic shape for the tub. Next, apply another coat of resin, and lay the next layer of cloth onto it. Allow this to harden, and repeat the process. At least three layers of fiberglass cloth or fiber will be needed. It is best to add several coats of resin after the last layer of cloth. Three layers of glass fiber generally are applied, followed by three more coats of resin. The last three coats of resin are sanded carefully to provide a smooth, curving surface in the exact shape desired.

Showers

Showers may be combined with the tub (Fig. 19-35) or may be separate (Fig. 19-36). In some smaller bathrooms, showers are the only bathing facility. They can

Fig. 19-35 *Combination tubs and showers are perhaps the most common unit. They can be made of almost any material or any combination of materials. (Owens-Corning Fiberglas.)*

Fig. 19-36 *Showers often are separate compartments for increased privacy.*

Fig. 19-37 *A number of accessory fittings such as handholds, soap dishes, and storage are available for showers.* *(NuTone.)*

be custom-made to fit an existing space, or standard units made from metal or fiberglass may be purchased. Shower stalls are available with the floor, three walls, and sometimes a molded ceiling as a single large unit. When made of fiberglass, they are molded as a single integral unit. When made of metal, they are joined by permanent joints or seams.

Standard showers are also available unassembled. This allows a unit to be brought in through halls and doors. The components consist of the drain mechanism, a floor unit, and wall panels.

Manufactured shower units often have handholds, rail ledges for shampoo, built-in soap dishes, and so forth molded into the walls. Custom-built units also can have conveniences molded into the walls but use separate pieces (Fig. 19-37).

Fiberglass units are usually more expensive than metal ones. Metal units have greater restraints on their design and appearance and are noisier than fiberglass units. Fiberglass should not be cleaned with abrasive cleaners.

Custom-built units are made from a variety of materials, including ceramic tile, wood, and laminated plastics. Tile is an ideal material but is relatively expensive. The grout between the tiles is subject to stains and is difficult to clean, but special grouts can be used to minimize these disadvantages. Floors for custom-made shower units must be carefully designed to include either a metal drain pan or special waterproofing membranes beneath the flooring.

Using laminated plastics for walls of showers provides several advantages. The material is almost impervious to stains and water, and a number of special moldings allow the materials to be used. The large size of the panels makes installation quick and easy.

The bottom surface of a shower should have a special nonslip texture; that is, it should be rough enough to prevent skidding but smooth enough to be comfortable. Neither raised patterns on the bottom of a bathtub nor stick-ons are very good.

Shelves, recessed handholds, and other surfaces in a shower area should be self-draining so that they will not hold accumulated water. Shelves approximately 36 to 42 inches above the floor of the shower are convenient for soap, shampoo, and other items. Handholds, vertical grab bars, and other devices used for support while entering or leaving the shower should be firmly anchored to wall studs.

FITTINGS

Fittings is the plumber's word for faucets, handles, and so forth. The available array of size, shape, and finish of fittings is almost endless. Both single and double faucets are obtainable with chrome, stainless, and gold-tone finishes. They can be operated by one or two handles that may be made of any material from metal to glass.

LIGHTING AND ELECTRICAL CONSIDERATIONS

Some of the work of the carpenter is influenced by the nature of the custom-made bathrooms. The following is provided to ensure proper installation of the bathroom.

Older bathrooms usually were lit with a single overhead light. Later, one or two lights were added near the mirrored medicine chest above the basin. Older bathrooms frequently have neither sufficient lighting nor enough electrical outlets for hair dryers, electric razors, electric toothbrushes, and water jets for dental hygiene.

Fig. 19-39 *Swag lamps are used in larger bathrooms for special lighting effects. (NuTone.)*

General lighting can be enhanced by using ceiling panels or wall lamps (Fig. 19-38). Another lighting idea is to use hanging swag lamps (Fig. 19-39). A large, free-hanging swag lamp would be inappropriate for a small bathroom. General lighting and lighting for the basin areas may be combined for smaller bathrooms.

Basin, or vanity, lamps should be placed above or to the sides of mirrors. They also can be placed in both locations. Light should not shine directly into the eyes but should come from above or to the side. One good idea is to use the special Hollywood makeup lights around mirrors (Fig. 19-40). They eliminate glare and give good light for grooming. They can be a single string of lights above the mirror or surround the mirror on the sides and top. They are best controlled near the basin area. Sepa-

Fig. 19-38 *Specialized lighting for vanity areas can be mounted on the walls. (NuTone.)*

Fig. 19-40 *Strips of lights surrounding a vanity mirror are both useful and popular. This Hollywood-style lighting can be controlled by a dimmer switch for increased flexibility. (NuTone.)*

rate controls may be desirable so that the user can adjust the lighting and reduce the number of bulbs lit. In addition, dimmer switches vary the intensity.

Special lights also may be desired for radiant heating and for keeping a suntan. The controls for these lights might be housed in several locations or with the general lighting switch. The controls for special areas such as bathing or toilet compartments are controlled inside the bathroom.

Bathroom outlets should be protected by a GFCI. This is a term used for the *ground-fault circuit interrupter*. If you are about to be shocked, it will turn off the circuit.

Newer bathrooms use skylights to give natural light for grooming. They also make good use of picture windows opening onto a patio or deck. Stained glass is also used in some bathroom designs.

The main thing in planning is to avoid a single light source and to use special area lights where needed. Also plan enough outlets to power everything that will be used.

BATHROOM BUILT-INS

Storage space is always needed in a bathroom. Tissues and towels are stored for instant use. Also, if several people use a bathroom, more towel space is needed. Clean towels and washcloths, soaps, shampoo, dental articles, grooming items, cleaning equipment, dirty clothes hampers, and even the family linens are all potential storage problems.

Bathrooms include deep shelves that are underused because no one can reach the back of them, particularly the top ones. Often, the items stacked in the front part of the shelf block the accessibility of the items in back. Sometimes built-ins are simply shelves or drawers that rotate, swing out, or move to allow better use of these back areas that cannot be reached easily. Figure 19-41 shows some ideas to improve efficiency. A closet or cabinet is a must for a bathroom. It should have shelves that are a variety of sizes and may incorporate a laundry hamper.

A laundry chute can be built into the space normally used for a hamper. It will take less space and perhaps save a lot of stair climbing. If a home has a basement laundry, a hole can be cut through the floor deck (but not the frame) so that soiled laundry can be dropped directly into the basement. It is not necessary for the laundry to drop directly into the laundry area, although that would be preferable.

Cabinets beneath the basins are extremely popular because they provide additional storage space housed

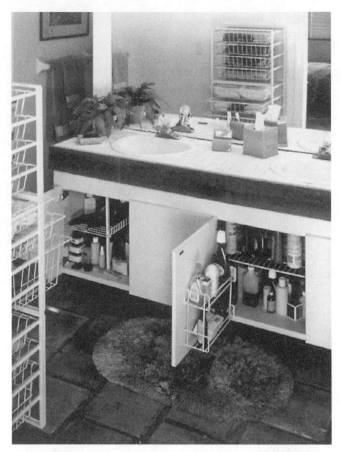

Fig. 19-41 *Bathroom storage and convenience can be increased by using racks and shelves that swing or roll out to save shelf space. (Closet Maid by Clairson.)*

in an attractive unit. In smaller bathrooms, this becomes more important because there is less space for closets or other built-ins.

People who have basins with wide counters may have to lean forward when using a mirror mounted on the wall. Medicine cabinets with mirrors (Fig. 19-42) hold the storage area out from the wall, which reduces the need to lean.

Where large mirrors are used behind basins, medicine cabinets can be built in on the ends of the counter. The doors may be either mirrors or wood. Mirrors provide more light and a more three-dimensional view, whereas the wood finishes may fit in with the décor (Fig. 19-43).

Building racks and shelves into walls is especially helpful when space is at a premium or when storage space is located in an area of high traffic flow. By using the space between the wall studs, items are recessed out of the way. Items such as toothbrushes, water jets, and electric razors can be kept readily accessible, hidden from view, and protected from splashing water by building them into the walls. Electrical

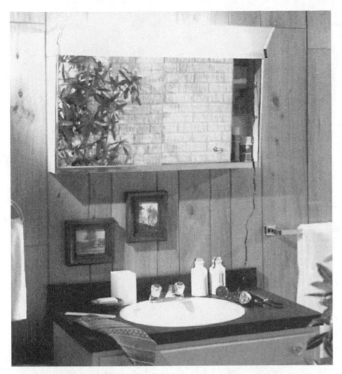

Fig. 19-42 *A cabinet that extends out from the wall minimizes the strain caused by leaning toward the mirror. (NuTone.)*

outlets also can be built into these areas to power these items. Sliding or hinged doors will hide them from view and keep water and dust off them.

Bathroom Layouts

Figure 19-44 shows bathroom plans that illustrate basic layouts and relationships, which can be adjusted to suit specific situations. Fixture spacing and clearances are important. Comfortable movement within a bathroom space is called for in being safe under bathroom circumstances. These recommended dimensions can be used in placing the fixtures on the plans and/or drawings of any bathroom. The bathroom will vary according to the sizes of the fixtures used.

Wheelchair Accessible

Notice that "wheelchair accessible" is shown in two of the drawings in Fig. 19-44. The doorway should have a minimum clear opening width of 32 inches. The door should not swing into the required clear floor space. The wheelchair should be able to make a 180-degree turn. Note that the turn should be either a 60-inch-diameter circle or a T-shaped area within a 60-inch cir-

Fig. 19-43 *Some examples of the many types of cabinets and doors available. (NuTone.)*

DOORWAY SHOULD HAVE A MINIMUM CLEAR OPENING WIDTH OF 32".
DOOR SHOULD NOT SWING INTO THE REQUIRED CLEAR FLOOR SPACE.

Fig. 19-44 *Bathroom layouts.*

cle or a square with arms 36 inches wide minimum and 60 inches long maximum. The clear floor spaces at fixtures, the accessibility route, and the wheelchair turning space are permitted to overlap.

FLOORS AND WALLS

Bathroom floors and walls are subject to water spills and splashes, heat, and high humidity. They should be capable of withstanding the heaviest wear under the most extreme conditions. Good flooring materials for bathrooms include ceramic and quarry tile, stone and brick, wood, resilient flooring, and special carpeting. Good wall materials include stone, tile, wood, and plastic laminates. Walls also can be painted, but regular flat paint is not advised around splash areas such as basins, tubs, and showers. If paint is to be used in those areas, use the best waterproof gloss or semigloss paint.

Environmental Considerations

Bathrooms are a source of water, and water can cause problems if it is allowed to accumulate and evaporate. Mildew and mold are two of the most undesirable effects of too much water in the wrong places. Keeping the bathroom dry and well ventilated are very important in maintaining a comfortable home with fewer germs and allergies.

If the house comes with a septic tank, it is a good idea to have it properly installed and pumped out on schedule. Keep the drainage field properly maintained to make sure that various associated germs and diseases are not allowed to proliferate.

Protect the well water location. Locate the septic tank where it is not a source of well water contamination. Test the water frequently.

CHAPTER 19 STUDY QUESTIONS

1. How do building codes influence bathroom design?
2. What are the new trends in bathrooms?
3. What is water hammer?
4. What is an air trap?
5. What are the advantages of small bathrooms?
6. What do building codes in local communities have to do with bathrooms?
7. How much water does it take to flush a new toilet today?
8. How is ventilation obtained for bathrooms with no windows?
9. Where are up-flush toilets used?
10. Where are wax seals used?
11. What are Japanese tubs called?

20
CHAPTER

Construction for Solar Heating

SOLAR SYSTEMS FOR HEATING AND COOLING A HOUSE are often thought of as a means of getting something for nothing. After all, the sun is free. All you have to do is devise a system to collect all this energy and channel it where you want it to heat the inside of the house or to cool it in the summer. Sounds simple, doesn't it? It can be done, but it is not cheap.

There are two ways to classify solar heating systems: active and passive. The *passive* type is the simplest. It relies entirely on the movement of a liquid or air by means of the sun's energy. You actually use the sun's energy as a method of heating when you open the curtains on the sunny side of the house during the cold months. In the summer, you can use curtains to block the sun's rays and try to keep the room cool. The passive system has no moving parts. (In the example, the moving of the curtains back and forth was an exception.) The design of the house can have a lot to do with this type of heating and cooling. It takes into consideration whether the climate requires the house to have an overhang to shade the windows or no overhang so that the sun can reach the windows and inside the house during the winter months.

The other type of system is called *active*. It has moving parts to add to the circulation of the heat by way of pumps to push hot water around the system or fans to blow the heated air and cause it to circulate.

Solar heating has long captivated people who want a carefree system to heat and cool their residences. However, as you read this chapter, you will find that there is no free source of energy. Some types of solar energy systems are rather expensive to install and maintain.

In this chapter you will learn

• How active and passive systems work

• How cooling and heating are accomplished

• Advantages and disadvantages of an underground house

• How various solar energy systems compare with conventional methods used for heating and cooling

PASSIVE SOLAR HEATING

Three concepts are used in passive heating systems: direct, indirect, and isolated gain (Fig. 20-1). Each of these concepts involves the relationship between the sun, storage mass, and living space.

Indirect Gain

In indirect gain, a storage mass is used to collect and store heat. The storage mass intercedes between the

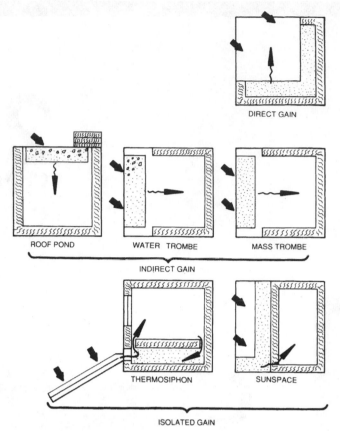

Fig. 20-1 *Three concepts for solar heating: direct gain, indirect gain, and isolated gain.*

sun and the living space. The three types of indirect-gain solar buildings are mass trombe, water trombe, and roof pond (see Fig. 20-1).

Mass trombe buildings involve only a large glass collector area with a storage mass directly behind it. There may be a variety of interpretations of this concept. One of the disadvantages is the large 15-inch-thick concrete mass constructed to absorb the heat during the day (Fig 20-2). The concrete has a black coating to aid the absorption of heat. Decorating around this gets to be a challenge. If it is a two-story house, most of the heat will go up the stairwell and remain in the upper bedrooms. Distribution of air by natural convection is viable with this system because the volume of air in the space between the glazing and storage mass is being heated to high temperatures and is constantly trying to move to other areas within the house. The mass also can be made of adobe, stone, or composites of brick, block, and sand.

Cooling is accomplished by allowing the 6-inch space between the mass and the glazed wall to be vented to the outside. Small fans may be necessary to move the hot air. Venting the hot air causes cooler air to be drawn through the house. This will produce some cooling during the summer. The massive wall and

Fig. 20-2 *Indirect-gain passive solar heating: mass trombe.*

ground-floor slab also maintain cooler daytime temperatures. Trees can be used for shade during the summer and then will drop their leaves during the winter to allow for direct heating of the mass.

A hot-air furnace with ducts built into the wall is used for supplemental heat. Its performance evaluation is roughly 75 percent passive solar heating contribution. Performance is rated as excellent. For summer, larger vents are needed, and in the winter, too much heat rises up the open stairwell.

Water trombe buildings are another of the indirect-gain passive solar heating types (Fig 20-3). The build-

ings have large glazed areas and an adjacent massive heat storage. The storage is in water or another liquid, held in a variety of containers, each with different heat-exchange surfaces to storage-mass ratios. Larger storage volumes provide greater and longer-term heat storage capacity. Smaller storage volumes provide greater heat-exchange surfaces and faster distribution. The tradeoff between heat-exchange surface and storage mass has not been fully developed. A number of different types of storage containers, such as tin cans, bottles, tubes, bins, barrels, drums, bags, and complete walls filled with water, have been used in experiments.

A gas-fired hot water heating system is used for backup purposes. This is done primarily because this system has a 30 percent passive solar heating contribution. When fans are used to force air past the wall to improve the heat circulation, it is classified as a *hybrid system.*

The *roof pond* type of building is exactly what its name implies. The roof is flooded with water (Fig. 20-4). It is protected and controlled by exterior movable insulation. The water is exposed to direct solar gain that causes it to absorb and store heat. Since the heat source is on the roof, it radiates heat from the ceiling to the living space below. Heat is by radiation only. The ceiling height makes a difference to the individual being warmed because radiation density drops off with distance. The storage mass should be uniformly spaced so that it covers the entire living area. A hybrid of the passive type must be devised if it is to be more efficient. A movable insulation has to be used on sunless winter days and nights to prevent unwanted heat

Fig. 20-3 *Indirect-gain passive solar heating: water trombe.*

Fig. 20-4 *Indirect-gain passive solar heating: roof pond.*

losses to the outside. It is also needed for unwanted heat gain in the summer.

This type of system does have some cooling advantages in summer. It works well for cooling in parts of the country where significant day-to-night temperature swings take place. The water is cooled down on summer evenings by exposure to the night air. The ceiling water mass then draws unwanted heat from the living and working spaces during the day. This takes advantage of the temperature stratification to provide passive cooling.

This type of system has not been fully tested, and no specifics are known at this time. It is still being tested in California.

Direct Gain

The direct-gain heating method uses the sun directly to heat a room or living space (Fig. 20-5). The area is open to the sun by using a large windowed space so that the sun's rays can penetrate the living space. The areas should be exposed to the south, with the solar exposure working on massive walls and floor areas that can hold the heat. The massive walls and thick floors are necessary to hold the heat. This is why most houses cool off at night even though the sun has heated the rooms during the day. Insulation has to be used between the walls and floors and the outside or exterior space. The insulation is needed to prevent the heat loss that occurs at night when the outside cools down.

Woodstoves and fireplaces can be used for auxiliary heat sources with this type of heating system. Systems of this type can be designed with 95 percent passive solar heating contribution. Overhangs on the south side of the building can be designed to provide shading against unwanted solar gain.

Isolated Gain

In this type of solar heating, the solar collection and storage are thermally isolated from the living spaces of the building. The *sunspace* isolated-gain passive solar building type collects solar radiation in a secondary space. The isolated space is separate from the living space. This design stores heat for later distribution.

This type of design has some advantages over the others. It offers separation of the collector-storage system from the living space. It is midway between the direct-gain system, where the living space is the collector of heat, and a mass or water trombe system, which collects heat indirectly for the living space.

Part of the design may be an atrium, a sun porch, a greenhouse, and a sunroom. The southern exposure of the house is usually the location of the collector arrangement. A dark tile floor can be used to absorb some of the sun's heat. The northern exposure is protected by a berm and a minimum of window area. The concrete slab is 8 inches thick and will store the heat for use during the night. During the summer, the atrium is shaded by deciduous trees. A fireplace and small central gas heater are used for auxiliary heat. Performance for the test run has reached the 75 to 90 percent level.

The *thermosiphon* isolated-gain passive solar building type is another type of solar-heated building (Fig. 20-6). In this type of solar heating system, the collector space is between the direct sunshine and the living space. It is not part of the building. A thermosiphoning heat flow occurs when the cool air or liquid naturally falls to the lowest point—in this case, the collectors. Once heated by the sun, the heated air or liquid rises up into an appropriately placed living space, or it can be moved to a storage mass. This causes a somewhat cooler air or liquid to fall again. This movement causes a continuous circulation to begin. Since the collector space is completely separate from the building, the thermosiphon system resembles an active system. However, the advantage is that no external fans or blowers are needed to move the heat-transfer medium. The thermosiphon principle has been applied in numerous solar domestic hot water systems. It offers good potential for space-heating applications.

Electric heaters and a fireplace with a heatilator serve as auxiliary sources. The south porch can be designed to shade the southern face with overhangs to protect the *clerestory* windows. Clerestory windows are those above the normal roofline, as in Fig. 20-6. Cross-ventilation is provided by these windows.

Time-Lag Heating

Time-lag heating was used by the Native American of the Southwest, where it got extremely hot during the day and cooled down at night. The diurnal (day-to-

Fig. 20-5 *Direct-gain passive solar heating.*

Fig. 20-6 *Isolated-gain passive solar heating system: thermosiphon.*

night) temperature conditions provided a clear opportunity for free or natural heating by delaying and holding daytime heat gain for use in the cool evening hours.

In the parts of the country where there are significant [20 to 35°F (–6.7 to 1.7°C)] day-to-night temperature swings, a building with thermal mass can allow the home itself to delay and store external daytime heating in its walls. The captured heat is then radiated to the building interior during the cool night. Internal heat gains come from people, lights, and appliances. This heat also can be absorbed and stored in the building structure.

Massive or heavy construction of walls, floors, and ceilings is used for this type of solar heating. Because it is dense, concrete, stone, or adobe has the capacity to hold heat. As the outside temperature rises during the day, so does the temperature of the building surface. The entire wall section heats up and gradually releases the stored heat to the room by radiation and convection. Two controls can make time-lag heating systems most effective for passive heating. In fact, 2, 4, 8, and even 12 hours of delay can be guaranteed by building walls of the right thickness and density. Choosing the right material and thickness can allow you to control what hour in the evening you begin heating. Exterior or sheathing insulation can prevent the heat-storage wall from losing its carefully gained heat to the outside, offering more passive heat to the inside. For additional heat, winter sunshine also can be collected and stored in massive walls, provided adequate shading is given to these walls to prevent overheating in the summer.

Uninsulated massive walls can cause problems with the auxiliary heating system if used improperly. In climates where there is no day-to-night temperature swing, uninsulated massive walls can cause problems. Continually cold or continually hot temperatures outside will build up in heavy exterior walls and will draw heat from the house for hours until these walls have been completely heated from the inside.

This type of house has been built and tested in Denver, Colorado, with a 65 percent passive solar heating contribution and 60 percent passive solar cooling contribution.

Underground Heating

The average temperature underground, below the frost line, remains stable at approximately 56°F (13.3°C). This can be used to provide effective natural heating as outside temperatures drop below freezing. The massiveness of the earth itself takes a long time to heat up and cool down. Its average annual temperature ranges between 55 and 65°F (12.8 and 18.3°C) with only slight increases at the end of summer and slight decreases at the end of winter. In climates with severe winter conditions or severe summer temperatures, underground construction provides considerably improved outside design temperatures. It also reduces wind exposure. The underground building method removes most of the heating load for maximum energy conservation.

One of the greatest disadvantages of the underground home is the humidity. If you live in a very humid region of the country, excessive humidity, moisture, and mildew can present problems. This is usually no problem during the winter season but can become serious in the summer. Underground buildings cannot take maximum advantage of comfortable outside temperatures. Instead, they are continuously exposed to 56°F ground temperatures. Thus, if you live in a comfortable climate, there is no need to build underground. You could, however, provide spring and fall living spaces outside the underground dwelling and use it only for summer cooling and winter heat conservation. Make sure that you do not build on clay that swells and slides. And do not dig deep into slopes without shoring against erosion.

An example is a test building that was constructed in Minneapolis. The building was designed to be energy efficient. It also was designed to use the passive cooling effects of the earth. Net energy savings over a conventional building are expected to be 80 to 100 percent during the heating period and approximately 45 percent during the cooling period.

PASSIVE COOLING SYSTEMS

There are six passive cooling systems. These include natural and induced ventilation systems, desiccant systems, and evaporative cooling systems, as well as the passive cooling that can be provided by night sky temperature conditions, diurnal (day-to-night) temperature conditions, and underground temperature conditions.

Natural Ventilation

Natural ventilation is used in climates where there are significant summer winds and sufficient humidity (>20 percent) so that the air movement will not cause dehydration.

Induced Ventilation

Induced ventilation is used in climate regions that are sunny but experience little summer wind activity.

Desiccant Cooling

Another name for desiccant cooling is *dehumidification*. This type of cooling is used in climates where high humidities are the major cause of discomfort. Humidities greater than 70 or 80 percent relative humidity (RH) will prevent evaporative cooling. Thus methods of drying out the air can provide effective summer cooling.

This type of cooling is accomplished by using two desiccant salt plates for absorbing water vapor and solar energy for drying out the salts. The two desiccant salt plates are placed alternately in the living space, where they absorb water vapor from the air, and in the sun, to evaporate this water vapor and return to solid form. The salt plates may be dried either on the roof or at the southern wall or, alternately, at east and west walls responding to morning and afternoon sun positions. Mechanized wheels transporting wet salt plates to the outside and dry salt plates to the inside also could be used.

Evaporative Cooling

This type of cooling is used where there are low humidities. The addition of moisture by using pools, fountains, and plants will begin an evaporation process that increases the humidity but lowers the temperature of the air for cooling relief.

Spraying the roof with water also can cause a reduction in the ceiling temperature and cause air movement to the cooler surface. Keep in mind that evaporative cooling is effective only in drier climates.

Water has to be available for makeup of the evaporated moisture. The atrium and the mechanical coolers should be kept out of the sun because it is the air's heat you are trying to use for the evaporation process, not solar heat. For the total system design, the pools of water, vegetation, and fountain court should be combined with the prevailing summer winds for efficient distribution of cool, humidified air.

Night-Sky Radiation Cooling

Night-sky radiation depends on clear nights in the summer. It involves the cooling of a massive body of water or masonry by exposure to a cool night sky. This type of cooling is most effective when there is a large day-to-night temperature swing. A clear night sky in any climate will act as a large heat sink to draw away the daytime heat that has accumulated in the building mass. A well-sized and properly exposed body of water or masonry, once cooled by radiation to the night sky, can be designed to act as a cold storage, draining heat away from the living space through the summer day and providing natural summer cooling.

The roof pond is one natural conditioning system that offers the potential for both passive heating and passive cooling. The requirements for this system involve the use of a contained body of water or masonry on the roof. This should be protected when necessary by moving insulation or by moving the water. The house has conventional ceiling heights for effective radiated heating and cooling. During the summer when it is too hot for comfort, the insulating panels are rolled away at night. This exposes the water mass to the clear night sky, which absorbs all the daytime heat from the water mass and leaves chilled water behind. During the day, the insulated panels are closed to protect the roof mass from the heat. The chilled storage mass below absorbs heat from the living spaces to provide natural cooling for most or all of the day.

This type of cooling offers up to 100 percent passive, nonmechanical air conditioning.

Time-Lag Cooling

This type of cooling already has been described in the section on time-lag heating. It is used primarily in climates that have a large day-to-night temperature swing. The well-insulated walls and floor will maintain the night temperature well into the day, transmitting little of the outside heat into the house. If 20-inch eaves over all the windows are used, they can exclude most of the summer radiation, thereby controlling the direct heat gain.

Underground Cooling

Underground cooling takes advantage of the fairly stable 56°F (13.3°C) temperature conditions of the earth below the frost line. The only control needed is the addition of perimeter insulation to keep the house temperatures above 56°F (13.3°C). In climates with severe summer temperatures and moderate to low humidity levels, underground construction provides stable and cool outside design temperatures as well as reduced sun exposure to remove most of the cooling load for maximum energy conservation.

ACTIVE SOLAR HEATING SYSTEMS

Active solar systems are modified systems that use fans, blowers, and pumps to control the heating process and distribution of the heat once it is collected. Active systems currently use the following six units to collect, control, and distribute solar heat.

Unit	Function
1. Solar collector	Intercepts solar radiation and converts it to heat for transfer to a thermal storage unit or to the heating load
2. Thermal storage unit	Can be either an air or liquid unit; if more heat than needed is collected, it is stored in this unit for later use; can be either liquid, rock, or a phase-change unit
3. Auxiliary heat source	Used as a backup unit when there is not enough solar heat to do the job
4. Heat-distribution system	Depending on the systems selected, could be the same as those used for cooling or auxiliary heating
5. Cooling-distribution system	Usually a blower and duct distribution capable of using air or liquid directly from either the solar collector or the thermal storage unit

Operation of Solar Heating Systems

It would take a book in itself to examine all the possibilities and maybe a couple of volumes more to present details of what has been done to date. Therefore, it is best to take a look at a system that is available commercially from a reputable firm that has been

Key components in a liquid air solar system: 1. Solar collectors, 2. Storage tank, 3. Hot water heat exchanger, 4. Hot water holding tank, 5. Space heating coil, 6. Purge coil (releases excess solar heat), 7. Expansion tank, 8. Heat exchanger.

Fig. 20-7 *Components of a liquid-to-air solar heating system.* (Lennox Furnace Co.)

making heating and cooling systems for years (Fig. 20-7).

Domestic Water-Heating System

The domestic water-heating system uses water heated with solar energy. It is more economically viable than whole-house space heating because hot water is required all year round. The opportunity to obtain a return on the initial investment in the system every day of the year is a distinct economic advantage. Only moderate collector temperatures are required to cause the system to function effectively. Thus domestic water can be heated during less than ideal weather conditions.

Indirect Heating/ Circulating Systems

Indirect heating systems circulate antifreeze solution or a special heat-transfer fluid through the collectors. This is done primarily to overcome the problem of draining liquid collectors during periods of subfreez-

ing weather (Fig. 20-8). Air collectors also can be used. As a result, there is no danger of freezing and no need to drain the system.

Circulating a solution of ethylene glycol and water through the collector and a heat exchanger is one means of eliminating the problem of freezing (Fig. 20-9). Note that this system requires a heat exchanger and an additional pump. The heat exchanger permits the heat in the liquid circulating through the collector to be transferred to the water in the storage tank. The extra pump is needed to circulate water from the storage tank through the heat exchanger.

The extra pump can be eliminated if

- The heat exchanger is located below the storage tank.

- The pipe sizes and heat-exchanger design permit thermosiphon action to circulate the water.

- A heat exchanger is used that actually wraps around and contacts the storage tank and transfers heat directly through the tank wall.

Fig. 20-8 *Collectors on the roof heat water that is circulated throughout the house.*

TC
TW
ON FOR T C 10° ABOVE T
OFF FOR Tw ABOVE 100° F

COLLECTORS

HEAT EXCHANGER

COLD SUPPLY   AUTOMATIC MIXING VALVE

TO H.W. SERVICE

STORAGE TANK

AUXILIARY HEATER

PUMP

Fig. 20-9 *Indirect solar water heating system.*

Safety is another consideration in the operation of this type of system. Two major problems might develop with liquid solar water heaters:

1. Excessive water may enter the domestic water service line.

2. High temperature and high pressure may damage collectors and the storage unit (Fig. 20-10).

If you want to prevent the first problem, you can add a mixing valve between the solar storage tank and the conventional water heater (Fig. 20-11).

Cold water is blended with hot water in the proper proportion to avoid excessive supply temperatures. The mixing valve is sometimes referred to as a *tempering valve*. Figure 20-12 shows the details of a typical connection for a tempering valve.

You can avoid excessive pressure in the collector loop carrying the antifreeze or heat-transfer solution by installing a pressure-relief valve in the loop. Set the valve to discharge at anything above 50 pounds per square foot. The temperature of the liquid may hit 200°F (93.3°C), so make sure that the relief valve is connected to an open drain. The fluid is unsafe and contaminated, so keep that in mind when disposing of it.

A temperature and pressure-relief valve is usually installed on the storage tank to protect it. Whenever water in the tank exceeds 210°F (98.9°C), the valve opens and purges the hot water in the tank. Cold water automatically enters the storage tank and provides a heating load for the collector loop. This cools down the system. Figure 20-13 gives examples of both safety devices installed in the system.

(A)

(B)

Fig. 20-10 *A. Liquid collector. B. Air collector.*

Fig. 20-11 *Schematic of the auxiliary heating equipment.*

Fig. 20-12 *A typical tempering valve.*

Fig. 20-13 *Note the locations of the safety valves on the heater tank and the expansion tank.*

- It does not have to be concerned with boiling fluid or freezing in the winter.

- It does not run the risk of losing the expensive fluid in the system.

It does have some disadvantages over the liquid type of system:

- It requires larger piping between the collector and heat exchanger.

- It requires more energy to operate the circulating fan than it does for the water pump.

- It needs a slightly larger collector.

Cycle Operation

The indirect and direct water-heating systems need some type of control. A differential temperature controller is used to measure the temperature difference between the collector and the storage. This controls pump operation.

The pump starts when there is more than a 10°F difference between the storage and collector temperatures. It stops when the differential drops to less than 3°F.

You can use two-speed or multispeed pumps in a system of this type to change the amount of water being circulated. As solar radiation increases, the pump is speeded up. This type of unit also improves the efficiency of the system.

The collector-loop expansion tank (Fig. 20-13) is required to absorb the expansion and contraction of the circulating fluid as it is heated and cooled. Any loop not vented to the atmosphere must be fitted with an expansion tank.

The heat exchanger acts as an interface between the toxic collector fluid and the *potable* (drinkable) water. The heat exchanger must be double-walled to prevent contamination of the drinking water if there is a leak in the heat exchanger. The shell and tube type (Fig. 20-14) often does not meet the local code or the health department requirements.

Air Transfer

Air-heating collectors can be used to heat domestic water (Fig. 20-15). The operation of this type of system is similar to that of the indirect liquid circulation system. The basic difference is that a blower or fan is used to circulate the air through the collector and heat exchanger rather than a pump to circulate a liquid.

The air-transfer method has advantages:

- It is not subject to damage owing to liquid leakage in the collector loop.

Fig. 20-14 *Heat-exchanger designs.*

Designing the Domestic Water-Heating System

Any heating job requires that you know the number of British thermal units (Btus) needed to heat a space. The type of heating system we are designing is no exception.

Table 20-1 shows minimum property standards for solar systems as designated by the U.S. Department of Housing and Urban Development. Note that the minimum daily hot water requirements for various residences and apartment occupancies are listed. For example, a two-bedroom home with three occupants should be provided with equipment that can provide 55 gallons per day of hot water. Many designers simply assume 20 gallons per day per person, which results in slightly higher requirements than those listed in the table.

TABLE 20-1 *Daily Hot Water Usage (140°F) for Solar System Design*

Category	One- and Two-Family Units and Apartments up to 20 Units				
Number of people	2	3	4	5	6
Number of bedrooms	1	2	3	4	5
Hot water per unit (gallons per day)	40	55	70	85	100

Another important consideration in sizing the solar domestic hot water system is the required change in temperature of incoming water. The water supplied by a public water system usually varies from 40 to 75°F (4.4 to 23.9°C) depending on location and season of the year. A telephone call to your local water utility will provide the water supply temperature in your area. Generally, the desired supply hot water temperature is from 140 to 160°F (60 to 71.1°C). Knowing these two temperatures and the volume of water required enables you to calculate the British thermal unit requirement for domestic hot water. Figure 20-16 shows how to calculate the British thermal unit requirements for heating domestic hot water.

To find the required collector area needed to provide some portion of the British thermal unit load, you can use a number of methods. Figure 20-17 shows the location of an add-on collector. One rule of thumb is that the amount of solar energy available at midaltitude in the continental United States is approximately equal to 2,000 British thermal units per square foot per day. Assuming a collector efficiency of 40 percent, 800 British thermal units per square foot per day can be collected with a properly installed collector. Using the example in Fig. 20-17, the collector should contain approximately 137 square feet. This is found by 109,956

Fig. 20-15 *Schematic of air-transfer-medium solar water-heating system.*

Example Problem:

Given that a family of six people live in a home where the incoming water temperature is 40°F and the requirement is for 150°F hot water, calculate the BTU requirement.

1. Compute: Water requirements = Number of people × 20 gallons per day
                                 = 6 × 20
                                 = 120 gallons of hot water used per day

Because a gallon of water weighs 8.33 pounds, the BTU requirement per day for hot water can be found by:

2. Compute: Heat Required = gallons per day × 8.33 × temperature rise
                           = (120 × 8.33) (150 − 40)
                           = 109,956 BTU per day

Fig. 20-16 *Circulating British thermal unit requirements.*

British thermal units per day divided by 800, which equals 137. The higher summer radiation levels and warmer temperatures would cause an excess capacity most of the time. A more practical approach is to provide nearly 100 percent solar hot water in July, which then might average out to 70 percent contribution for the year. Thus a collector area of 0.7 × 137, or about 96 square feet, might be a more realistic installation (Fig. 20-18).

The rule of thumb sizing procedures used here assume that the collector is installed facing due south and inclined at an angle equal to the local latitude plus 10 degrees. Modification of these optimal collector installation procedures will reduce the effectiveness of the collector. In case the ideal installation cannot be achieved, it will be necessary to increase the size of the collector to compensate for the loss in effectiveness.

Other Components

The components for solar domestic water-heating systems are available in kits prepackaged with instructions (Fig. 20-19). These eliminate the need to size the storage tank, expansion tank, and pump. If you wish to select individual components, it will be necessary to make the same types of calculations for whole-house heating to determine the sizes of such components. Tank storage typically would be based on one day's supply of energy, which is based on the daily British thermal unit load. Figure 20-20 illustrates a typical piping and wiring arrangement for a solar water-heating system.

IS THIS FOR ME?

The basic question for everyone is: Is this for me? What are the economics of the system? Most manufacturers of packaged solar water-heating systems provide some type of economic analysis to assist installation contractors in selling their customers. For example, payback time for fuel savings to equal the total investment may be as little as 6 to 9 years.

A computer service called *SOLCOST* provides a complete analysis based on the information supplied to the computer service. It includes a collector size optimization calculation that will provide the customer the optimal savings over the life of the equipment. For full details, contact:

International Business Services, Inc.
Solar Group
1010 Vermont Avenue
Washington, DC 20005

(A) BERM UNIT

(B) PORCH COVER

(C) CAR PORT SCREEN UNIT

SOLAR PANEL (TYP)

(D) CAR PORT COVER

SOLAR PANEL

(E) CAR PORT COVER

Fig. 20-17 *Collector placement for remodeling jobs.*

BUILDING MODIFICATIONS

In order to make housing more efficient, it is necessary to make some modifications in present-day carpentry practices. For instance, the following must be done to make room for better and more efficient insulation:

1. Truss rafters are modified to permit stacking of two 6-inch-thick batts of insulation over the wall plate. The truss is hipped by adding vertical members at each end and directly over the 2 × 6 studs of the outer wall.

2. All outside walls use thicker (2 × 6) studs on 2-foot centers to accommodate the thicker insulation batts.

3. Ductwork is framed into the living space to reduce heat loss as warmed air passes through the ducts to rooms.

4. Wiring is rerouted along the soleplate and through notches in the 2 × 6 studs. This leaves the insulation cavity in the wall free of obstructions.

Elevations B, C, D, K, and L are probably the most economical methods of accommodating a steep tilt angle for solar collectors.

Elevations E, H, M, and O are probably the most economical.

Elevations I, J, M, and possibly O, depending on proportions, must be treated carefully.

Elevations G, N, P, Q, and S are possibilities but are more costly than the first group and especially more costly if clearstory space and/or fenestration are to be provided.

Elevations R and T are of some interest in that they provide a method of hiding the collectors completely, but their comparative cost needs special analysis.

Elevation F is obviously very expensive and only applicable under certain conditions. It does provide a three-story opportunity.

Fig. 20-18 *Collector roof configurations.* (National Association of Home Builders.)

5. Partitions join outside walls without creating a gap in the insulation. Drywall passes between the soleplate and abutting studs of the interior walls. The cavity is fully insulated.

6. The window area is reduced to 8 percent of the living area.

7. Box headers over the door and window openings receive insulation. In present practice, the space is filled with 2 × 10s or whatever is needed. Window header space can be filled with insulation. This reduces heat loss through the wood that normally would be located there.

Fig. 20-19 *Prepackaged solar-assisted domestic hot water unit.*

BUILDING UNDERGROUND

A number of methods are being researched for possible use in solar heating applications. One of the most inexpensive ways of obtaining insulation is building underground. However, one problem with underground living is psychological. People do not like to live where they can't see the sun or outside. The idea of building underground is not new. The Chinese have done it for centuries. But the problem comes in selling the public the idea. It will take a number of years of research and development before a move is made in this direction.

Advantages

There are a number of advantages to going underground. By using a subterranean design, the builder can take advantage of the earth's insulative properties.

The ground is slow to react to climatic temperature changes. It is a perfect year-round insulator. There is a relatively constant soil temperature at 30 feet below the surface. This could be ideal for moderate climates because the temperature would be a constant 68°F.

By building underground, it would be possible to use the constant temperature to reduce heating and cooling costs. A substantial energy reduction or savings could be realized in the initial construction also.

Figure 20-21 shows a roof-suspended earth home. This is one of the designs being researched at Texas A&M University. It uses the earth as a building material and uses the wind, water, vegetation, and sun to modify the climate.

A suitable method of construction uses beams to support the walls. This means that structural beams could be stretched across the hole, and walls could be suspended from the beams (Fig. 20-22). The inside and outside walls also would be dropped from the beams. This would allow something other than wood to be used for walls because they would be non-weight-bearing partitions. The roof would be built several feet above the surface of the ground to allow natural lighting through the skylights and provide a view of outside. This would get rid of some of the feeling of living like a mole.

Another design being researched involves tunneling into the side of a hill (Fig. 20-23). This still uses the insulative qualities of the earth and allows a southern exposure wall. In this way, windows and a conventional-type front door could be used. Exposing only the southern wall also helps to reduce heating costs. The steep hills in some areas could be used for this

Fig. 20-20 *Piping and wiring diagram for solar water-heating system.*

Fig. 20-21 *Roof-suspended earth home.*

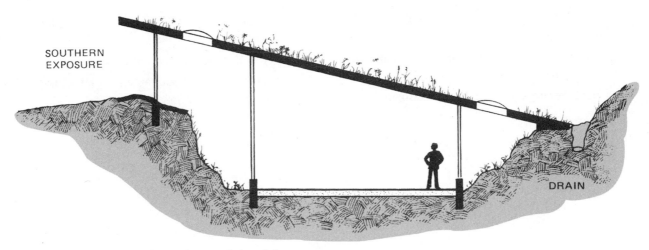

Fig. 20-22 *Another variation of a roof-suspended earth home.*

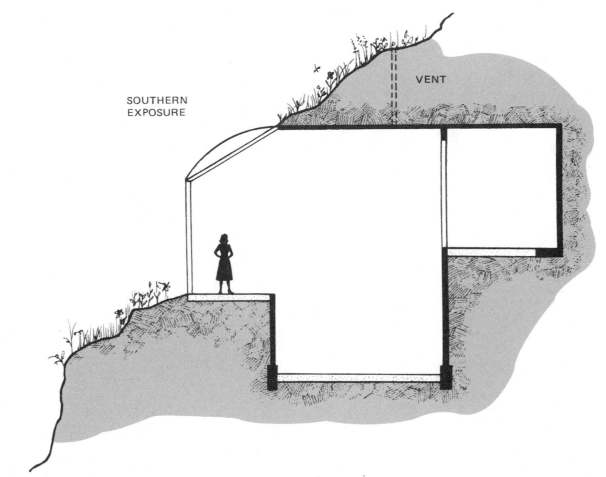

Fig. 20-23 *Hillside earth home.*

type of housing, where it would be impossible to build conventional-type housing.

Many more designs will be forthcoming as the need to conserve energy becomes more apparent. The carpenter will have to work with materials other than wood. The new methods and new materials will demand a carpenter willing to experiment and apply known skills to a rapidly changing field.

CHAPTER 20 STUDY QUESTIONS

1. Where are solar heated houses built?

2. Explain how the Tech House air-conditioning system works.

3. Explain how the fireplace works in Tech House.

4. How much can exterior shutters cut heat loss through windows?

5. How does the entry vestibule contribute energy saving in Tech House?

6. What temperature does the air in the attic of a Tech House reach?

7. How much energy do our homes consume each year?

8. What are the advantages of building a house underground?

9. How does a solar collector work?

10. What are the disadvantages of building a house underground?

21
CHAPTER

Alternative Framing Methods

THIS CHAPTER ILLUSTRATES NEW METHODS THAT are presently being used to frame houses. These newer methods are becoming increasingly popular as the price of lumber continues to increase — while its quality decreases. This chapter covers the following aspects of the most popular alternative framing methods:

- The advantages and disadvantages to each of these methods
- The tools that are associated with each of these methods
- The basic sequence for framing a house with these methods
- The installation of utilities inside houses built with these framing methods

WOOD FRAMES PREDOMINATE

Approximately 90 percent of homes in the United States are built with wood-frame walls. Because of this heavy use, almost all the old-growth forests have been harvested. New-growth lumber is typically the only type available to builders. This presents a present-day problem because new-growth lumber is fertilized and watered to make it grow quickly. This produces a larger cell structure, the end result of which is wood that is not as dense or as strong and is more susceptible to warping, cracking, and other defects. In addition, the price of lumber has tripled over the past decade and will continue to rise as quality, old-growth lumber becomes less available.

Because of this trend, many home builders use alternative methods for framing houses. These methods are variations of methods used in commercial buildings and therefore meet most local residential building code requirements. In most cases, these framing methods exceed code specifications and make a home more resistant to hurricanes, earthquakes, termites, and other common nuisances that afflict traditionally framed homes.

STEEL FRAMING

Steel-framed homes are not a new phenomenon. Their debut was at the 1933 Chicago World's Fair. However, it wasn't until the 1960s that the leading steel manufacturers entered the residential housing market. Steel homes have not gained wide popularity in the residential market because of their initial cost. Steel framing had always been more expensive than wood framing. With the rapid increase in the cost of lumber, steel studs are now, in most cases, less expensive than

wooden 2 × 4s. In most areas of the country, a steel-framed home costs as much as a wood-framed home or only 15 percent more. If the price of lumber continues to rise at the same rate as it has in the past, then steel-framed homes or other methods of framing will become more economical and commonplace.

Advantages and Disadvantages of Steel-Framed Homes

Today, most commercial and industrial buildings are constructed with steel as part of the primary structure. Steel is used because it is the strongest, safest, and most cost-effective building material available. When steel framing is used instead of wood, its advantages include the following:

- There is no shrinking, warping, or twisting, which allows walls, floors, and ceilings to be straight and square, thereby reducing finish work.
- There is a higher strength-to-weight ratio so that homes can withstand hurricanes, earthquakes, high winds, heavy snow loads, torrential rains, and termites and can span greater distances with less support.
- The house is assembled with screws, which eliminates nail pops and squeaks.
- Steel studs, joists, and trusses are quicker to install.
- Steel-framed homes weigh as much as 30 percent less, so foundations can be lighter.
- Steel-framed homes provide a better path to ground when struck by lightning, reducing the likelihood of explosions or secondary fires.
- Most steel framing is made to exact size in a factory, so less waste is produced.
- Steel is 100 percent recyclable.

The main disadvantage to residential steel framing is its initial cost. Other disadvantages include the following:

- Local building code officials might be unfamiliar with its use.
- Different tools are required.
- Carpenters are not familiar with the building practices associated with steel framing.

Types of Steel Framing

The two main types of steel framing are *red iron* and *galvanized*. Red-iron framing is based on the standard com-

Fig. 21-1 *Red-iron framework at 8-foot intervals with galvanized-steel framing in between to support sheathing.* (Tri-Steel Structures.)

mercial application in which the structure is built out of a thick red-iron frame at 8-foot spans to distribute the load (Fig. 21-1). Galvanized-sheet metal studs, joists, and other framing members are installed at 24 inches on center (O.C.) for a screwing base for sheathing.

Galvanized-steel framing is based on the traditional wooden "stick built" method. In this method, the entire home is built using galvanized sheet-metal framing members, as illustrated in Fig. 21-2. This method is preferred by some builders because its similarity to traditional wood-framing methods makes it easier to retrain carpenters, and cranes typically are not needed to lift heavy red-iron framing members. However, exterior walls require additional horizontal bracing spaced 16 to 24 inches O.C., and roof framing requires more rafters and purlins.

Tools Used in Steel Framing

Because most metal-framed homes are engineered and sold in kits, most of the framing members are cut to length. However, tin snips or aviation snips are frequently used to cut angles and notches in base track for plumbing manifolds. A chop saw or circular saw with a metal cutting blade is used to make most straight cuts.

Screw guns are used instead of hammers. Powder-actuated nail guns are used to attach base track to concrete foundations. Wrenches and ratchets are used to secure nuts and bolts in red-iron connections. A crane is needed for one day to erect the red-iron framing.

Tape measurers are used to lay out all framing connections, and speed squares are used to transfer measurements around corners. Levels and chalk lines are used frequently to keep everything plumb and square. Vice grips are used to clamp framing members together while they are being attached.

Sequence

Most residential steel suppliers are eager for builders to convert to steel framing and will provide extensive training as well as videotapes on framing with steel. In addition, they sell framing kits of precut steel members for standard model floor plans. Detailed plans are

Fig. 21-2 *Galvanized-steel framing. Note the horizontal wall bracing and additional rafters that are required for homes made without a red-iron frame.* (Tri-Steel Structures.)

provided with each kit and include specifications for stud spacing and other framing members depending on the size of the structure. Therefore, only the general sequence for framing with steel is covered in this chapter:

1. Anchor bolts for red-iron framing can only deviate ⅛ inch to ensure proper alignment and accuracy. Therefore, they are installed after the concrete has hardened. This is done by drilling holes for each anchor bolt with a hammer-drill. Anchor bolts range in length and diameter depending on load specifications. Two anchor bolts are used to fasten each red-iron girder. Anchor bolts are packaged in glass tubes, which break when they are inserted into the hole. When the tube breaks, a two-part adhesive mixes together and bonds the anchor bolt to the concrete foundation. Usually, 24 hours is required for the adhesive to cure before any red-iron framing members are attached.

2. Red-iron framing members are assembled with nuts and bolts on a flat, level surface. Each section usually consists of two vertical supports with rafters and bracing. They are erected using a crane.

3. Once the red-iron framing members are attached to the anchor bolts, they are straightened by attaching them to come-alongs that are attached to stakes around the perimeter of the home. Come-alongs are tightened and loosened until the framing members are plumb and square. Galvanized sheet-metal purlins then are attached every 2 feet between the 8-foot span of the red-iron rafters to keep them square.

4. Galvanized framing for walls is usually secured to concrete foundations by running two beads of sealant (sill seal) between the concrete and the galvanized base track. A powder-actuated nail gun then is used to drive fasteners every 12 inches O.C. in the base track.

5. Studs are secured by one screw on each side of the base track. Typically, ½-inch No. 8 self-drilling low-profile wafer, washer, or pan-head screws are used to secure all galvanized framing members. MIG welders also can be used to attach framing members; however, screws are preferred. Studs are attached to red-iron or any other thicker framing members by ¾-inch No. 12 self-drilling low-profile head screws. Vice grips always should be used to clamp the framing members together while they are being attached.

6. Furring or hat channel is spaced 16 to 24 inches O.C. for ceiling framing or for horizontal rein-

Part Identification Guide

	STUD - C Channel Type, formed with knurled outer flanges to prevent screw ride, hollow partition framing.
	TRACK - Channel configuration formed with unhemmed legs sized to receive the corresponding size stud.
	FURRING - HAT CHANNEL - Section designed for screw alignment of finished surface material in wall and ceiling furring.
	FURRING - RESILIENT CHANNEL - is one of the most effective methods of improving sound transmission loss through wood framed partitions and ceilings.
	FURRING - Z CHANNEL - Designed for vertical placement in wall furring for rigid insulation.
	ANGLE RUNNER - Angle sections used to secure and brace finish surface material.
	FLAT STRAPPING - Strap bracing is for use on internal and external framing to prevent racking and hold proper alignment.
	BRIDGING - Metal stud wall bridging is used as flush mount backing for mounting cabinets, handrails, grab bars, and most casework.

Fig. 21-3 *Steel-frame part identification guide.* (*Southeastern Metals Manufacturing Co.*)

forcement of exterior walls in the galvanized framing method. Steel framing members are identified in Fig. 21-3.

7. Plywood and oriented-strand board (OSB) are used commonly as exterior sheathing for steel-framed homes. Exterior sheathing is staggered

1 The Foundation (slab, pier & beam, pilings or basement) is prepared and exterior wall sections are bolted together on the ground.

2 Frame sections are raised, joined together and anchored to the foundation.

3 Roof trusses are assembled on the ground, lifted into place, attached to the side walls, and braced with furring.

4 Sheathing is attached to the roof and walls. Doors and windows are installed and dormers are created. Interior framing and insulation are installed.

5 Exterior finish of your choice is applied and the shell is now complete. The interior can be completed at your own pace.

Fig. 21-4 *Basic sequence for framing and sheathing using the red-iron method.* (Tri-Steel Structures.)

every 4 feet on the roof and on the walls of the galvanized-framing method. Plywood and OSB can be laid horizontally on top of each other without staggering for walls in the red-iron method. The basic sequence for framing and sheathing a red-iron home is shown in Fig. 21-4. Sheathing should be attached by screws every 24 inches O.C., with five screws securing a 48-inch width. This equates to a minimum of 25 screws per 4- × 8-foot sheet.

8. Six screws are used at web-stiffening areas such as window sills supported by 6-inch cripple studs. Most other connections require only two screws (one on each side).

9. Utilities, such as electrical wiring and plumbing, can run through the prepunched holes in the studs and joists. However, plastic grommets must first be inserted in the holes so that the sharp steel edges of the framing members will not damage the insulation on the wiring. The remaining electrical work usually involves the same techniques used in commercial steel-framing applications.

GALVANIZED FRAMING

When galvanized steel studs are used to frame the entire home, heavier-gauge studs (12 to 18 gauge) are used for the exterior load-bearing walls, and lighter-gauge studs (20 to 25 gauge) are used for the interior curtain walls. Typically, most houses are engineered by a company in advance; depending on the weight of the roof, tile, or composition shingles, or if there is a second story, the gauge of the studs will vary from home to home.

There are three main ways to assemble a steel-stud wall. One way is to have all the walls fabricated into panels in a factory that is off site (Fig. 21-5). This method, typically called *panelization*, is the fastest-growing segment of residential new construction. The advantages to panelization include

1. Less material is needed because human errors on the job site are prevented.

2. There is no lost time on the job site because of inclement weather.

3. Fewer skilled laborers are needed on the job site. This means a reduction in the builder's liability and in workers' compensation insurance.

4. Panelized walls produce consistent quality. This is so because the factory-controlled conditions include preset jigs and layout tables.

5. Panelized walls can be ordered in advance and shipped to the job site for easy installation. This significantly reduces the project construction time (Fig. 21-6).

Fig. 21-5 *Panelized walls and trusses save material and waste on the job site.* (Steel Framing, Inc.)

Fig. 21-6 *Factory-made panels being unloaded at the construction site.* (Steel Framing, Inc.)

The second way to assemble a steel wall is called *in-place framing*. This method requires the carpenter to first secure the steel base track to the foundation. It can be done by bolting it down with anchor bolts or by using powder-actuated nailers. When connecting the base track to a slab using a powder-actuated nailer, two low-velocity fasteners are spaced a minimum of every 24 inches. Typically, most exterior walls are connected to the slab with anchor bolts or J-bolts. The base track usually is reinforced with an additional piece of track or stud where the anchor bolt extends through the base track, as shown in Fig. 21-7.

Steel studs are attached to the base track every 16 or 24 inches O.C. with a No. 8 self-tapping screw on either side. Spacing depends on the load placed on each wall and the gauge of the stud being used. The corner stud is braced to the ground to prevent movement from the wind or from the stresses during assembly. A spacer stud is placed between each stud thereafter to keep the studs in place. Steel channel (which comes in 16-foot lengths) can be placed between the *cutouts* of each stud to brace the wall instead of the usual spacers, as shown

Fig. 21-7 *J-bolt foundation connection.* (Steel Framing, Inc.)

WALL STUD

1 1/2" CRC 16 GA.

150CLA150–43 @ 3"/5" ANGLE
(WIDTH IS 1/2" LESS THAN STUD DEPTH)
(2) #10x3/4" HEX SCREWS TO STUD
(1) #10x3/4" HEX SCREWS TO 1 1/2" CRC 16 GA.

NOTES:
FOR STUD HEIGHTS 10'–0" AND LESS, INSTALL CRC AT MID HEIGHT.
FOR STUD HEIGHTS GREATER THAN 10'–0" INSTALL AT THIRD POINTS.

Fig. 21-8 *Wall bridging detail.* (Nuconsteel Commercial Corp.)

in Fig. 21-8. Additional steps in fabricating the wall will be discussed in the following paragraphs.

Tilt-up framing is another way to assemble a steel-stud wall. In this method, the wall is assembled flat (horizontally). Then it is lifted into a vertical position. This method makes it easier for carpenters to assemble a flat, straight wall—that is, if the surface they are assembling the wall on is flat. Torpedo levels, which are magnetic, can be placed on the steel studs to ensure that each is level and square before being installed into the base track with No. 8 self-taping screws.

Steel-strap blocking, wood blocking, or track blocking can be used between studs for reinforcing. This is particularly helpful where overhead cabinets or other heavy fixtures may be attached. Steel straps are also fastened diagonally in the corners of the framing. This keeps the walls square and protects them from shear forces. However, steel strapping is not necessary if OSB or plywood is fastened to the corners of the steel framing.

Details for corners, window and door openings, wall-to-wall panels, intersecting wall panels, and wall-to-roof trusses are shown in Figs. 21-9 through 21-15.

Fig. 21-9 *Corner framing.* (Nuconsteel Commercial Corp.)

INSULATED CONCRETE FORMS

One of the fastest-growing framing methods is the insulated concrete form (ICF). In this method, walls are made of concrete reinforced by rebar surrounded by Styrofoam (expanded polystyrene beads). Most ICFs are hollow Styrofoam blocks that are stacked together like building blocks with either tongue-and-groove joints or finger joints, as shown in Fig. 21-16. Each manufacturer makes the block a different size. Some manufacturers have ties molded in their blocks, whereas others have separate ties that are inserted into the block at the job site. Once the blocks are assembled, reinforced, and braced, concrete is poured into the blocks using a concrete pumper truck. An example of what the concrete and rebar would look like inside the foam is shown in Fig. 21-17.

Advantages and Disadvantages to ICFs

Insulated concrete form homes are framed in concrete, which allows them to have the following advantages over conventional wood-frame homes:

Fig. 21-10 *Typical structural wall framing.* (Nuconsteel Commercial Corp.

TOP TRACK

THIS INDICATES
A SCREW

WALL STUDS AT
24" O.C. SPACING

TYPICAL DOOR OPENING

BASE TRACK

- WALL STUDS SPACING @ 24" O.C.
- USE #6 x $^{7}/_{16}$" SHARP POINT SCREWS FOR ALL CONNECTIONS

Fig. 21-11 *Typical nonstructural wall framing.* (Nuconsteel Commercial Corp.)

- The home can withstand hurricane-force winds (200 miles per hour).

- The home is bullet-resistant.

- Termites cannot harm its structural integrity.

- The home exceeds building code requirements.

- Heating and cooling costs are lowered by 50 to 80 percent.

- Load-bearing capacity is higher (27,000 versus 4,000 plf).

- Outside noise reduction is greater.

- Homeowner's insurance rates should decrease by 10 to 25 percent.

- Fire rating of the walls is measured in hours versus minutes.

- Pollen is reduced inside the home.

An ICF home is much stronger and more energy efficient than a wood-framed home. However, its major disadvantage is cost. The average ICF home costs about twice as much to frame as its chief competitor. Because the framing of a home is just one of its costs, this figure really amounts to only a 5 to 15 percent increase in the total cost of the home.

Another disadvantage to an ICF is that there is no room for error. If a window or door opening is off, it is much more difficult to cut reinforced concrete than

ALTERNATE

1-24V

PANEL — PANEL

1-24V

(1) #8 x 1/2" TRUSS HEAD SCREW @ 24" O.C.

THIS INDICATES TOP & BOTTOM TRACK

LOAD BEARING STUDS

STUD MATERIAL (ONE SIZE LARGER THAN INTERIOR WALL STUD) (2) #8 x 1/2" TRUSS HEAD SCREWS @ TOP TRACK & BOTTOM TRACK

wood. Carpenters also must be retrained to use this framing method correctly.

Tools Used in Insulated Concrete Form Framing

Because concrete can be molded into any shape, it is essential to mold the frame of a home straight and square so that walls are plumb and level. In order to achieve this, a carpenter should have a framing square, 2- and 4-foot levels, a chalk line, and 30- and 100-foot tape measures. The foam forms can be cut easily with a handsaw or sharp utility knife. In most cases, the foam forms are glued together, so a caulking gun is required.

All the foam forms need to be braced until the concrete hardens. If the construction crew is not using

(2) #8 x 1/2" TRUSS SCREW @ 24" O.C. (TYP)

PANEL — PANEL

THIS INDICATES TOP & BOTTOM TRACK

LOAD BEARING STUDS

200STR-33 CONTINUOUS STAMP
(1) #10 x ³/₄" HEX TO EACH TRUSS
(1) #10 x ³/₄" @ 3" O.C. TO BLOCKING

NUTRUSS™ TOP CHORD

1¹/₂"

VANES

NUTRUSS™ BOTTOM CHORD

S/H1 TYP
W/(X) #10 x 1" HEX SCREWS
TO TRUSS AND (X) #8 x ¹/₂"
TRUSS HEAD TO WALL

18 GA. BLOCKING AT
EVERY OTHER SPACE
W/#10 x ³/₄" PAN PH
@ 3" O.C. TO TOP TRACK
BLOCKING PROVIDED IN
STOCK LENGTH FOR
FIELD CUTTING

WALL STUD

ATTACH ROOF DECK TO EACH
TRUSS W/(1)#10 x ³/₄" HEX
AND TO BLOCKING W/(1)
#10 x ³/₄" HEX @ 3" O.C.

NUTRUSS™ TOP CHORD

METAL ROOF DECK

NUTRUSS™ BOTTOM CHORD

BLOCKING AT
EVERY OTHER BAY
W/#10 x ³/₄" PAN PH
SCREWS @ 3" O.C.
TO TOP TRACK

STUD WALL AND
TOP TRACK

SIDE VIEW

Fig. 21-15 *Truss to stud wall with blocking.* (Nuconsteel Commercial Corp.)

Centerline of cores

Furring strips

6"
12"
12"
12"
6"

POLYSTEEL® FORM
(U.S.PAT. 4879853)
THE EASY
STAY-IN-PLACE
INSULATED CONCRETE
BUILDING SYSTEM

16"

48"

9-1/4" (6" core)
11" (8 core)

Fig. 21-16 *Insulated concrete form.* (American Polysteel Forms.)

metal braces, then wooden braces need to be constructed. Therefore, a framing hammer, nails, circular saw, and crowbar are needed.

Rebar is run horizontally and vertically throughout the walls of an ICF home. Tools used to work with rebar are rebar cutting and bending tools, a metal cutoff saw, a rebar twist tie tool (pigtail), 8-inch dikes, 9-inch lineman's pliers, a hack saw, tin snips, and wire-cutting pliers. A hot knife (Fig. 21-18) or router can be used to cut notches in the foam to place electrical wiring.

Sequence

Insulated concrete form framing is a radically new approach for carpenters who are familiar with the traditional wooden "stick built" method. The only wood used in this wall-framing method is pressure-treated

Fig. 21-17 *Concrete and rebar inside an insulated concrete form.*

HOT KNIFE

Fig. 21-18 *A hot knife used to cut foam for electrical wiring and plumbing.* (Avalon Concepts.)

lumber, which is used for sealing rough door and window openings. Because this method is different and each manufacturer has specific guidelines that must be followed to ensure proper installation, training is offered to all individuals constructing ICFs. Many ICFs also send representatives to the job site to supervise those who are building their first home using this method. Therefore, only general guidelines are covered here:

1. Before starting, rebar should extend vertically 2 to 6 feet from the foundation every 1 to 2 linear feet depending on building code requirements.

2. Most manufacturers suggest placing 2 × 4s or some type of bracing around the perimeter of the foundation as a guide for setting the foam forms.

3. Once the first course of foam blocks is laid, continue to do so, placing rebar horizontally every 1 to 2 feet (every or every other course of block) as required.

4. Vertical bracing should be tied to the perimeter bracing at 6-foot intervals. Corners should be braced on each intersecting edge with additional diagonal bracing spaced at 4-foot intervals. Figure 21-19 shows a typical corner bracing diagram. As forms are stacked, vertical bracing can be screwed into the metal or plastic ties built or inserted into the foam blocks.

5. Openings for windows and doors should be blocked with pressure-treated lumber.

6. After the foam blocks are stacked to the correct height, place bracing on the top, and secure it to the side bracing. Because foam is very light, it will float. Proper bracing is essential to prevent the foam walls from floating and causing

CORNER BRACE WITH FORMS

2 X 4 vertical brace on inside of corner.

Double tie wire around 2 X 4 and 2 X 6s above every course.

Mitered corner

Fig. 21-19 *Corner bracing technique for insulated concrete forms. (American Polysteel Forms.)*

blowouts when the concrete is poured inside them. Additional bracing techniques are illustrated in Fig. 21-20.

7. Walls usually are poured with a boom pump truck in 4-foot increments. Anchor bolts are set in the top of the wall to secure the top plate as a nailing base for roof construction (Fig. 21-21).

8. Sheathing can be screwed into the metal or plastic ties located every foot in the foam block, as shown in Fig. 21-22.

9. Electrical wiring and boxes can be installed by gouging out a groove with a router or hot knife. Figure 21-23 illustrates electrical wiring tips.

Types of Foam

The most commonly used types of foam in ICFs are *expanded polystyrene* (EPS) and *extruded polystyrene* (XPS). EPS consists of tightly fused beads of foam. Vending-machine coffee cups, for example, are made of EPS. XPS is produced in a different process and is more continuous, without beads or the sort of "grain" of EPS. The trays in prepackaged meat at the grocery store are made of XPS. The two types can differ in cost, strength, R value, and water resistance. EPS varies somewhat. It comes in various densities; the most common are 1.5 and 2 pounds per cubic foot. The denser foam is a little more expensive but is a little stronger and has a slightly higher R value.

Some stock EPS is now available with insect-repellent additives. Although few cases of insect penetration into the foam have been reported to date, some ICF manufacturers offer versions of their product made of treated material. Whether you buy your own foam or choose a preassembled system, you might want to check with the manufacturer about this.

Three Types of ICF Systems

The main difference in ICF systems is that they vary in their unit sizes and connection methods. They can be divided into three types: *plank*, *panel*, and *block systems*, as shown in Fig. 21-24. The panel system is the largest. It is usually 4 × 8 feet in size. This means that the wall area can be erected in one step but may require more cutting. These panels have flat edges and are connected one to the other with fasteners such as glue, wire, or plastic channel.

Plank systems are usually 8 feet long with narrow (8- or 12-inch) planks of foam. These pieces of foam are held at a constant distance of separation by steel or plastic ties. The plank system has notched, cut, or drilled edges. The edges are where the ties fit. In addi-

2 X 4 TOP RAIL SUPPORT

2 X 4s

Fig. 21-20 *Additional bracing techniques.* (American Polysteel Forms.)

HORIZONTAL TOP BRACE

◆ A horizontal top brace secured to the steel furring strips and supported by a diagonal brace staked to the ground will also provide the alignment and security necessary to keep the Forms in place during the pouring of concrete.

POLYSTEEL FORM® TOP BRACE

Polysteel Forms®

Screw a continuous horizontal 2 X 4 to the top course of Polysteel Form® wall with 2 1/2" deck screws at 4'–0" O.C., a 2 X 4 diagonal brace is screwed to the horizontal 2 X 4 and is nailed to a bracing stake driven into the ground.

Fig. 21-21 *Anchor bolt placement for roof framing.* (AFM.)

Fig. 21-22 *Attaching sheathing directly to plastic ties in the concrete forms.* (AFM.)

tion to spacing the planks, the ties connect each course of planks to the one above and below.

Block systems include units ranging from the standard concrete block (8 × 16 inches) size to a much larger 16-inch-high × 4-foot-long unit. Along their edges are teeth or tongues and grooves for interlock-

Fig. 21-23 *Installing electrical wiring and boxes in insulated concrete forms.* (American Polysteel Forms.)

FOR THIN PORTIONS OF EPS:
4"x4"x1-1/2" shallow box with single gang 1/2" raised cover.

Fig. 21-24 *Diagrams of ICF formwork made with the three basic units: panel on the left, plank in the center, and block on the right.* (PCC.)

Fig. 21-25 *Cutaway diagrams of ICF walls with the three basic cavity shapes: flat, grid, and post-and-beam.* (PCC.)

ing; they stack without separate fasteners on the same principle as children's Lego blocks.

Another difference is the shape of the cavities. Each system has one of three distinct cavity shapes. The shapes are flat, grid, or post-and-beam. These produce different shapes of concrete beneath the foam, as shown in Fig. 21-25.

Note how the flat cavities produce a concrete wall of constant thickness, just like a conventionally poured wall that was made with plywood or metal forms. Cavities are usually "wavy," both horizontally and verti-

cally. If the forms are removed, it can be seen just how the walls resemble a breakfast waffle. The post-and-beam cavities are cavities with concrete only every few feet, horizontally and vertically. In the most extreme post-and-beam systems, there is a 6-inch-diameter concrete "post" formed every 4 feet and a 6-inch concrete "beam" at the top of each story.

Keep in mind that no matter what the shape of the cavity, all systems have ties. These are the cross-pieces that connect the front and back layers of foam that make up the form. If the ties are plastic or metal, the concrete is not affected significantly. However, in some grid systems, they are foam and are much larger, forming breaks in the concrete about 2 inches in diameter every foot or so. Figure 21-26 illustrates the differences.

Table 21-1 shows eight different systems used in ICF construction with their dimensions, fastening surface, and various notes providing additional detail.

Another difference is that many of the systems also have a fastening surface that is some material other than foam embedded into the units that crews can sink screws or nails into the same way as fastening to a stud. Often this surface is simply the ends of the ties; however, other systems have no embedded fastening surface. These units are all foam, including the ties. This generally makes them simpler and less expensive but requires crews to take extra steps to con-

TABLE 21-1 *Available ICF Systems**

	Dimensions² (width x height x length)	Fastening surface	Notes
Panel systems			
Flat panel systems			
R-FORMS	8" x 4' x 8'	Ends of plastic ties	Assembled in the field; different lengths of ties available to form different panel widths.
Styroform	10" x 2' x 8'	Ends of plastic ties	Shipped flat and folded out in the field; can be purchased in larger/smaller heights and lengths.
Grid panel systems			
ENER-GRID	10" x 1'3" x 10'	None	Other dimensions also available; units made of foam/cement mixture.
RASTRA	10" x 1'3" x 10'	None	Other dimensions also available; units made of foam/cement mixture.
Post-and-beam panel systems			
Amhome	9 3/8" x 4' x 8'	Wooden strips	Assembled by the contractor from foam sheet. Includes provisions to mount wooden furring strips into the foam as a fastening surface.
Plank systems			
Flat plank systems			
Diamond Snap-Form	1' x 1' x 8'	Ends of plastic ties	
Lite-Form	1' x 8" x 8'	Ends of plastic ties	
Polycrete	11" x 1' x 8'	Plastic strips	
QUAD-LOCK	8" x 1' x 4'	Ends of plastic ties	
Block systems			
Flat block systems			
AAB	11.5" x 16⅝" x 4'	Ends of plastic ties	
Fold-Form	1' x 1' x 4'	Ends of plastic ties	Shipped flat and folded out in the field.
GREENBLOCK	10" x 10" x 3'4"	Ends of plastic ties	
SmartBlock Variable Width Form	10" x 10" x 3'4"	Ends of plastic ties	Ties inserted by the contractor; different length ties available to form different block widths.
Grid block systems with fastening surfaces			
I.C.E. Block	9 1/4" x 1'4" x 4'	Ends of steel ties	
Polysteel	9 1/4" x 1'4" x 4'	Ends of steel ties	
REWARD	9 1/4" x 1'4" x 4'	Ends of plastic ties	
Therm-O-Wall	9 1/4" x 1'4" x 4'	Ends of plastic ties	
Grid block systems without fastening surfaces			
Reddi-Form	9 5/8" x 1' x 4'	Optional	Plastic fastening surface strips available
SmartBlock Standard Form	10" x 10" x 3'4"	None	
Post-and-beam block systems			
ENERGYLOCK	8" x 8" x 2'8"	None	
Featherlite	8" x 8" x 1'4"	None	
KEEVA	8" x 1' x 4'	None	

¹ All systems are listed by brand name.

² "Width" is the distance between the inside and outside surfaces of foam of the unit. The thickness of the concrete inside will be less, and the thickness of the completed wall with finishes added will be greater.

Fig. 21-26 *Cutaway diagrams of ICF grid walls with steel/plastic ties and foam ties. (PCC.)*

nect interior wallboard, trim, exterior siding, and so on to the walls.

Cutaway views of the wall panels are shown in Figs. 21-27 through 21-34. In Fig. 21-35, an R-Forms panel is being assembled on site. In Fig. 21-36, there is

Fig. 21-29 *Cutaway diagram of a post-and-beam panel wall. (PCC.)*

Fig. 21-27 *Cutaway diagram of a flat panel wall. (PCC.)*

Fig. 21-30 *Cutaway diagram of a flat plank wall. (PCC.)*

Fig. 21-28 *Cutaway diagram of a grid panel wall. (PCC.)*

Fig. 21-31 *Cutaway diagram of a flat block wall. (PCC.)*

Fig. 21-32 *Cutaway diagram of a grid block wall with fastening surfaces. (PCC.)*

Fig. 21-33 *Cutaway diagram of a grid block wall without fastening surfaces.(PCC.)*

Fig. 21-34 *Cutaway diagram of a post-and-beam block wall. (PCC.)*

Fig. 21-35 *Site-assembly of R-Forms panels. (PCC.)*

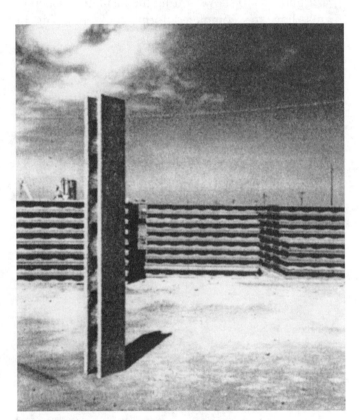

Fig. 21-36 *An ENER-GRID panel. (PCC.)*

an on-site pile of Ener-Grid panels. Figure 21-37 shows the top view of an Amhome panel that has furring strips embedded. Note how light the panels are. They are easily handled by one person. In Fig. 21-38, the worker is folding out a Fold-Form block before use.

Two varieties of SmartBlock are shown in Fig. 21-39. Three of them, A, C, and D, are assembled with plastic ties. A grid block without fastening surfaces is shown in Fig. 21-39B.

Fig. 21-37 *Top view of an Amhome panel with embedded furring strips.* (PCC.)

Fig. 21-39 *The two varieties of SmartBlock: a flat block assembled with plastic ties (A, C, and D) and a grid block without fastening surfaces (B).* (PCC.)

Figure 21-40 shows setting the insulating concrete forms, how the completed forms look, pumping in concrete, and the completed house.

Foam Working Tools

Some of the tools needed for working with foam are not familiar to the usual framer of houses. These tools may include a thermal cutter. This cutter is a new tool that cuts a near-perfect line through foam and plastic

units in one pass. Figure 21-41 shows this device. It is made up of a taut resistance-controlled wire mounted on a bench frame. It is heated with electricity and drawn though the unit while the wire is red hot. It melts a narrow path through the foam and plastic ties. It is a worthwhile tool to have if you are building a high-volume ICF house. It will not cut rough metal ties or the foam-and-cement material of grid panel systems. However, companies selling thermal cutters are usually located in every community. Grid panel sys-

A

B

Fig. 21-38 *Folding out a Fold-Form block before use.* (PCC.)

Fig. 21-40 *A. Bracing corners with window and door inserts. B. Garage doors framed, poured, and roof décor applied. C. More form-work in place with window backs. D. Side walls in place with framing to support the concrete pouring operation. E. Garage door framing. Note the steel rebar through the foam blocks and the corner bracing. F. Bracing for a window. Note rebar.*

tems can be cut with any of the bladed tools used by a carpenter, but sometimes a chain saw is a handy tool. It goes quickly through the heavier material of these systems and cuts through in one pass, whether cutting on the ground or in place. Tools needed for cutting and working with foam are shown in Table 21-2.

Gluing and Tying Units

Insulated concrete form units are frequently glued at the joints to hold them down, hold them together, and prevent concrete leakage. Common wood glue and most construction adhesives do the job well. Popular

Fig. 21-40 *G. Note treatment of the bay window shown here without siding. Also, note the foam being coated where backfill will be placed. H. Foam blocks filled with concrete; roof it attached and shingled. I. Foam block in place. Roof framing is complete and ready for roofing. J. Finished house looks like any other type of construction. K. Blocked-in area for basement entrance.*

brands are Liquid Nail, PL 200, and PL 400. Some of these can dissolve the foam, but if they are applied in a thin layer, the amount of foam lost is usually insignificant. Look for an adhesive that is "compatible with polystyrene." Figure 21-42 shows an industrial foam gun.

Rebar is often precut to length and prebent, but even if it is, workers generally have to process a few bars in the field. Most ICF systems also have cradles that hold the bars in place for the pour, but a few bars need to be wired to one another or to ties to keep them in position.

L

M

N

Fig. 21-40 *L. Finished house with foam foundation and basement to be finished with stucco. This will protect the foam blocks. M. Another foam-concrete home with a stucco finish. N. A smaller foam-concrete home with conventional appearance.*

It is possible to bend rebar with whatever tools are handy and cut it with a hacksaw; however, if you process large quantities, you might prefer buying or renting a cutter-bender. A large manual cutter-bender is pictured in Fig. 21-43. It makes the job faster and

Fig. 21-41 *A thermal cutter. Note the white line that is part of the cutting device. It is attached to the transformer on the left of the framework. (PCC.)*

Fig. 21-42 *An industrial foam gun can be used as a glue gun. (PCC.)*

easier. Cutters-benders are available at steel-supply, concrete-supply, and masonry-supply houses. Almost any steel wire can hold rebar in place, but most efficient are rolls of precut wire, as shown in Fig. 21-44. These wires make the job faster and easier. Wire coils and belt-mounted coil holders are both sold by suppliers of concrete products, masonry, and steel.

Pouring Concrete

Concrete is best poured at a more controlled rate into ICFs than it is into conventional forms. An ordinary chute can be used for foundation walls (basement or

TABLE 21-2 *Useful Tools and Materials*

Operation or Class of Material	
Tool or Material	**Comments**
Cutting and shaving foam	
Drywall or keyhole saw	For small cuts, holes, and curved cuts.
PVC or mitre saw	For small, straight cuts and shaving edges.
Coarse sandpaper or rasp	For shaving edges.
Bow saw or garden pruner	For faster straight cuts.
Circular saw	For fast, precise, straight cuts. For cutting units with steel ties, reverse the blade or use a metal-cutting blade.
Reciprocating saw	For fast cuts, especially in place.
Thermal cutter	For fast, very precise cuts on a bench. Not suitable for steel ties or grid panel units.
Chain saw	For fast cuts of grid panel units.
Lifting units	
Forklift, manual lift, or boom or crane truck	For carrying large grid panel units and setting them in place. For upper stories, a truck is necessary.
Gluing and tying units	
Wood glue, construction adhesive, or adhesive foam	
Small-gage wire	For connecting units of flat panel systems.
Bending, cutting, and wiring rebar	
Cutter-bender	
Small-gage wire or precut tie wire or wire spool	
Filling and sealing formwork	
Adhesive foam	
Placing concrete	
Chute	For below-grade pours.
Line pump	Use a 2-inch hose.
Boom pump	Use two "S" couplings and reduce the hose down to a 2-inch diameter.
Evening concrete	
Mason's trowel	
Dampproofing walls below grade	
Nonsolvent-based dampproofer or nonheat-sealed membrane product	
Surface cutting foam	
Utility knife, router, or hot knife	Heavier utility knives work better. Use a router with a half-inch drive for deep cutting.
Fastening to the wall	
Galvanized nails, ringed nails, and drywall screws	For attaching items to fastening surfaces. Use screws only for steel fastening surfaces.
Adhesives	For light and medium connections to foam.
Insulation nails and screws	For holding lumber inside formwork.
J-bolt or steel strap	For heavy structural connections.
Duplex nails	For medium connections to lumber.
Small-gage wire	For connecting to steel mesh for stucco.
Concrete nails or screw anchors	For medium connections to lumber after the pour.
Flattening foam	
Coarse sandpaper or rasp	For removing small high spots.
Thermal cutter	For removing large bulges.
Foam	
Expanded polystyrene or extruded polystyrene	Consider foam with insect-repellent additives
Concrete	
Midrange plasticizer or superplasticizer	For increasing the flow of concrete without decreasing its strength. Can also be accomplished by changing proportions of the other ingredients.
Stucco	
Portland cement stucco or polymer-based stucco	

Fig. 21-43 *This is a rebar cutter-bender.* (PCC.)

Fig. 21-45 *Line pump for concrete.* (PCC.)

Fig. 21-44 *Roll of precut tie wires.* (PCC.)

stem); this is the least expensive option because it comes free with the concrete truck. Precise control is more difficult with a chute. You must pour more slowly than with conventional forms, and you must move the chute and truck frequently to avoid overloading any one section of the formwork.

The smaller line pump pushes concrete through a hose that lies on the ground (Fig. 21-45). The crew holds the end of the hose over the formwork to drop concrete inside. If possible, use a 2-inch hose. One or two workers can handle it, and it generally can be run at full speed without danger. If only a 3-inch hose is available, you can use it, but pump slowly until you learn how much pressure the forms can take.

Boom pumps are mounted on a truck that also holds a pneumatically operated arm (the boom). The hose from the pump causes the concrete to move along the length of the boom and then hang loose from the end

(Fig. 21-46). By moving the boom, the truck's operator can dangle the hose wherever the crew calls for it. One worker holds the free end to position it over the formwork cavities. The standard hose diameter is 4 inches.

You will need to have the hose diameter reduced to 2 or 3 inches with tapered steel tubes called *reducers*. The narrower diameter slows down and smoothes out the flow of the concrete. Figure 21-47 shows how reducers are arranged on a boom with hoses. Also, ask for two 90-degree elbow fittings on the end of the hose assembly, as seen in the figure. These form an S in the line that further breaks the fall of the concrete. The concrete can be leveled off on top of the form by using a mason's trowel.

CONCRETE BLOCK

The second most popular residential framing method is concrete block. Making up less than 5 percent of the residential framing market, concrete block is another traditional method that has been around for a long time. It is used mainly when homeowners are interested in a framing method that is resistant to inclement weather and termites. This type of construction is much more costly and time-consuming. Because the techniques associated with masonry and concrete work are numerous and detailed, only the major steps are reviewed here:

1. Snap a chalk line on the footing or foundation where the concrete block should be placed.
2. Spread mortar 1 or 2 inches thick on the footing or foundation and 1 inch or so wider and longer than the block.
3. Place a concrete block on each corner of the wall. Make sure that both blocks are level and straight.

Fig. 21-46 *Boom pump for pouring concrete.* (PCC.)

Boom

5" to 4" Reducer

4" Hose

4" to 3" Reducer
(or 4" to 2")

Two 90-degree elbows

Fig. 21-47 *Diagram of boom pump fittings that are suitable for pouring ICF walls.* (PCC.)

4. Stretch a string line tightly between the front top edge of both blocks. This can be done by wrapping a string around a brick and placing it on top of the concrete block.

5. Place concrete blocks evenly for about seven blocks (if using 8- × 16-inch blocks) from the corner. Do this for both corners.

6. Continue to build up the corners by backing up each level of block (course) by half a block. When finished, one block should be on the top of each corner. Each corner is called a *lead*.

7. Place concrete blocks between each lead using a string to keep the top and front edge of the block straight and level. Mortar should be placed on one end of each block. A ⅜-inch-thick layer of mortar is common because the actual size of most 8- × 16-inch concrete block is 7⅝ × 15⅝ inches.

8. Allowances should be made for rough openings of windows and doors. Lintels or metal ledges are used to support block over door and window openings.

9. The top course of concrete block is filled with mortar, and anchor bolts are inserted at certain intervals according to building code requirements.

10. Mortar joints are usually smoothed (tooled) with a round bar, square bar, or S-shaped tool.

CHAPTER 21 STUDY QUESTIONS

1. Why are many builders beginning to use alternative methods for framing a house?

2. What are the main advantages and disadvantages of steel framing?

3. List the two types of steel framing, and describe their differences.

4. How are anchor bolts attached to a concrete foundation using the red-iron framing method?

5. How is electrical wiring run through a steel-frame home?

6. What does ICF mean, and how is it constructed?

7. List at least two advantages and disadvantages to ICFs.

8. How is sheathing installed in an ICF home?

9. How is electrical wiring installed in an ICF home?

10. Define the following terms: *lead*, *course*, *lintel*, and *tooled*.

11. What are three ways to build a steel-framed wall?

12. List two advantages of a panelized steel-framed wall.

13. What type of fastener is used to assemble steel and stud walls?

14. What gauge steel studs are used for interior and exterior walls?

15. What gauge steel is used for studs of weight-bearing walls?

Permanent Wood Foundation System

THE PERMANENT WOOD FOUNDATION (PWF) IS AN innovative building system that saves builders time. It creates comfortable, warm living areas that enhance a home's value. A PWF system consists of a load-bearing frame made of pressure-treated lumber and sheathed with pressure-treated plywood. This type of construction is suitable for crawl spaces as well as for split-entry or full basements (Fig. 22-1).

Footing plate may extend beyond foundation to avoid cutting treated lumber

Optional stud for panelized foundation wall sections

Footing plate

Optional stud for crawl space foundation wall sections

Plan View Typical Corner Detail

Space panels ⅛" and caulk along vertical joints

Field applied 2x top plate (stagger joints in top plates)

1½" plywood lap to cover field applied top plate

2x top plate end-nailed to studs

APA Rated Plywood Sheathing

Optional 2x4 blocking between studs

Caulk along horizontal joint

2x studs

Plywood face grain across studs (shown) or parallel to studs

2x bottom plate end-nailed to studs

2x footing plate (stagger joints in bottom plate and footing plate)

Fig. 22-1A *Foundation panel construction details.*

FLOOR JOIST

FLASHING
PLYWOOD MAY OVERLAP FIELD
APPLIED TOP PLATE FOR SHEAR TRANSFER

8" MIN.

FIELD APPLIED 2 × ___ TOP PLATE
2 × ___ TOP PLATE
CAULK
FINISH GRADE SLOPE ¹/₂" PER FOOT
MIN. 6' FROM WALL
2 × ___ STUD WALL
INSULATION AS APPROPRIATE
1 × ___ OR PLYWOOD STRIP PROTECTING
TOP OF POLYETHYLENE FILM

WARM SIDE
VAPOR BARRIER

PLYWOOD
ASPHALT OR POLYETHYLENE FILM STRIPS
3" OR 4" CONC. SLAB
4" GRAVEL OR CRUSHED STONE FILL
1 × ___ SCREED BOARD (OPTIONAL)
POLYETHYLENE FILM
___ × ___ BOTTOM PLATE
2 × ___ FOOTING PLATE

BELOW FROST LINE

OPTIONAL INTERIOR
FINISH

BACKFILL W/CRUSHED STONE
OR GRAVEL (SEE TEXT FOR HGT.)

³/₄ d

d
2d

WARM SIDE
VAPOR BARRIER

Fig. 22-1B *How the basement wall is placed in reference to the base and floor joists.*

Builders and homeowners have chosen the PWF system for good reasons, including design flexibility, faster construction, and larger, more comfortable living space.

The *International Building Code* recognizes both the one- and two-family dwellings with a PWF foundation. The code recognizes this type of construction, as do many local and national organizations.

This system is a proven one that has been demonstrated by long-term ground tests conducted by the U.S. Department of Agriculture's Forest Service. In these tests, pressure-treated wood withstood severe decay and termite conditions over decades of exposure. The walls are designed to resist and distribute the earth, wind, and seismic loads and stresses that may crack other types of foundations. The PWF is accepted by the major building codes, by federal agencies, and by lending, home warranty, and fire insurance institutions. This building system has been proven by years of success in more than 300,000 homes and other structures throughout the United States. As with conventional wood frame walls, the wood foundation is adaptable to virtually any design. The PWF fits a variety of floor plans. It can be used for both level and sloping sites (Fig. 22-2).

Virtually any home design—traditional or contemporary—can benefit from the many advantages of a PWF. Room additions are also simplified.

Panel Foundations

Panel foundations can be site built or made in a shop. Third-party inspection may be required; look for grade stamps, treatment stamps, stud spacing, insulation, nailing, depth of saw cuts, and plywood requirements, which are some of the items that a third-party inspector will note. All PWF's must be designed and installed in accordance with current building code standards.

Wood foundations are easy to build. This is only true if one is building from an accurately designed plan. When such a plan is incorrect, or if something is left out of the plan, or if a design is made using a guide manual, keep in mind that these are not design manuals. Major mistakes can be made during the construction process. These errors cause problems for the owner, builder, and building department.

The wood used in the wood foundation should be waterproofed by using chemically saturated wood to prevent rot and attack by various insects, termites being the most bothersome. Waterborne preservatives are

Fig. 22-1C *Applying glue to the floor joists before laying the plywood subfloor.*

applied to the plywood and lumber in a high-pressure chamber, where the preservatives are driven deep into the wood fiber and permanently bonded in the cellular structure of the wood. After treatment, the wood is dried to normal moisture levels. Pressure-preserved wood retains all its stiffness, strength, and workability and has the added value of permanent protection against mold, decay, and insect infestation.

Building Materials

High-quality building materials are used for the PWF. Plywood must meet the strict requirements of U.S. Product Standard PS 1-83 for construction and industrial plywood. Lumber must be of a species for which allowable unit stresses are given in the *National Design Specification for Wood Construction* and must be grade-marked by an approved inspection agency. Both plywood and lumber must be pressure-treated in accordance with the penetration, retention, and drying requirements of the American Wood Preservers Bureau AWPB-FDN Standard. Required preservative retention is 50 percent higher than codes require for normal ground-contact application.

The PWF is the result of extensive design and engineering analyses by the U.S. Department of Agriculture Forest Service, the National Forest Products

Window rough opening

Field applied 2x top plate*

PWF stud wall

Insulation

4-mil poly. sheeting

Door rough opening

Pipe drain

Interior finish

6-mil poly. sheeting

Sump

8" min. between fin. grade and sheathing

Gravel base

Concrete slab

Plywood protection strip

APA Rated Plywood Sheathing

Plywood strip protecting top of polyethylene sheeting (12" nom. width)

6-mil polyethylene sheeting

6-mil polyethylene sheeting or 30 lb roofing felt

2x footing plate

Gravel footing

* Not required to be treated if exterior finish grade is 8″ or more below bottom of plate.

Fig. 22-2 *This is a typical wood foundation basement.*

Association, and the National Association of Home Builders Research Foundation.

Additional in-ground structural testing has been conducted by the American Plywood Association. Durability of the system was demonstrated in long-term tests conducted by the Forest Service. Pressure-preservative-treated wood in these tests has withstood severe decay and termite conditions with treatment levels lower than required by the AWPB standard for PWFs. As a result the PWF basement has all the comforts of an above-ground room. Real-wood construction lends a feeling of warmth—not the musty, damp feeling usually associated with basements. The wood

basement is an integral part of the home. This type of basement allows for every square foot of it to become living space—below ground and above.

Energy Considerations

The wood framing of the PWF makes it easy to install thick, economical batt-type insulation. This means less heat loss through the walls and greater long-term energy savings. The cost of installing insulation is less, too, because the wood-frame walls are already in place. Consider this comparison: In a concrete foundation, ¾- or 1½-inch foam sheathing typically is installed for insulation. This gives an R rating of R-3 to R-6. In the wood foundation, it's easy to fit 3½-inch-thick insulation between the 2 × 4 studs, producing an R-11 rating. If 5½-inch insulation is installed between 2 × 6 studs, the energy rating increases to R-19.

Finishing

The PWF is easier to finish inside than conventional foundations because the wooden studs are already in place. Plumbing and wiring are simplified and concealed. No furring strips are needed for installing insulation and gypsum board or paneling because it's so easy; many PWF homeowners do the finishing themselves. In this way, it can be done according to their own tastes and imagination and at less cost.

Adding Living Space

PWFs provide more actual living space in basements than concrete or masonry houses of the same dimensions. This is so because the wood foundation walls don't need to be as thick as concrete or masonry walls. Insulation takes less space in the PWF because it fits into the cavities of the wood-frame wall. This means that extra furring strips aren't necessary. This, in turn, produces more living area (Fig. 22-3).

Remodeling

PWFs are easy to remodel or modify. You can cut out window or door openings or add a whole room. Simple finishing touches such as pictures and shelves can be hung without special tools or fasteners. When adding onto your home, you won't have to worry about providing access for a concrete truck. PWFs have even been used to replace old and weakened mason and concrete foundations—with considerable savings in time and expense.

Flexibility

Flexibility is one of the primary considerations when selecting a PWF. The system can be used in both single- and multistory structures and for both site-built and manufactured homes. The system can be adapted to almost any home design and site plan. It can be engineered for a variety of soil conditions, even those with high water tables. It is as easy to install for crawl-space designs as for split-entry and full-basement buildings.

Pressure-Treated Wood Concerns and Considerations

When most wood is exposed to the elements, excessive moisture, or contact with the ground, it will decompose. This is so because four conditions are required for the decay and insect attack to occur:

- Moisture
- Favorable temperature (approximately 50 to 90°F)
- Oxygen
- A source of food (wood fiber)

Fig. 22-3 *A cutaway showing the comfortable room addition possible by using a PWF for a house.*

If any of these conditions is removed, infestation and decomposition will not occur.

Southern pine has long been a preferred species when pressure treatment with preservatives is required. This is so because of the ease with which it can be treated. The unique cellular structure of southern pine permits deep, uniform penetration of preservative chemicals, rendering the wood useless as a food source for fungi, termites, and microorganisms. Some 85 percent of all pressure-treated wood is southern pine. See Table 22-1 for softwood lumber and plywood standards.

Why design and build with pressure-treated wood? Figure 22-4 shows, by zones, the level of wood deterioration throughout the United States. As shown, deterioration zones from moderate to severe cover most of the country.

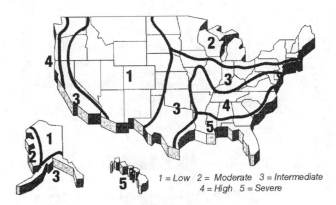

1 = Low 2 = Moderate 3 = Intermediate
4 = High 5 = Severe

Fig. 22-4 *Wood deterioration zones in the United States.*

Modern science has developed preservative treatments that are odorless and colorless and leave the wood preserved and dry to the touch. Treatment with chemical preservatives protects wood that is exposed to the elements, is in contact with the ground, or is used in high-humidity areas.

However, beware—all treated wood is not considered equal. Most wood species do not readily accept chemical preservatives and must first be incised or perforated by a series of small slits along the grain of the wood. Incising allows sufficient penetration of the preservatives to meet American Wood Preserver's Association standards. Southern pine doesn't require incising.

In addition, the use of treated southern pine provides no measurable risk to humans, animals, or plant or other life. Scientific research studies have shown the following:

- Preservative-treated wood products last longer than alternative materials, conserving a renewable natural resource.

- Wood preservatives do not aggressively leach into the ground or waterways or drinking water or adversely affect marine life.

- Preservative-treated wood products have been tested extensively and have proven to be more durable than alternative products that require more energy to produce and also may be aesthetically unacceptable to consumers.

How long does pressure-treated wood last? Currently available research shows that wood that has been properly treated and installed for its intended use can be expected to last for many decades.

Ongoing tests are sponsored and monitored by the U.S. Forest Services' Forest Products Laboratory. Test stakes of treated wood have been buried in the ground at various locations stretching from the Mississippi Delta to the Canadian border. Data analysis indicate that CCA-treated southern pine stakes, still in place since 1938, have shown no failures at chemical retention levels of 0.29 pounds of preservatives per foot of wood or higher.

Types of Wood Preservative

Several waterborne preservatives are used commonly, including

- Chromated copper arsenate (CCA)
- Alkaline copper quat (ACQ)
- Copper azole (CA)
- Sodium borate (SBX)

As of December 31, 2003, CCA has been withdrawn for most residential consumer-use treated-lumber applications. CCA treatment will continue to be allowed for certain industrial, agricultural, foundation, and marine applications. More information on alternative preservatives and links to preservative manufacturers can be found on the Southern Pine Council Web site.

In consultation with the Environmental Protection Agency, manufacturers have made a transition to alternative wood preservatives for the residential and outdoor market. Leading preservative manufacturers include

- ArchWood Protection, Inc.
- Chemical Specialties, Inc.
- Osmose, Inc.

These manufacturers have amended their respective registrations for CCA, limiting the use of CCA to approved industrial and commercial applications.

TABLE 22-1 *AWPA Standards for Softwood Lumber and Plywood*

		USE CATEGORY STANDARDS				COMMODITY STANDARDS

SELECTION GUIDE

Use Category	Service Conditions	Use Environment	Common Agents of Deterioration	Typical Applications
UC1	Interior construction Above Ground Dry	Continuously protected from weather or other sources of moisture	Insects only	Interior construction and furnishings
UC2	Interior construction Above Ground Damp	Protected from weather, but may be subject to sources of moisture	Decay fungi and insects	Interior construction
UC3A	Exterior construction Above Ground Coated & rapid water runoff	Exposed to all weather cycles, not exposed to prolonged wetting	Decay fungi and insects	Coated millwork, siding and trim
UC3B	Exterior construction Above Ground Uncoated or poor water runoff	Exposed to all weather cycles including prolonged wetting	Decay fungi and insects	Decking, deck joists, railings, fence pickets, sill plates, uncoated millwork
UC4A	Ground contact or fresh water Non-critical components	Exposed to all weather cycles, normal exposure conditions	Decay fungi and insects	Fence, deck, and guardrail posts, crossties & utility poles (low decay areas)
UC4B	Ground contact or fresh water Critical components or difficult replacement	Exposed to all weather cycles, high decay potential including salt water splash	Decay fungi and insects with increased potential for biodeterioration	Permanent wood foundations, building poles, horticultural posts, crossties & utility poles (high decay areas)
UC4C	Ground contact or fresh water Critical structural components	Exposed to all weather cycles, severe environments, extreme decay potential	Decay fungi and insects with extreme potential for biodeterioration	Land & fresh water piling, foundation piling, crossties & utility poles (severe decay areas)
UC5A	Salt or brackish water and adjacent mud zone Northern waters	Continuous marine exposure (salt water)	Salt water organisms	Piling, bulkheads, bracing
UC5B	Salt or brackish water and adjacent mud zone New Jersey to Georgia, South of San Francisco	Continuous marine exposure (salt water)	Salt water organisms including Limnoria tripunctata	Piling, bulkheads, bracing
UC5C	Salt or brackish water and adjacent mud zone South of Georgia and Gulf Coast	Continuous marine exposure (salt water)	Salt water organisms including Martesia, Sphaeroma	Piling, bulkheads, bracing
UCFA	Fire protection as required by codes Above Ground Interior construction	Continuously protected from weather or other sources of moisture	Fire	Roof sheathing, roof trusses, studs, joists, paneling
UCFB	Fire protection as required by codes Above Ground Exterior construction	Subject to wetting	Fire	Vertical exterior walls, inclined roof surfaces or other construction which allows water to quickly drain

COMMODITY STANDARDS

C1 – All Timber Products – Preservative Treatment by Pressure Processes

C2 – Lumber, Timber, Bridge Ties and Mine Ties

C3 – Piles

C4 – Poles

C5 – Fence Posts

C6 – Crossties and Switch Ties

C9 – Plywood

C11 – Wood Blocks for Floors and Platforms

C14 – Wood for Highway Construction

C15 – Wood for Commercial – Residential Construction

C16 – Wood Used on Farms

C17 – Playground Equipment

C18 – Marine Construction

C20 – Structural Lumber: Fire Retardant Treatment by Pressure Processes

C22 – Permanent Wood Foundations

C23 – Round Poles and Posts used in Building Construction

C24 – Sawn Timber used to Support Residential & Commercial Structures

C25 – Sawn Crossarms

C27 – Plywood Fire Retardant Treatment by Pressure Processes

C28 – Glued-Laminated Members

C29 – Lumber and Plywood to be used for the Harvesting, Storage and Transportation of Food Stuffs

C30 – Lumber, Timbers and Plywood for CoolingTowers

C31 – Lumber used Out of Contact with the Ground and Continuously Protected from Liquid Water

C32 – Glue Laminated Poles

C33 – Structural Composite Lumber

C34 – Shakes & Shingles

NOTE: The Standards are not in consecutive order, as some have been deleted over time. Only those pertaining to Southern Pine have been included here.

SPECIAL NOTE: The American Wood-Preservers' Association (AWPA) developed the Use Category System (UCS) to provide a simple way to use AWPA Standards. The UCS defines a series of different exposures for treated wood products. Each exposure has a different degree of biodegradation hazard and/or product service life expectation. All treated wood commodities can be placed into one of the Use Categories (see table above). There are five Use Categories based on exposures and expected product performance, ranging from weather protected to salt water marine. A separate Use Category is provided for fire retardant applications. The Use Category number relates to the hazard associated with certain use environments, while the letter following the number (if present) relates to the risk. In general, as the Use Category number increases, there is a consequent increase in the required preservative retention. The depth of penetration requirements may also increase.

The UCS was developed as a format change for the AWPA Commodity (or "C") Standards and is not intended to make substantive technical changes to those Standards. Inconsistencies in the UCS and the C Standards may exist due to the enormity of the task to construct the UCS. Therefore, in the event of conflicts between the UCS and the C Standards during the interim phase-in period, the C Standards will govern until such time as the C Standards may be deleted from a future edition of the AWPA Book of Standards. For user convenience, treated wood specification guidelines in this publication include both the appropriate Use Category Standard and a cross reference to its corresponding Commodity Standard.

To specify a treated wood commodity using the UCS, use the AWPA Book of Standards. The user should first find the appropriate Use Category for the expected service conditions in Section 2 "Service Conditions for Use Category Designations", and a definite application in Section 3 "Guide to Treated Wood End Uses". The user may then refer to the appropriate product or application in Section 7 "Commodity Specification." The Table of Contents for Commodity Specifications using AWPA's Use Category System is shown to the right.

COMMODITY SPECIFICATIONS
USE CATEGORY SYSTEM

Sawn Products	A
Permanent Wood Foundations (PWF)	A
Posts	B
Playground Material	B
Round Building Poles	B
Crossties	C
Utility Poles	D
Round Timber Piling	E
Wood Composites	F
Marine Applications (Salt Water)	G
Fire Retardants	H
Non-Pressure Applications	I

Constructing the PWF

PWF construction is similar to wood-frame exterior wall construction with some exceptions. Because PWF walls are used in below-grade applications, all lumber and plywood is pressure-treated with preservatives for decay and termite resistance. These have already been discussed. Other differences include the use of stainless steel nails, an offset footing plate, and framing anchors to connect foundation studs and floor joists to the top plates of the foundation walls in high-backfill conditions.

As with conventional wood-frame walls, the wood foundation is adaptable to virtually any design. It fits a variety of floor plans. This type of foundation can be used for both level and sloping sites.

Radon

In certain localities where emission of radon gas from the soil or groundwater is prevalent, a plastic pipe and tee can be installed through the basement floor for basement-type PWFs. For crawl space PWFs, a perforated plastic pipe can be installed on the ground inside the crawl space beneath the vapor retarder. In both applications, the pipe is connected to a vent pipe and exhaust fan to depressurize the soil under the basement floor or crawl space vapor retarder, removing radon gas from the soil under and around the building. If a sump is used, the sump cover should be sealed and connected to the vent pipe and exhaust fan to remove radon gas from the sump pit.

Advantages of the PWF

Builders and home buyers across the country are choosing the PWF. Here are just a few of the features that are making PWFs increasingly popular.

Flexibility PWFs can he used in a variety of building types and sizes, including both single- and multi-story houses, condominiums and apartments, and site-built and manufactured houses. PWFs are suitable for crawl-space, split-entry, and full-basement designs. Remodeling contractors have found the PWF ideal for room additions, especially where site access is limited.

Offices or other commercial and nonresidential buildings also can be built on a PWF. The PWF can be engineered for almost any large or complex building design or to satisfy special site constraints and can be adapted for a variety of soil conditions, including low-bearing-capacity soils, expansive soils, and high water tables. The system even can be adapted for such uses as retaining walls and swimming pools.

Scheduling The builder's or subcontractor's carpentry crews install the PWF, reducing the need for scheduling other trades. The PWF can be installed under nearly any weather conditions, even below freezing, so the building season is extended. On remote sites, high delivery costs and delays for concrete are eliminated. There's no need to wait for setting and stripping concrete forming or allowing concrete to cure.

The PWF is easily installed by a small crew, often in less than a day or even in just a few hours. As soon as the foundation is framed and sheathed, construction of floors and walls can proceed. Shorter construction time means savings in interim construction financing and greater productivity.

Comfortable basement Several features of the PWF make a home attractive to buyers. First, there is comfort. PWF basements have all the livability of above-ground rooms. Wood construction lends a feeling of warmth. A damp feeling is usually associated with masonry basements. PWFs incorporate superior drainage features that prevent moisture problems typical of ordinary foundations. The result is a warm, dry, below-grade living space.

Second, the wood-framed walls of the PWF make it easy to install thick, economical insulation. This means less heat loss through the foundation wall and greater long-term savings. The cost of installing insulation is less because the wood-framed walls are already in place.

Consider this comparison: In a concrete foundation, ¾- or 1½-inch foam sheathing typically is installed for insulation. This gives an R rating of R-3 to R-6. In a wood foundation, it's easy to fit 3½-inch-thick insulation between PWF studs, producing an R-11 to R-15 rating. If 5½-inch insulation is installed between 2 × 6 or 2 × 8 PWF studs, the R rating increases to R-19 or R-21. The National Energy Policy Act mandates that the basement of a new home must be insulated properly. Several states already have adopted this code.

The economical answer to meeting state energy code requirements begins with a PWF. Research has found that to build an 8-inch basement wall with an insulation value of R-19, the concrete costs some 30 percent more than the PWF. Cement block can cost up to 62 percent more. Energy savings and an incomparably dry basement are possible. Comfortable living areas below grade are also possible.

Finishing Another advantage of PWFs is the ease of finishing. The easily worked-with wood studs are already in place. This means that plumbing, wiring, and interior wall installation are simplified. Because it's so easy, many PWF homebuyers elect to do the finishing themselves according to their own tastes and imagination— often at less cost.

More space PWFs also can mean added living space. Wood foundation walls need not be as thick as comparable concrete or masonry walls. Less space is needed for insulation, too, because it fits into the cavities of the wood-framed wall, and extra furring strips or wall studs aren't necessary. PWFs are easy to remodel or modify. Window or door openings can be cut out or whole rooms can be added. Additional structural engineering may be required for certain remodeling projects.

Radon gas PWF systems, both new construction and retrofits, have definite advantages for radon gas resistance. The gravel layer beneath the basement floor serves as a collection system for soil gas. The gas is easily vented to the outside.

Soil Conditions

The type of soil and general grading conditions at the building site are factors in determining foundation construction details such as footing design and backfill and drainage provisions. Soils are classified by their composition and how they drain. Table 22-2 lists common soil types and their properties. Soil classifications for most areas are listed in the standard series of soil surveys published by the U.S. Department of Agriculture's Soil Conservation Service.

PWFs may be built in group I, II, or III soils. In poorly drained group III soils, granular fill under the slab for basement-type foundations must be at least 8 inches deep as opposed to the 4-inch minimum for group I and II soils.

In such soil conditions, it may be more practical to build an above-grade crawl-space foundation/floor system, especially for sites having a high water table or where extreme amounts of rain often fall in short periods. Regardless of soil type, above-grade crawl-space foundation/floor systems have the cost benefit of minimum excavation and backfill.

Group IV soils generally are unsatisfactory for wood foundations, that is, unless special measures are taken.

For building sites in regions where expansive clay soils in groups II, III, or IV occur, a licensed soils engineer should be consulted to determine modifications required for foundation footings, drainage, soil moisture control, and backfill around the foundation. In such cases, special design considerations and construction details may be needed to avoid soil expansion or shrinkage that otherwise might affect foundation and floor performance.

For basement-type foundations, a sump draining to daylight or into a storm sewer or other stormwater drainage system is recommended for all soil groups. In addition, for all types of foundations in all soil groups, the ground surface around the foundation should be graded to slope ½ inch per foot away from the structure. The backfill should be free of organic material, voids, or chunks of clay. It should be compacted and no more permeable than the surrounding soil.

Site Preparation

Site clearing and excavation methods for a PWF are the same as for conventional foundation systems. Organic materials, including tree stumps and other vegetation, should be removed and topsoil separated from excavated earth, which may be used later for backfilling or grading.

After clearing the site, it's wise to use a plot plan to locate foundation footings and trenches for plumbing, sewer, gas, and electrical lines and drainage trenches. Excavation for foundation footings, plumbing and other services, and drainage trenches must be completed before soil treatment for termite protection, if required. Otherwise, retreatment is necessary.

After utility trenches are dug to the desired level, they can be lined with fine gravel or sand before pipes and conduit are set in place. Then the trenches are filled the rest of the way with gravel, coarse sand, or crushed rock (Fig. 22-5).

Footings and Backfill

Granular materials are recommended footings under foundation walls and for fill under basement slabs or treated-wood basement floors. They are also good for a portion of the backfill to provide an optimal drainage system to keep the under-floor area and foundation walls dry. The granular material may be crushed stone, gravel, or sand. They must be clean and free of silt, clay, and organic material. The size limitations are

- Maximum of ½ inch for crushed stone
- Maximum of ¾ inch for gravel
- Minimum of ¹⁄₁₆ inch for sand

Continuous poured concrete may be used for the footings beneath foundation walls. If a concrete footing is used, it should be placed on gravel to maintain

TABLE 22-2 *Types of Soils and Related Design Properties*

Soil Group	Unified Soil Classification Symbol	Soil Description	Allowable Bearing in Pounds per Square Foot with Medium Compaction or Stiffness[1]	Drainage Characteristic[2]	Frost Heave Potential	Volume Change Potential Expansion[3]
Group I Excellent	GS	Well-graded gravels, gravel-sand mixtures, little or no fines.	8000	Good	Low	Low
	GP	Poorly graded gravels or gravel-sand mixtures, little or no fines.	8000	Good	Low	Low
	SW	Well-graded sands, gravelly sands, little or no fines.	6000	Good	Low	Low
	SP	Poorly graded sands or gravelly sands, little or no fines.	5000	Good	Low	Low
	GM	Silty gravels, gravel-sand-silt mixtures.	4000	Medium	Medium	Low
	SM	Silty sand, sand-silt mixtures.	4000	Medium	Medium	Low
Group II Fair to Good	GC	Clayey gravels, gravel-sand-clay mixtures.	4000	Medium	Medium	Low
	SC	Clayey sands, sand-clay mixture.	4000	Medium	Medium	Low
	ML	Inorganic silts and very fine sands, rock flour, silty or clayey fine sands or clayey silts with slight plasticity	2000	Medium	High	Low
	CL	Inorganic clays of low to medium plasticity, gravelly clays, sandy clays, silty clays, lean clays.	2000	Medium	Medium	Medum
Group III Poor	CH	Inorganic clays of high plasticity, fat clays.	2000	Poor	Medium	High
	MH	Inorganic silts, micaceous or diatomaceous fine sandy or silty soils, elastic silts.	2000	Poor	High	High
Group IV Unsatisfactory	OL	Organic silts and organic silty clays of low plasticity.	400	Poor	Medium	Medium
	OH	Organic clays of medium to high plasticity, organic silts	–0–	Unsatisfactory	Medium	High
	Pt	Peat and other highly organic soils.	–0–	Unsatisfactory	Medium	High

[1]Allowable bearing value may be increased 25 percent for very compact, coarse-grained gravelly or sandy soils or very stiff fine-grained clayey or silty soils. Allowable bearing value may be decreased 25 percent for loose, coarse-gained gravelly or sandy soils, or soft, fine-grained clayey or silty soils.
[2] The percolation rate for good drainage is over 4 inches per hour, medium drainage is 2 to 4 inches per hour, and poor is less than 2 inches per hour.
[3] For expansive soils, contact local soils engineer for verification of design assumptions.
[4] Dangerous expansion might occur if these soil types are dry but subject to future wetting.

continuity of the drainage system. Otherwise, drains through the concrete footing must be provided.

All types of foundations need footings to be placed on undisturbed soil. The footing excavation should extend below the frost line. The footing trench depth and width, as with conventional systems, depend on the loads to be carried by the foundation (Fig. 22-6).

Excavations must be wide and deep enough into undisturbed soil that the footings will be centered under the foundation walls. Use of granular or concrete footings distributes vertical loads from the structure and foundation walls to the soil. Footings are required under perimeter and interior load-bearing walls.

Site Drainage

Site drainage is an important feature in keeping any type of foundation dry and trouble-free. Drainage systems have been developed for wood foundations to keep crawl spaces and basements dry under virtually any condition. Granular footings and backfill are key elements in the PWF drainage system. They provide an unobstructed path for the water to flow away from the foundation or into a sump for a basement house. This prevents a buildup of pressure against the foundation and helps to avoid leaks (Fig. 22-7).

Following are several methods of providing drainage around excavated foundations. In permeable

Fig. 22-5 *A. Typical excavation for footings and under-floor tile, pipes, conduit, and sumps. B. Above-grade crawl-space foundation for high-water-table or soil-drainage problems.*

Fig. 22-6 *Footing drainage trench to daylight.*

group I soils, such as gravel, sand, silty gravels, and gravel-silt-sand mixtures with a percolation rate of more than 4 inches per hour, the gravel footing of the PWF provides the drainage trench for sites with good surface water runoff (Figs. 22-8 and 22-9).

For continuous concrete footings, place gravel under and around the outside of the footing, at least 12 inches wide by 12 inches deep. Water will collect in the gravel beneath the footing and then can be drained away from the foundation. Cover the gravel with 6-mil polyethylene sheeting or wrap gravel with water-permeable filter fabric to prevent soil from washing into the footing (Fig. 22-10).

Group II soils such as gravel-sand-clay mixtures, clay, gravels or sands, inorganic silts, and fine sand with medium to poor percolation characteristics re-

Fig. 22-7 *Footing drainage trench to sump or storm sewer.*

3/4" or larger gravel or crushed stone (must not plug perforated pipe)

Cover

Concrete slab

Polyethylene sheeting

2'-0" min.

Perforated 4" diameter plastic or fiber drain pipe

Excavated sump pit

90° ell-4" diameter plastic or fiber pipe (P-trap optional)

2'-6" min.

NOTE: Vertical pipe may be extended through slab with a clean-out plug in floor.

Fig. 22-8 *Sump for medium wet-drained soils.*

quire a trench sloping away from the gravel foundation footing. If it is a sloping site, it may be practical to dig a drainage trench to daylight, where the site slop intersects the drainage trench (Fig. 22-13). Place about 6 inches of gravel in the trench, and cover it with 6-mil polyethylene sheeting. An alternative is to place a drain tile or soil pipe into the trench from the gravel footing to the point where the trench emerges.

Foundation footings and basement sumps, when used, should be drained to a storm sewer, drainage swale, or to daylight. On level sites where direct drainage to daylight or a storm sewer is not practical, it may be necessary to dig drywells or sump pits at several locations around the outside of the foundation (Fig. 22-14). Rely on drywells only in areas of well-drained group I soils with high sand or gravel content. The bottom of the drywell should project into undisturbed, porous soil at a level above the highest seasonal groundwater table.

The sump pit, about 3 feet in diameter, should be at least 2 feet deeper than the base of the gravel footing. The top of the sump should be at a lower elevation than the footing. Connect the sump pit to the footing either with a gravel-filled trench or with a drainage tile or soil pipe.

In medium-drained soils, a gravel-filled sump should be sufficient to provide proper drainage. In poorly drained soils, either provide drainage from the sump pit to a storm sewer or, in extreme drainage problem cases, install a sump pump inside a prefabricated sump tile (see Fig. 22-14).

For crawl-space foundations, where the interior ground level is below the outside finish grade, granular drainage trenches or drain pipes are recommended for draining footings or perimeter drains by gravity to daylight, storm sewers, or other approved stormwater drainage system.

On sites where proper drainage may be expensive or troublesome, consider using an above-grade foundation/floor system, and make sure that the finish grade slopes away from the foundation.

In rainy climates, provide for drainage inside the foundation. This can be done by grading to a low spot on the ground inside such that the under-floor area will drain to it.

After the building is complete, make sure that the foundation and under-floor areas remain dry by providing for adequate drainage of stormwater. Most important is the use of gutters, downspouts, and splash blocks or drainpipes to direct water runoff away from the building. Also, slope adjacent porches or patios to drain away from the building.

Polyethylene sheeting

Concrete slab

Pressure treated laminated plywood manhole cover (two 23/32" pieces)

Precast concrete pipe, 24" dia. x 30" high OR 20"x20"x30" high treated plywood and lumber sump crock

2" min. gravel fill on all sides

90° ell - 4" diameter pipe (trap required when drained to sewer)

Fig. 22-9 *Sump for poorly drained soils.*

PWF stud wall

6-mil poly.
sheeting

2x bottom plate

Concrete slab

2x footer plate

NOTES:

(1) In basement PWF constructions, gravel or crushed
rock backfills half way up the foundation wall are
recommended and typical, as shown.

(2) In poorly drained soils, gravel backfill can be extended
up to just below grade line for optimum perimeter
drainage.

Gravel or
crushed rock

Fig. 22-10 *Basement PWF with perimeter gravel drainage.*

h_i (in.)	Max. h_o (in.)
6	17
8	23
10	28
12	33
14	37

Subfloor

Insulation

Sheathing or siding

Field applied 2x top plate

2x top plate

Floor joist

Plywood

Exterior finish grade —
slope 1/2" per foot for
minimum 6' from wall

8" min.

2x stud wall

18" min.

Polyethylene
sheeting

Polyethylene
sheet
(optional in
crawl space
construction)

h_o
(see table)

2x bottom plate

h_i (6" min.)

2x footing plate

3/4d

Below frost line

Gravel or crushed
stone footing

d

2d

Fig. 22-11 *Crawl-space foundation.*

716   Permanent Wood Foundation System

Fig. 22-12 *Crawl-space foundation on concrete footing.*

Labels in figure 22-12:
- Sheathing or siding
- Subfloor
- Floor joist
- PWF wall panel
- 8" min.
- Note: Inside fill may be required with high backfills.
- 18" min. to joists 12" min. to girders
- 9" min.
- Galvanized anchor bolts*(size and spacing as required)
- 7"
- Polyethylene sheeting
- Below frost line
- 12" (1 story) 15" (2 story)
- Concrete footing (refer to building code for reinforcing requirements)

Fig. 22-13 *Deep gravel footing for PWF.*

Labels in figure 22-13:
- Subfloor
- Sheathing or siding
- Floor joist
- PWF wall panel
- 8" min.
- 18" min. to joists 12" min. to girders
- Polyethylene sheeting
- 9" min.
- 6" min.
- 6-mil poly. sheeting
- d
- 2d
- Deep gravel footing, drained to sump or daylight
- Below frost line

Figure 22-14 labels (lower left):
- Field applied 2x top plate
- Floor joist
- Framing anchors for deep backfill
- 2x stud wall
- Required insulation
- Vapor retarder
- Interior finish
- Concrete slab
- Polyethylene sheeting
- 1x screed board (optional)
- Sheathing or exterior siding may overlap field applied top plate for shear transfer (Flashing not required if siding overlaps). See Note
- 8" min. between grade and 2x top plate
- 2x top plate*
- Caulk
- 1x or plywood strip protecting top of polyethylene sheeting (12" nom.)
- Finish grade slope ½" per foot min. 6' from wall
- Backfill**
- Asphalt or polyethylene strips
- Polyethylene sheeting
- 2x bottom plate
- 2x footing plate
- ¾d
- Below frost line
- Gravel, coarse sand, or crushed stone fill (4" for Group I and II soils, 6" for Group III)
- d
- 2d

* Not required to be treated if backfill is more than 8″ below bottom of plate. Typical for all details

** Backfill with crushed stone or gravel 12″ for Group I soils, and half the backfill height for Groups II and III soils.

NOTE: For daylight basement foundations, use double header joists (stagger end joints) or splice header joist for continuity on uphill and daylight sides of building.

Fig. 22-14 *Basement foundation.*

Building the PWF Step by Step

There are a number of steps in building the PWF. Many of them should be studied before undertaking the project, either to build it or to have it built. In fact, it would be best to access the entire plans in all their details from the Web site of the Southern Pine Council at www.southernpinecouncil.com. A complete 55-page manual is available at no charge for one copy.

The information shown here is to give you enough detail to make a decision to go ahead with this alternative type of foundation. As you can see, there are a number of advantages to this type of construction for homes, offices, and stores. The following figures and their captions will walk you through the entire process of building a PWF.

Installation of a PWF is speedy. On a prepared site, a PWF typically can be installed in less than 1 day. This makes it ideal for use where there is a problem with the weather. Experienced workers can install one in just a few hours (Figs. 22-15 through 22-25).

The next 17 steps shown in Fig. 22-26A illustrate the rapidity with which the PWF can be completed. Figure 22-26B shows a completed basement. Note the placement of treated-plywood sheathing.

Figures 22-27 and 22-28 show the advantage of the PWF when it comes to finishing the interior and exterior foundation walls to blend with the rest of the house.

Finishing a PWF House

It is easy to finish the interior and exterior foundation walls of a PWF structure. The walls can be finished to blend with the rest of the house. The exterior wood foundation walls can be overlaid with APA-rated siding. The siding can be patterned southern pine exterior siding or other types of code-approved siding that blend with the architectural style of the structure (see Fig. 22-28). The wood siding should be at least 6 inches from the ground. Applying the siding is done, as with all types of housing, as shown in Fig. 22-27.

The exterior foundation walls can be stained, painted, or covered with a code-approved stucco finish. Film-forming oil-based finishes are not recommended. They tend to crack and flake quickly in areas such as knotholes.

However, penetrating semitransparent oil-based stains generally perform well over CCA-treated plywood. This is so because they allow some of the wood to show through. The finished color may be affected by the color of the panel surface. Earth tones or a green stain usually will mask any discoloration from the chemical treatment.

Latex solid-color stains or paint systems also show excellent performance over CCA-treated plywood. Earth tones usually provide the best appearance. However, if pastels or white finishes are used, a stain-blocking all-acrylic latex primer, followed by a compatible all-acrylic latex topcoat, produces optimal results.

Plywood treated with ACA may contain blotchy deposits of residual surface salts. Thorough brushing to remove these excess surface salts prior to finishing is essential. Because of the blotchy appearance of the ACA treatment, only an all-acrylic latex paint system is recommended. It should consist of at least one coat

NOTES:

(1) Provide drains through footing at 6′ on center.

(2) 4″ layer of gravel, crushed stone or coarse sand under and along the sides of the concrete footing.

Concrete Footing Detail at Foundation Anchor Bolts

Fig. 22-15 Basement foundation on concrete footing.

of stain-blocking acrylic latex primer, followed by a companion acrylic latex topcoat. Medium to darker colors are suggested.

The Southern Pine Council provided the illustrations for the last part of this chapter.

Joist Perpendicular to Foundation Wall

Subfloor

Floor joist

PWF side wall

Fig. 22-16   Fastening foundation sidewalls to floor system.

Joists Parallel to Foundation Wall

Blocking @ 24″ or 48″ o.c.

Floor joist

PWF end wall

Fig. 22-17   Fastening foundation end walls to floor system.

Caulk

1x or plywood strip protecting polyethylene sheeting (12" nom.)

8" min.

Note: Nailing to be as normally required for shallow or no fill. Lateral soil forces transferred through 2 x 3 thrust plate.

Plywood

Polyethylene sheeting

Subfloor

Floor joist

2x8 joists notched $1^3/4$"
2x10 and wider joists notched 2"

Interior finish (optional)

Field applied 2x3 thrust plate

2x top plate

2x PWF stud wall

Interior finish (optional)

Fig. 22-18   Fastening foundation sidewalls to floor system.

Fig. 22-19 *Daylight basement end walls.*

Labels for Fig. 22-19:
- 16d nail between each stud
- Concrete slab keyed between studs
- Polyethylene sheeting
- 2x stud
- Gravel or crushed rock
- 2x footing plate
- Gravel footing
- Sheathing or exterior siding
- Plywood 3/8" min.
- Rigid insulation (thickness as required)
- h_1 (8" min.)
- h_2 (9" min.) h_2 greater than $h1$
- 3/4d
- Below frost line
- d
- 2d

Fig. 22-21 *Basement foundation stub wall with low backfill.*

Labels for Fig. 22-21:
- Subfloor
- Floor joist
- Treated stub wall
- Sheathing or exterior siding
- Untreated bearing wall above treated stub wall
- Caulk or "Z" flashing
- 8" min.
- 1_1 (1_1 less than 1_2)
- 1_2 (min. 18")

Fig. 22-20 *PWF stem wall, concrete slab on grade.*

Labels for Fig. 22-20:
- Subfloor
- Floor joist
- Blocking in outer bay as required
- Blocking between studs
- Concrete "keys" between studs
- Concrete slab
- Sheathing or exterior siding
- 8" min.
- All plywood edges fastened to blocking or framing

Fig. 22-22 *Basement foundation wall with partial backfill.*

Labels for Fig. 22-22:
- Subfloor
- Floor joist
- Untreated 2x top plates
- Untreated blocking between studs
- Untreated sheathing or exterior siding
- Caulk or "Z" flashing (for siding)
- Plywood (applied horizontally)
- 8" min.

Field applied 2x top plate

Note: Floor system must be in place before backfill.

2x top plate

Interior foundation wall (required between basement and crawlspace). Construction of this wall is the same as for full height foundation wall.

Crawl space or elevated foundation wall

Plywood

2x stud

Below frost line

4" min.

Full-height foundation wall

Gravel or crushed stone

Bottom plate

Footing plate

Support frame made of 2x members with bottom plate bearing on extended footing plate, sheathed with plywood

Backfill with gravel or crushed stone

Field cut extended footing plate at rise of excavation

Extended footing plate

FIELD INSTALLATION SEQUENCE

(1) Install full-height foundation wall and extended footing plate.

(2) Install interior foundation wall (required between basement and crawl space areas).

(3) Install support frame.

(4) Cover support frame with plywood sheathing.

(5) Place gravel or crushed stone around support frame.

(6) Install crawl space wall.

Full-height foundation wall

Interior foundation wall

Crawl space wall

Section View At "A"

Fig. 22-23 *Stepped footing, crawl space, or basement foundation.*

Fig. 22-24 *Knee wall with brick veneer.*

Labels (Fig. 22-24):
Subfloor
Floor joist
2x top plate
Required insulation
2x PWF stud wall
Vapor retarder
Optional interior finish
Plywood
1x screed board (optional)
Concrete slab
Polyethylene sheeting
1" air space
Veneer tie
Field applied 2x top plate
Brick veneer
Finish grade slope 1/2" per foot for min. 6' from wall
2x brick ledger
8" min.
Install 1x4 to provide expansion joint to prevent frost heave.
2x knee wall
Asphalt or polyethylene strips
Polyethylene sheeting
1x bottom plate
2x footing plate
3/4d
Below frost line
Gravel or crushed stone fill
d
2d

Fig. 22-25 *Alternate brick veneer detail.*

Labels (Fig. 22-25):
APA Rated Plywood Sheathing
Optional interior finish
Subfloor
Floor joist
Joist hanger
Ledger
Brick
1" air space
Building paper
Flashing
PWF wall panel

722   Permanent Wood Foundation System

Installation of a Permanent Wood Foundation is speedy. On a prepared site, a PWF can typically be installed in less than one day, but it's not unusual to see one installed in just hours. Here's the installation sequence:

1 The building site is prepared. Topsoil is removed and all excavation and trenching completed. Utility and drainage lines have been installed.

2 A minimum of 4″ of gravel, coarse sand, or crushed rock is laid as a base for the concrete slab or the wood floor to be installed later. Thickness of gravel under footings is relative to their width.

3 Gravel is leveled, extending several inches beyond where treated wood footing plates will rest. The gravel under the wood footing plates performs the same function as a conventional concrete spread footing, receiving and distributing loads from foundation walls.

4 Restaking the house is next, after drainage system and gravel footing installation.

5 Foundation sections can be built at the jobsite, or be prefabricated in panelized sections for accurate and rapid installation. Each section is composed of a footing plate, bottom plate, wood studs, plywood, and single top plate. Here, prefabricated sections are delivered ready for installation.

6 The first section goes up. 8′ x 8′ panels can be easily set without mechanical assistance. It is recommended that the first two panels installed be located at a corner, because a corner is self-bracing. Check the level of first sections

7 Caulking is applied between plywood edges of adjoining foundation sections. In full-basement construction, the plywood joints must be sealed full-length with caulking compound. Caulking is not required in crawl space construction.

8 Additional panels of the foundation are attached. Using these preframed sections, windows and door openings are already cut and framed, reducing onsite labor costs.

9 Bracing supports foundation panels while additional sections are installed.

Fig. 22-26A *The PWF can be completed very quickly.*

10 Remaining sections of foundation walls are erected. Extended footing plate automatically offsets wall section joints.

11 Panels are plumbed and aligned. When an accurate line has been established, sections can be shifted inward or outward as needed from the line of reference.

12 Stakes along the footing plate keep the panels from sliding until permanent attachment of all sections is completed.

13 The last section is installed. Once all the foundation wall sections are in place, the entire structure is rechecked to be level and square.

14 The second top plate (typically untreated lumber) is attached. Top plate joints are staggered so that they do not fall directly over joints between foundation wall sections. At corners, joints in the double top plates overlap as in conventional wall construction.

15 Prior to backfilling, 6-mil polyethylene sheeting is attached at the grade line and is draped over the portion of the foundation wall that will be below grade. The top edge is protected with a treated wood strip that is caulked. This strip is a guide for backfilling.

16 Installation of a treated wood floor will optimize the comfort of below-ground living areas.

17 Backfilling takes place after the basement floor is poured and cured, and the first floor framing and floor sheathing are installed. These steps give the PWF lateral restraint for backfill loads. To avoid excessive deflection, backfill in layers of 6″ to 8″ and tamp to compact. Avoid operating heavy equipment near walls during backfilling.

In Group I soils, the first 12″ or more of backfill is the same material as used for footings. For Group II and III soils, backfill with the same materials as footings, for half the height of the backfill. This portion of the fill is covered with strips of 30-pound asphalt paper or 6-mil polyethylene to permit water seepage, yet prevent infiltration of fine soils. Verify that no polyethylene sheeting is exposed below the grade strip.

Fig. 22-26A *The PWF can be completed very quickly.*

Fig. 22-26B *A complete basement. Note the placement of treated-plywood sheathing.*

Fig. 22-28 *The PWF exterior foundation walls blend with the rest of the house.*

Fig. 22-27 *Application of siding.*

23
CHAPTER

Private
Water
Systems

WATER SYSTEMS CAN BE PUBLIC OR PRIVATE. The private water system is the responsibility of the homeowner. It is installed according to the latest rules and regulations furnished by the local health department.

PUBLIC WATER SUPPLIES

The water main is thought of as having always been there. It has, of course, if you live within a city's limits or in a municipal utility district (MUD) in some states. The water main is the conduit or large pipe through which a public or community water system conveys water to all service connections. Figure 23-1 shows how the typical home installation is made. The water main is usually located in the street—in front of the house, having been located there before the building permits were issued.

Once tapped, the water main furnishes water to the house by way of builder-installed plumbing, a service pipe. This plumbing involves a stop valve or a place to turn off the water to the house. From there, it goes to the curb box that provides access to a water meter that measures and records the quantity of water that passes through the service pipe. There is usually a control valve for shutting off the water supply to a building in case of emergency. The service pipe connects the house to the water main. The water main usually is installed by or under the jurisdiction of a public utility. Once the service has been extended inside the build-

ing, a shutoff valve is installed. From there, all the water in the house is controlled by the valve's on/off function.

In this chapter you will learn

- The difference between the public and private water supply systems
- What types of maintenance are required for obtaining your own water supply
- Where a well may be placed
- Where to obtain detailed information on water wells
- How pressure is regulated on private systems
- How pumps are installed and water is chlorinated
- How water tanks are operated and maintained
- How the Environmental Protection Agency (EPA) protects your private water supply

PRIVATE WATER SYSTEMS

In order to obtain water for a rural location, out of the reach of city water, a constantly flowing stream must be found, or there must be a reliable source of well water.

Drilling a Well, Boring a Well, or Driving a Well

A well must be drilled and located at least 100 feet away from the house's sewers, septic tanks, and

- Water main is the conduit through which a public or community water system conveys water to all service connections.

- Service pipe connects a building to a water main, usually installed by or under the jurisdiction of a public utility.

- Building shutoff valve

- Corporation stop is a valve controlling the flow of water from a main to a service pipe.

- Curb box provides access to a water meter that measures and records the quantity of water that passes through a service pipe, and the control valve for shutting off the water supply to a building in case of an emergency.

Fig. 23-1 *Public water supply.*

sewage disposal fields and should be accessible so as to permit removal of the well casings or pump for maintenance or repair. Codes are written for well locations and installations and should be checked before hiring someone to produce a well (Fig. 23-2).

Groundwater is the drinking water source for more than 50 percent of the U.S. population and more than 90 percent of its rural population. In New York State, for example, approximately 3 million rural residents rely on groundwater to supply their drinking water. Most residents with private water supplies also have onsite wastewater treatment systems, commonly called *septic systems*. Approximately 1.5 million households in New York have some kind of septic system. They are in most instances satisfied with their water supplies—82 percent are "somewhat satisfied" or "very satisfied."

Drinking Water

It is interesting to note that a study of the rural homeowners found that most of the people are satisfied with their water supplies and do not perceive any problems. However, many do not

- Take steps to actively protect their water supplies
- Test their drinking water
- Pump their septic tanks on a regular basis

The study showed that upstate New York residents do not employ even low-cost maintenance practices for their water supplies.

Private water supply protection is the responsibility of individual homeowners. However, only a few local programs sponsor regular testing of private water supplies. This leads to incomplete information and assumptions about the quality of rural drinking water.

Water from wells, springs, lakes, and so forth should be tested for purity. This should be done before the water is used for human consumption. State and/or local boards of health or private laboratories will furnish sterile containers for water samples and test the water for purity. If the water is contaminated, filters, chlorinators, or iodine feeders can be installed in the system to make the water safe for human consumption.

Well Water

Well water, if the source is deep enough, is usually pure, cool, and free of discoloration and taste or odor problems. A sample should be checked for bacteria and chemical content by the local health department before a well is put into operation.

Water must be supplied to a house

- In the correct quantity
- At the proper flow rate, pressure, and temperature

Total operating head = service pressure minus friction head loss.

Pressure tank maintains service pressure. It requires electrical power and a fused disconnect switch.

Install supply pipe below frost line.

(A)

(B)

Fig. 23-2 *Private well water supply.*

For human consumption, water must be potable—free of harmful bacteria—and palatable. To avoid the clogging or corrosion of pipes and equipment, water may have to be treated for hardness or excessive acidity.

Keep in mind that water is consumed by drinking, cooking, cleaning, clothes washing, and bathing. Some heating, ventilation, and air-conditioning systems also use water for cooling, heating, and controlling humidity. Fire-protection systems store water for extinguishing fires.

The EPA has a Web site to furnish information to those using private drinking water wells. It is www.epa.gov/safewaterwells/booklet/whatshould do.html.

Listed below are the eight basic steps you should take to maintain the safety of your drinking water once a system has been installed:

1. Identify potential problems sources.
2. Talk with local experts.
3. Have your water tested periodically.
4. Have the test results interpreted and explained clearly.
5. Set a regular maintenance schedule for your well.
6. Do the scheduled maintenance.
7. Keep accurate, up-to-date records.
8. Remedy any problems.

Water Pressure

Pressure tanks can be used to maintain service pressure. They require electrical power to the pump and a fused disconnect switch. For proper operation, the water level in a pressure tank should be based on two-thirds water and one-third air. With the tank top third of the tank filled with air, water is pumped into the tank, and the air is compressed until the top (or high) setting of the pressure switch is reached, usually 40 pounds. The pressure switch then turns off the pump. Water then can be used from the tank. As the water is pushed out of the tank, the pressure drops until the low setting of the pressure switch is reached (usually 20 pounds), and the pump starts, beginning the cycle over again. Occasionally, owing to the failure of an air volume control or a leak in the pressure tank, the air cushion (the top third of the tank) is lost or diminished.

If there is only a very small area of air at the top of the tank, the pressure will drop from 40 to 20 pounds, and immediately on opening a faucet or valve, the pump will start. If there is no air at the top of the tank, the pump will operate continuously because the pump will be unable to build up enough pressure to cause the

pressure switch to turn the pump off. When these conditions occur, the tank is waterlogged. A waterlogged tank should be corrected as soon as possible; the frequent on/off operation of the pressure switch will result in burned contacts on the switch and possible damage to the pump motor (Fig. 23-3).

Fig. 23-3 *Sequence of operation for a typical pressure tank.*

Pressure Tanks

Pressure tanks with plastic pipe installations require a torque arrester as well as cable guards. The purpose of the pressure tank is to allow an amount of water to be drawn before the pressure drops enough to cause the pump to start. Without a pressure tank, the pump would start and stop continuously when water is drawn. There are two types of pressure tanks: the standard tank that requires an air volume control and the precharged tank.

- On a standard pneumatic tank system, air is introduced to compensate for that which is absorbed by the water. Each time the pump cycles, air is added to the tank through a bleeder and snifter valve. The excess air is released by a float assembly (air volume control) in the upper side tapping of the tank (see Fig. 23-6).

- In a precharged tank, a flexible diaphragm or bladder separates the air and water areas of the tank. The air chamber is precharged by means of a tire valve with pressure of 2 pounds per square foot less than the cutout pressure of the pump. Because this is not in contact with the water, it cannot be absorbed by the water. Therefore, the original charge of air is never lost (see Fig. 23-7).

- In a precharged tank system, none of the fittings for air introduction or air level control are required (see Fig. 23-5). The piping in the well is also different for the two systems. The precharged tank system does not require a bleeder orifice assembly, which simplifies the installation.

The tank size should be selected to keep the pump starts per day as low as practical for maximum life. Excessive cycling of the motor accelerates motor bearing wear. It also affects spline wear, pump wear, and contact erosion. On single-phase motors, use as a guide 100 starts per day (24 hours). Motors operating on three-phase current should use 300 starts per day as a rule of thumb.

Pressure tanks that use a permanently sealed-in air charge, similar to the tank shown in Fig. 23-3, can be used instead of the older models. The permanently sealed tank has a number of advantages. Installation is much simpler. There is but one opening in the tank.

Waterlogging is eliminated by the sealed-in air cushion. By eliminating waterlogging, pump starts are kept to a minimum. This adds to the life of the pump and the control switch. The water reservoir is coated, and inside the tank is corrosion-free.

Operation of the Pressure Tank

The four steps in the operation of a pressurized tank are illustrated in Fig. 23-3: The tank is precharged with air, and the pump is off (A). The pump is on, and water enters the reservoir (B). This compresses the air cushion and raises the air-cushion pressure. Figure 23-3C shows the tank in its filled position and a pressure of 50 pounds per square foot. This pressure operates a pressure-sensitive switch and turns off the pump. In Fig. 23-3D, the water is being used, and the pressure in the air cushion drops. Once it drops to the cut-in point, the pump starts. This is the beginning of a repeating the cycle. Pressurized tanks are available in sizes from 2.0 gallons up to 410 gallons. If more capacity is needed, it is possible to add more tanks. Figure 23-4 shows the many configurations of the single-tank installation.

Pressure Switch

The heart of automatic operation is the pressure switch. The pressure switch provides automatic opera-

Fig. 23-4 *Typical installation with standard pneumatic tank.*

tion by starting when the pressure drops to the switch cut-in setting and stops when the pressure reaches the switch cutout setting. The pressure switch must be installed as close to the tank as possible (Figs. 23-5 through 23-7).

Fig. 23-5 *Bladder operation.*

Relief Valve (Pressure)

A properly sized pressure-relief valve must be installed on any system where the pump pressure can exceed the pressure tank's maximum working pressure or on systems where the discharge line can be shut off or obstructed (see Fig. 23-8). The relief-valve drain port should be piped to a drain.

Keep in mind that not providing a relief valve can cause extreme overpressure, which could result in personal and/or property damage.

Pump Installation

After the pump has been tested and wires spliced and wrapped, it is time to drop the pump into the well. By following the installation instructions for the pump chosen, use Schedule 80 polyvinyl chloride (PVC) pipe or galvanized pipe. If either of these two types is used, a foot clamp or vise will be required to hold the pipe when connecting the next pipe length.

Install the pump in a well that is sand-free, straight, and has sufficient flow of water to supply the pump. Clear the well of sand and any other foreign matter with a test pump before installing the submersible pump.

Fig. 23-6 *Typical installation with precharged tank.*

Fig. 23-7 *Typical installation of a single tank.*

Fig. 23-8 *Smart pressure package installations. Note the location of the relief valve.* (Flexcon Industries.)

Note: Using the submersible pump to clean the well will void the warranty. When drilling a new well in an area where sand is a problem, a sand screen must be installed to protect the pump and motor.

Chlorinate the well first. This is done by dropping 24 to 48 HTH chlorine tablets into the well before lowering the pump. This prevents contamination and the growth of iron bacteria. The bacteria later could plug the well and the pump. The chlorinated water is pumped out of the system when testing the pump flow.

Be sure the top edge of the well casing is perfectly smooth. This is necessary because a sharp or jagged edge can cut or scrape the cable and cause a short.

Install a line check valve within 25 feet of the pump and below the drawdown level of the water supply. The check valve should be the same size as the discharge outlet of the pump or larger.

Note: Use of pipe that is smaller than the discharge tapping of the pump will restrict the capacity of the pump and lower its operating performance.

When connecting the first length of pipe and placing the pump in the well casing, take care to maintain a centered pump in the well. It is easier to handle the pump if a short piece of pipe is installed first rather than a long piece. Install the check valve at the end of the first piece of pipe. Do this prior to lowering the pump into the well. Maintain the alignment as the pump is placed and lowered into the well; a torque arrester is recommended. Position the torque arrestor to within 6 inches of the pump discharge, and clamp the arrestor to the pipe. Wrap the pipe with enough tape at the top and bottom of the torque arrester to keep it from sliding up the pipe while the pump is being lowered into the well.

If not already done, splice the electrical cable to the motor leads (Fig. 23-9). The cable and ground wire should be taped to the discharge pipe. Tape the cable about 5 feet above the discharge and every 20 feet thereafter. Install cable guards if required. This is to eliminate rubbing against the well casing. Do not let the cable drop over the edge of the well casing. Never allow the weight of the pump to hang on the cable alone.

Warning: Since most submersible pump problems are electrical, it is very important that all electrical work be done properly. Therefore, all electrical hookup work or electrical service work should be done by a qualified electrician or serviceperson only!

Once the motor (pump) is connected to power, it may be necessary to disconnect the power source before working on or near the motor. Be aware of the shock potential of the control box and wiring. If the power disconnect is out of sight, be sure to lock it in the open (or off) position and tag it to prevent unexpected application of power by someone not familiar with the repair work going on.

Lower the pump into the well slowly without forcing. Use a vise or foot clamp to hold the pipe while

(A)

(B)

Fig. 23-9 *A. Making an electrical splice. B. Water pump.*

connecting the next length. A boom, tripod, or pump-setting rig is recommended. Lower the pump to approximately 10 feet below maximum drawdown of the water, if possible, and keep it approximately 10 feet from the bottom. Do not set the pump on the bottom of the well. Before each new length of pipe is added, attach the coupling to the top of that length of pipe. This will provide a stop for the foot clamp to hold while the next section of pipe is being installed.

On a standard tank with air volume control, a bleeder orifice is required. Install the bleeder orifice in the discharge pipe 5 feet or more below the snifter valve. See Fig. 23-4 and Table 23-1 for well seal and/or pitless adapter installations.

All installations should have a well seal. Make sure that the seal is seated, and tighten the bolts evenly. Be sure to assemble the tee to the pipe above the well seal to prevent dropping the pipe and pump down the

TABLE 23-1 *Well Seal/Pitless Adapter Installation*

Distance Table	
Tank Size, Gallons	Depth from Horizontal Check Valve to Bleeder Orifice, Feet
42	5
82	10
120	15
220	15
315	20
525	20–35

Note: Installations that use a precharged pressure tank do not require a bleeder orifice.

well as you lower it. It is important that the well seal and piping be protected from freezing.

On a pitless adapter installation, the connection to the system supply line is made below ground. Install the pitless adapter following the instructions included with a particular brand or design being used in the installation. Follow all applicable state and local plumbing codes.

Test Run

Once the installation is complete, there is the next step—usually the one you have been waiting for in this project—the test run. The preliminary test run is done when the pump is at a desired depth, and a throttle valve is installed for the test run. Wire the single-phase motor though the control box. Follow instructions in the box regarding color coding of the wires and so forth. Wire the three-phase motor through a magnetic starter. Test the cable for continuity with an ohmmeter. With the pump discharge throttled, run the pump until the water is clear of sand or other impurities. Gradually open the discharge.

Be sure that you do not stop the pump before the water runs clear. This may take several hours. If the pump stops with sand in it, it will lock.

If the pump lowers water in the well far enough to lose prime, either lower the pump in the well (if possible) or throttle the discharge to the capacity of the well. If the well is low capacity, use a low well level control. On three-phase units, establish the correct motor rotation by running in both directions. Change rotating direction by exchanging any two of the three motor leads. The rotation that gives the most water flow is the correct rotation.

Pressure Tank Installation

On a new installation, you have to install the pressure tank along with the pressure switch, pressure gauge, pressure-relief valve, check valve, gate valves, and unions as shown in Figs. 23-5 and 23-6.

On replacement-pump installations, you have to make sure that the tank system is in good operating condition. A waterlogged tank may cause pump failure.

Water Conditioning Equipment

Some wells produce water that is usable only with filtering and conditioning. For instance, a well with a high iron content in the water can cause depositions of iron oxide to form in the piping, tanks, water heaters, and water closet tanks. Once these deposits are formed, it is almost impossible to get rid of them. Water pressure tanks, water heaters, and piping may have to be replaced. However, you can install an iron filter to remove the iron before it causes damage in the piping system. Filters also can be installed to remove any objectionable taste or odor from so-called sulfur water. Well water, if the source is deep enough, is usually pure and cool with no discoloration, taste, or odor problems. A sample should be checked for bacteria and chemical content by the local health department before a well is put into permanent operation.

Locating the Equipment

Water conditioning equipment should have its location well planned. It depends on the location and type of house being served. There are a number of ways to connect the conditioner to the filter and to condition the water. Figure 23-10 shows a few of the possible locations for the water conditioner.

Select the location of the water softener with care. Various conditions that contribute to proper location include the following:

* Locate as close as possible to the water supply source.
* Locate as close as possible to a floor or laundry tube drain.

(A)

(B)

(C)

(D)

Fig. 23-10 *Typical installations and equipment locations.*

- Locate unit in correct relationship to other water conditioning equipment.

- Select a location where the floor is level. If the floor is rough and/or uneven, you can level it by placing the cabinet or tanks on ¾-inch plywood and shim to level as needed.

- Locate the softener in the supply line before the hot water heater. Temperatures above 100°F (38°C) will damage the softener and void the warranty.

- Allow sufficient space around the softener installation for easy servicing.

- Provide a nonswitched 110/120-volt, 60-hertz power source for the control valve.

- Complete instructions come with the installation whether you buy it at Sears or it is installed by a professional plumber.

The EPA regulates public water systems. It does not have the authority to regulate private drinking water wells. Approximately 15 percent of Americans rely on their private water supplies, and these supplies are not subject to EPA standards, although some states and local governments do set rules to protect users of these wells. Unlike public drinking water systems serving many people, they do not have experts regularly checking the water's source and quality before it is sent to the tap. These households must take special precautions to protect and maintain their drinking water supplies and sources.

CHAPTER 23 STUDY QUESTIONS

1. Whose responsibility is a private water system?
2. What is the difference between a public and private water system?
3. What is the difference between drilled, bored, and driven wells?
4. What steps do you need to take to obtain and keep your private water supply safe?
5. Describe the water that is supplied by a well and its necessary characteristics.
6. What is the EPA?
7. How does the EPA figure into your private water supply?
8. What is the purpose of a water pressure tank?
9. How does the pressure tank operate?
10. What is the difference between a pressure switch and a relief valve?
11. What does adding chlorinating tablets to a well prevent?

24
CHAPTER

Private
Sewage
Facilities

MANY PEOPLE LIVE IN THE COUNTRY. THIS choice of living location leads to another expense in the building of a house. Every domicile must have a sewage system as well as a source of water to serve the needs of the occupants. Chapter 23 covered private water supplies. Inasmuch as there is no city sewage system to serve those outside the limits of most cities, this means that some method of disposal of human wastes must be found. The traditional outhouse is no longer acceptable. Disposal must be done in an approved manner to prevent the spread of disease and odor.

Things you should know about the private sewage system include

- Septic tanks are needed where public sewage lines are not available
- Septic tanks and disposal fields require attention
- Septic tank systems usually need periodic maintenance
- Locating a septic tank system is very important
- The location and operation of the disposal field are very important
- Grease traps are needed sometimes
- Maintenance and repair of the system should be recorded
- Newer private disposal systems are now available

One of the most readily acceptable means of fulfilling the requirements of modern living and meeting local health department sanitation regulations is the installation of a septic tank. Local (county) regulations probably will govern the minimum size of the septic tank. One rule is commonly used to figure the storage capacity of the tank: The tank should equal the number of gallons of sewage entering the tank in a 24-hour period. At the rate of 100 gallons per person per day, a septic tank for a four-person household, usually a two-bedroom house, should have a minimum capacity of 400 gallons storage. If a garbage disposer is used, the capacity should be increased by 50 percent. This means that a 400- to 600-gallon tank is called for when a four-person household is used in the equation. However, in actual practice, the minimum size of any septic tank should be 1,000 gallons.

The actual size of the disposal system depends on the number of fixtures served and the permeability of the soil, as determined by a percolation test. Sewage disposal systems are designed by sanitary engineers and must be approved and inspected by the health department before being put in use. Consult your local building and health codes for specific regulations and requirements.

SEPTIC TANKS AND DISPOSAL FIELDS

The following information is provided for your guidance and planning. An understanding of the septic tank system and its disposal field will aid in living with this type of system.

Septic tanks and disposal fields are used by many kinds of suburban or rural areas that have no service or treatment facilities. Septic tanks come in a variety of sizes and shapes. They can be made of precast concrete, fiberglass, or steel. Each type has its advocates, but the precast concrete tank seems to be the longest lasting if installed properly.

Septic Tank Operation

A septic tank is a covered watertight tank for receiving the discharge from a building sewer, separating out the solid organic matter (which is decomposed and purified by anaerobic bacteria), and allowing the clarified liquid to discharge for final disposal (Fig. 24-1).

The liquid effluent, which is about 70 percent purified, may flow into one of the following systems:

- A *drain field*, which is an open area containing an arrangement of absorption trenches through which effluent from a septic tank may seep or leach into the surrounding soil.
- A *seepage pit*, which is lined with a perforated masonry or concrete wall and sometimes used as a substitute for a drain field when the soil is absorbent and the highest level of the water table is at least 2 feet below the bottom of the pit (Fig. 24-2).
- A *subsurface sand filter*, which consists of distribution pipes surrounded by graded gravel in an intermediate layer of clean coarse sand, and a system of underdrains to carry off the filtered effluent (Fig. 24-3). (Sand filters are used only where other systems are not feasible.)

As mentioned previously, local regulations probably will govern the minimum size of the septic tank. Septic tanks function by a combination of bacterial action and gases. Solids entering the tank drop to the bottom. Bacteria and gases cause decomposition to take place, breaking down the solids into liquids, and in the course of this process, the indissoluble solids, or sludge, settle to the bottom of the tank. Decomposition in an active tank takes about 24 hours. The sludge

(A)

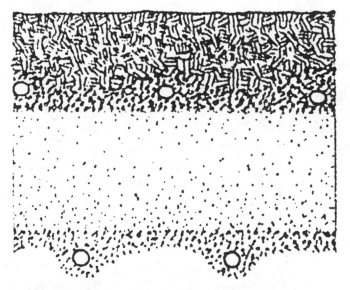

(B)

Fig. 24-1 *A. Cast concrete tank. B. Fiberglass tanks.*

Fig. 24-2 *Pit used for sewage disposal purposes.*

Fig. 24-3 *Sand filter. Note the two layers of pipes.*

builds up on the bottom of the tank so that periodically it must be pumped out (commonly once a year). However, in other instances, it is only once in 10 years, depending on the usage of the tank. In the cleaning or pumping process, only the sludge on the bottom of the tank should be removed. The crust, formed at the water level near the top of the tank should not be disturbed. The only exception to this rule is if the crust has become coated with grease, and the bacterial action of the tank has been destroyed. If this happens, the crust on top will have to be removed. The bacterial action will begin again when the top is placed on the tank and the tank is sealed. Special compounds can be purchased to hasten the resumption of bacterial action. However, these compounds are rarely, if ever, needed. Figure 24-4A shows a rough sketch of a septic tank system.

Septic Tank Location

The top of the septic tank should be located at a minimum depth of 12 inches below ground level. The actual depth probably will be somewhat greater owing to the depth of the sewer entering the septic tank. The septic tank also must be located at least 100 feet away from any well and downhill so that all drainage is away from the well. Local regulations must be followed as to placement of a septic tank. The system should be located 50 feet from streams and 10 feet from buildings and property lines. The dosing chamber of a large septic tank employs siphonic action to automatically discharge a large volume of effluent when a predetermined quantity has accumulated (see Fig. 24-4B).

Septic Tank Disposal Field

The disposal field must be sized using the rate of absorption established in the percolation tests (Fig. 24-5). If the absorption rate is 1 inch in 60 minutes, then a factor 2.35 times the number of gallons of sewage entering the septic tank per day will determine the number of square feel of trench bottom area needed. Using the recommended figure of 100 gallons per day per person, if the absorption rate is 1 inch in 60 minutes and there are four persons in the household, the formula is 2.35 × 400 = square feet of trench bottom needed for the disposal field. If the disposal field has five fingers (trenches), each trench would need 188 square feet of trench bottom; if the trench is 2½ feet (30 inches) wide, then each trench would be 75 feet long.

If the absorption rate established in the tests is 1 inch in 10 minutes, then the factor 0.558 × 100 gallons per person per day would be used. Again, using the base of four persons in the household, 0.558 × 400 = square feet of trench bottom needed. With five fingers or trenches installed in the disposal field, each trench should have 45 square feet of trench bottom. If the trench is 2½ feet wide, then each trench should be 18 feet long.

The figures used in these examples may be helpful, but local regulations in your area can vary somewhat. The rules and regulations governing sizes of disposal fields in your area should be followed.

Disposal fields serve two purposes. The fingers of a disposal field provide storage for the discharge of a septic tank until the discharge can be absorbed by the earth. The fingers also serve as a further step in the purification of the discharge through the action of bacteria in the earth. Liquids discharged into the disposal field are disposed of in two ways. One is by evaporation into the air. Sunlight, heat, and capillary action draw the subsurface moisture to the surface in dry weather (Fig. 24-6).

In periods of wet or extremely cold weather, the liquids must be absorbed into the earth. The discharge from the septic tank should go into a siphon chamber. It is then siphoned into a distribution box. The distribution box directs the discharge so that approximately the same amount of liquid goes into each trench. The

HOUSE

GREASE TRAP

SEPTIC TANK

DISTRIBUTION BOX IS EQUIPPED WITH BAFFLES TO DIRECT EFFLUENT TO VARIOUS PARTS OF THE DISPOSAL FIELD.

DRAINAGE FIELD

Fig. 24-4A Sewage disposal system.

Labels in figure:
SEALED JOINTS — PRECAST REINFORCED CONCRETE SLABS — 8" MIN. — GROUND LEVEL — 3"
INLET — 1" VENT SPACE — APPROX. 20% OF LIQUID DEPTH — OUTLET
LIQUID LEVEL — SCUM — 40% OF LIQUID DEPTH — 4" SOIL PIPE
HOUSE SEWER — 3" — 6"
COMMON OR SANITARY TEE — EFFLUENT — 30" - 60" — SHORT RADIUS ELBOW
SLUDGE
6" REINFORCED CONCRETE WALLS AND FLOOR

Fig. 24-4B *Sectional view of a typical septic tank.*

drainage field should not run more than 60 feet. Figure 24-5B shows the sewage septic tank disposal field layout.

A siphon, also referred to as a *dosing siphon*, is desirable for this reason: The siphon does not operate until the water level in the siphon chamber reaches a predetermined point. When this level is reached, the siphonage action starts, and a given number of gallons are discharged into the disposal field. The sudden rush of this liquid into the distribution box and then into the separate fingers of the disposal system ensures that each finger will receive an equal share of the discharge. The discharge thus received by the fingers will have time to be absorbed before another siphon action occurs.

A siphon is not essential to the operation of a septic tank and disposal field. However, a siphon will improve the efficiency of the system. The lateral distance between trenches in a disposal field will be governed by ground conditions and local regulations. Construction of the disposal field varies owing to soil conditions; basically, a layer of gravel is placed in the bottom of the trench; field tile or perforated drain tile then is laid on the gravel. If field tile (farm tile) is used, the tiles should be spaced ¼ inch apart and should slope away from the distribution box at the rate of 4 inches fall per 100 feet. The space between the tile should be covered by a strip of heavy asphalt-coated building paper or by a strip of asphalt roofing shingle (Fig. 24-7).

The trench then should be filled to within 6 inches of ground level with gravel. Topsoil then should be added to bring the finished trench to ground level. Since some settling of the topsoil will occur, it is wise to mound the earth slightly over the trenches to allow for settling.

Absorption trenches are 18 to 30 inches wide and 30 inches deep. They contain coarse aggregate and a perforated distribution pipe through which the effluent from the septic tank is allowed to seep into the soil. The distribution pipes should run perpendicular to the slop and at least 2 feet minimum to the water table (see Fig. 24-7).

The Grease Trap

Grease that is present in dishwashing water can destroy the action of a septic tank, and if allowed to get into the disposal field, it will coat the surface of the earth in the disposal field and prevent absorption. Grease is normal in sewage; the septic tank can dispose of the grease from the kitchen. If it is trapped before it reaches the septic tank, the life of the septic tank and the disposal field will be greatly prolonged.

A grease trap must be large enough that the incoming hot water will be cooled off as soon as it reaches the grease trap. A 400-gallon septic tank is ideal for use as a grease trap. The water present in this size tank always will be relatively cool, and the grease present in the dishwashing water will congeal and rise to the top of the tank. The inlet side and the discharge side of the tank have battles, or fittings, turned down.

Fig. 24-5A *Distribution box and drainage field.*

Thus the water entering and leaving the tank will be trapped. The grease will congeal and float to the top of the tank, and relatively clear water will flow out of the grease trap into the septic tank. Drain-cleaning compounds such as sodium hydroxide and those that contain lye never should be used with a septic tank installation. The drainage piping from the kitchen sink will not be connected to the drainage system of the

HOUSE

PRECAST CONCRETE GREASE TRAP

SEPTIC TANK

DISTRIBUTION BOX

DRAINAGE FIELD

60' (18m) MAXIMUM

ABSORPTION TRENCHES

house when a grease trap is used. The kitchen sink drainage should be piped out separately, and the grease trap should be located as near as possible to the point where the sink drain line exits the house. The top of the grease trap should be within 12 inches of ground level. A manhole can be extended up to ground level with a lightweight locking-type manhole cover; this will provide easy access for skimming off the grease. A periodic check should be done and the grease skimmed off when it reaches a depth of 3 to 4 inches. The tile in the disposal field will vary in depth according to local conditions. The septic tank supplier, if unfamiliar with a siphon and siphon chamber, can get this information. If a siphon is not used, the outlet of the septic tank is piped directly to the distribution box. The siphon outlet, if a siphon is used, must be lower than the inlet to the siphon chamber.

Newer Wastewater Treatments

Newer technology provides for a cleaner fluid emission from the wastewater treatment unit. The septic tank has some limitations, and some areas, especially heavily populated zones, no longer allow them to be installed.

One of the newer systems reduces normal household wastewater to a clear, odorless liquid in just 24 hours. It uses the same process as central treatment plants (Fig. 24-8).

The primary treatment compartment receives the household wastewater and holds it long enough to allow solid matter to settle to the sludge layer at the tank's bottom. Organic solids are here broken down physically and biochemically by anaerobic bacteria, the bacteria that live and work without oxygen. Grit and other untreatable materials are settled out and held back. The partially broken down, finely divided material that is passed on to the aeration compartment is much easier to treat than raw sewage.

It can be seen in part (2) of Fig. 24-8 that the aeration compartment takes the finely divided, pretreated material from the primary compartment and mixes it with activated sludge. It is then aerated. The aerator injects larger quantities of fresh air into the compartment to mix the compartment's entire contents and to provide oxygen for the aerobic digester process.

The aerator is mounted in a concrete housing that rises to the ground level to give it access to fresh outside air (Figs. 24-9 and 24-10).

As air is injected into the liquid, the aerator breaks up this air into tiny bubbles so that more air meets the liquid, thus hastening the aerobic digestion process. Aerobic bacteria, which are bacteria that live and work

Fig. 24-5B *Sewage septic tank disposal field layout.*

(A) Trench in clay soil

(B) Trench in loose soil

Fig. 24-6 *Trench layouts for various soil conditions.*

Fig. 24-7 *Absorption trench.*

Fig. 24-8 *Home wastewater treatment plant.* (Jet, Inc.)

in the presence of oxygen, then use the oxygen in the solution to break down the wastewater completely and convert it to odorless liquids and gases.

The final phase is shown in part (3) of Fig. 24-8. The settling/clarification compartment has a tube settler that eliminates currents and encourages the settling of any remaining particulate material that is returned, by way of the tank's sloping end wall, to the aeration compartment for further treatment. A nonmechanical surface skimmer, operated by hydraulics, skims floating materials from the compartment. The remaining odorless, clarified liquid flows into the final discharge line through the baffled outlet.

Fig. 24-9 *Aerator and control panel.* (Jet, Inc.)

This type of home treatment system is often mandated by local health officials where there is a high water table or the soil has poor percolation.

Environmental Concerns

By paying attention to recommended maintenance and repair of the septic system, the environment is protected from untreated sewage and the well water supply is kept free of contamination. The main concern is protection of the health and well-being of occupants of a house.

CHAPTER 24 STUDY QUESTIONS

1. How many people in the United States have a private sewage system (septic tank)?

2. How does a septic tank operate? Does it have to be maintained or serviced?

3. Where are sand filters used?

4. How is the effluent dissipated into the ground?

5. What type of soil is best for a septic tank?

Fig. 24-10 *Home water treatment plant serving a single home.* (Jet, Inc.)

6. How is the size of a septic system determined?

7. What is sludge? Describe it.

8. Where is the grease trap located? Why?

9. How does newer technology make it easier to own and operate a septic system?

10. What is an aerator, and why do you need one?

11. How is wastewater turned to odorless, colorless liquid?

12. What is aerobic bacteria, and how are they used in a septic system?

Glossary

access Access refers to the freedom to move to and around a building, or the ease with which a person can obtain admission to a building site.

acoustical This refers to the ability of tiles on a ceiling to absorb or deaden sound.

aggregate Aggregate refers to sand, gravel, or both in reference to concrete mix.

air-drying Method of removing excess moisture from lumber using natural circulation of air.

alligatoring This occurs when paint cracks and resembles the skin of an alligator.

anchor bolt An anchor bolt is a steel pin that has a threaded end with a nut and an end with a 90° angle in it. The angled end is pushed into the wet concrete and becomes part of the foundation for anchoring the flooring or sill plates.

annular ring There are two colored rings that indicate the growth of a tree. The colors indicate the growth of springwood and summerwood.

apron A piece of window trim that is located beneath the window sill. Also used to designate the front of a building such as the concrete apron in front of a garage.

arbor An axle on which a cutting tool is mounted. It is a common term used in reference to the mounting of a circular saw blade.

asphalt shingle This is a composition-type shingle used on roof. It is made of a saturated felt paper with ground-up pieces of stone embedded and held in place by asphaltum.

asphalt shingles These are shingles made of asphalt or tar-impregnated paper with a mineral material embedded; they are very fire resistant.

auxiliary locks Auxiliary locks are placed on exterior doors to prevent burglaries.

awl An awl is a tool used to mark wood with a scratch mark. It can be used to produce pilot holes for screws.

awning picture window This type of window has a bottom panel that swings outward; a crank operates the moving window. As the window swings outward, it has a tendency to create an awning effect.

backsplash A backsplash is the vertical part of a countertop that runs along the wall to prevent splashes from marring the wall.

backsaw This saw is easily recognized since it has a very heavy steel top edge. It has a fine-tooth configuration.

balloon frame This type of framing is used on two-story buildings. Wall studs rest on the sill. The joists and studs are nailed together, and the end joists are nailed to the wall studs.

balustrade A complete handrail assembly. This includes the rails, the balusters, subrails, and fillets.

baluster The baluster is that part of the staircase which supports the handrail or bannister.

bannister The bannister is that part of the staircase which fits on top of the balusters.

baseboard Molding covering the joint between a finished wall and the floor.

base shoe A molding added at the bottom of a baseboard. It is used to cover the edge of finish flooring or carpeting.

batten A batten is the narrow piece of wood used to cover a joint.

batter boards These are boards used to frame in the corners of a proposed building while the layout and excavating work takes place.

batts Batts are thick pieces of fiberglass that can be inserted into a wall between the studs to provide insulation.

bay window Bay windows stick out from the main part of the house. They add to the architectural qualities of a house and are used mostly for decoration.

beam A horizontal framing member. It may be made of steel or wood. Usually the term is used to refer to a wooden beam that is at least 5 inches thick and at least 2 inches wider than it is thick.

bearing partition An interior divider or wall that supports the weight of the structure above it.

bearing wall A bearing wall has weight-bearing properties associated with holding up a building's roof or second floor.

benchmark A point from which other measurements are made.

bevel A bevel is a tool that can be adjusted to any angle. It helps make cuts at the number of degrees that is desired and is a good device for transferring angles from one place to another.

bifold A bifold is a folding door used to cover a closet. It has two panels that hinge in the middle and fold to allow entrance.

blistering Blistering refers to the condition that paint presents when air or moisture is trapped underneath and makes bubbles that break into flaky particles and ragged edges.

blocking Corners and wall intersections are made the same as outside walls. The size and amount of blocking can be reduced. The purpose of blocking is to provide nail surfaces at the corners. These are needed at inside and outside nail surfaces. They are a base for nailing wall covering.

blockout A form for pouring concrete is blocked out by a frame or other insertion to allow for an opening once the concrete has cured.

blocks This refers to a type of flooring made of wood. Wide pieces of boards are fastened to the floor, usually in squares and by adhesives.

board and batten A finished wall surface consisting of vertical board with gaps between them. The battens or small strips of wood cover the gaps.

board foot A unit of lumber measure equaling 144 cubic inches. The base unit (B.F.) is 1 inch thick and 12 inches square, or $1 \times 12 \times 12 = 144$ cubic inches.

bonding This is another word for gluing together wood or plastics and wood.

bottom or heel cut This refers to the cutout of the rafter end which rests against the plate. The bottom cut is also called the foot or seat cut.

bow A term used to indicate an upward warp along the length of a piece of lumber that is laid.

bow window A window unit that projects from an exterior wall. It has a number of windowpanes that form a curve.

brace A brace is an inclined piece of lumber applied to a wall or to roof rafters to add strength.

brace scale A brace scale is a table that is found along the center of the back of the tongue and gives the exact lengths of common braces.

bridging Bridging is used to keep joists from twisting or bending.

buck A buck is the same as a blockout.

builder's level This is a tripod-mounted device that uses optical sighting to make sure that a straight line is sighted and that the reference point is level.

building codes Building codes are rules and regulations that are formulated in a code by a local housing authority or governing body.

building paper Also called tar paper, roofing paper, and a number of other terms. It has a black coating of asphalt for use in weatherproofing.

building permits Most incorporated cities or towns have a series of permits that must be obtained for building. This allows for inspections of the work and for placing the house on the tax roles.

built-ins This is a term used to describe the cabinets and other small appliances that are built into the kitchen, bathroom, or family room by a carpenter. They may be custom cabinets or may be made on the site.

bundle This term refers to the packaging of shingles. A bundle of shingles is usually a handy method for shipment.

butt A term that can be used a couple of ways: A *butt hinge* is one where the two parts meet edge-to-edge, allowing movement of the two parts when held together by a pin; to *butt* means to meet edge to edge, such as in a joining of wooden edges.

cantilever Overhangs are called *cantilevers*; they are used for special effects on porches, decks, or balconies.

carpenter's square This steel tool can be used to check for right angles, to lay out rafters and studs, and to perform any number of measuring jobs.

carpet strips These are wooden or metal strips with nails or pins sticking out. They are nailed around the perimeter of a room, and the carpet is pulled tight and fastened to the exposed nails.

carpet tape This is a tape used to seam carpet where it fits together.

carriage A notched stair frame is called a carriage.

casement This is a type of window hinged to swing outward.

casing A door casing refers to the trim that goes on around the edge of a door opening and also to the trim around a window.

cathedral ceiling A cathedral ceiling is not flat and parallel with the flooring; it is open and follows the shape of the roof. The open ceiling usually precludes an attic.

caulk Caulk is any type of material used to seal walls, windows, and doors to keep out the weather. Caulk is usually made of putty or some type of plastic material, and it is flexible and applied with a caulking gun.

cellulose fiber Insulation material made from cellulose fiber. Cellulose is present in wood and paper, for example.

cement Cement is a fine-powdered limestone that is heated and mixed with other minerals to serve as a binder in concrete mixes.

ceramic tile Ceramic tile is made of clay and fired to a high temperature; it usually has a glaze on its surface. Small pieces are used to make floors and wall coverings in bathrooms and kitchens.

certificate of occupancy This certificate is issued when local inspectors have found a house worthy of human habitation. It allows a contractor to sell a house. It is granted when the building code has been complied with and certain inspections have been made.

chair A chair is a support bracket for steel reinforcing rods that holds the rods in place until the concrete has been poured around them.

chalk line A chalk line is used to guide a builder. It is snapped, causing the string to make a chalk mark on the surface being worked on. For example, the builders can follow it with their shingles.

chipboard Chipboard is used as an underlayment. It is constructed of wood chips held together with different types of resins.

chisel A wood chisel is used to cut away wood for making joints. It is sharpened on one end, and the other is hit with the palm of the hand or with a hammer to cut away wood for door hinge installation or to fit a joint tightly.

claw hammer This is the common hammer used by carpenters to drive nails. The claws are used to extract nails that bend or fail to go where they are wanted.

cleat Any strip of material attached to the surface of another material to strengthen, support, or secure a third material.

clerestory A short exterior wall between two sections of roof that slope in different directions. The term is also used to describe a window that is placed in this type of wall.

closed-cut For a closed-cut valley, the first course of shingles is laid along the eaves of one roof area up to and over the valley. It is extended along the adjoining roof section. The distance is at least 12 inches. The same procedure is followed for the next courses of shingles.

cold chisel This chisel is made with an edge that can cut metal. It has a one-piece configuration, with a head to be hit by a hammer and a cutting edge to be placed against the metal to be cut.

common rafter A common rafter is a member that extends diagonally from the plate to the ridge.

concrete Concrete is a mixture of sand, gravel, and cement in water.

condensation The process by which moisture in the air becomes water or ice on a surface (such as a window) whose temperature is colder than the air's temperature.

contact cement Contact cement is the type of glue used in applying countertop finishes. Both sides of the materials are coated with the cement, and the cement is allowed to dry. The two surfaces are then placed in contact, and the glue holds immediately.

contractor A contractor is a person who contracts with a firm, a bank, or another person to do a job for a certain fee and under certain conditions.

contractor's key This is a key designed to allow the contractor access to a house while it is under construction. The lock is changed to fit a pregrooved key when the house is turned over to the owner.

convection Transfer of heat through the movement of a liquid or gas.

coped joint This type of joint is made with a coping saw. It is especially useful for corners that are not square.

coping saw This saw is designed to cut small thicknesses of wood at any curve or angle desired. The blade is placed in a frame, with the teeth pointing toward the handle.

corner beads These are metal strips that prevent damage to drywall corners.

cornice The cornice is the area under the roof overhang. It is usually enclosed or boxed in.

course This refers to alternate layers of shingles in roofing.

cradle brace A cradle brace is designed to hold Sheetrock or drywall while it is being nailed to the ceiling joists. The cradle brace is shaped like a T.

crawl space A crawl space is the area under a floor that is not fully excavated. It is only excavated sufficiently to allow one to crawl under it to get at the electrical or plumbing devices.

cricket This is another term for the *saddle*.

cripple jack A cripple jack is a jack rafter with a cut that fits in between a hip and a valley rafter.

cripple rafter A cripple rafter is not as long as the regular rafter used to span a given area.

cripple stud This is a short stud that fills out the position where the stud would have been located if a window, door, or some other opening had not been there.

crisscross wire support This refers to chicken wire that is used to hold insulation in place under the flooring of a house.

crosscut saw This is a handsaw used to cut wood across the grain. It has a wooden handle and a flexible steel blade.

cup To warp across the grain.

curtain wall Inside walls are often called curtain walls. They do not carry loads from roof or floors above them.

dado A rectangular groove cut into a board across the grain.

damper A damper is an opening and closing device that will close off the fireplace area from the outside by shutting off the flue. It can also be used to control draft.

deadbolt lock A deadbolt lock will respond only to the owner who knows how to operate it. It is designed to keep burglars out.

deck A deck is the part of a roof that covers the rafters.

decorative beams Decorative beams are cut to length from wood or plastic and mounted to the tastes of the owner. They do not support the ceiling.

diagonals Diagonals are lines used to cut across from adjacent corners to check for squareness in the layout of a basement or footings.

dividers Dividers have two points and resemble a compass. They are used to mark off specific measurements or transfer them from a square or a measuring device to the wood to be cut.

dormer Dormers are protrusions that stick out from a roof. They may be added to allow light into an upstairs room.

double-hung windows Double-hung windows have two sections, one of which slides past the other. They slide up and down in a prearranged slot.

double plate This usually refers to the practice of using two pieces of dimensional lumber for support over the top section or wall section.

double trimmer Double joists used on the sides of openings are called double trimmers. Double trimmers are placed without regard to regular joist spacings for openings in the floor for stairs or chimneys.

downspouts These are pipes connected to the gutter to conduct rainwater to the ground or sewer.

drain tile A drain tile is usually made of plastic. It generally is 4 inches in diameter, with a number of small holes to allow water to drain into it. It is laid along the foundation footing to drain the seepage into a sump or storm sewer.

drawer guide The drawers in cabinets have guides to make sure that the drawer glides into its closed position easily without wobbling.

drop siding Drop siding has a special groove, or edge cut into it. The edge lets each board fit into the next board. This makes the boards fit together and resist moisture and weather.

drywall This is another name for panels made of gypsum.

ductwork Ductwork is a system of pipes used to pass heated air along to all parts of a house. The same ductwork can be used to distribute cold air for summer air conditioning.

dutch hip This is a modification of the hip roof.

eaves Eaves are the overhang of a roof projecting over the walls.

eaves trough A gutter.

elevation Elevation refers to the location of a point in reference to the point established with the builder's level or transit. Elevation indicates how high a point is. It may also refer to the front elevation or front view of a building or the rear elevation or what the building looks like from the rear. Side elevations refer to a side view.

energy Energy refers to the oil, gas, or electricity used to heat or cool a house.

essex board measure This is a table on the back of the body; it gives the contents of any size lumber. The table is located on the steel square used by carpenters.

excavation Excavate means to remove. In this case, excavation refers to the removal of dirt to make room for footings, a foundation, or the basement of a building.

expansion joint This is usually a piece of soft material that is about 1 inch thick and 4 inches wide and is placed between sections of concrete to allow for expansion when the flat surface is heated by the sun.

exposure Exposure refers to the part of a shingle or roof left to the weather.

extension form An extension form is built inside the concrete outer form. It forms a stepped appearance so that water will not drain into a building but drain outward from the slab or foundation slab.

faced insulation This insulation usually has a coating to create a moisture barrier.

faced stapling Faced stapling refers to the strip along the outer edges of the insulation that is stapled to the outside or 2-inch sides of the 2 × 4 studs.

facing Facing strips give cabinets a finished look. They cover the edges where the units meet and where the cabinets meet the ceiling or woodwork.

factory edge This is the straight edge of linoleum made at the factory. It provides a reference line for the installer.

factory-produced housing This refers to housing that is made totally in a factory. Complete units are usually trucked to a place where a basement or slab is ready.

false bottom This is a system of 1- × 6-inch false-bottom or box beams that provide the beauty of beams without the expense. The false beams are made of wood or plastic materials and glued or nailed in place. They do not support any weight.

fascia Fascia refers to a flat board covering the ends of rafters on the cornice or eaves. The eave troughs are usually mounted to the fascia board.

FHA The Federal Housing Administration.

fiberglass Fiberglass is insulation material made from spun resin or glass. It conducts little heat and creates a large dead air space between layers of fibers. It helps conserve energy.

firebrick This is a special type of brick that is not damaged by fire. It is used to line the firebox in a fireplace.

fire stops Fire stops are short pieces nailed between joists and studs.

flaking This refers to paint that falls off a wall or ceiling in flakes.

flashing Flashing is metal used to cover chimneys or other things projecting through the roofing. It keeps the weather out.

floating The edges of drywall sheets are staggered, or floated. This gives more bracing to the wall since the whole wall does not meet at any one joint.

floating Floating refers to concrete work; it lets the smaller pieces of concrete mix float to the top. Floating is usually done with a tool moved over the concrete.

floorboards This refers to floor decking. Floorboards may be composed of boards or may be a sheet of plywood used as a subfloor.

flue The flue is the passage through a chimney.

flush This term means to be even with.

folding rule This is a device that folds into a 3- × 6-inch rectangle. It has the foot broken into 12 inches. Each inch is broken into 16 points. Snap joints hinge the rule every six inches. It will spread to as much as 6 feet, or 72 inches.

footings Footings are the lowest part of a building. They are designed to support the weight of the building and distribute it to the earth or rock formation on which it rests.

form A form is a structure made of metal or wood used as a mold for concrete.

Formica Formica is a laminated plastic covering made for countertops.

foundation The foundation is the base on which a house or building rests. It may consist of the footings and walls.

framing Roof framing is composed of rafters, ridge board, collar beams, and cripple studs.

framing square This tool allows a carpenter to make square cuts in dimensional lumber. It can be used to lay out rafters and roof framing.

French doors This usually refers to two or more groups of doors arranged to open outward onto a patio or veranda. The doors are usually composed of many small glass panes.

frostline This is the depth to which the ground freezes in the winter.

furring strips These are strips of wood attached to concrete or stone. They form a nail base for wood or paneling.

gable This is the simplest kind of roof. Two large surfaces come together at a common edge, forming an inverted V.

galvanized iron This material is usually found on roofs as flashing. It is sheet metal coated with zinc.

gambrel roof This is a barn-shaped roof.

girder A girder is a support for the joists at one end. It is usually placed halfway between the outside walls and runs the length of the building.

grade The grade is the variation of levels of ground or the established ground-line limit on a building.

grid system This is a system of metal strips that support a drop ceiling.

grout Grout is a white plaster-like material placed into the cracks between ceramic tiles.

gusset A gusset is a triangular or rectangular piece of wood or metal that is usually fastened to the joint of a truss to strengthen it. It is used primarily in making roof trusses.

gutter This is a metal or wooden trough set below the eaves to catch and conduct water from rain and melting snow to a downspout.

gypsum Gypsum is a chalk used to make wallboard. It is made into a paste, inserted between two layers of paper, and allowed to dry. This produces a plastered wall with certain fire-resisting characteristics.

handsaw A handsaw is any saw used to cut wood and operated by manual labor rather than electricity.

hang a door This term refers to the fact that a door has to be mounted on hinges and aligned with the door frame.

hangers These are metal supports that hold joists or purlins in place.

hardware In this case hardware refers to the metal parts of a door. Such things as hinges, locksets, and screws are hardware.

hardwood The wood that comes from a tree that sheds its leaves. This doesn't necessarily mean the wood itself is hard. A poplar has soft wood, but it is classified as a hardwood tree. An oak has hard wood and is also classified as a hardwood tree.

hardboard A type of fiberboard pressed into thin sheets. Usually made of wood chips or waste material from trees after the lumbering process has been completed.

header A header is a board that fits across the ends of joists.

head lap This refers to the distance between the top and the bottom shingle and the bottom edge of the shingle covering it.

hearth A hearth is the part of a fireplace that is in front of the wood rack.

hex strips This refers to strips of shingles that are six-sided.

hip rafters A hip rafter is a member that extends diagonally from the corner of the plate to the ridge.

hip roof A hip roof has four sides, all sloping toward the center of the building.

hollow-core doors Most interior doors are hollow and have paper or plastic supports for the large surface area between the top and bottom edge and the two faces.

honeycomb Air bubbles in concrete cause a honeycomb effect and weaken the concrete.

insert stapling This refers to stapling insulation inside the 2 × 4 stud. The facing of the insulation has a strip left over. It can be stapled inside the studs or over the studs.

insulation Insulation is any material that offers resistance to the conduction of heat through its composite materials. Plastic foam and fiberglass are the two most commonly used types of insulation in homes today.

insulation batts These are thick precut lengths of insulation designed to fit between studs.

interlocking Interlocking refers to a type of shingle that overlaps and interlocks with its edges. It is used in high winds.

in the white This term is used to designate cabinets that are assembled but unfinished.

jack rafter Any rafter that does not extend from the plate to the ridge is called a jack rafter.

jamb A jamb is the part that surrounds a door window frame. It is usually made of two vertical pieces and a horizontal piece over the top.

joist Large dimensional pieces of lumber used to support the flooring platform of a house or building are called joists.

joist hangers These are metal brackets that hold up the joist. They are nailed to the girder, and the joist fits into the bracket.

joist header If the joist does not cover the full width of the sill, space is left for the joist header. The header is nailed to the ends of the joists and rests on the sill plate. It is perpendicular to the joists.

kerf The cut made by a saw blade.

key A key is a depression made in a footing so that the foundation or wall can be poured into the footing, preventing the wall or foundation from moving during changes in temperature or settling of the building.

kicker A kicker is a piece of material installed at the top or side of a drawer to prevent it from falling out of a cabinet when it is opened.

kiln dried Special ovens are made to dry wood before it is used in construction.

king-post truss This is the type of roof truss used to support low pitch roofs.

ladder jack Ladder jacks hang from a platform on a ladder. They are most suitable for repair jobs and for light work where only one carpenter is on the job.

landing A landing is the part of a stairway that is a shaped platform.

lap This refers to lap siding. Lap siding fits on the wall at an angle. A small part of the siding is overlapped on the preceding piece of siding.

laths Laths are small strips of wood or metal designed to hold plastic on the wall until it hardens for a smooth finish.

ledger A ledger is a strip of lumber nailed along an edge or bottom of a location. It helps support or keep from slipping the girders on which the joists rest.

left-hand door A left-hand door has its hinges mounted on the left when viewed from the outside.

level A level is a tool using bubbles in a glass tube to indicate the level of a wall, stud, or floor. Keeping windows, doors, and frames square and level makes a difference in their fit and operation.

level-transit This is an optical device that is a combination of a level and a means for checking vertical and horizontal angles.

load Load refers to the weight of a building.

load conditions These are the conditions under which a roof must perform. The roof has to support so much wind load and snow. Load conditions vary according to locale.

lockset The lockset refers to the doorknob and associated locking parts inserted in a door.

mansard This type of roof is popular in France and is used in the United States also. The second story of the house is covered with the same shingles used on the roof.

manufactured housing This term is used in reference to houses that are totally or partially made within a factory and then trucked to a building site.

mastic Mastic is an adhesive used to hold tiles in place. The term also refers to adhesives used to glue many types of materials in the building process.

military specifications These are specifications that the military writes for the products it buys from the manufacturers. In this case, the term refers to the specifications for a glue used in making trusses and plywood.

miter box The miter box has a hacksaw mounted in it. It is adjustable for cutting at angles such as 45 and 90°. Some units can be adjusted by a level to any angle.

modular homes These houses are made in modules or small units which are nailed or bolted together once they arrive at the foundation or slab on which they will rest.

moisture barrier A moisture barrier is some type of material used to keep moisture from entering the living space of a building. Moisture barrier, vapor seal, and membrane mean the same thing. It is laid so that it covers the whole subsurface area over sand or gravel.

moisture control Excess moisture in a well-insulated house may pose problems. A house must be allowed to breathe and change the air occasionally, which in turn helps remove excess moisture. Proper ventilation is needed to control moisture in an insulated house. Elimination of moisture is another method but it requires the reduction of cooking vapors and shower vapors, for example.

moldings Moldings are trim mounted around windows, floors, and doors as well as closets. Moldings are usually made of wood with designs or particular shapes.

monolithic slab Mono means "one." This refers to a one-piece slab for a building floor and foundation all in one piece.

nail creep This is a term used in conjunction with drywall, the nails pop because of wood shrinkage. The nailheads usually show through the panel.

nailers These are powered hammers that have the ability to drive nails. They may be operated by compressed air or by electricity.

nailhead stains These occur whenever the iron in the nailhead rusts and shows through the paint.

nail set Finish nails are driven below the surface of the wood by a nail set. The nail set is placed on the head of a nail, and the large end of the nail set is struck with a hammer.

newels The end posts of a stairway are called newels.

octagon scale This "eight square" scale is found on the center of the face of the tongue of a steel square. It is used when timber is cut with eight sides.

open This refers to the type of roofing that allows a joint between a dormer and the main roof. It is an open valley type of roofing. The valley where two roofs intersect is left open and covered with flashing and roofing sealer.

orbital sander This power sander will vibrate, but in an orbit. Thus causes the sandpaper to do its job better than it would if used in only one direction. An orbital sander can be used to finish off windows, doors, counters, cabinets, and floors.

panel This refers to a small section of a door that takes on definite shape, or to the panel in a window made of glass.

panel door This is a type of door used for the inside of a house. It has panels inserted in the frame to give it strength and design.

parquet Parquet is a type of flooring made from small strips arranged in patterns. It must be laminated to a base.

partition A partition is a divider wall or section that separates a building into rooms.

peeling This is a term used in regard to paint that will not stay on a building. The paint peels and falls off or leaves ragged edges.

penny (d) This is the unit of measure of the nails used by carpenters.

perimeter The perimeter is the outside edges of a plot of land or building. It represents the sum of all the individual sides.

perimeter insulation Perimeter insulation is placed around the outside edges of a slab.

pile A pile is a steel or wooden pole driven into the ground sufficiently to support the weight of a wall and building.

pillar A pillar is a pole or reinforced wall section used to support the floor and consequently the building. It is usually located in the basement, with a footing of its own to spread its load over a wider area than the pole would normally occupy.

pitch The pitch of a roof is the slant or slope from the ridge to the plate.

pivot This refers to a point where the bifold door is anchored and allowed to move so that the larger portion of the folded sections can move.

plane Planes are designed to remove small shavings of wood along a surface. One hand holds the knob in front, and the other hand holds the handle in the back of the plane. A plane is used to shave off door edges to make them fit properly.

planks This refers to a type of flooring usually made of tongue-and-groove lumber and nailed to the subflooring or directly to the floor joists.

plaster This refers to plaster of Paris mixed with water and applied to a lath to cover a wall and allow for a finished appearance that will take a painted finish.

plaster grounds A carpenter applies small strips of wood around windows, doors, and walls to hold plaster. These grounds may also be made of metal.

plastic laminates These are materials usually employed to make countertops. Formica is an example of a plastic laminate.

plate The plate is a roof member that has the rafters fastened to it at their lower ends.

platform frame This refers to the flooring surface placed over the joists; it serves as support for further floor finishing.

plenum A plenum is a large chamber.

plumb bob This is a very useful tool for checking plumb, or the upright level of a board, stud, or framing member. It is also used to locate points that should be directly under a given location. It hangs free on a string, and its point indicates a specific location for a wall, a light fixture, or the plumb of a wall.

plumb cut This refers to the cut of the rafter end which rests against the ridge board or against the opposite rafter.

post and beam Posts are used to support beams, which support the roof decking. Regular rafters are not used. This technique is used in barns and houses to achieve a cathedral-ceiling effect.

prehung This refers to doors or windows that are already mounted in a frame and are ready for installation as a complete unit.

primer This refers to the first coat of paint or glue when more than one coat will be applied.

pulls The handle or the part of the door on a cabinet or the handle on a drawer that allows it to be pulled or opened is called a pull.

purlin Secondary beams used in post-and-beam construction are called purlins.

rabbet A groove cut in or near the edge of a piece of lumber to fit the edge of another piece.

radial-arm saw This type of power saw has a motor and blade that moves out over the table which is fixed. The wood is placed on the table and the blade is pulled through the wood.

rafter scales This refers to a steel square with the rafter measurements stamped on it. The scales are on the face of the body.

rail The vertical facing strip on a cabinet is the stile. The horizontal facing strip is the rail.

rake On a gabled roof, a rake is the inclined edge of the surface to be covered.

random spacing This refers to spacing that has no regular pattern.

rebar A rebar is a reinforcement steel rod in a concrete footing.

reinforcement mesh Reinforcement mesh is made of 10-gage wires spaced about 4 to 6 inches apart. It is used to reinforce basements or slabs in houses. The mesh is placed so that it becomes a part of the concrete slab or floor.

remodeling This refers to changing the looks and function of a house.

residential building A residential building is designed for people to live in.

resilient flooring This type of flooring is made of plastics rather than wood products. It includes such things as linoleum and asphalt tile.

ridge board This is a horizontal member that connects the upper ends of the rafters on one side to the rafters on the opposite side.

right-hand door A right-hand door has the opening or hinges mounted on the right when viewed from the outside.

right-hand draw This means that the curtain rod can be operated to open and close the drapes from the right-hand side as one faces it.

rise In roofing, rise is the vertical distance between the top of the double plate and the center of the ridge board. In stairs, it is the vertical distance from the top of a stair tread to the top of the next tread.

riser The vertical part at the edge of a stair is called a riser.

roof brackets These brackets can be clamped onto a ladder used for roofing.

roof cement A number of preparations are used to make sure that a roof does not leak. Roof cement also can hold down shingle tabs and rolls of felt paper when it is used as a roof covering.

roofing This term is used to designate anything that is applied to a roof to cover it.

rough line A rough line is drawn on the ground to indicate the approximate location of footing.

rough opening This is a large opening made in a wall frame or roof frame to allow the insertion of a door or window or whatever is to be mounted in the open space. The space is shimmed to fit the object being installed.

router A router will cut out a groove or cut an edge. It is usually powered and has a number of different shaped tips that will carve its shape into a piece of wood. It can be used to take the edges off countertops.

run The run of a roof is the shortest horizontal distance measured from a plumb line through the center of the ridge to the outer edge of the plate.

R values This refers to the unit that measures the effectiveness of insulation. It indicates the relative value of the insulation for the job. The higher the number, the better the insulation qualities of the materials.

saber saw The saber saw has a blade that can be used to cut circles in wood. It can cut around any circle or curve. The blade is inserted in a hole drilled previously and the saw will follow a curved or straight line to remove the block of wood needed to allow a particular job to be completed.

saddle A saddle is the inverted V-shaped piece of roof inserted between the vertical side of a chimney and the roof.

saturated felt Other names for this material are tar paper and builder's felt. It is roll roofing paper and can be used as a moisture barrier and waterproofing material on roofs and under siding.

scabs Scabs are boards used to join the ends of a girder.

scaffold A scaffold is a platform erected by carpenters to stand on while they work on a higher level. Scaffolds are supported by tubing or 2 × 4s. Another name for scaffolding is *staging*.

screed A screed may be a board or pipe supported by metal pins. The screed is leveled with the tops of the concrete forms. It is removed after the section of concrete is leveled.

scribing Scribing means marking.

sealant A sealant is any type of material that will seal a crack. This usually refers to caulking when carpenters use the term.

sealer coat The sealer coat ensures that a stain is covered and the wood is sealed against moisture.

shakes This is a term used for shingles made of handsplit wood, in most cases western cedar.

sheathing This is a term used for the outside layer of wood applied to studs to close up a house or wall. It is also used to cover the rafters and make a base for the roofing. It is usually made of plywood today. In some cases, sheathing is still used to indicate the 1- × 6-inch wooden boards used for siding undercoating.

shed In terms of roofs, this is the flat sloping roof used on some storage sheds. It is the simplest type of roof.

Sheetrock This is another name for panels made of gypsum.

shim To shim means to add some type of material that will cause a window or door to be level. Usually wood shingles are wedge-shaped and serve this purpose.

shingles This refers to material used to cover the outside of a roof and take the ravages of weather. Shingles may be made of metal, wood, or composition materials.

shingle stringers These are nailing boards that can have cedar shingles attached to them. They are spaced to support the length of the shingle that will be exposed to weather.

shiplap An L-shaped edge, cut into boards and some sheet materials to form an overlapping joint with adjacent pieces of the same material.

side lap The side lap is the distance between adjacent shingles that overlap.

siding This is a term used to indicate that the studs have been covered with sheathing and the last covering is being placed on it. Siding may be made of many different materials—wood, metal, or plastic.

sill This is a piece of wood that is anchored to the foundation.

sinker nail This is a special nail for laying subflooring. The head is sloped toward the shank but is flat on top.

size Size is a special coating used for walls before the wallpaper is applied. It seals the wall and allows the wallpaper paste to attach itself to the wall and paper without adding undue moisture to the wall.

skew-back saw This saw is designed to cut wood. It is hand-operated and has a serrated steel blade that is smooth on the non-cutting edge of the saw. It is 22 to 26 inches long and can have 5½ to 10 teeth per inch.

skilled worker A skilled worker is a person who can do a job well each time it is done, or the person who has the ability to do the job a little better each time. Skilled means the person has been at it for some time, usually 4 to 5 years at the least.

sliding door This is usually a large door made of glass, with one section sliding past the other to create a passageway. A sliding door may be made of wood or glass and can disappear or slide into a wall. Closets sometimes have doors that slide past one another to create an opening.

sliding window This type of window has the capability to slide in order to open.

slope Slope refers to how fast the roof rises from the horizontal.

soffit A covering for the underside of the overhang of a roof.

soil stack A soil stack is the ventilation pipe that comes out of a roof to allow the plumbing to operate properly inside the house. It is usually made of a soil pipe (cast iron). In most modern housing, the soil stack is made of plastic.

soleplate A soleplate is a 2 × 4 or 2 × 6 used to support studs in a horizontal position. It is placed against the flooring and nailed into position onto the subflooring.

span The span of a roof is the distance over the wall plates.

spreader Special braces used across the top of concrete forms are called spreaders.

square This term refers to a shingle-covering area. A square consists of 100 square feet of area covered by shingles.

square butt strip This refers to shingles for roofing purposes that were made square in shape but produced in strips for ease in application.

staging This is the planking for ladder jacks. It holds the roofer or shingles.

stain Stain is a paint-like material that imparts a color to wood. It is usually finished by a clear coating of shellac, varnish or Satinlac, or brush lacquer.

stapler This device is used to place wire staples into a roof's tar paper to hold it in place while the shingles are applied.

steel square The steel square consists of two parts—the blade and tongue or the body.

stepped footing This is footing that may be located on a number of levels.

stile A stile is an upright framing member in a panel door.

stool The flat shelf that rims the bottom of a window frame on the inside of a wall.

stop This applies to a door. It is the strip on the door frame that stops the door from swinging past the center of the frame.

storm door A storm door is designed to fit over the outside doors of a house. It may be made of wood, metal, or plastic, and it adds to the insulation qualities of a house. A storm door may be all glass, all screen, or a combination of both. It may be used in summer, winter, or both.

storm window Older windows have storms fitted on the outside. The storms consist of another window that fits over the existing window. The purpose is to trap air that will become an insulating layer to prevent heat transfer during the winter. Newer windows have thermopanes, or two panes mounted in the same frame.

stress skin panels These are large prebuilt panels used as walls, floors, and roof decks. They are built in a factory and hauled to the building site.

strike-off After tamping, concrete is leveled with a long board called a strike-off.

strike plate This is mounted on the door frame. The lock plunger goes into the hole in the strike plate and rests against the metal part of the plate to hold the door secure against the door stop.

striker This refers to the strike plate. The striker is the movable part of the lock that retracts into the door once it hits the striker plate.

stringer A carriage is also called a stringer.

strip flooring Wooden strip flooring is nothing more than the wooden strips that are applied perpendicular to the joists.

strongbacks Strongbacks are braces used across ceiling joints. They help align, space, and strengthen joists for drywall installation.

stucco Stucco is a type of finish used on the outside of a building. It is a masonry finish that can be put on over any type of wall. It is applied over a wire mesh nailed to the wall.

stud This refers to the vertical boards (usually 2×4 or 2×6) that make up the walls of a building.

stump A stump is that part of a tree which is left after the top has been cut and removed. The stump remains in the ground.

subfloor The subfloor is a platform that supports the rest of the structure. It is also referred to as the underlayment.

sump pump This refers to a pump mounted in a sump or well created to catch water from around the foundation of a house. The pump takes water from the well and lifts it to the grade level or to a storm sewer nearby.

surveying Surveying means taking in the total scene. In this case, it refers to checking out the plot plan and the relationship of the proposed building with others located within eyesight.

suspended beams False beams may be used to lower a ceiling like a grid system; these beams are suspended. They use screw eyes attached to the existing ceiling joists.

sway brace A sway brace is a piece of 2×4 or similar material used to temporarily brace a wall from the wind until it is secured.

swinging door A swinging door is mounted so that it will swing into or out of either of two rooms.

table saw A table saw is electrically powered, with a motor-mounted saw blade supported by a table that allows the wood to be pushed over the table into the cutting blade.

tail The tail is the portion of a rafter that extends beyond the outside edge of the plate.

tail joist This is a short beam or joist supported in a wall on one end and by a header on the other.

tamp To tamp means to pack tightly. The term usually refers to making sand tightly packed or making concrete mixed properly in a form to get rid of air pockets that may form with a quick pouring.

taping and bedding This refers to drywall finishing. Taping is the application of a strip of specially prepared tape to drywall joints; bedding means embedding the tape in the joint to increase its structural strength.

team A team is a group of people working together.

terrazzo This refers to two layers of flooring made from concrete and marble chips. The surface is ground to a very smooth finish.

texture paint This is a very thick paint that will leave a texture or pattern. It can be shaped to cover cracked ceilings or walls or beautify an otherwise dull room.

thermal ceilings These are ceilings that are insulated with batts of insulation to prevent loss of heat or cooling. They are usually drop ceilings.

tie A tie is a soft metal wire that is twisted around a rebar or reinforcement rod and chair to hold the rod in place till concrete is poured.

tin snips This refers to a pair of scissors-type cutters used to cut flashing and some types of shingles.

tongue-and-groove Roof decking may have a groove cut in one side and tongue led in the other edge of the piece of wood so that the two adjacent pieces will fit together tightly.

track This refers to the metal support system that allows the bifold and other hung doors to move from closed to open.

transit-mix truck In some parts of the country, this is called a Redi-Mix truck. It mixes the concrete on its way from the source of materials to the building site where it is needed.

traverse rod This is another name for a curtain rod.

tread The part of a stair on which people step is the tread.

trestle jack Trestle jacks are used for low platforms both inside and outside. A ledger, made of 2×4 lumber, is used to connect two trestle jacks. Platform boards are then placed across the two ledgers.

trimmer A trimmer is a piece of lumber, usually a 2×4, that is shorter than the stud or rafter but is used to fill in where the longer piece would have been normally spaced except for the window or door opening or some other opening in the roof or floor or wall.

trowel A trowel is a tool used to work with concrete or mortar.

truss This is a type of support for a building roof that is prefabricated and delivered to the site. The W and King trusses are the most popular.

try square A try square can be used to mark small pieces for cutting. If one edge is straight and the handle part of the square is placed against this straightedge, the blade can be used to mark the wood perpendicular to the edge.

underlayment This is also referred to as the subfloor. It is used to support the rest of the building. The term may also refer to the sheathing used to cover rafters and serve as a base for roofing.

unfaced insulation This type of insulation does not have a facing or plastic membrane over one side of it. It has to be placed on top of existing insulation. If used in a wall, it has to be covered by a plastic film to ensure a vapor barrier.

union A union is a group of people with the same interests and with proper representation for achieving their objectives.

utilities Utilities are the things needed to make a house a home. They include electricity, water, gas, and phone service. Sewage is a utility that is usually determined to be part of the water installation.

utility knife This type of knife is used to cut the underlayment or the shingles to make sure they fit the area assigned to them. It is also used to cut the saturated felt paper over a deck.

valley This refers to the area of a roof where two sections come together and form a depression.

valley rafters A valley rafter is a rafter which extends diagonally from the plate to the ridge at the line of intersection of two roof surfaces.

vapor barrier This is the same as a moisture barrier.

veneer A veneer is a thin layer or sheet of wood.

vent A vent is usually a hole in the eaves or soffit to allow the circulation of air over an insulated ceiling. It is usually covered with a piece of metal or screen.

ventilation Ventilation refers to the exchange of air, or the movement of air through a building. This may be done naturally through doors and windows or mechanically by motor-driven fans.

vernier This is a fine adjustment on a transit that allows for greater accuracy in the device when it is used for layout or leveling jobs at a construction site.

vinyl Vinyl is a plastic material. The term usually refers to polyvinyl chloride. It is used in weather stripping and in making floor tile.

vinyl-asbestos tile This is a floor covering made from vinyl with an asbestos filling.

water hammer The pounding sound produced when the water is turned off quickly. It can be reduced by placing short pieces of pipe, capped off at one end, above the most likely causes of quick turnoffs, usually the dishwasher and clothes washing machine.

water tables This refers to the amount of water that is present in any area. The moisture may be from rain or snow.

weatherstripping This refers to adding insulating material around windows and doors to prevent the heat loss associated with cracks.

winder This refers to the fan-shaped steps that allow the stairway to change direction without a landing.

window apron The window apron is the flat part of the interior trim of a window. It is located next to the wall and directly beneath the window stool.

window stool A window stool is the flat narrow shelf which forms the top member of the interior trim at the bottom of a window.

wrecking bar This tool has a number of names. It is used to pry boards loose or to extract nails. It is a specially treated steel bar that provides leverage.

woven This refers to a type of roofing. Woven valley-type shingling allows the two intersecting pieces of shingle to be woven into a pattern as they progress up the roof. The valley is not exposed to the weather but is covered by shingles.

zoning laws Zoning laws determine what type of structure can be placed in a given area. Most communities now have a master plan which recognizes residential, commercial, and industrial zones for building.

Index